Software-Defined Data Infrastructure Essentials

Cloud, Converged, and Virtual
Fundamental Server Storage I/O Tradecraft

Software-Defined Data Infrastructure Essentials

Cloud, Converged, and Virtual
Fundamental Server Storage I/O Tradecraft

Greg Schulz

CRC Press
Taylor & Francis Group
Boca Raton London New York

CRC Press is an imprint of the
Taylor & Francis Group, an **informa** business
AN AUERBACH BOOK

CRC Press
Taylor & Francis Group
6000 Broken Sound Parkway NW, Suite 300
Boca Raton, FL 33487-2742

First issued in paperback 2021

© 2017 by Taylor & Francis Group, LLC
CRC Press is an imprint of Taylor & Francis Group, an Informa business

No claim to original U.S. Government works

ISBN-13: 978-1-4987-3815-6 (hbk)
ISBN-13: 978-1-03-209676-6 (pbk)

Visit the Taylor & Francis Web site at
http://www.taylorandfrancis.com

and the CRC Press Web site at
http://www.crcpress.com

Contents

Preface

This book follows from my previous books, *Resilient Storage Networks: Designing Flexible Scalable Data Infrastructures* (aka "The Red Book"), *The Green and Virtual Data Center: Enabling Sustainable Economic, Optimized, Efficient and Productive IT Environments* (aka "The Green Book"), and *Cloud and Virtual Data Storage Networking: Your Journey to Efficient and Effective Information Services* (aka "The Yellow, or Gold, Book"). *Software-Defined Data Infrastructures Essentials* is more than a follow-up to these previous works; it looks in various directions—up, down, left, right, current and emerging—and extending into various adjacent data infrastructure topic areas.

Software-Defined Data Infrastructures Essentials provides fundamental coverage of physical, cloud, converged, and virtual server storage I/O networking technologies, trends, tools, techniques, and tradecraft skills. Software-defined data centers (SDDC), software data infrastructures (SDI), software-defined data infrastructures (SDDI, and traditional data infrastructures support business applications including components such as a server, storage, I/O networking, hardware, software, services, and best practices, among other management tools. Spanning cloud, virtual, container, converged (and hyper-converged) as well as legacy and hybrid systems, data infrastructures exist to protect, preserve, and serve data and information.

With a title containing terms such as tradecraft, essentials, fundamentals, and advanced emerging topics, some will assume that the material in this book is for the beginner or newbie, which it is. However, being focused on fundamentals and essentials of data infrastructure topics, where there is constant change (some evolutionary, some revolutionary), there are also plenty of "new" essential fundamentals to expand or refresh your tradecraft (i.e., "game"). By "game" I refer to your skills, experience, and abilities to play in and on the data infrastructure game field, which also means being a student of the IT data infrastructure game.

Regardless of whether you are a new student of IT or a specific focus area in a college, university, or trade school, or are going through a career transition, or simply moving from one area of IT to another, this book is for you. Why? Because it converges various data infrastructure topics, themes, trends, techniques, and technologies that can be used in various ways.

If, on the other hand, you are a seasoned pro, veteran industry expert, guru, or "unicorn" with several years or decades of experience, you will find this book useful as well. From web-scale, software-defined, containers, database, key-value store, cloud, and enterprise to small or medium-size business, there are plenty of themes, technologies, techniques, and tips to help develop or refine your server storage I/O tradecraft (game). Thus, there is plenty of new material or new ways to use new and old things even if you are a new or old expert ;).

One of the main and recurring themes in this book is the importance of understanding and recognizing context about server and storage I/O fundamentals. This means gaining as well as expanding (and sharing) your experience with technologies, tools, techniques, and trends in what is also known as your tradecraft.

To some people, the phrase "server storage I/O network tradecraft and fundamentals" will mean how I/O is done, from applications to servers to storage devices. Other people may assume a hardware focus or a software focus. Still others may see a focus on physical machines (PM) or bare metal, virtual, container, cloud, or converged solutions. Context is a recurring theme throughout this book; after all, sometimes words have two or more meanings.

"Tradecraft" refers to the skills and experiences needed and obtained from practicing or being a part of a particular trade. For example, if you were (or are) a successful intelligence spy,

Key Themes and What You Will Learn in This Book Include:

Theme and Topic	What You Will Learn
Data infrastructure supporting information infrastructure	Server, storage, I/O networking hardware, software, services, resources defined by policies and best practices
SDDC, SDI, SDDI Data infrastructure essentials Converged CI/HCI	Server, storage, I/O networking hardware, software, services, processes, procedures, and best practice fundamentals along with advanced and emerging topics for today and tomorrow
Context matters	Clarify different terms that have various meanings
Decision making, strategy, planning, and management	How you can create a strategy, plan, or manage and decide what you do not know. Gain insight into applications and technology to avoid "flying blind."
Everything is not the same	Different environments have various attributes and needs
Fast applications need fast data resources	Fast applications need robust underlying data infrastructures: fast servers, storage, I/O resources, and policies
Hybrid home run	Hybrid tools, technology, techniques, and services are the IT home run as they adapt to environments different needs
Knowing your toolbox	Know the tools and technologies in your toolbox as well as techniques on when and how to use them in new ways
Protect, preserve, and serve IT information and assets	Data infrastructure exists in physical, virtual, cloud, and containers to support business and information applications
Server storage I/O tradecraft	Enhancing, expanding, or refreshing your skills and "game"
Software defined	Hardware, software, and services all get defined by software algorithms and data structures (programs) that require some hardware existing somewhere in the technology stack
Various usage examples	While everything is not the same, there are some common usage and deployment scenarios to use as examples
What's in your toolbox	Understanding various hardware, software, and services tools, technologies, and techniques and where to use them

your tradecraft would be the skills, techniques, and experiences of using different tools and techniques to accomplish your job.

Other examples of tradecraft are if you are or were a medical doctor, your tradecraft would be that of providing health care, or if you are an airline pilot, your tradecraft would include flying the plane but also navigating, managing systems, and knowing the various technologies, procedures, routes, and tricks to get your job done. If your field is sales, marketing, finance, or engineering (among others), you possess fundamental tradecraft skills, knowledge, experiences, and practices from those disciplines that can be leveraged while learning new techniques, trends, and topics.

Regarding trends, I often hear people tell me that this year (whenever you read this) is the most exciting time ever, with more happening in the server, storage I/O networking, and related technologies. I agree, as my usual response is that every year for the past several decades has been exciting, each with something new. Since becoming involved with servers, storage, I/O hardware, software, and services, from applications to systems to components, I have found that there is always something new. Some things are evolutionary and prompt a sense of *déja vu*—of having seen or experienced them in the past. Some are revolutionary, new or first-time experiences, while others can be technolutionary (a blend of new, revolutionary along with evolutionary).

While there are plenty of new things, sometimes those new things get used in old ways; and sometimes old things can get used in new ways. As you have probably heard before, the one thing that is constant is change, yet something else that occurs is that as things or technologies change, they get used or remain the same. A not-so-bold prophecy would be to say that next year will see even more new things, not to mention old things being used in new ways.

For example, many technology changes or enhancements have occurred from the time I started writing this book until its completion. There will be more from the time this goes to the publisher for production, then until its release and you read it in print or electronically. That is where my companion website, www.storageio.com, along with my blog, www.storageioblog.com, and Twitter @StorageIO come into play. There you can further expand your tradecraft, seeing what's current, new, and emerging, along with related companion content to this book.

In terms of buzzwords, let's play some buzzword bingo. Here are some (among others) of the trends, technologies, tools, and techniques that are covered in this book: software-defined, containers, object, cloud, and virtual, physical, virtual server infrasture (VSI), virtual desktop infrasture (VDI) and work spaces, emerging, legacy server, micro-servers (and services), along with context matters. This includes server, storage I/O networking hardware as well as software, and services, tips, and techniques.

I also discuss converged, CI, HCI, SDDC, VMware, Hyper-V, KVM, Xen converged and hyper-converged, cluster-in-box or cloud-in-box, Azure and Azure Stack, AWS, OpenStack, Ceph, Mesos, Kubernetes, Hadoop, Hortonworks, and hive, as well as "big data" items. Let's not forget little data, big fast data, structured, unstructured, SQL, and NoSQL Cassandra, MongoDB, and other database or key-value stores. Also scale-out object storage, S3, Swift, backup/data protection as well as archiving, NFV and SDN, MySQL, SQL Server, benchmarking, capacity planning, IoT, artificial intelligence (AI), BC/BR/DR, strategy and acquisition decision making.

In terms of context, there is SRM, which can stand for storage or system resource management and monitoring as well as VMware Site Recovery Manager; telemetry, system resource analysis (SRA) and analytics along with CDM; Splunk, Kudu, Windows Nano, Linux LXC, CoreOS, Docker and other containers, S2D, VHDX, VMDK, VVOL, TBW and DWPD, Reed-Soloman (RS), erasure codes, and LRC, along with mirroring and replication. How about storage class memory (SCM), Flash SSD, NBD, NVM, NVMe, PCIe, SAS, SATA, iSCSI, RoCE, InfiniBand, 100 GbE, IP, MPLS, file systems, NFS, HDFS, SDI, SMB/CIFS, Posix, de-dupe, thin provisioning, compression, and much more.

Say "BINGO" for your favorite buzzwords; that is how buzzword bingo is played!

In case you have not connected the dots yet, the cover ties in themes of using new and old things in new ways, existing and emerging technology spanning hardware, software, services, and techniques. Also, for fun, the cover color combined with my other books represents the primary colors and wavelengths of the rainbow (Red, Green, Blue and Yellow) that are also leveraged in modern high-density fiber optic communications, and high-definition video. With that being said, let's get into *Software-Defined Data Infrastructure Essentials: Cloud Converged Virtual Fundamental for Server Storage I/O Networking.*

Who Should Read This Book

Software-Defined Data Infrastructure Essentials: Cloud, Converged, and Virtual Fundamental Server Storage I/O Tradecraft is for people who are currently involved with or looking to expand their knowledge and tradecraft skills (experience) of data infrastructures. Software-defined data centers (SDDC), software data infrastructures (SDI), software-defined data infrastructure (SDDI) and traditional data infrastructures are made up of software, hardware, services, and best practices and tools spanning servers, I/O networking, and storage from physical to software-defined virtual, container, and clouds. The role of data infrastructures is to enable and support information technology (IT) and organizational information applications.

Everything is not the same in business, organizations, IT, and in particular servers, storage, and I/O. This means that there are different audiences who will benefit from reading this book. Because everything and everybody is not the same when it comes to server and storage I/O along with associated IT environments and applications, different readers may want to focus on various sections or chapters of this book.

If you are looking to expand your knowledge into an adjacent area or to understand what's "under the hood," from converged, hyper-converged to traditional data infrastructures topics, this book is for you. For experienced storage, server, and networking professionals, this book connects the dots as well as provides coverage of virtualization, cloud, and other convergence themes and topics.

This book is also for those who are new or need to learn more about data infrastructure, server, storage, I/O networking, hardware, software, and services. Another audience for this book is experienced IT professionals who are now responsible for or working with data infrastructure components, technologies, tools, and techniques.

For vendors, there are plenty of buzzwords, trends, and demand drivers as well as how things work to enable walking the talk as well as talking the talk. There is a mix of Platform 2 (existing, brownfield, Windows, Linux, bare metal, and virtual client-server) and Platform 3 (new, greenfield, cloud, container, DevOp, IoT, and IoD). This also means that there is a Platform 2.5, which is a hybrid, or in between Platforms 2 and 3, that is, existing and new emerging. For non-vendors, there is information on different options for usage, and the technologies, tools, techniques, and how to use new and old things in new ways to address different needs.

Even if you are going to a converged or hyper-converged cloud environment, the fundamental skills will help you connect the dots with those and other environments. Meanwhile, for those new to IT or data infrastructure-related topics and themes, there is plenty here to develop (or refresh) your skillsets, as well as help you move into adjacent technology areas.

Student of IT	New to IT and related topics, perhaps a student at a university, college, or trade school, or starting a new or different career.
Newbie	Relatively new on the job, perhaps first job in IT or an affiliated area, as well someone who has been on the job for a few years and is looking to expand tradecraft beyond accumulating certificates of achievement and expand knowledge as well as experiences for a current or potential future job.
Industry veteran	Several decades on the job (or different jobs), perhaps soon to retire, or simply looking to expand (or refresh) tradecraft skills in a current focus area or an adjacent one. In addition to learning new tradecraft, continue sharing tradecraft experiences with others.
Student of the game	Anyone who is constantly enhancing game or tradecraft skills in different focus areas as well as new ones while sharing experiences and helping others to learn.

How This Book Is Organized

There are four parts in addition to the front and back matter (including Appendices A to G and a robust Glossary). The front matter consists of Acknowledgments and About the Author sections; a Preface, including Who Should Read This Book and How This Book Is Organized; and a Table of Contents. The back matter indicates where to learn more along with my companion sites (www.storageio.com, www.storageioblog.com, and @StorageIO). The back matter also includes the Index.

Figure 1 illustrates the organization of this book, which happens to align with typical data infrastructures topics and themes. Using the figure as a guide or map, you can jump around to different sections or chapters as needed based on your preferences.

In between the front and back matter exists the heart of this book: the fundamental items for developing and expanding your server storage I/O tradecraft (experience). There are four parts, starting out with big picture fundaments in Chapter 1 and application, data infrastructures and IT environment items in Chapter 2.

Part Two is a deep dive into server storage I/O, covering from bits and bytes, software, servers, server I/O, and distance networking. Part Three continues with a storage deep dive, including a storage medium and device components, and data services (functionality). Part Four puts data infrastructure together and includes server storage solutions, managing data infrastructures, and deployment considerations, tying together various topics in the book.

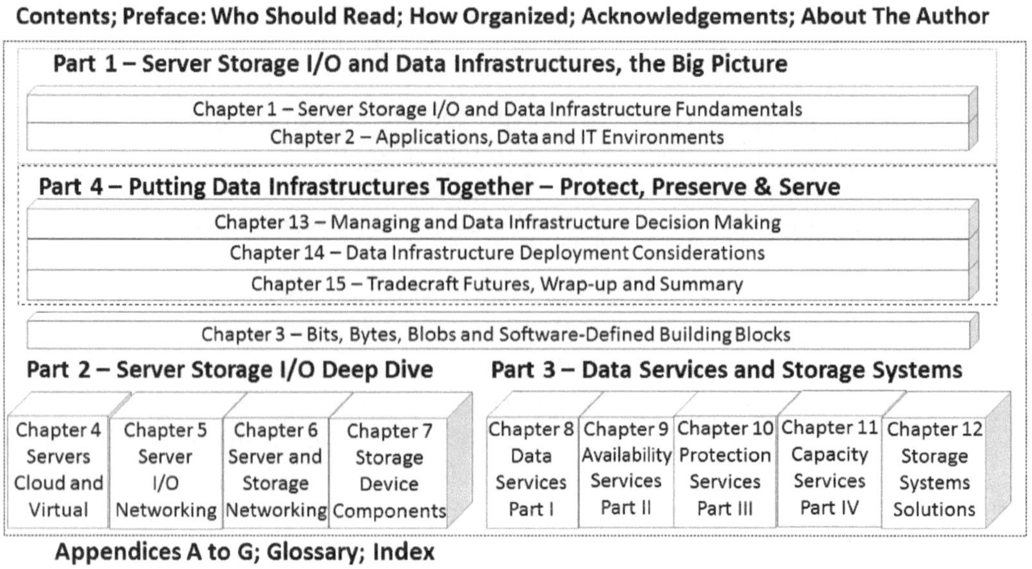

Contents; Preface: Who Should Read; How Organized; Acknowledgements; About The Author
Part 1 – Server Storage I/O and Data Infrastructures, the Big Picture
Chapter 1 – Server Storage I/O and Data Infrastructure Fundamentals
Chapter 2 – Applications, Data and IT Environments
Part 4 – Putting Data Infrastructures Together – Protect, Preserve & Serve
Chapter 13 – Managing and Data Infrastructure Decision Making
Chapter 14 – Data Infrastructure Deployment Considerations
Chapter 15 – Tradecraft Futures, Wrap-up and Summary
Chapter 3 – Bits, Bytes, Blobs and Software-Defined Building Blocks
Part 2 – Server Storage I/O Deep Dive
Part 3 – Data Services and Storage Systems
Chapter 4 Servers Cloud and Virtual
Chapter 5 Server I/O Networking
Chapter 6 Server and Storage Networking
Chapter 7 Storage Device Components
Chapter 8 Data Services Part I
Chapter 9 Availability Services Part II
Chapter 10 Protection Services Part III
Chapter 11 Capacity Services Part IV
Chapter 12 Storage Systems Solutions
Appendices A to G; Glossary; Index

Figure 1 The organization of the book.

In each chapter, you will learn as part of developing and expanding (or refreshing) your data infrastructures tradecraft, hardware, software, services, and technique skills. There are various tables, figures, screenshots, and command examples, along with who's doing what. You will also find tradecraft tips, context matters, and tools for your toolbox, along with common questions as well as learning experiences. Figure 2 shows common icons used in the book.

Feel free to jump around as you need to. While the book is laid out in a sequential hierarchy "stack and layer" fashion; it is also designed for random jumping around. This enables you

Mobile Devices
Management Tools
Database, Key-Value Meta-data Repository
Audio
Photo Image
Video
File System Object / Blob Hierarchy
Documents
Wireless
Hypervisor
SAN, NAS Object
Standalone Bare Metal
Single and Dual Server With Hypervisors
Scale-Out SAN, NAS, Object
Cloud (Public, Private, Hybrid)
Memory NVM, SCM, Flash SSD
Storage NVM, SSD, Hybrid, HDD
Tape
Adapters
HBA, RAID, NIC Card
Switch / Router
Cloud

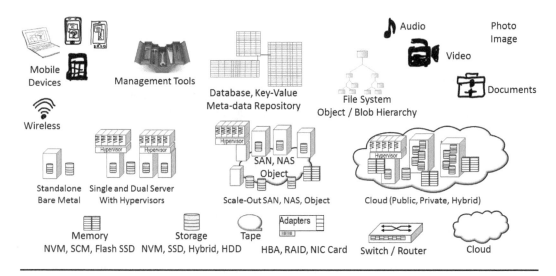

Figure 2 Common icons used in this book.

to adapt the book's content to your needs and preferences, which may be lots of small, quick reads, or longer, sustained deep reading. Appendix F provides a guide on how to use this book for different audiences who have various focus, interests, and levels of experience.

In case you did not pick up on it, I just described the characteristics of software-defined data infrastructures leveraging server storage I/O technologies for different applications spanning cloud, virtual, container, legacy, and, of course, software—all of which are defined to your needs.

Acknowledgments

Writing a book is more than putting pen to paper—or, in this case, typing on a computer—it includes literally hundreds of hours working behind the scenes on various activities. In some ways, writing a book is similar to a technology development project, whether that be hardware, software, or a service, in that there is the initial assessment of the need, including make (do it yourself, DIY) or buy (have someone write it for you).

For this, like my other solo book projects, I went with make (DIY), using a publisher as a service for the postproduction and actual printing as well as distribution. Behind-the-scenes activities included research and discussions with practitioners (new and old). Other activities included plenty of hands-on behind-the-wheel lab time and learning. That was in addition to actual content generation, editing, reviewing, debugging, more editing, working with text as well as graphics, administrative project management, contracts, marketing, and production, among other activities.

Thanks and appreciation to all of the vendors, vars, service providers, press and media, freelance writers as well as reporters, investors and venture capitalists, bloggers, and consultants, as well as fellow Microsoft MVPs and VMware vExperts. Also thanks to all Twitter tweeps and IT professionals around the world that I have been fortunate enough to talk with while putting this book together.

I would also like to thank all of my support network as well as others who were directly or indirectly involved with this project, including Chad Allram, Annie and Steve (and Teddy) Benjamin, Gert and Frank Brouwer, Georgiana Comsa, Rob Dombrowsky, Jim Dyer, Carl Folstad, Steve Guendert, Mark Hall, TJ Hoffman, Nate Klaphake, Anton Kolomyeytsev, Kevin Koski, Ray Lucchesi, Roger Lund, Corey Peden, Bruce Ravid, Drew Robb, and George Terwey, among many others.

Special thanks to Tom Becchetti, Greg Brunton, Mark McSherry, and Dr. "J" Metz. Thanks to John Wyzalek, my publisher, along with everyone else at CRC/Taylor & Francis/ Auerbach, as well as a big thank you to Theron Shreve at DerryField Publishing Services and his associates, Lynne Lackenbach and Marje Pollack, for working their magic.

Finally, thanks to my wife Karen (www.karenofarcola.com) for having the patience to support me while I worked on this project.

To all of the above and, to you the reader, thank you very much.

About the Author

Greg Schulz is Founder and Senior Analyst of the independent IT advisory and consultancy firm Server StorageIO (www.storageio.com). He has worked in IT at an electrical utility and at financial services and transportation firms in roles ranging from business applications development to systems management and architecture planning.

Greg is the author of the Intel Recommended Reading List books *Cloud and Virtual Data Storage Networking* (CRC Press, 2011) and *The Green and Virtual Data Center* (CRC Press, 2009) as well as *Resilient Storage Networks* (Elsevier, 2004), among other works. He is a multi-year VMware vSAN and vExpert as well as a Microsoft MVP and has been an advisor to various organizations including CompTIA Storage+ among others.

In addition to holding frequent webinars, on-line, and live in-person speaking events and publishing articles and other content, Greg is regularly quoted and interviewed as one of the most sought-after independent IT advisors providing perspectives, commentary, and opinion on industry activity.

Greg has a B.A. in computer science and a M.Sc. in software engineering from the University of St. Thomas. You can find him on Twitter @StorageIO; his blog is at www.storageioblog.com, and his main website is www.storageio.com.

About the Author

Part One

Server Storage I/O, Software-Defined and Data Infrastructures

Part One includes Chapters 1 and 2, and provides an overview of the book as well as key concepts including industry trends, different environments, and applications that rely on data infrastructures. Software-defined data infrastructures (SDDI), also known as software-defined data centers (SDDC), span from legacy to virtual, containers, cloud, converged, and hybrid solutions.

Buzzword terms, trends, technologies, and techniques include application, big data applications, cloud, landscapes, little data, performance, availability, capacity, and economics (PACE), server storage I/O networking, software-defined, structured and unstructured, among others.

Part One

Drivers of Tourist Satisfaction:
Demand and Destination Attributes

Chapter 1

Server Storage I/O and Data Infrastructure Fundamentals

What good is data if you can't process it into information?

What You Will Learn in This Chapter

- How/why everything is not the same in most IT environments
- What are data infrastructures and their fundamental components
- Server and storage I/O tradecraft and basic terminology
- IT industry trends and server storage I/O demand drivers
- How to articulate the role and importance of server storage I/O
- How servers and storage support each other connected by I/O networks

This opening chapter kicks off our discussion of server storage input/output (I/O) and data infrastructure fundamentals. Key themes, buzzwords, and trends addressed in this chapter include server storage I/O tradecraft, data demand drivers, fundamental needs and uses of data infrastructures, and associated technologies.

Our conversation is going to span hardware, software, services, tools, techniques, and industry trends along with associated applications across different environments. Depending on when you read this, some of the things discussed will be more mature and mainstream, while others will be nearing or have reached the end of their usefulness.

On the other hand, there will be some new things emerging while others will have joined the "Where are they now list?" similar to music one-hit wonders or great expectations that don't pan out. Being a book about fundamentals, some things change while others remain the same, albeit with new or different technologies or techniques.

In other words, in addition to being for information technology (IT) beginners and new-bies to servers, storage, and I/O connectivity hardware and software, this book also provides fundamental technology, tools, techniques, and trends for seasoned pros, industry veterans, and other experts. Thus, we are going to take a bit of a journey spanning the past, present, and future to understand why things are done, along with ideas on how things might be done differently.

Key themes that will be repeated throughout this book include:

- There is no such as thing as an information recession.
- There is more volume of data, and data is getting larger.
- Data is finding new value, meaning that it is living longer.
- Everything is not the same across different data centers, though there are similarities.
- Hardware needs software; software needs hardware.
- Servers and storage get defined and consumed in various ways.
- Clouds, public, private, virtual private and hybrid servers, services and storage.
- Non-volatile memory (NVM) including NAND flash solid-state devices (SSD).
- Context matters regarding server, storage I/O, and associated topics.
- The fundamental role of data infrastructure: Protect, preserve, and serve information.
- Hybrid is the IT and technology home-run into the future.
- The best server, storage, I/O technology depends on what are your needs.

1.1. Getting Started

Server storage I/O data infrastructure fundamentals include hardware systems and components as well as software that, along with management tools, are core building blocks for converged and nonconverged environments. Also fundamental are various techniques, best practices, policies, and "tradecraft" experience tips that will also be explored. Tradecraft is the skills, experiences, insight, and tricks of the trade associated with a given profession.

When I am asked to sum up, or describe, server, storage, and I/O data infrastructures in one paragraph, it is this: Servers need memory and storage to store data. Data storage is accessed via I/O networks by servers whose applications manage and process data. The fundamental role of a computer server is to process data into information; it does this by running algorithms

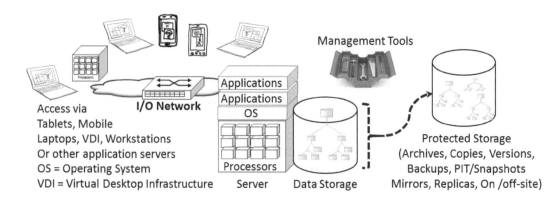

Figure 1.1   Server storage I/O fundamentals: the "big picture."

(programs or applications) that must be able to have data to process. The sum of those parts is the software-defined data infrasture enabled by hardware, software, and other services.

Likewise, the fundamental role of data storage ("storage") is to provide persistent memory for servers to place data to be protected, preserved, and served. Connectivity for moving data between servers and storage, from servers to servers, or from storage to storage is handled via I/O networks (internal and external). There are different types of servers, storage, and I/O networks for various environments, functionalities, as well as application or budget needs.

Figure 1.1 shows a very simplistic, scaled-down view of servers, storage, and I/O resources supporting applications being accessed via tablets, mobile devices, laptops, virtual desktop infrastructure (VDI), workstations, and other servers. Also shown in Figure 1.1 is storage (internal or external, dedicated and shared) being protected by some other storage system (or service). We will be "right-clicking" or "drilling down" (i.e., going into more detail) about each of the above as well as other areas concerned with server, storage, and I/O data infrastructure fundamentals throughout this chapter and the rest of the book.

Keeping in mind that Figure 1.1 is the "big picture" and a simple one at that means we could scale it down even further to a laptop or tablet, or, in the opposite direction, to a large web-scale or cloud environment of tens of thousands of servers, storage, and I/O components, hardware, and software.

A fundamental theme is that servers process data using various applications programs to create information; I/O networks provide connectivity to access servers and storage; storage is where data gets stored, protected, preserved, and served from; and all of this needs to be managed. There are also many technologies involved, including hardware, software, and services as well as various techniques that make up a server, storage, and I/O enabled data infrastructure.

Server storage I/O and data infrastructure fundamental focus areas include:

- Organizations—Markets and industry focus, organizational size
- Applications—What's using, creating, and resulting in server storage I/O demands
- Technologies—Tools and hard products (hardware, software, services, packaging)
- Tradecraft—Techniques, skills, best practices, how managed, decision making
- Management—Configuration, monitoring, reporting, troubleshooting, performance, availability, data protection and security, access, and capacity planning

Applications are what transform data into information. Figure 1.2 shows how applications, which are software defined by people and software, consist of algorithms, policies, procedures, and rules that are put into some code to tell the server processor (CPU) what to do.

Application programs include data structures (not to be confused with infrastructures) that define what data looks like and how to organize and access it using the "rules of the road" (the algorithms). The program algorithms along with data structures are stored in memory, together with some of the data being worked on (i.e., the active working set).

Additional data is stored in some form of extended memory—storage devices such as non-volatile memory (NVM), solid-state devices (SSD), hard disk drives (HDD), or tape, among others, either locally or remotely. Also shown in Figure 1.2 are various devices that perform input/output (I/O) with the applications and server, including mobile devices as well as other application servers. In Chapter 2 we take a closer look at various applications, programs, and related topics.

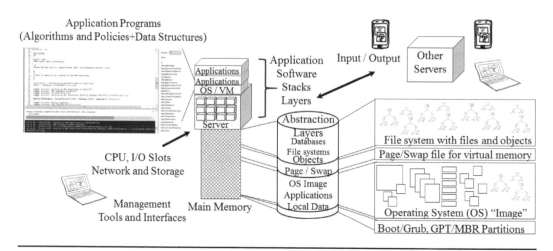

Figure 1.2   How data infrastructure resources transform data into information.

1.2. What's the Buzz in and around Servers, Storage, and I/O?

There is a lot going on, in and around data infrastructure server, storage, and I/O networking connectivity from a hardware, software, and services perceptive. From consumer to small/medium business (SMB), enterprise to web-scale and cloud-managed service providers, physical to virtual, spanning structured database (aka "little data") to unstructured big data and very big fast data, a lot is happening today.

Figure 1.3 takes a slightly closer look at the server storage I/O data infrastructure, revealing different components and focus areas that will be expanded on throughout this book.

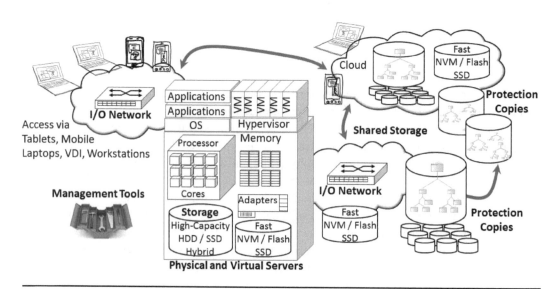

Figure 1.3   What's inside the data infrastructure, server, storage, and I/O resources.

Some buzz and popular trending topics, themes, and technologies include, among others:

- Non-volatile memory (NVM), including NAND flash SSD
- Software defined data centers (SDDC), networks (SDN), and storage (SDS)
- Converged infrastructure (CI), hyper-converged infrastructure (HCI)
- Cluster-in-Box and Cloud-in-Box (CiB) along with software stack solutions
- Scale-out, scale-up, and scale-down resources, functional and solutions
- Virtualization, containers, cloud, and operating systems (OS)
- NVM express (NVMe) and PCIe accessible storage
- Block, file, object, and application program interface (API) accessed storage
- Data protection, business resiliency (BR), archiving, and disaster recovery (DR)

In Figure 1.3 there are several different focus topics that enable access to server and storage resources and services from various client or portable devices. These include access via server I/O networks, including:

- Local area networks (LAN)
- Storage area networks (SAN)
- Metropolitan area networks (MAN), and wide area networks (WAN)

Also, keep in mind that hardware needs software and software needs hardware, including:

- Operating systems (OS) and Docker, Linux, as well as Windows containers
- Hypervisors (such as Xen and KVM, Microsoft Hyper-V, VMware vSphere/ESXi)
- Data protection and management tools
- Monitoring, reporting, and analytics
- File systems, databases, and key-value repositories

All of these are critical to consider, along with other applications that all rely on some underlying hardware (bare metal, virtual, or cloud abstracted). There are various types of servers, storage, and I/O networking hardware as well as software that have various focus areas or specialties, which we will go deeper into in later chapters.

There is a lot of diversity across the different types, sizes, focus, and scope of organizations. However, it's not just about size or scale; it's also about the business or organizational focus, applications, needs (and wants), as well as other requirements and constraints, including a budget. Even in a specific industry sector such as financial services, healthcare or life science, media and entertainment, or energy, among others, there are similarities but also differences. While everything that comes to server storage I/O is not the same, there are some commonalities and similarities that appear widely.

Figure 1.4 shows examples of various types of environments where servers, storage, and I/O have an impact on the consumer, from small office/home office (SOHO) to SMB (large and small), workgroup, remote office/branch office (ROBO), or departmental, to small/medium enterprise (SME) to large web-scale cloud and service providers across private, public, and government sectors.

In Figure 1.4 across the top, going from left to right, are various categories of environments also known as market segments, price bands, and focus areas. These environments span

Different Industries, Database, Web, Application Development/DevOpp, Big Data and Analytics, Networking and CDN. Research, Medical/Life Science, Financial Video, VDI, High Performance Compute (HPC), General Purpose, Public and Private Clouds, Data Protection, Rich Content, File and Data Serving

Various Environments

Consumer, SOHO, ROBO

Workgroup, Departmental, SMB

SME, Enterprise and Cloud Scale

Different Attributes

Local and Remote

Scale-up, Scale-down, Scale-out

Physical, Virtual, Cloud

Software and Hardware Defined

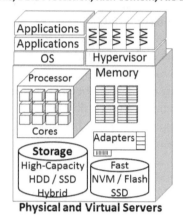

Applications | VM VM VM VM
Applications |
OS | Hypervisor
Processor | Memory
Cores | Adapters
Storage
High-Capacity HDD / SSD Hybrid | Fast NVM / Flash SSD

Physical and Virtual Servers

Management Tools

Management Functions

Data Protection

Provision and Repair

Analyze and Remediate

Monitor and Report

Figure 1.4   Different environments with applications using servers, storage, and I/O.

different industries or organizational focus from academic and education to healthcare and life sciences, engineering and manufacturing to aerospace and security, from media and entertainment to financial servers, among many others.

Also shown in Figure 1.4 are some common functions and roles that servers, storage, and I/O resources are used for, including network routing and access and content distribution networks (CDN). Other examples include supercomputing, high-performance computing (HPC), and high-productivity computing. Some other applications and workloads shown in Figure 1.4 include general file sharing and home directories, little data databases along with big data analytics, email and messaging, among many others discussed further in Chapter 2.

Figure 1.4 also shows that the role of servers (processors, memory, and I/O connectivity) combined with some storage can be deployed in various ways from preconfigured engineered packaged solutions to build your own leveraging open source and available (aka commodity or white box) hardware.

Besides different applications, industry, and market sectors concerned with server, storage, and I/O topics, there are also various technology and industry trends. These include, among others, analytics, application-aware, automation, cloud (public, private, community, virtual private, and hybrid) CiB, HCI, CI, containers, and micro services.

Other data infrastructure and applications include data center infrastructure management (DCIM), data lakes, data ponds, data pools and data streams, data protection and security, HCI, insight and reporting, little data, big data, big fast data, very big data, management, orchestration, policies, software defined, structured data and unstructured data, templates, virtual server infrastructures (VSI), and virtual desktop infrastructure (VDI), among others.

Additional fundamental buzz and focus from a technology perspective include:

- Server-side memory and storage (storage in, adjacent to, or very close to the server)

- NVM including 3D NAND flash, 3D XPoint, and others such as phase change memory (PCM), DRAM, and various emerging persistent and nonpersistent memory technologies
- Storage class memories (SCM), which have the persistence of NVM storage and the performance as well as durability of traditional server DRAM
- I/O connectivity including PCIe, NVMe, SAS/SATA, InfiniBand, Converged Ethernet, RDMA over Converged Ethernet (RoCE), block, file, object, and API-accessed storage
- Data analytics including Hadoop, Hortonworks, Cloudera, and Pivotal, among others
- Databases and key-value repositories including SQL (AWS RDS, IBM DB2, Microsoft SQL Server, MariaDB, MemSQL, MySQL, Oracle, PostgresSQL, ClearDB, TokuDB) and NoSQL (Aerospike, Cassandra, CouchDB, HBASE, MongoDB, Neo4j, Riak, Redis, TokuDB) as well as big data or data warehouse (HDFS based, Pivotal Greenplum, SAP HANA, and Teradata), among others

1.2.1. Data Infrastructures—How Server Storage I/O Resources Are Used

Depending on your role or focus, you may have a different view than somebody else of what is infrastructure, or what an infrastructure is. Generally speaking, people tend to refer to infrastructure as those things that support what they are doing at work, at home, or in other aspects of their lives. For example, the roads and bridges that carry you over rivers or valleys when traveling in a vehicle are referred to as infrastructure.

Similarly, the system of pipes, valves, meters, lifts, and pumps that bring fresh water to you, and the sewer system that takes away waste water, are called infrastructure. The telecommunications network—both wired and wireless, such as cell phone networks—along with electrical generating and transmission networks are considered infrastructure. Even the planes, trains, boats, and buses that transport us locally or globally are considered part of the transportation infrastructure. Anything that is below what you do, or that supports what you do, is considered infrastructure.

This is also the situation with IT systems and services where, depending on where you sit or use various services, anything below what you do may be considered infrastructure. However, that also causes a context issue in that infrastructure can mean different things. For example in Figure 1.5 the user, customer, client, or consumer who is accessing some service or application may view IT in general as infrastructure, or perhaps as business infrastructure.

Those who develop, service, and support the business infrastructure and its users or clients may view anything below them as infrastructure, from desktop to database, servers to storage, network to security, data protection to physical facilities. Moving down a layer in Figure 1.5 is the information infrastructure which, depending on your view, may also include servers, storage, and I/O hardware and software.

For our discussion, to help make a point, let's think of the information infrastructure as the collection of databases, key-value stores, repositories, and applications along with development tools that support the business infrastructure. This is where you may find developers who maintain and create actual business applications for the business infrastructure. Those in the information infrastructure usually refer to what's below them as infrastructure. Meanwhile, those lower in the stack shown in Figure 1.4 may refer to what's above them as the customer, user, or application, even if the actual user is up another layer or two.

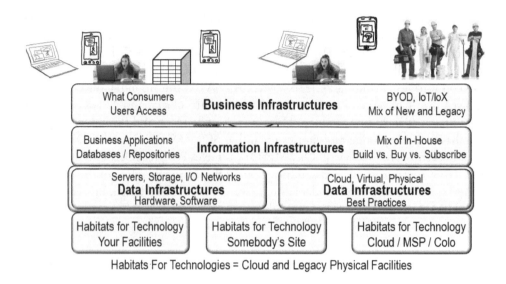

Figure 1.5   IT Information and data infrastructure.

Context matters in the discussion of infrastructure. So for our of server storage I/O funda-mentals, the data infrastructures support the databases and applications developers as well as things above, while existing above the physical facilities infrastructure, leveraging power, cool-ing, and communication network infrastructures below.

Figure 1.6 shows a deeper look into the data infrastructure shown at a high level in Figure 1.5. The lower left of Figure 1.6 shows the common-to-all-environments hardware, software, people, processes, and practices that comprise tradecraft (experiences, skills, tech-niques) and "valueware." Valueware is how you define the hardware and software along with

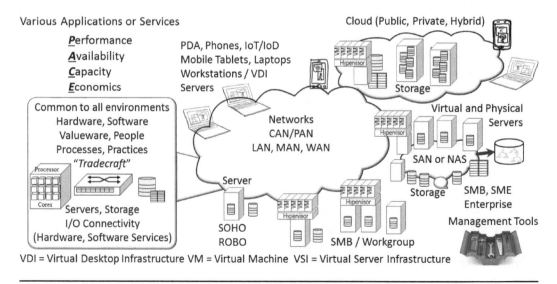

Figure 1.6   Data infrastructures: server storage I/O hardware and software.

any customization to create a resulting service that adds value to what you are doing or supporting. Also shown in Figure 1.6 are common application and services attributes including performance, availability, capacity, and economics (PACE), which vary with different applications or usage scenarios.

Common server storage I/O fundamentals across organizations and environments include:

- While everything is not the same, there are similarities.
- One size, technology, or approach does not apply to all scenarios.
- Some things scale up, others scale down; some can't scale up, or scale down.
- Data protection includes security, protection copies, and availability.
- The amount (velocity), as well as size (volume) of data continues to grow.
- Servers generate, collect, and process data into information.
- I/O networks move data and information results between servers, storage, and users.
- Storage is where data is stored temporarily or long-term.

Figure 1.7 shows the fundamental pillars or building blocks for a data infrastructure, including servers for computer processing, I/O networks for connectivity, and storage for storing data. These resources including both hardware and software as well as services and tools. The size of the environment, organization, or application needs will determine how large or small the data infrastructure is or can be.

For example, at one extreme you can have a single high-performance laptop with a hypervisor running OpenStack and various operating systems along with their applications leveraging flash SSD and high-performance wired or wireless networks powering a home lab or test environment. On the other hand, you can have a scenario with tens of thousands (or more) servers, networking devices, and hundreds of petabytes (PBs) of storage (or more).

In Figure 1.7 the primary data infrastructure components or pillar (server, storage, and I/O) hardware and software resources are packaged and defined to meet various needs.

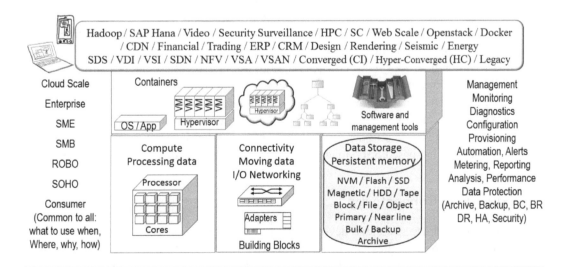

Figure 1.7   Server storage I/O building blocks (hardware, software, services).

Software-defined storage management includes configuring the server, storage, and I/O hardware and software as well as services for use, implementing data protection and security, provisioning, diagnostics, troubleshooting, performance analysis, and other activities. Server storage and I/O hardware and software can be individual components, prepackaged as bundles or application suites and converged, among other options.

Servers and Memory

Fast applications need faster software (databases, file systems, repositories, operating systems, hypervisors), servers (physical, virtual, cloud, converged), and storage and I/O networks. Servers and their applications software manage and process data into information by leveraging local as well as remote storage accessed via I/O networks.

Application and computer server software is installed, configured, and stored on storage. That storage may be local or external dedicated, or shared. Servers leverage different tiers of memory from local processor cache to primary main dynamic random access memory (DRAM). Memory is storage, and storage is a persistent memory.

Memory is used for holding both applications and data, along with operating systems, hypervisors, device drivers, as well as cache buffers. Memory is staged or read and written to data storage that is a form of persistent memory ranging from non-volatile memory (NVM) NAND flash SSD to HDD and magnetic tape, among others. Bare-metal or physical machine (PM) servers can be virtualized as virtual machines (VM) or as cloud instances or cloud machines (CM).

Data Storage

Servers or other computers need storage, storage needs servers, and I/O networks tie the two together. The I/O network may be an internal PCIe or memory bus, or an external Wi-Fi network for IP connection, or use some other interface and protocol. Data and storage may be coupled directly to servers or accessed via a networked connection to meet different application needs. Also, data may be dedicated (affinity) to a server or shared across servers, depending on deployment or application requirements.

While data storage media are usually persistent or non-volatile, they can be configured and used for ephemeral (temporary) data or for longer-term retention. For example, a policy might specify that data stored in a certain class or type of storage does not get backed up, is not replicated or have high-availability (HA) or other basic protection, and will be deleted or purged periodically. Storage is used and consumed in various ways. Persistent storage includes NVM such as flash and SCM-based SSD, magnetic disk, tape, and optical, among other forms.

I/O Networking Connectivity

Servers network and access storage devices and systems via various I/O connectivity options or data highways. Some of these are local or internal to the server, while others can be external over a short distance (such as in the cabinet or rack), across the data center, or campus, or metropolitan and wide area, spanning countries and continents. Once networks are set up, they typically are used for moving or accessing devices and data with their configurations stored in some form of storage, usually non-volatile memory or flash-based. Networks can be wired using copper electrical cabling or fiber optic, as well as wireless using various radio frequency (RF) and other technologies locally, or over long distances.

Packaging and Management

There are also various levels of abstraction, management, and access, such as via block, file, object, or API. Shared data vs. data sharing can be internal dedicated, external dedicated, external shared and networked. In addition to various ways of consuming, storage can also be packaged in different ways such as legacy storage systems or appliances, or software combined with hardware ("tin wrapped").

Other packaging variations include virtual storage appliance (VSA), as a cloud instance or service, as well as via "shrink wrap" or open-source software deployed on your servers. Servers and storage hardware and software can also be bundled into CI, HCI, and CiB, similar to an all-in-one printer, fax, copier, and scanner that provide converged functionality.

1.2.2. Why Servers Storage and I/O Are Important (Demand Drivers)

There is no information recession; more and more data being generated and processed that needs to be protected, preserved, as well as served. With increasing volumes of data of various sizes (big and little), if you simply do what you have been doing in the past, you better have a big budget or go on a rapid data diet.

On the other hand, if you start using those new and old tools in your toolbox, from disk to flash and even tape along with cloud, leveraging data footprint reduction (DFR) from the application source to targets including archiving as well as deduplication, you can get ahead of the challenge.

Figure 1.8 shows data being generated, moved, processed, stored, and accessed as information from a variety of sources, in different locations. For example, video, audio, and still image data are captured from various devices, copied to local tablets or workstations, and uploaded to servers for additional processing.

Data is also captured from various devices in medical facilities such as doctors' offices, clinics, labs, and hospitals, and information on patient electronic medical records (EMR) is accessed. Digital evidence management (DEM) systems provide similar functionalities, supporting devices such as police body cameras, among other assets. Uploaded videos, images, and photos are processed; they are indexed, classified, checked for copyright violations using waveform analysis or other software, among other tasks, with metadata stored in a database or key-value repository.

The resulting content can then be accessed via other applications and various devices. These are very simple examples that will be explored further in later chapters, along with associated tools, technologies, and techniques to protect, preserve, and serve information.

Figure 1.8 shows many different applications and uses of data. Just as everything is not the same across different environments or even applications, data is also not the same. There is "little data" such as traditional databases, files, or objects in home directories or shares along with fast "big data." Some data is structured in databases, while other data is unstructured in file directories.

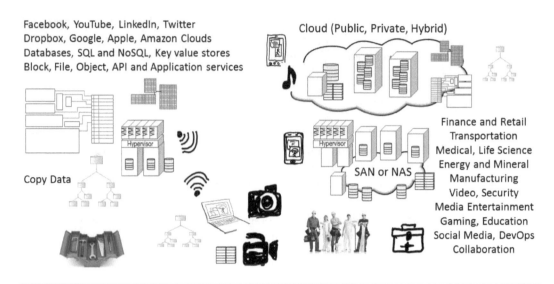

Figure 1.8   Examples of server storage I/O demand drivers and influences.

1.2.3. Data Value

Data also may have different values at different times. Very often, context is needed: Data can represent a value or a piece of information, but one must also consider the value or worth of that data item to the organization. The basic three types of data value are no value, has value, or has an unknown value.

Data that has unknown value may eventually have some value or no value. Likewise, data that currently has some form of value may eventually have no value and be discarded, or may be put into a pool or parking place for data with known value until its new value is determined or the time has come to delete the data. In addition to the three basic types of data value, there can be different varieties of data with a value, such as high-value or low-value.

Besides the value of data, access is also part of its life cycle, which means the data may be initially stored and updated, but later becomes static, to be read but not modified. We will spend more time in Chapter 2 looking at different types of applications in terms of data values, life cycles, and access patterns.

General demand drivers of data value include:

- Creation—The Internet of Things (IoT) and the Internet of Devices (IoD), such as sensors and other devices including camera, phones, tablets, drones, satellites, imagery, telemetry, and log detail data, legacy creation, and machine-to-machine (M2M).
- Curation, Transformation—Transformation, analytics, and general processing, across different sectors, industries, organizations, and focus areas. Also tagging and adding metadata to data where applicable to add value and context.
- Consumption—Content distribution, cache, collaboration, downloads, various consumers

- Preservation—With more data being stored for longer periods of time, combined with other factors mentioned above, simply preserving data is in itself a large demand driver.

Additional demand drivers include:

- Increased number of devices creating and consuming data
- More volumes of data being created, processed, and stored, faster
- Log data (call detail, transactions, cookies, and tracking)
- Larger data items (files, images, videos, audio)
- More copies of items being kept for longer periods of time
- Data finding new value now, or in the future
- Ability to process data faster results in more information quicker
- Old and new applications creating, curating, consuming data
- Mobile and consumerized access of information services
- Easier-to-use applications and services to create and consume data
- Virtual machines and virtual disks creating virtual data sprawl

In addition to the above drivers, there are also various industry- or organization-focused applications including government, police and military, healthcare and life sciences, security and surveillance video, media and entertainment, education, research, manufacturing, energy exploration, transportation, finance, and many others.

1.3. Server Storage I/O Past, Present, and Future

There is always something new and emerging, including the next shiny new thing (SNT) or shiny new technology. Likewise, there are new applications and ways of using the SNT as well as legacy technologies and tools.

If you are like most people, your environment consists of a mix of legacy applications (and technology) as well as new ones; or perhaps everything is new, and you are starting from scratch or "greenfield" (i.e., all new). On the other hand, you might also be in a situation in which everything is old or legacy and is simply in a sustaining mode.

In addition to new technology, some existing technology may support both new and old or legacy applications. Some of these applications may have been re-hosted or moved from one operating hardware and operating system environment to another, or at least upgraded. Others may have been migrated from physical to virtual or cloud environments, while still others may have been rewritten to take advantage of new or emerging technologies and techniques.

1.3.1. Where Are We Today? (Balancing Legacy with Emerging)

A trend has been a move from centralized computer server and storage to distributed, then to centralized via consolidation (or aggregation), then back to distributed over different generations from mainframe to time-sharing to minicomputers to PCs and client servers, to the web and virtualized to the cloud. This also includes going from dedicated direct attached storage to

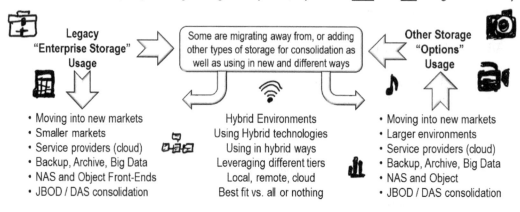

Figure 1.9   Server and storage yesterday and today.

clustered and shared storage area networks (SAN) or network attached storage (NAS) to object storage, virtual and cloud storage, and back to direct attached storage, among other options.

What this all means is that from a server storage I/O hardware, software, and services perspective, consideration is needed for what is currently legacy, how to keep it running or improve it, as well as what might be a shiny new thing today but legacy tomorrow, while also keeping an eye on the future.

This ties into the theme of keeping your head in the clouds (looking out toward the future) but your feet on the ground (rooted in the reality of today), and finding a balance between what needs to be done and what you want to do. Figure 1.9 shows, for example, how enterprise and other storage options are evolving, taking on new or different roles while also being enhanced.

Some of the changes that are occurring include traditional enterprise-class storage systems being used for secondary roles or combined with software-defined storage management tools, virtualization, cloud, and other technologies. Also shown in Figure 1.9 from a high level are how traditional lower-end or open technologies are handling more enterprise workloads.

Common to what is shown in Figure 1.9 as well as in other scenarios is that the trend of using commercial off-the-shelf (COTS) servers continues to evolve. COTS are also known as commodity, industry standard, and white box products, with motherboards, processors, and memory. This means that servers are defined to be data and storage systems or appliances, as well as being application servers with different hardware configurations to support those needs. Besides physical servers, many cloud providers also offer cloud machines (instances) with various standard configurations ranging from high-performance computing to memory- or I/O-intensive, among other options.

Another trend centers around convergence (all-in-ones and others), where server, storage, and I/O hardware is designed with software to support applications as well as storage and management functions as turnkey solutions. We shall have more to say about these and other trends later.

1.3.2. Where Are We Going? (Future Planning, Leveraging Lessons Learned)

To see and understand where we are going will help us to see where we have been and where we are currently. At first glance, Figure 1.10 may look similar to what you have seen in the past, but what's different and will be covered in more depth in subsequent chapters is the role of servers being defined as storage. In addition to being defined as media content servers for video, audio, or other streaming applications, as a database, email, the web, or other servers, servers are also increasingly being leveraged as the hardware platform for data and storage applications.

A closer look at Figure 1.10 will also reveal new and emerging non-volatile memory that can be used in hybrid ways, for example, as an alternative to DRAM or NVRAM for main memory and cache in storage systems or networking devices, as well as as a data storage medium. Also shown in Figure 1.10 are new access methods such as NVM Express (NVMe) to reduce the latency for performing reads and writes to fast NVM memory and storage. NVMe, while initially focused for inside the server or storage system, will also be available for external access to storage systems over low-latency networks.

Building off of what is shown in Figure 1.10, Figure 1.11 provides another variation of how servers will be defined to be used for deployments as well as how servers provide the processing capability for running storage application software. In Figure 1.11 the top center is a shrunk-down version of Figure 1.10, along with servers defined to be storage systems or appliances for block SAN, NAS file, and Object API including converted, nonconverged, and virtual and cloud based applications.

Figure 1.11 shows how the primary data infrastructure building block resources, including server, storage, I/O networking hardware, software, and services, combine to support various

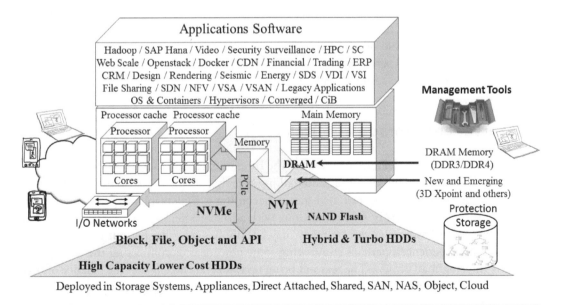

Figure 1.10   Evolving server and storage I/O leveraging new technologies.

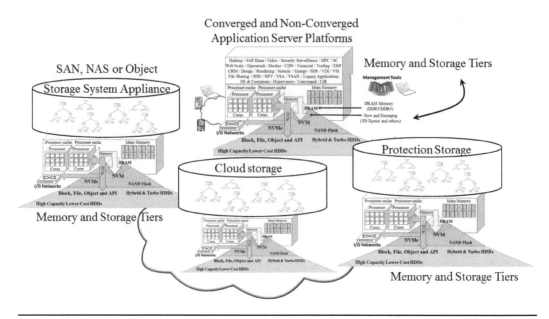

Figure 1.11 Servers as storage and storage as servers, converged and nonconverged.

needs. Not shown in detail is the trend toward protecting data at ingest vs. the past trend of protecting after the fact. This also means collecting metadata and other information about data when it is being stored, to reduce the overhead complexity of searching or scanning data after the fact for management and protection purposes.

Figure 1.11 also illustrates that fundamentals for server, storage, and I/O networking need to have a focus spanning hardware, software, services, tools, processes, procedures, techniques, and tradecraft, even if the resulting solutions are cloud, converged, virtual, or nonconverged. Remember, the primary role of data infrastructures and servers, storage, and I/O resources is to protect, preserve, and serve existing as well as new applications and their data in a cost-effective, agile, flexible, and resilient way.

1.4. Server and Storage I/O Tradecraft

Server and storage I/O tradecraft encompasses the fundamental and as well as advanced skills, techniques, and expertise for managing and making sure that a data infrastructure can support the organization's needs. Tradecraft also means becoming a student of the game, where the game is your particular craft or focus area along with adjacent topics.

For some, what has been covered already may be simply a refresher of your existing tradecraft skills and experiences, perhaps stimulating some new ideas or connecting the dots, while for others this might all be new. The reason I bring up tradecraft here is that the balance of this book will help you expand or refresh your specific tradecraft as we continue to explore server storage I/O fundamentals.

There are many different tasks, activities, and functions concerned with servers, storage, and I/O hardware, software, and services, depending on your environment. Likewise, depending on your focus area or job responsibility, you might be focused on lower-level detailed tasks, or higher-level vision, strategy, architecture, design, and planning, or some combination of these tasks.

In general, common tasks and activities include, among others:

- Vision, strategy, architecture design, project planning, and management
- Configuration of hardware and software; change management
- Troubleshooting, diagnostics, and resolution
- Remediation; applying hardware or software updates
- Maintaining inventory of hardware and software resources
- New installations, configuration changes, updates
- Disposition of old items, resources, life-cycle management
- Migrations, conversions, movement, media maintenance
- What to buy, lease, rent, subscribe, or download
- Data protection including archiving, backup, high availability, business continuity, disaster recovery, and security
- Reporting, analysis, forecasting, chargeback or show-back
- Defining processes, policies, templates, and workflows for automation
- Service-level agreements (SLA), service-level objective (SLO) management
- Learning new things; learning how to use old things in new ways
- Capturing and preserving knowledge, skills, and tradecraft to be shared with others
- Expanding focus to comprehending vs. simply memorizing to pass a test or certification
- Moving beyond memorizing things to pass a test or get a certificate

If you have not noticed yet, my perspective of servers is that they include a conversation of storage and I/O networking from a hardware and software perspective. Likewise, storage is more than HDD or SSD hardware or management software; there are also the applications, servers, and I/O networks along with management tools. I/O networks are more than cables, wires, switches, routers, and software, because all of these are important and converge to support data infrastructures.

This means that for your tradecraft, you need to acquire skills in these different yet related areas. Even if your focus is to be a hardware or software or services person from a particular resource standpoint (e.g., server, virtualization, cloud, networking, storage, performance, data protection, or security, among others), knowing how the other pieces work together and interact to support applications is important.

Trends involving tradecraft include capturing existing experiences and skills from those who are about to retire or simply move on to something else, as well as learning for those new to IT or servers, storage, I/O, and data infrastructure hardware, software, and services. This means being able to find a balance of old and new tools, techniques, and technologies, including using things in new ways for different situations.

Part of expanding your tradecraft skillset is knowing when to use different tools, techniques, and technologies from proprietary and closed to open solutions, from tightly integrated to loosely integrated, to bundled and converged, or to a la carte or unbundled components, with do-it-yourself (DIY) integration. Tradecraft also means being able to balance when to make a change of technology, tool, or technique for the sake of change vs. clinging to something comfortable or known, vs. leveraging old and new in new ways while enabling change without disrupting the data infrastructure environment or users of its services.

A couple of other trends include the convergence of people and positions within organizations that may have been in different silos or focus areas in the past. One example is the rise of Development Operations (also known as DevOps), where instead of separate development, administration, and operations areas, they are a combined entity. This might be *déja vu* for some of you who grew up and gained your tradecraft in similar types of organizations decades ago; for others, it may be something new.

Regarding fundamental tradecraft skills, if you are a hardware person it's wise to learn software; if you are a software person, it's wise to acquire some hardware experience. Also, don't be afraid to say "I don't know" or "it depends on" when asked a question. This also means learning how information technology supports the needs of the business, as well as learning the technology the business uses.

Put another way, in addition to learning server storage I/O hardware and software tradecraft, also learn the basic tradecraft of the business your information systems are supporting. After all, the fundamental role of IT is to protect, preserve, and serve information that enables the business or organization; no business exists just to support IT.

1.5. Fundamental Server and Storage I/O Terminology (Context Matters)

For some of you this will be a refresher; for others, an introduction of server and storage I/O talk (terms and expressions). There are many different terms, buzzwords, and acronyms about IT and particularly server storage I/O–related themes, some of which have more than one meaning.

This is where context matters; look for context clarification in the various chapters. For example, SAS can mean serial attached SCSI in the context of physical cabling, or the protocol command set, as a disk or SSD device, as a server to storage, or storage system to storage device connectivity interface. SAS can also mean statistical analysis software for number crunching, statistics, analysis, big data, and related processing.

Context also matters in terms that are used generically, similar to how some people refer to making a copy as making a Xerox, or getting a Kleenex vs. a tissue. Thus, a "mainframe" could be legacy IBM "big iron" hardware and software or a minicomputer or large server running Windows, Unix, Linux, or something other than a small system; others might simply refer to that as a server or a server as the mainframe, as the cloud, and vice versa (Hollywood and some news venues tend to do this).

The following is not an exhaustive list nor a dictionary, rather some common terms and themes that are covered in this book and thus provide some examples (among others). As examples, the following also include terms for which context is important: a bucket may mean one thing in the context of cloud and object storage and something else in other situations. Likewise, a container can mean one thing for Docker and micro-services and another with reference to cloud and object storage or archives.

Buckets	Data structures or ways of organizing memory and storage, because object storage buckets (also known as containers) contain objects or items being stored.

	There are also non-object or cloud storage buckets, including database and file system buckets for storing data.
Containers	Can refer to database or application containers, data structures such as archiving and backup "logical" containers, folders in file systems, as well as shipping containers in addition to Docker or computer micro-services.
CSV	Comma Separated Variables is a format for exporting, importing, and interchange of data such as with spreadsheets. Cluster Shared Volumes for Microsoft Windows Server and Hyper-V environments are examples.
Flash	Can refer to NAND flash solid-state persistent memory packaged in various formats using different interfaces. Flash can also refer to Adobe Flash for video and other multimedia viewing and playback, including websites.
Objects	There are many different types of server and storage I/O objects, so context is important. When it comes to server, storage, and I/O hardware and software, objects do not always refer to object storage, as objects can refer to many different things.
Partitions	Hard disk drive (HDD), solid-state device (SSD), or other storage device and aggregate subdividing (i.e., storage sharing), database instance, files or table partitioned, file system partition, logical partition or server tenancy.
PCI	Payment Card Industry and PCIe (Peripheral Computer Interconnect Express)
SMB	Small/Medium Business, refers to the size or type of an organization or target market for a solution (or services). SMB is also the name for Server Message Block protocols for file and data sharing, also known as CIFS (Common Internet File System) aka Windows file sharing.
SME	Small/Medium Enterprise, a type or size of organization; also refers to a Subject Matter Expert skilled in a given subject area.
Virtual disk	Abstracted disks such as from a RAID controller, software-defined storage such as Microsoft Storage Spaces, as well as virtual machine including Hyper-V VHD/VHDX, VMware VMDK, and qcow, among others.
VM	Virtual Memory (Operating System Virtual Memory, Hypervisor Virtual Memory, Virtual Machine Memory), Virtual Machine, Video Monitor.

You can find a more comprehensive glossary at the end of this book. One of the links at the end of this book is to the SNIA (Storage Network Industry Association) Dictionary available in several languages, including English and Portuguese.

1.6. What's in Your Fundamental Toolbox?

Do you have a virtual or physical server storage I/O toolbox? If not, now is a good time to create one. On the other hand, if you already have a toolbox, now is also a good time to start updating it, refamiliarizing yourself with what you have, or replacing broken or worn-out items. Keep in mind that a toolbox for server storage I/O can contain physical items such as hardware adapters, cables, connectors, drives, or spare parts, along with actual tools (e.g., screwdrivers, magnifying glass, flashlight).

Note that your toolbox should also include items on your physical or virtual bookshelf, a collection of videos, white papers, reports, tutorials, notes, and tip sheets.

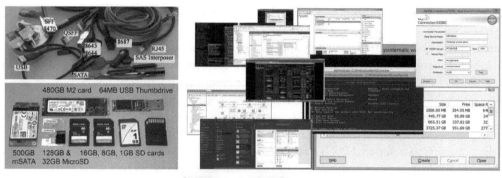

Physical, Virtual, Containers
Cloud and Software Defined

Hardware, Physical Tools, Cables
Connectors, Adapters, Storage
Servers and Management Software

Data Infrastructure Tool Box

Figure 1.12   Server storage I/O fundamental toolbox.

A server storage I/O toolbox should also exist in a virtual form, that is, software tools for configuring, managing, monitoring, troubleshooting, and reporting to support your various environments. For example, I have a physical toolbox that consists of parts and tools (tweezers, screwdrivers, flashlights, magnifying glass, pickups, among other things (see Figure 1.12), along with plastic tie straps, cable labels, screws, and related items.

I also have a virtual toolbox that consists of various software tools that I use for servers, storage, and I/O physical, virtual, and cloud solutions. Note that in addition to hardware and software tools, I also have notes and tip sheets with commands, scripts, and other hints to help me remember how or what to do across different systems including physical, cloud, virtual, container, and hypervisor operating systems, as well as database, data protection, and other focus areas.

> *Tip:* Knowing about the tool is important; however, so too is knowing how to use a tool and when along with where or for what. This means knowing the tools in your toolbox, but also knowing when, where, why, and how to use a given tool (or technology), along with techniques to use that tool by itself or in conjunction with multiple other tools.

Part of server storage I/O data infrastructure tradecraft is understanding what tools to use when, where, and why, not to mention knowing how to improvise with those tools, find new ones, or create your own.

Remember, if all you have is a hammer, everything starts to look like a nail. On the other hand, if you have more tools than you know what to do with, or how to use them, perhaps fewer tools are needed along with learning how to use them by enhancing your skillset and tradecraft.

1.7. Common Questions and Tips

A common question I get asked is *"Why can't we just have one type of data storage such as NAND flash SSD, or put everything in the cloud?"* My usual answer is that if you can afford to buy, rent, or subscribe to a cloud or managed service provider (MSP) all-flash array (AFA) solution, then do so.

While some organizations are going the all-flash storage approach, most cannot afford to do so yet, instead adopting a hybrid approach. The cloud is similar: Some have gone or will be going all-cloud, others are leveraging a hybrid approach at least near-term. What about virtual storage and storage virtualization for the software-defined data center? Sure, some environments are doing more with software-defined storage (depending on what your definition is) vs. others.

What is the best storage? That depends on how you will be using the storage and with what applications. Will the storage be shared across multiple servers or hypervisors? Do you need to share files and data? What are the Performance, Availability, Capacity, and Economic (PACE) requirements? With what does the storage need to coexist within your environment?

Do cloud and virtual servers need the cloud and virtual storage? If your cloud or virtual server is running at or on a public cloud, such as Amazon or rack space, you may want to access their cloud and virtual storage. If your virtual server is running on one of your own servers, you may be fine using traditional shared storage.

Are clouds safe? They can be if you use and configure them in safe ways, as well as do your due diligence to assess the security, availability, as well as the privacy of a particular cloud provider. Keep in mind that cloud data protection and security are a shared responsibility between you and your service provider. There are things that both parties need to do to prevent cloud data loss.

Why not move everything to the cloud? The short answer is data locality, meaning you will need and want some amount of storage close to where you are using the data or applications—that might mean a small amount of flash or RAM cache, or HDD, Hybrid, or similar. Another consideration is available and effective network bandwidth or access.

How to develop tradecraft? There are many ways, including reading this book along with the companion websites as well as other books, attending seminars and webinars, participating in forums and user groups, as well as having a test lab in which to learn and try things. Also, find a mentor you can learn from to help capture some of his or her tradecraft, and if you are experienced, become a mentor to help others develop their tradecraft.

1.8. Learning Experience

In subsequent chapters, in addition to the "What's in Your Toolbox?" and "Common Questions and Tips" sections, there are also some learning experiences. These will vary by chapter, depending on the topic focus. To get started and see what you have learned (or refreshed), practice the following and see how you do. Of course, as with many server storage I/O fundamental topics, the actual answer may depend on what your real needs and requirements are.

Understand the context and comprehend the question; don't just try to memorize the answers.

- Can you clarify storage characteristics besides cost per capacity?

- What are some things that are different in most IT environments?
- Where does a server store things, including an operating system, applications, and data?
- What are some common similarities between different applications?
- What is server storage and I/O tradecraft? How do you acquire and maintain it?
- Why is context important for server storage I/O topics, trends, terms, and technologies?
- What is causing the need for more servers, storage, and I/O resources?
- Articulate the role and importance of server storage I/O hardware and software

1.9. Chapter Summary

Servers and other devices need some form of temporary and persistent storage. Likewise, storage exists to provide temporary and persistent memory to services and other devices. There are many different types of devices spanning from large to small servers, cloud, virtual, and physical. There are also various devices that rely on storage. Figure 1.11 summarizes what we have discussed in this chapter, tying together the themes of various environments, different applications, server storage I/O hardware and software spanning legacy, physical, software-defined, virtual, and cloud, along with demand drivers and associated management.

Just as there are many different types of devices that access storage, there are also various types of storage devices, systems, solutions, hardware, software, and services. A key theme for all storage is to protect, preserve, and service data that gets transformed into information.

General action items include:

- Avoid treating all data and applications the same.
- Look at data storage from more than a cost-per-space capacity basis.
- Storage can be complex. However, it can also be simplified, depending on your needs.
- Know your applications PACE criteria and requirements as well as needs vs. wants.
- Don't be afraid of cloud and virtualization.

Bottom line: Show up for your server storage I/O and data infrastructure activities informed, prepared to play an effective game. The fundamental role of IT data infrastructures including server, storage I/O connectivity hardware, and software and management tools is to support protecting, preserving, and serving data and information across various types of organizations.

Chapter 2

Application and IT Environments

Algorithms + Data Structures = Programs

– Niklaus Wirth (Prentice-Hall, 1976)

What You Will Learn in This Chapter

- Application PACE characteristics, attributes, and tradecraft
- Different types of data and applications
- Structured and unstructured data, little data, and big data
- Why everything is not the same in data centers or applications
- Where applications and data get processed
- How applications are supported by data infrastructures

In this chapter, we build on the fundamentals that were laid down as a foundation in Chapter 1, expanding the focus to various applications and environments. Keep in mind that everything is not the same when it comes to server and storage I/O hardware, software, cloud, and managed services tools and technology. Likewise, there are different types of environments and applications that we will now take a closer look at in this chapter. Key themes, buzzwords, and trends in this chapter include little data, big data, structured and unstructured data, application performance, availability, capacity, and economics where applications and data get processed.

2.1. Getting Started

Many storage discussions focus on just the data. However, we are going to expand the discussion since we are covering data infrastructures including servers, storage, and I/O networking hardware as well as software tools. We will extend the theme that applications (programs) that run on servers have algorithms and data structures that reside in memory and data storage. Where we are going with this discussion and why it is important is that different applications rely on servers and storage in various ways; however, storage systems are storage applications deployed on some server platform. Data infrastructure components are converging, as are applications, people, processes, techniques, services, and technologies.

2.1.1. Tradecraft and Context for the Chapter

Tradecraft means more than just servers and storage performance and cost per capacity, and, as we mentioned in Chapter 1, server storage and I/O fundamentals include more than just hardware, software, and services technologies or tools.

Part of advancing your tradecraft is understanding the interrelationships of various applications and their data dependencies (or vice versa), along with corresponding resource characteristics. This means learning about the business and organizations that IT systems and applications support, beyond just the basic characteristics of associated technologies. These characteristics include Performance, Availability, Capacity, and Economics (PACE), which are common to all applications and which we will be discussing in greater detail later in this chapter.

Often when it comes to servers and storage I/O in particular, the focus is mostly on data or how it is accessed. This makes sense because the fundamental role of data infrastructure and storage is to protect, preserve, and serve data so that applications can access or transform it into useful information. However, when it comes to servers, storage I/O networking hardware, software, and services, it is also about the applications that handle the processing of data.

Keep in mind that applications consist of one or more programs on one or more servers (or computers) that interact with data. Also, keep in mind that programs are the algorithms, logic, rules, and policies that determine what to do with data, when, where, why, and how. In addition to algorithms, programs also rely on some data that is defined as data structures and stored in memory or persistent storage.

Hence the theme at the top of this chapter: *Algorithms + Data Structures = Programs,* which is not coincidentally the title of a book by Niklaus Wirth that I highly recommend reading (see Appendix for more information). Even if you are not planning on ever doing any programming or system design, simply reading it will give you some different perspectives and appreciation for the relationship between software that are algorithms and their associated data structures.

The importance is understanding what data is being stored and why, how it is used by programs that run on processors (computers or servers), interacting with storage devices via I/O paths. Thus the essence of server storage and I/O convergence relates back to the fundamental premise that applications algorithms require computer processors and storage, including memory for storing data structures.

2.2. Everything Is Not the Same with Servers Storage and I/O

Just as there are many different types of environments, from small to large web-scale and cloud or managed service providers, along with various types of organization and focus, there are also diverse applications and data needs. Keep in mind that a fundamental role of data infra-structures, and IT in general, is to protect, preserve, and serve information when and where it is needed. What this means is that while everything is not the same, from types and sizes of organization to business and services functionality focus or application needs, there are some common similarities or generalities.

Some of these commonalities come from an organizational or business industry focus, such as defense and security, e-commerce and retail, education, energy, finance, government, legal and law enforcement, healthcare and life sciences, manufacturing, media and entertainment, scientific research, social media and networking, and transportation, among many others. Other commonalities found across different organizations include sales, marketing and public relations, accounting and payroll, human resources, back-office functions and file sharing, among others.

2.2.1. Various Types of Environments (Big and Small)

There are broad generalities—for example, types of applications, business focus or industry sector, small, medium, large—but there are differences. Likewise, there are technology products, tools, and services that can fit different needs; they may also be optimized for certain things. Some environments are more complex due to their size, others may be small but also complex.

Everything is not the same, although there are similarities and generalities:

- Different environments (industry focus and size)
- Applications and how they are used, together with their characteristics
- Various workloads and size or amount of data
- Requirements and constraints to support information needs

There are various sizes, types, and focuses for environments:

- Uber-large web-scale, cloud, and managed service providers
- Ultra-large enterprise, government, or other organizations and small service providers
- Large enterprise, government, education, and other organizations
- Small/medium enterprise (SME) and other organizations
- Small/medium business (SMB) enterprises, from small office/home office (SOHO) to SME
- Workgroup, departmental, remote office/branch office (ROBO)
- Prosumer (e.g., power user) and consumer

Note that technology solutions and services are also often aligned, positioned, or targeted toward the above general categories by vendors, as well as for consumption by those using the services. What defines the size or category of an organization varies. Organization size can be defined by the number of employees, revenue, business ranking, the number of offices or locations, the number of servers, desktops, or amount of storage or budget spent on hardware and software, among other considerations.

Do you treat everything the same?

- From a performance perspective?
- In terms of availability, including data protection?

- About efficiency (usage) and effectiveness (productivity)?
- Regarding space capacity or type of storage?
- Do you know your costs for servers, storage, I/O, and relates services?
- How you align applicable technology to your business or organization needs?

Understanding the organization and its focus as well as its information needs helps to know the characteristics of the applications and data that is needed. This knowledge is useful for making decisions about configuration, upgrades, performance, availability, capacity, or other changes as well as troubleshooting data infrastructure on a cloud, virtual, or physical basis.

2.2.2. Gaining Data and Application Insight

Gathering insight and awareness into your applications, along with their associated data and characteristics, involves a mix of technology tools, reading what is available in any internal knowledge base as well as external general available information.

There is another valuable resource that is part of expanding your tradecraft, which is learning and sharing experiences with others in your environment as well as peers in other organizations. From a technology tools perspective, there are many different tools for conducting discovery of what applications and data, along with resource usage and activity, occurs.

An example about expanding tradecraft is, for example, an automobile technician who has learned how to change oil, check tire air pressure, or other basic entry-level functions. On the other hand, a master mechanic knows how to perform more involved tasks, from engine to transmission repair or rebuilding, bodywork, along with troubleshooting. A master mechanic not only knows what buttons, knobs, tools, and techniques to use for different tasks, he also knows how to diagnose problems, as well as what usually causes those problems to occur.

For IT and data infrastructure tradecraft this means expanding from basic tasks to being able to do more advanced things—for example, expanding tradecraft from knowing the different hardware, software, and services resources as well as tools, to what to use when, where, why, and how. Another dimension of expanding data infrastructure tradecraft skills is gaining the experience and insight to troubleshoot problems, as well as how to design and manage to eliminate or reduce the chance of things going wrong.

Note that while expanding your awareness of new technologies along with how they work is important, so too is understanding application and organization needs. Expanding your tradecraft means balancing the focus on new and old technologies, tools, and techniques with business or organizational application functionality.

This is where using various tools that themselves are applications to gain insight into how your data infrastructure is configured and being used, along with the applications they support, is important.

Some of these tools may be application-specific, while others look at data stored and correlate back to applications, and still others may be performance- or availability-centric. Figure 2.1 shows a simple discovery and assessment of one of my systems using a tool called Spotlight on Windows (via Dell, there is a free version).

Spotlight (on Windows) provides me with insight into how data infrastructure resources are being used. There are many other tools available, some of which are mentioned later in the

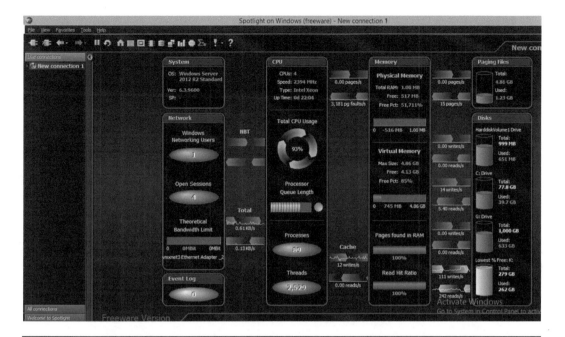

Figure 2.1   Gaining insight into application activity using Spotlight.

toolbox section of this chapter. Which tool is best for your needs will depend on what you need it to do, as well as the budget, size, and scope of the environment. After all, everything is not the same, and what is good for you or me may be different for somebody else.

In addition to using various tools including those that you can install on your server and storage systems (cloud, virtual or legacy), there are also cloud services that can collect information and then analyze and provide reports and insights back to you. Besides cloud-based tools and those you can install and use yourself, there are also service providers, vendors, value-added resellers (VARs), and consultants that can also help with assessing your environment.

These services range from high-level, showing you what and how to collect the information yourself, to turnkey, in-depth data discovery and application assessment project engagements. What is going to be best for you and your environment will depend on your needs. When in doubt, ask your peers in other organizations or those you know via various venues, user groups, and sites.

2.2.3. Various Types of Applications

Recall from Chapter 1 (Figure 1.4) that there are different layers of infrastructure depending on where you sit (or stand) when accessing and using information services. Each of the different layers of infrastructure and information services is built on and deployed via underlying layers of applications also known as infrastructure. Some of these applications are also known as tools, utilities, plug-ins, agents, drivers, middleware, or valueware, among other terms, and they may be internal to your organization or delivered via a cloud or managed services provider.

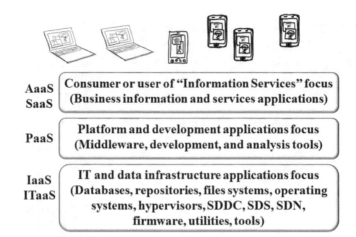

AaaS SaaS	**Consumer or user of "Information Services" focus** **(Business information and services applications)**
PaaS	**Platform and development applications focus** **(Middleware, development, and analysis tools)**
IaaS ITaaS	**IT and data infrastructure applications focus** **(Databases, repositories, files systems, operating** **systems, hypervisors, SDDC, SDS, SDN,** **firmware, utilities, tools)**

Figure 2.2   Stacking and application layers.

In general, software applications include, among others:

- Frameworks and application suites such as Microsoft Office or Office 365
- Cloud-based software as a service (SaaS) and applications as a service (AaaS)
- Plug-ins, apps, or applets for mobile devices and web browsers
- Agents, drivers, tools, and utilities for workstations and servers
- Application development and business logic tools
- Middleware, productivity, and development platforms
- Databases, file systems, data repositories, and utilities
- Operating systems, hypervisors, and associated management tools
- Server, storage, and I/O networking hardware or appliance firmware

All of these applications can be packaged, delivered, and consumed in various ways.

Depending on your focus, you might be the consumer of some information services for work or entertainment. Shown in Figure 2.2 are examples such as if you are a systems or applications developer, traditional or DevOps-focused, you may be using platform as a service (PaaS) for development along with databases, tools, containers, code management, testing, and middleware, either on-site or via the cloud.

If you are using a traditional desktop environment installed on a workstation, laptop, or tablet, you are using a collection of applications including an operating system, office, and other tools. On the other hand, if you are using a virtual desktop infrastructure (VDI), you are using another collection of applications.

The important takeaway is that there are various layers of infrastructure and applications that in turn support upper- or higher-level applications and functionalities. These applications extend to those that define hardware to be a web server, compute server, storage system, data or database server, content server, or a network router, switch, or cache appliance, to a virtual data center or simply a virtual server among other functions.

What this means is that storage and data management software are installed on a cloud, virtual, or physical server and that storage system can also host different applications—hence the importance of understanding various applications along with their characteristics and data activity.

Business Information and Services Applications

There are many different types of "information services" focused applications ranging from consumer entertainment to social media/networking (YouTube, Facebook, Pinterest, LinkedIn, Instagram, Twitter, Yammer, and many others), education, e-commerce, information sites as well as organizational and business-related sites, among others.

Examples include emerging applications that support or enable the Internet of Things (IoT), Internet of Devices (IoD), the web, cloud- and mobile-to-legacy human resources, content management systems (CMS) including Drupal, Axis video, digital asset management (DAM), and digital evidence managers (DEMS) used for managing police body cameras. Others include content delivery networks (CDN) and web cache, as well as customer relationship management (CRM) and sales/marketing-focused applications such as Salesforce.com or Office 365, ArcGIS, Hive data warehouse, Nifi for distributed data distribution, spark and Splunk, among others.

There are also computer-assisted design (CAD) applications and applications for controlling manufacturing devices, enterprise resource planning (ERP) systems such as M1, Epicor, Oracle, and SAP, financial accounting, payroll, banking, insurance, investments, and trading, medical and healthcare, life sciences including pharmaceuticals, as well as research.

More examples include video and entertainment, gaming, security and surveillance, photo sharing such as Shutterfly and Shutterstock, and other content services. Other applications and functionalities include electronic data interchange (EDI), payments and supply-chain management to support manufacturing and other activity, as well as messaging such as email, texting, workflow routing, and approvals.

Content solutions span from video (4K, HD, SD resolution, and legacy streaming video, pre/post-production, and editing), audio, imaging (photo, seismic, energy, healthcare, etc.) to security surveillance (including intelligent video surveillance [ISV] as well as intelligence surveillance and reconnaissance [ISR]).

These and other applications are packaged, delivered, and consumed in various ways from "shrink-wrapped," where the software is bought or leased installed on your data infrastructure servers, storage, I/O networks, and associated software, or provided as a service accessible via a cloud, managed service provider, or other web means. The software may also be made available on a services basis, that is, SaaS or AaaS, among other variations and pricing models.

Platform and Development Applications (Tools)

Other applications include those used for creating the previous (among others) applications by traditional developers or DevOps. These tools sit on and rely on the services and capabilities provided by underlying data infrastructures that are also made up of various applications, software tools, and utilities. Platform and development applications, tools, utilities, plug-ins, frameworks, middleware, and servers have various functionalities, from code creation to debugging, management libraries, and release coordination tracking, to testing tools.

Other platform and development applications tools include scripts, application program interface (API) extensions or customization capabilities, languages, and automation and code generators. Environments may consist of Java, .Net, Pivotal, and Cloud Foundry, among others, as well as leveraging and building on applications such as statistical analysis software (SAS), Hadoop and its variations including Cloudera, Oracle ERP, SAP, and a long list of others.

Some other tools, sites, and services to add to your vocabulary, some of which will be discussed later in addition to those already mentioned, include Ansible, Chef, Docker, GitHub, Puppet, Salt, object-oriented framework tools such as Visual Studio, Hibernate for Java, and databases, among others.

There are even some legacy tools supporting decades-old applications based on Basic, COBOL, and C+, among others. The above and others can be delivered via platform as a service (PaaS) managed service and cloud providers, or on-premises and include proprietary as well as open-source-based tools.

IT and Data Infrastructure Applications (Tools)

In addition to business information and services applications, along with platform and development applications tools and utilities, there are also IT and data infrastructure applications including tools, utilities, SQL and NoSQL databases, repositories, volume managers and file systems, operating systems, containers, hypervisors, and device drivers.

There is also other software-defined data center (SDDC), software-defined network (SDN), and network function virtualization (NFV), along with software-defined storage (SDS) software tools or applications. Additional data infrastructure applications include cloud platforms such as OpenStack, database tools, monitoring and management, help desk and IT service management (ITSM), analytics, inventory and asset tracking, data center infrastructure management (DCIM), performance management database (PMDB), configuration management database (CMDB), data protection including security, backup/restore, disaster recovery (DR) and high-availability (HA), key management, along with data discovery and search.

Others include workstation and desktop (physical and virtual desktop infrastructure), file and services such as SharePoint, email and messaging, to name a few. Note that hardware is further defined as firmware, microcode, and basic input/output system (BIOS) software, and many storage and I/O networking systems or appliances are in fact servers with storage software installed on a basic operating system.

2.2.4. Various Types of Data

There are many different types of applications to meet various business, organization, or functional needs. Keep in mind that applications are based on programs which consist of algorithms and data structures that define the data, how to use it, as well as how and when to store it. Those data structures define data that will get transformed into information by programs while also being stored in memory and on data storage in various formats.

Just as various applications have different algorithms, they also have different types of data. Even though everything is not the same in all environments, or even how the same applications get used across various organizations, there are some similarities. Likewise, even though there are different types of applications and data, there are also some similarities and general characteristics. Keep in mind that information is the result of programs (applications and their algorithms) that process data into something useful or of value.

Data typically has a basic life cycle of:

- Creation and some activity, including being protected
- Dormant, followed by either continued activity or going inactive
- Disposition (delete or remove)

The above is a simple life cycle, and given that everything is not the same, for some environments there may be additional activities beyond the three mentioned.

In general, data can be

- Temporary, ephemeral or transient
- Dynamic or changing ("hot data")
- Active static on-line, near-line, or off-line ("warm data")
- In-active static on-line or off-line ("cold data")

General data characteristics include:

- Value—From no value to unknown to some or high value
- Volume—Amount of data, files, objects of a given size
- Variety—Various types of data (small, big, fast, structured, unstructured)
- Velocity—Data streams, flows, rates, load, process, access, active or static

Figure 2.3 shows how different data has various values over time. Data that has no value today or in the future can be deleted, while data with unknown value can be retained.

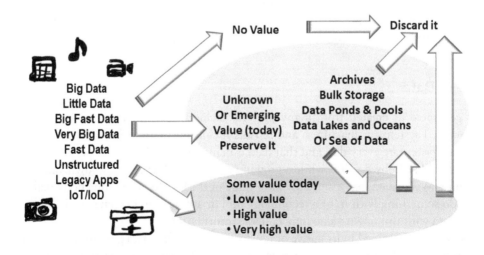

Figure 2.3   Different data with various values over time.

General characteristics include the value of the data which in turn determines its performance, availability, capacity, and economic considerations. Also, data can be ephemeral (temporary) or kept for longer periods of time on persistent, non-volatile storage (you do not lose the data when power is turned off). Examples of temporary include work and scratch areas such as where data gets imported into, or exported out of, an application or database.

Data can also be little, big, or big and fast, terms which describe in part the size as well as volume along with the speed or velocity of being created, accessed, and processed. The importance of understanding characteristics of data and how their associated applications use them is to enable effective decision making about performance, availability, capacity, and economics of data infrastructure resources.

Data Value

There is more to data storage than how much space capacity per cost. All data has one of three basic values:

- No value = ephemeral/temp/scratch = Why keep?
- Some value = current or emerging future value, which can be low or high
- Unknown value = protect until value is unlocked

In addition to the above basic three, data with some value can also be further subdivided into little value, some value, or high value. Of course, you can keep subdividing into as many more or different categories as needed—after all, everything is not always the same across environments.

Besides data having some value, that value can also change by increasing or decreasing in value over time, or even going from unknown to known value, known to unknown, or to no value. Data with no value can be discarded, if in doubt, make and keep a copy of that data somewhere safe until its value (or lack of value) is fully known and understood.

The importance of understanding the value of data is to enable effective decision making on where and how to protect, preserve, and store the data in a cost-effective way. Note that cost-effective does not necessarily mean the cheapest or lowest-cost approach, rather it means the way that aligns with the value and importance of the data at a given point in time.

Volume of Data

Data is generated or created from many sources, including mobile devices, social networks, web-connected systems or machines, and sensors including IoT and IoD. There is more data growing at a faster rate every day, and that data is being retained for longer periods. Some data being retained has known value, while a growing amount of data has unknown value.

Unknown-value data may eventually have value in the future, when somebody realizes that they can do something with it, or a technology tool or application becomes available to transform the data with unknown value into valuable information.

Some data gets retained in its native or raw form, while other data get processed by application program algorithms into summary data, or is curated and aggregated with other data to be

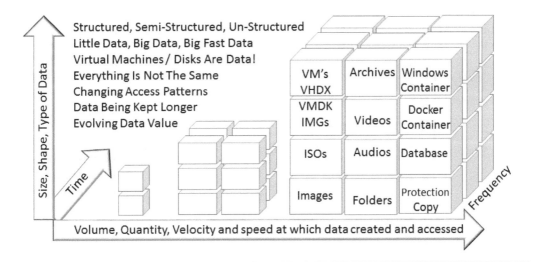

Structured, Semi-Structured, Un-Structured
Little Data, Big Data, Big Fast Data
Virtual Machines / Disks Are Data!
Everything Is Not The Same
Changing Access Patterns
Data Being Kept Longer
Evolving Data Value

Size, Shape, Type of Data

Time

VM's VHDX | Archives | Windows Container

VMDK IMGs | Videos | Docker Container

ISOs | Audios | Database

Images | Folders | Protection Copy

Frequency

Volume, Quantity, Velocity and speed at which data created and accessed

Figure 2.4   Increasing data velocity and volume, more data and data getting larger.

transformed into new useful data. Figure 2.4 shows, from left to right and front to back, more data being created, and that data also getting larger over time. For example, on the left are two data items—objects, files, or blocks representing some information.

In the center of Figure 2.4 are more columns and rows of data, with each of those data items also becoming larger. Moving farther to the right, there are yet more data items stacked up higher, as well as across and farther back, with those items also being larger. Figure 2.4 could represent blocks of storage, files in a file system, rows and columns in a database or key-value repository, or objects in a cloud or object storage system.

In addition to more data being created, some of that data is relatively small in terms of the records or data structure entities being stored. However, there can be a large quantity of those smaller data items. In addition to the amount of data, as well as the size of the data, protection or overhead copies of data are also kept. Another dimension is that data is also getting larger where the data structures describing a piece of data for an application have increased in size. For example, a still photograph taken with a digital camera, cell phone, or other mobile hand-held device, drone, or other IoT device, increases in size with each new generation of cameras as there are more megapixels.

Variety of Data

In addition to having value and volume, there are also different varieties of data, including ephemeral (temporary), persistent, primary, metadata, structured, semistructured, unstruc-tured, little, and big data. Keep in mind that programs, applications, tools, and utilities get stored as data, while they also use, create, access, and manage data.

There is also primary data and metadata, or data about data, as well as system data that is also sometimes referred to as metadata. Here is where context comes into play as part of trade-craft, as there can be metadata describing data being used by programs, as well as metadata about systems, applications, file systems, databases, and storage systems, among other things, including little and big data.

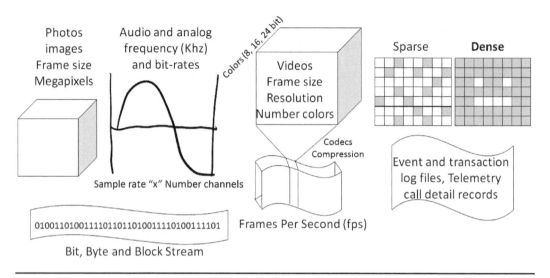

Figure 2.5   Varieties of data (bits, bytes, blocks, blobs, and bitstreams).

Context also matters regarding big data, as there are applications such as statistical analysis software, Splunk and Hadoop, among others, for processing (analyzing) large amounts of data. The data being processed may not be big in terms of the records or data entity items, but there may be a large volume such as with telemetry. In addition to big data analytics, data, and applications, there is also data that is very big (as well as large volumes or collections of data sets).

For example video and audio, among others, may also be referred to as big fast data, or very big data. A challenge with larger data items is the complexity in moving over the distance promptly, as well as processing requiring new approaches, algorithms, data structures, and storage management techniques.

Likewise, the challenges with large volumes of smaller data are similar in that data needs to be moved, protected, preserved, and served in a cost-effective manner for long periods of time. Both large and small data are stored (in memory or storage) in various types of data repositories.

In general, data in repositories is accessed locally, remotely, or via a cloud using:

- Object and Application Programming Interface (API)
- File-based using local or networked file systems
- Block-based access of disk partitions, LUNs (logical unit numbers), or volumes

Figure 2.5 shows varieties of data including (left) photos or images, audio, videos, and various log, event, and telemetry data, as well as (right) sparse and dense data.

Velocity of Data

Data, in addition to having value (known, unknown, or none), volume (size and quantity), and variety (structured, unstructured, semistructured, primary, metadata, small, big), also has velocity. Velocity refers to how fast (or slowly) data is accessed, including being stored, retrieved, updated, scanned, or if it is active (updated, or fixed static) or dormant and inactive.

In addition to data access and life cycle, velocity also refers to how data is used, such as random or sequential or some combination. Think of data velocity as how data, or streams of data, flow in various ways.

Velocity also describes how data is used and accessed, including:

- Active (hot), static (warm and WORM), or dormant (cold)
- Random or sequential, read or write-accessed
- Real-time (on-line, synchronous) or time-delayed

Why this matters is that by understanding and knowing how applications use data, or how data is accessed via applications, you can make informed decisions as to how to design, configure, and manage servers, storage, and I/O resources (hardware, software, services) to meet various needs. Understanding the velocity of the data both for when it is created as well as when used is important for aligning the applicable performance techniques and technologies.

Active (Hot), Static (Warm and WORM), or Dormant (Cold)

In general, the data life cycle (called by some cradle to grave, birth or creation to disposition) is create, save and store, perhaps update and read with changing access patterns over time, along with value. During that time, the data (which includes applications and their settings) will be protected with copies or some other technique, and perhaps eventually disposed of.

Between the time when data is created and when it is disposed of, there are many variations of what gets done and needs to be done. Considering static data for a moment, some applications and their data, or data and their applications, create data which is active for a short period, then goes dormant, then is active again briefly before going cold (see the left side of Figure 2.6). This is a classic application, data, and information life-cycle model (ILM), and tiering or data movement and migration that still applies for some scenarios.

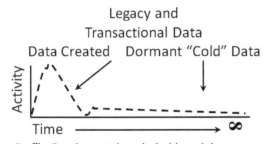

Profile: Data is created, worked with, and then goes dormant after some period of time with a probability of little to no future access or use. Data is updated, read and write vs. write once read many or append. Block and File.
Examples: Databases, general file serving, email, transactional, project-oriented data among others

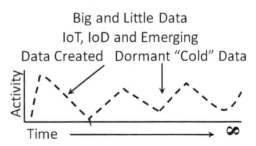

Profile: Data is created, worked with, and then may go idle briefly, then accessed, then idle, then active, then idle, then active, then idle, then active, many reads, new writes or appends, more meta data.
Examples: EMR, PACs, energy, exploration, Web, reference and lookup, meta data, fixed content, CDN, Big Data, Social Media, Video, Security, DEMs, research, media, entertainment, analytics, seasonal applications. Block, File, Object, API, and Bulk.

Figure 2.6   Changing data access patterns for different applications.

However, a newer scenario over the past several years that continues to increase is shown on the right side of Figure 2.6. In this scenario, data is initially active for updates, then goes cold or WORM (Write Once Read Many); however, it warms back up as a static reference, on the web, as big data, and for other uses where it is used to create new data and information.

Data, in addition to its other attributes already mentioned, can be active (hot), residing in a memory cache, buffers inside a server, or on a fast storage appliance or caching appliance. Hot data means that it is actively being used for reads or writes: This is what the term *heat map* pertains to in the context of the server, storage data, and applications. The heat map shows where the hot or active data is along with its other characteristics.

Context is important here, as there are also IT facilities heat maps, which refer to physical facilities including what servers are consuming power and generating heat. Note that some current and emerging data center infrastructure management (DCIM) tools can correlate the physical facilities power, cooling, and heat to actual work being done from an applications perspective. This correlated or converged management view enables more granular analysis and effective decision making on how to best utilize data infrastructure resources.

In addition to being hot or active, data can be warm (not as heavily accessed) or cold (rarely if ever accessed), as well as on-line, near-line, or off-line. As their names imply, warm data may occasionally be used, either updated and written, or static and just being read. Some data also gets protected as WORM data using hardware or software technologies. WORM data, not to be confused with warm data, is fixed or immutable: To preserve its context for legal or other reasons, it cannot be modified.

Also, note that some WORM data can also be warm data in that it is read-only, yet read frequently, perhaps residing in a read cache for fast lookup or access.

When looking at data (or storage), it is important to see when the data was created as well as when it was modified. However, you should avoid the mistake of looking only at when it was created or modified: Instead, also look to see when it was last read, as well as how often it is read. You might find that some data has not been updated for several years, but it is still accessed several times an hour or minute. Also, keep in mind that the metadata about the actual data may be being updated, even while the data itself is static.

Also, look at your applications characteristics as well as how data gets used, to see if it is conducive to caching or automated tiering based on activity, events, or time. For example, there is a large amount of data for an energy or oil exploration project that normally sits on slower lower-cost storage, but that now and then some analysis needs to run on.

Using data and storage management tools, given notice or based on activity, which large or big data could be promoted to faster storage, or applications migrated to be closer to the data to speed up processing. Another example is weekly, monthly, quarterly, or year-end processing of financial, accounting, payroll, inventory, or enterprise resource planning (ERP) schedules. Knowing how and when the applications use the data, which is also understanding the data, automated tools, and policies, can be used to tier or cache data to speed up processing and thereby boost productivity.

Random or Sequential, Read or Write Accessed

One aspect of data velocity is order access, including random, sequential, as well as reads and writes that can be large, small, or mixed. Sequential refers to contiguous (adjacent) data

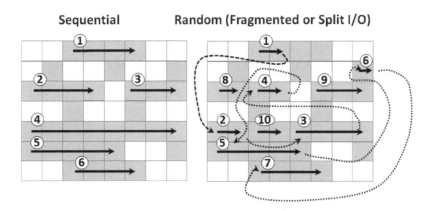

Figure 2.7 Sequential and random access, large and small I/O.

accessed, such as on the left side of Figure 2.7, while random access is shown on the right in the figure. Also, note the flat address space shown in Figure 2.7. In addition to random I/O, the right side of Figure 2.7 also represents fragmented, split, or noncontiguous I/O and data placement.

In Figure 2.7 (left), sequential, also known as streaming I/O operations, are shown in order of access (the numbers), with the dark squares representing the size of or how much data is being read or written. For example, (1) shows four data blocks being read or written, then four more (2), followed by three more (3), then eight more (4) in a large (or long) access, then five in (5), finally four in (6). In Figure 2.7 (right), random access is shown by jumping around with a mix of small and large operations.

Note that random and sequential can both have large or small access that are reads, writes, or a mix. However, often sequential access is associated with large reads or writes, while random access can be smaller—although note that this is not a hard-and-fast rule, as I/O sizes are getting larger. Likewise, applications may access data using both random and sequential access, for example, randomly accessing some data and then sequentially streaming other data.

Why this matters is that knowing data access patterns of how applications use data, as well as insight into data value, velocity, and other attributes, are important for managing servers, storage, and I/O resources for performance and other activities.

Real-Time (On-Line, Synchronous) or Time-Delayed

Real-time and time-delayed refer generally to how data is accessed and used by applications. Some data may be accessed in real time, on-line with no delays, including reads and updates or writes locally. That same data may be accessed in a time-delayed mode remotely, perhaps by accessing a read or buffered cache for collaboration or sharing purposes. Likewise, data written or updated from one location to another might be done asynchronously, in which case the application keeps running while data is written in the background, with eventual consistency (e.g., write behind).

The benefit is that applications can keep running without causing delays to users while data is written at a slower pace. The caveat is that data integrity means that either the application or data infrastructure needs to provide consistency safeguards to protect against data loss. With synchronous access, applications wait for notification that data has been safely written to ensure strong consistency and protect against data loss.

Note that for reads, instead of waiting for data to be retrieved, it can be cached in a local buffer to speed up access. However, cache consistency and coherency needs to exist to prevent stale reads, which mean that if the underlying data changes, the cache or buffers need to be upgraded to avoid using old data.

Habitats for Data

Another aspect of data (along with applications) is where they exist and reside. Data repositories are habitats for where data lives and is stored. These repositories exist in physical facilities either on your own site, colocated (colo), or on shared hosting or cloud managed services. Repositories can be as simple as a file system dedicated to a standalone device, external and shared via portable media, network, or cloud services, among other variations. Other data repositories include structured and unstructured databases and data warehouses for little and big data, both temporary and permanent, near- or long-term.

Unstructured data repositories include object and file systems, solutions, or services that may also have marketing or functional names such as data ponds, data pools, data lakes, data reservoirs, or data hotels, among others. Different data types may have different velocity, but, in general, as it flows downstream from its source to other destinations, data also gets larger, just like water in a river.

Similarly, data repositories can also overflow, resulting in a flood and causing disruptions if not properly managed—prompting the question of whether it is easier and more cost-effective to take a proactive stand to find and fix problems at the source, or to have insurance to clean them up after the fact. Keep that thought in mind as we go further along our journey in this book.

Some applications store and access data in highly structured databases, such as Intersystem's Caché, Microsoft SQL Server, IBM DB2, MySQL, or variations including ClearDB, TokuDB, and Aerospike, as well as in memory-based solutions such as MemSQL, Oracle, SAP, HANA, and Sybase, among others. In addition to SQL databases, there are also NoSQL databases and repositories, including Cassandra, Mongo, HBase, CouchDB, Kudo, and Riak, among others.

In additional to traditional databases, there are also repositories optimized for data warehouse and big data analytics, including Teradata, IBM Netezza, Oracle ERP, Pivotal Greenplum, Hadoop, and HDFS, among others. In the case of block-based access, the application, database, or file system works with applications to know what files to access.

In other scenarios, applications access a file by name in a folder or directory that is part of a mount point or share in a file system, either locally or on a remote file server or NAS device. Another means of data access is via an object API, whereby a client requests information from a server via a defined mechanism.

In the case of a database, applications only need to know the schema or how the database is organized to make queries with tools such as Structured Query Language (SQL); the database handles read or writing of data either in a block or file system mode. For block mode, the database is assigned LUNs or storage space where it creates its files or data sets that it manages. If a file system is being used, the database leverages the underlying file system to handle some of the storage management tasks.

Structured data has defined attributes, making searching or other functions relatively easy. This structure, however, can make adding or changing the organization more complex or costly. As a result, there is the growing category of unstructured data, also known as file accessed data. The value proposition of unstructured data is that there is no formal organization other than files stored in a folder or directory in a file system.

Some file systems and files can support additional metadata or properties. The flexibility of unstructured data causes challenges, however, when it comes to being able to search or determine what the files contain. With a database, the schema or organization makes it relatively easy to search and determine what is stored.

With unstructured data, additional metadata needs to be discovered via tools including search and classification tools. Metadata includes information about the contents of files along with dependencies on other information or applications. The use of structured or unstructured data depends on preference, the performance of the specific file system or storage being used, desired flexibility, and other criteria.

2.3. Common Applications Characteristics

Different applications will have various attributes, in general, as well as how they are used, for example, database transaction activity vs. reporting or analytics, logs and journals vs. redo logs, indices, tables, indices, import/export, scratch and temp space. Performance, availability, capacity, and economics (PACE) describes the applications and data characters and needs (Figure 2.8).

All applications have PACE attributes, however:

- PACE attributes vary by application and usage
- Some applications and their data are more active than others
- PACE characteristics may vary within different parts of an application

Think of applications along with associated data PACE as its personality or how it behaves, what it does, how it does it, and when, along with value, benefit, or cost as well as quality-of-service (QoS) attributes. Understanding applications in different environments, including data values and associated PACE attributes, is essential for making informed server, storage, and I/O decisions from configuration to acquisitions or upgrades—when, where, why, and

Figure 2.8 Application PACE attributes.

how to protect, and how to optimize performance including capacity planning, reporting, and troubleshooting, not to mention addressing budget concerns.

Primary PACE attributes for active and inactive applications and data are

P—Performance and activity (how things get used)
A—Availability and durability (resiliency and protection)
C—Capacity and space (what things use or occupy)
E— Economics and Energy (people, budgets, and other barriers)

Some applications need more performance (server computer, or storage and network I/O), while others need space capacity (storage, memory, network, or I/O connectivity). Likewise, some applications have different availability needs (data protection, durability, security, resiliency, backup, business continuity, disaster recovery) that determine the tools, technologies, and techniques to use.

Budgets are also nearly always a concern, which for some applications means enabling more performance per cost while others are focused on maximizing space capacity and protection level per cost. PACE attributes also define or influence policies for QoS (performance, availability, capacity), as well as thresholds, limits, quotas, retention, and disposition, among others.

2.3.1. Performance and Activity (How Resources Get Used)

Some applications or components that comprise a larger solution will have more performance demands than others. Likewise, the performance characteristics of applications along with their associated data will also vary. Performance applies to the server, storage, and I/O networking hardware along with associated software and applications.

For servers, performance is focused on how much CPU or processor time is used, along with memory and I/O operations. I/O operations to create, read, update, or delete (CRUD) data include activity rate (frequency or data velocity) of I/O operations (IOPs), the volume or amount of data being moved (bandwidth, throughput, transfer), response time or latency, along with queue depths. Activity is the amount of work to do or being done in a given amount of time (seconds, minutes, hours, days, weeks), which can be transactions, rates, IOPs, latency, bandwidth, throughput, response time, queues, reads or writes, gets or puts, updates, lists, directories, searches, pages views, files opened, videos viewed, or downloads.

Server, storage, and I/O network performance include:

- Processor CPU usage time and queues (user and system overhead)
- Memory usage effectiveness including page and swap
- I/O activity including between servers and storage
- Errors, retransmissions, retries, and rebuilds

Figure 2.9 shows a generic performance example of data being accessed (mixed reads, writes, random, sequential, big, small, low- and high-latency) on a local and a remote basis. The example shows how for a given time interval (see lower right), applications are accessing

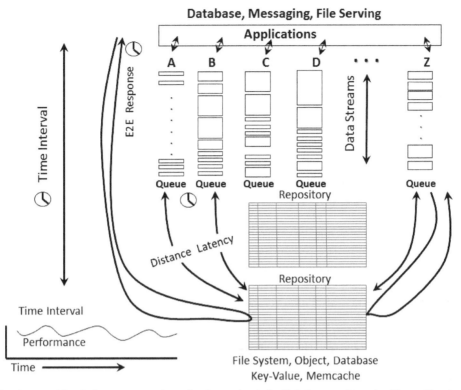

Figure 2.9 Server, storage, I/O performance fundamentals.

and working with data via different data streams in the larger image left center. Also shown are queues and I/O handling along with end-to-end (E2E) response time.

Also shown on the left in Figure 2.9 is an example of E2E response time from the application through the various data infrastructure layers, as well as, lower center, the response time from the server to the memory or storage devices. Various queues are shown in the middle of Figure 2.9 which are indicators of how much work is occurring, if the processing is keeping up with the work or causing backlogs. Context is needed for queues, as they exist in the server, I/O networking devices, and software drivers, as well as in storage among other locations.

Some basic server, storage, I/O metrics that matter include:

- Queue depth of I/Os waiting to be processed and concurrency
- CPU and memory usage to process I/Os
- I/O size, or how much data can be moved in a given operation

- I/O activity rate or IOPs = amount of data moved/I/O size per unit of time
- Bandwidth = data moved per unit of time = I/O size × I/O rate
- Latency usually increases with larger I/O sizes, decreases with smaller requests
- I/O rates usually increase with smaller I/O sizes and vice versa
- Bandwidth increases with larger I/O sizes and vice versa
- Sequential stream access data may have better performance than some random access data
- Not all data is conducive to being sequential stream, or random
- Lower response time is better, higher activity rates and bandwidth are better

Queues with high latency and small I/O size or small I/O rates could indicate a performance bottleneck. Queues with low latency and high I/O rates with good bandwidth or data being moved could be a good thing. An important note is to look at several metrics, not just IOPs or activity, or bandwidth, queues, or response time. Also, keep in mind that metrics that matter for your environment may be different than those for somebody else.

Something to keep in perspective is that there can be a large amount of data with low performance, or a small amount of data with high-performance, not to mention many other variations. The important concept is that as space capacity scales, that does not mean performance also improves or vice versa—after all, everything is not the same.

2.3.2. Availability (Accessibility, Durability, Consistency)

Just as there are many different aspects and focus areas for performance, there are also several facets to availability. Note that applications performance requires availability and availability relies on some level of performance. Availability is a broad and encompassing area that includes data protection to protect, preserve, and serve (backup/restore, archive, BC, BR, DR, HA) data and applications. There are logical and physical aspects of availability including data protection as well as security including key management (manage your keys or authentication and certificates) and permissions, among other things.

Availability = accessibility (can you get to your application and data) + durability (is the data intact and consistent). This includes basic Reliability, Availability, Serviceability (RAS), as well as high availability, accessibility, and durability. "Durable" has multiple meanings, so context is important. Durable means how data infrastructure resources hold up to, survive, and tolerate wear and tear from use (i.e., endurance), for example, Flash SSD or mechanical devices. Another context for durable refers to data, meaning how many copies in various places.

Server, storage, and I/O network availability topics include:

- Resiliency and self-healing to tolerate failure or disruption
- Hardware, software, and services configured for resiliency
- Accessibility to reach or be reached for handling work
- Durability and consistency of data to be available for access
- Protection of data, applications, and assets including security
- Backup/restore, replication, snapshots, sync, and copies
- Basic Reliability, Availability, Serviceability, HA, failover, BC, BR, and DR
- Alternative paths, redundant components, and associated software

- Applications that are fault-tolerant, resilient, and self-healing
- Nondisruptive upgrades, code (application or software) loads, and activation
- Immediate data consistency and integrity vs. eventual consistency
- Virus, malware, and other data corruption or loss prevention

From a data protection standpoint, the fundament rule or guideline is *4 3 2 1,* which means having at least four copies consisting of at least three versions (different points in time), at least two of which are on different systems or storage devices and at least one of those is off-site (online, off-line, cloud, or other). There are many variations of the *4 3 2 1* rule (Figure 2.10) along with approaches on how to manage technology to use. We will go deeper into data protection and related topics in later chapters. For now, remember:

4 At least four copies of data (or more)—Enables durability in case a copy goes bad, deleted, corrupted, failed device, or site.

3 The number (or more) versions of the data to retain—Enables various recovery points in time to restore, resume, restart from.

2 Data located on two or more systems (devices or media/mediums)—Enables protection against device, system, server, file system, or other fault/failure.

1 With at least one of those copies being off-premise and not live (isolated from active primary copy)—Enables resiliency across sites, as well as space, time, distance gap for protection.

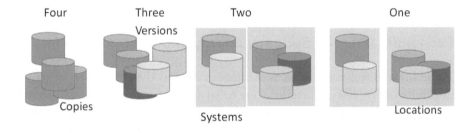

Figure 2.10   *4 3 2 1* data protection.

2.3.3. Capacity and Space (What Gets Consumed and Occupied)

In addition to being available and accessible in a timely manner (performance), data (and applications) occupy space. That space is memory in servers, as well as using available consumable processor CPU time along with I/O (performance) including over networks. Data and applications also consume storage space where they are stored. In addition to basic data space, there is also space consumed for metadata as well as protection copies (and overhead), application settings, logs, and other items. Another aspect of capacity includes network IP ports and addresses, software licenses, server, storage, and network bandwidth or service time.

Server, storage, and I/O network capacity topics include:

- Consumable time-expiring resources (processor time, I/O, network bandwidth)
- Network IP and other addresses

- Physical resources of servers, storage, and I/O networking devices
- Software licenses based on consumption or number of users
- Primary and protection copies of data and applications
- Active and standby data infrastructure resources and sites
- Data footprint reduction (DFR) tools and techniques for space optimization
- Policies, quotas, thresholds, limits, and capacity QoS
- Application and database optimization

DFR includes various techniques, technologies, and tools to reduce the impact or overhead of protecting, preserving, and serving more data for longer periods of time. There are many different approaches to implementing a DFR strategy, since there are various applications and data. Some common DFR techniques and technologies include archiving, backup modernization, copy data management (CDM), clean up, compress, and consolidate, data management, deletion and dedupe, storage tiering, RAID (including parity-based, erasure codes, local reconstruction codes [LRC], and Reed-Solomon, among others), and protection configurations along with thin-provisioning, among others.

DFR can be implemented in various complementary locations from row-level compression in database or email to normalized databases, to file systems, operating systems, appliances, and storage systems using various techniques. Also, keep in mind that not all data is the same; some is sparse, some is dense, some can be compressed or deduped while others cannot. Likewise, some data may not be compressible or dedupable. However, identical copies can be identified with links created to a common copy.

2.3.4. Economics (People, Budgets, Energy and other Constraints)

If one thing in life and technology that is constant is change, then the other constant is concern about economics or costs. There is a cost to enable and maintain a data infrastructure on premise or in the cloud, which exists to protect, preserve, and serve data and information applications. However, there should also be a benefit to having the data infrastructure to house data and support applications that provide information to users of the services. A common economic focus is what something costs, either as up-front capital expenditure (CapEx) or as an operating expenditure (OpEx) expense, along with recurring fees.

In general, economic considerations include:

- Budgets (CapEx and OpEx), both up front and in recurring fees
- Whether you buy, lease, rent, subscribe, or use free and open sources
- People time needed to integrate and support even free open-source software
- Costs including hardware, software, services, power, cooling, facilities, tools
- People time includes base salary, benefits, training and education

2.4. Where Applications and Data Get Processed and Reside

Data habitats or repositories reside in various locations involving different software along with hardware and services spanning cloud, virtual, and physical environments. A trend today is to place data either closer to the server or where the applications run, or move the applications close to where the data is. For some, this may be a new concept, for others evolutionary, yet for some it will also be *déjà vu* from "back in the day."

What is common across those scenarios is a locality of reference. The locality of reference involves keeping data close to the application and server, or applications as close to the data as possible, to reduce overhead and the time delay of moving or accessing data. Keep in mind that the best I/O is the one that you do not have to do: More data in main memory, larger capacity and faster main memory, as well as faster access to NVM and other fast storage, all help to reduce the overhead of I/Os and subsequent application impact.

Leveraging information about applications and data helps to determine where best to house data and applications, including data infrastructure resources, as well as on-site or on-premise, off-site, cloud, managed server, or other options. Some applications and data lend themselves to being centralized, while others are distributed and rely on collaboration and data synchronization or coordination tools.

Where is the best place to locate applications and data, along with the type of servers, storage, I/O networking hardware, and software, depends on your organization needs and preferences. Some applications and environments have shifted over time from centralized to distributed, to centralized, and so forth, following technology trends, while others adapt to organizational needs and preferences.

Applications and data can reside and be processed on or at:

- Dedicated or shared resources (server, storage, I/O networking)
- Converged, hyperconverged, cluster-in-box, or unbounded solutions
- Physical (bare metal), virtual machine, or container-based solutions
- Legacy mainframe and proprietary, open systems, *nix or Windows
- Best-of-breed components converged or hyperconverged solutions
- Locally on-site, on-premise, private or hybrid cloud
- Remote at a secondary site, colocation facility
- Hosted, managed services, or cloud provider
- Tightly or loosely coupled or decoupled compute and storage

2.4.1. The Past, Today, and Tomorrow

Part of developing and expanding your tradecraft about servers, storage, and I/O networking hardware, software, and services is to understand the past. By understanding the past to leverage lessons and techniques learned from others, common mistakes can be minimized or avoided. Likewise, leveraging prior art or experiences as part of your tradecraft can help when looking at new or emerging technologies, techniques, tools, and trends to make use of new things in new ways.

If we know where we have been, we can learn from past lessons what to avoid or prevent or design around for the future. If you know where you have been and where you are going, you should know when you are arriving or at your destination, or the next stop on your journey.

A little bit of history can help to put things into perspective as well as help explain why things are what they are today, why some see new and revolutionary while others see *déjà vu*.

This also explains why some things cannot simply change overnight, particularly considering existing investment in products, hardware, software, services, tools, applications, and skill sets as well as why operating systems or other technologies cannot be changed instantly. Part of developing or expanding your tradecraft means understanding business or organizational needs along with how the applications that support them use and rely on data and data infrastructures.

2.5. What's in Your Application Toolbox?

Tools in my toolbox include Acronis image, Caspa network sniffer and analyzer, Cisco networking tools, Clonezilla, cloud service providers (AWS, Bluehost, Google, Microsoft, and VMware vCloud, among others) along with access management tools, Cloudberry, Dell Toad tools including Benchmark Factory, Foglight, Spotlight, and others.

There are also Fio, VNC, Putty, iperf, Gpart, Grub, Hadoop, and HDFS-related tools, Hammerdb, Iperf, JAM Treesize Pro along with Space sniffer, Jungledisk, Lenovo, Microsoft (Diskpart, Disk2vhd, VMM, Diskspd, Exchange, Hyper-V, SyncToy, Windows desktop and servers), SQL Server along with Perfmon), MySQL and other SQL as well as NoSQL databases or repositories, Retrospect, Rsync, Rufus, S3fs, S3motion, and Seagate Seatools, among others. Also, there are various Linux operating systems, containers, and open-source tools along with OpenStack and Ceph, Vdbench, Wordpress plug-ins, and VMware tools, among others.

2.6. Common Questions and Tips

Part of developing or expanding your tradecraft is practicing and applying what you learn.

Why not keep all data? If the data has value, and you have a large enough budget, why not? On the other hand, most organizations have a budget and other constraints that determine how much and what data to retain.

Isn't the objective to keep less data to cut costs? If the data has no value, then get rid of it. On the other hand, if data has value or unknown value, then find ways to remove the cost of keeping more data for longer periods of time so its value can be realized.

On a server in an environment where there is a good response time and high I/O rate, a large or high queue depth and a relatively low amount of server CPU time being spent on system time or overhead would indicate what? There is a lot of work being done with a large arrival rate, yet the system can keep up with the work. However, if the response time increased, that could indicate a bottleneck somewhere. An increase in CPU time might occur if there are many small I/Os being handled.

An application is doing very large sequential reads (256 KB or larger) across a 15-TB device, and the system is showing a low CPU usage with a high queue depth and large response time. The amount of data being transferred is indicated as 21 Gbps; is this a problem? It depends on what the storage system and its I/O network are capable of, but at 21 Gbps that equates to about 2.6 GB or about half of what a 40-GbE network line rate is capable of. Keep in mind that with large data transfer, your I/O rate or activity will tend to decrease and response time increase. Likewise, less CPU is needed to handle a few larger I/Os than many smaller ones.

2.7. Learning Experience

For this chapter's learning experience, practice with the following and see how you do. Of course, as with many server, storage, I/O fundamental topics, the actual answer may depend on the context. Hence, it's important to understand the context and comprehend the question rather than simply memorize an answer.

- Explain the impact of increasing data volume.
- What is the difference between data volume and velocity?
- How do you determine the value of data?
- What do programs and applications have in common?
- Who is Niklaus Wirth, and why is he relevant to this chapter?
- What is the relationship between applications and data?
- Why is understanding applications and data important?
- What's the impact on servers, storage, and I/O of latency and bandwidth?
- What is the importance of sparse vs. dense, random vs. sequential data?
- How do you decide what data infrastructure resources to use when and where?

2.8. Chapter Summary

Everything is not the same, from businesses, organizations, or institutions (as well as consumers) to their applications and data. Having insight into the applications, data, and their characteristics help to determine where they should run and on what, vs. simply based on what is the newest trend or technology. The data infrastructure might be more flexible to change as long as it can support the upper-level applications and information infrastructure. Be mindful of how or what the upper layers depend on. The upper level might be more agile than the lower-level data infrastructures; the key is knowing what to do when, where, why, and how for different situations.

General action items include:

- Expand your tradecraft, balancing technology and business insight.
- Gain insight into the organization, its applications, and the data characteristics.
- Learn about applications and their PACE characteristics.
- Use tools to understand different attributes of data including changing value.
- Understand how to align technology with business needs and vice versa.
- Explore new technologies as well as habitats for technologies.
- Use new and old things in new ways.

That wraps up this chapter and Part One. Next we will be going deeper, including bits, bytes, blocks, blobs, and much more. The bottom line for now is to keep in mind that applications are programs and data structures, and programs = algorithms plus data structures that rely on a server, storage, I/O hardware, and software.

Part Two

Server I/O and Networking Deep Dive

Part Two, including Chapters 3 through 6, goes into more detail about data infrastructure components spanning hardware, software, and services. In addition to basic bits and bytes, we look at server compute along with I/O on a local and wide-area basis—that is, how your data infrastructure components communicate among themselves and with others.

Buzzword terms, trends, technologies, and techniques include abstraction, API, architecture, block, converged, cloud, container, DAS, ephemeral, file, Flash, HCI, I/O, IP, LAN, MAN, physical, NAS, NFV, NVMe, object, PCIe, SAN, SDN, SSD, software-defined, stateful, stateless, virtual, VM, and WAN, among others.

Chapter 3

Bits, Bytes, Blobs, and Software-Defined Building Blocks

Bits, bytes, blocks, buckets, and blobs get software defined into applications and data infrastructures.

What You Will Learn in This Chapter

- Stateful vs. stateless, ephemeral vs. persistent
- Storage vs. data sharing
- Server and storage I/O counting
- Server and storage I/O architectures and access
- Block, LUN, volume, DAS, and SAN
- Files, file systems, global namespace, and NAS
- Objects, buckets, containers, and cloud storage
- Layers of abstraction and added functionality

In this chapter, the theme is bits, bytes, blocks, blobs, and software-defined data infrastructure building blocks as we start to go deeper into server and storage I/O fundamentals. This means looking at some bits, bytes, blocks, blobs, and more, along with tools, technologies, tradecraft techniques, and trends. Each of the subsequent chapters in this section drill down further, going deeper into the various related topics. Key themes, buzzwords, and trends addressed in this chapter include physical, virtual, cloud, DAS, SAN, and NAS, along with application and storage access (block, file, object, and API), stacks, and layers, among other related themes.

3.1. Getting Started

There are many types of environments, applications, and data, along with their corresponding attributes and characteristics. Remember from Chapter 2 that applications are programs, which are algorithms plus their data structures defined to accomplish some task or function. Those applications are in turn supported by data infrastructures that are made up of servers, storage, I/O networking hardware, software, and services that get configured via software-defined management tools.

The theme of the chapter is bits, bytes, blobs, and software-defined building blocks, as those are the hardware and software items that get defined to support applications (e.g., data infrastructures), as well as the applications themselves. While some may view this as a hardware-centric discussion, others will see it as software-focused. Still others will see this as a converged discussion where hardware and software get defined into data infrastructures and the applications they support.

Building from Chapter 2 and our discussion of applications, some applications are stateless, meaning they do not store any local data and can be shut down or terminated without concern about data loss. Other applications are stateful, meaning that they have some data whose context and consistency (i.e., state) as well as content needs to be persistent. Persistent data means data is stored and preserved (stateful), able to survive system (hardware or software) shutdowns, crash, or applications problems.

Crash consistency means being able to go back to a point in time before a system crash or shutdown to the state of the system as of that point in time. Application consistency means being able to go back to a particular time and state the application along with its data were in when something happened. Think of application and transactional consistency as being more fine-grain or detailed vs. more coarse system or device consistency.

For example, if you do a financial transaction such as a banking deposit, you want absolute statefulness and transactional consistency (along with privacy security), where the data as well as transaction logs are stored on persistent (lossless), protected storage. Granted, some people may be more concerned about persistence, lossless, and statefulness of deposits vs. withdrawals.

On the other hand, consider watching a video or movie via Netflix, Hulu, or Amazon Prime/Fire, among others services. While it might be an inconvenience, you may not be as concerned if some bits or bytes are dropped (lossy), resulting in a blurred or less-than-perfect display.

Also, to improve performance over slow networks, the resolution can be lowered, removing some bits and bytes (i.e., lossy) as opposed to lossless. Lossless means all of the bits are preserved even though they may take up more space when they are stored or sent over a network. Likewise, the local video cache kept on persistent storage contains ephemeral data that can be refreshed if needed.

The opposite of persistent is ephemeral or temporary and stateless. Note that persistent storage can be configured and used as ephemeral, that is, as a temporary scratch or workspace, but only during the life cycle of what is using it or for some period. We'll have more to say about persistence, ephemeral, stateless, stateful, lossless, lossy, and consistency later.

Realizing that data infrastructures exist to support applications and their data needs, let's look into the bits and bytes that span hardware, software, clouds, and virtual from the server to storage and I/O networking. This includes how bits and bytes are organized via software into blocks, buckets, and blobs, along with other data structures that get stored. Figure 3.1 shows a data infrastructure architecture that aligns with how this book is organized.

Keep in mind that applications are programs and programs are made up of algorithms and data structures. Those data structures range from database records to files in a file system or objects stored in a cloud or object repository. Objects and files may, in turn, contain (encapsulate) other files, objects, or entities, including databases, operating systems, backups

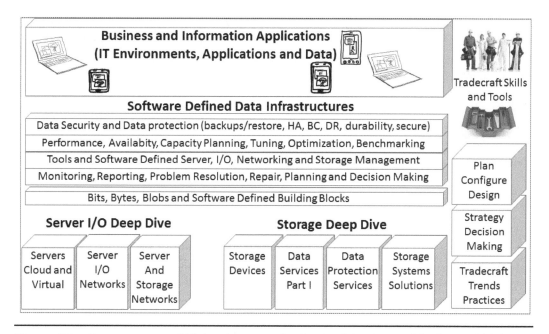

Figure 3.1   Server storage I/O hardware and software bits, bytes, blocks, and blobs.

or archives, snapshots, presentations, word documents, videos, photos, audio, among many other items.

Other examples include low-level data structures such as how an HDD or SSD is partitioned and formatted for use, including adding a file system, image, or other organization of bits and bytes. Remember that servers save their software including hypervisors, operating systems, tools, utilities, databases, files, and data along with settings for upper-level applications on some form of persistent memory or storage.

Looking at Figure 3.1, our focus shifts from the upper-level applications and data along with their general attribute characteristics to the bits and bytes of data infrastructures. Regarding the basic bits and bytes of server storage I/O hardware and software, we are going to span, or if you prefer, converge some themes. First, as a reminder, hardware needs software, software needs hardware; even virtual, cloud, and software-defined environments require hardware.

Also, in Figure 3.1, data infrastructures include not only hardware that gets defined by software; software and services also get defined by other software as part of creating, configuring, customizing, or deploying and using data infrastructures and their resources.

3.1.1. Tradecraft and Context for the Chapter

Some of what is covered in this chapter is computer science, general IT, systems administration, database administration (DBA) or network administration, systems programming, general programming and software-defined development, along with cloud, virtual, operating systems, and file systems, among other tradecraft disciplines. Also, this is the basis or foundation for discussions of software-defined storage (SDS), software-defined data centers (SDDC), software-defined networking (SDN), and software-defined data infrastructures (SDDI).

One thing that I find missing from software-defined marketing (SDM) discussions are software-defined data structures which get defined by the software to use, as well as reside in the cloud, virtual, and physical systems (and hardware). Another is software-defined storage (or data center or network) management.

I mention data structures and management in the context of bits, bytes, blocks, and blobs because they are both software-defined to support storage, servers, and data infrastructure applications. Data infrastructure applications that have bits and bytes defined into data structures include storage partitions, volumes, logical unit numbers (LUNs), file systems, databases, VMs, as well as data and object repositories that support storing and using data.

Server, storage, I/O, and network bits and bytes tradecraft topics include:

- Servers store their data in some form of persistent storage (memory).
- There are various software layers (stacks), abstractions, and functionalities.
- Block, file, object, API, and programmatic binding provide access to data.
- All applications do some form of create, read, update, and delete (CRUD).
- I/O operations include Read, Get, Write, Put, List, Head, Delete, Create, and Close.

Some bits and bytes tradecraft terms and buzzwords include:

Altitude	Relative height layer where a piece of software, application, driver, protocol, or function exists. Firmware and BIOS are low altitudes, hypervisors are higher, with operating systems even further up.
Cluster	Collection of things or resources, clustered file system or database spanning two or more servers, groups of data blocks allocated within a file system also known as extents. Clusters can be used for scaling-out (horizontal).
Fragmentation	Files in file systems along with data in databases get fragmented when their data is no longer able to be stored contiguous to each other. The result is that data is spread out using available space, which impacts performance. This is like trying to find a parking spot in a full parking lot (takes more time).
Contiguous	Data is organized, grouped, and stored logically or physically adjacent or sequential to each other in a nonfragmented manner.
Image	Refers to a photo or video, as well as an instance of an operating system, hypervisor, virtual machine, or state of a copied application.
Nesting	Refers to encapsulating or wrapping something inside of something else. For example, nesting a hypervisor such as Hyper-V running on a Windows Server that in turn runs on top of a nested VMware ESXi VM. A file system can be nested inside of, or on top of, another file system, in different layers of virtual disks, among other forms of encapsulation.
Stack	Refers to a data structure in a CPU (physical or virtual) for keeping track of data and code being worked on. Also refers to different layers of software that sit on top of each other to create a data or information infrastructure and application. This includes lower-level BIOS, firmware, device drivers, hypervisors, operating systems, agents, tools, networking, and file system databases, among other utilities or software.

3.1.2. Removing Object Obstacles: Context Matters

The word *object* has many different meanings depending on context, both inside the IT world and outside it. As a noun, the term can refer to a thing that can be seen or touched as well as a person or thing of action or feeling.

In addition to being a person, place, or physical thing, an object can be a software-defined data structure that describes something—for example, a database record describing somebody's contact or banking information, or a file descriptor with name, index ID, date and time stamps, permissions, and access control lists, as well as other attributes or metadata. Other examples are objects or blobs stored in a cloud or object storage system repository, or items in a hypervisor, operating system, container image, or other application.

Besides being a noun, *object* can also be a verb, expressing disapproval or disagreement with something or someone. In an IT context, *object* can also refer to a programming method (e.g., object-oriented programming [OOP], or Java [among other environments] objects and classes) and systems development, in addition to describing entities with data structures.

In other words, a data structure describes an object that can be a simple variable, constant, complex descriptor of something being processed by a program, as well as a function or unit of work. There are also objects unique or context-specific to environments besides Java or databases, operating systems, hypervisors, file systems, cloud, and other things.

3.2. Server and Storage I/O Basics

Server (or other computer) CPUs move data in and out of different tiers of memory including processor-level caches, DRAM main memory, and persistent and nonpersistent memory and storage.

Algorithms, that is, program code instructions, are what tell the CPUs to do as well as how to define and use bits and bytes organized into various data structures that make up the data infrastructure. Data infrastructures include databases, repositories such as file systems, operating systems, utilities and tools, hypervisor and device driver software, along with associated server and storage I/O networking hardware or services.

Remember that memory is storage and storage is persistent memory. Memory and storage hold data structures and computer server processors run algorithms that when combined result in programs to produce information. Applications and data have various attributes and characteristics, including Performance, Availability, Capacity, and Economics (PACE). Likewise, data is organized and accessed via various management abstraction layers defined by software.

Note that both hardware and software get defined or configured to take on different roles as part of supporting applications via data infrastructures. For example, servers get defined by various application-supporting infrastructure layers, including storage servers or data repositories as well as networking devices. Data repositories include databases and file systems, object stores, operating systems, and hypervisors, among other resources.

Software and data-defining servers, storage, and I/O resources:

- Transform bits and bytes into blocks, files, blobs, and objects of information.
- Compute (do work) at a given speed (e.g., gigahertz).
- Hold application algorithms and data structures (e.g., programs) in memory.

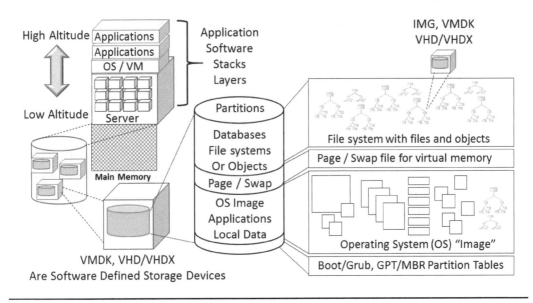

Figure 3.2 Data infrastructure layers and software-defining resources.

In this regard,

- Memory and storage capacity is measured in bytes including gigabytes (binary or decimal).
- Storage access is measured in activity, data moved, response time, and queues per time.

Other aspects to consider include:

- I/O and networking access as well as data movement
- Software and licenses for various tools and applications
- IP addresses, endpoints, and other cloud (or local) consumable resources

Storage is often divided into storage pools or spaces, files, object systems or other databases, key-value and metadata repositories. These repositories are implemented in hypervisors for virtual server infrastructures (VSI) and virtual desktop infrastructures (VDI), operating systems and databases that support other applications and functions. These different software-defined entities are layered on top of others, building up a stack that collectively provides services, management, and abstraction of underlying hardware as well as software resources.

Figure 3.2 looks at how bits and bytes get defined by software applications (utilities, tools, device drivers, agents, plug-ins) into device partitions, LUNs, volumes, and VMs.

3.2.1. From Bits to Bytes, Blocks to Blobs (Server Storage I/O Counting)

Storage is an extension of memory, and cache is a convergence of memory and external media. Digital data is stored as 1's and 0's, binary bits indicating on or off, implemented using different physical techniques depending on the physical media (disk, tape, optical, solid-state memory).

The bits are grouped into bytes (1 byte = 8 bits) and subsequently organized into larger groups for different purposes. Note that computer memory is counted in binary, while storage is often referred to in decimal. Likewise, memory and storage are usually referred to in bytes, whereas I/O networking is often discussed in terms of bits moved.

IT counting values include:

- Time (duration and speed), quantity (bits, bytes, blocks, and blobs)
- SI/International numbering scheme, big B, little b, bits and bytes
- Base 2—e.g., 2 ^ 2 = 4, 2 ^ 4 = 16, 2 ^ 8 = 256
- Binary (base 2)—e.g., 0, 1, 10, 11, 100, 101, 110, 111, 1000, 1001, 1010, 1011, etc.
- Octal (base 8)–e.g., 0, 1, 2, 3, 4, 5, 6, 7, 10, etc.
- Decimal (base 10)—e.g., 0, 1, 2, 3, 4, 5, 6, 7, 8, 9, 10, 11, 12, 13, etc.
- Hexadecimal (base 16) e.g. 0, 1, 2, 3, 4, 5, 6, 7, 8, 9, A, B, C, D, E, F, 10, 11, etc.

In addition to base 2, base 8, base 10, and base 16 numbering schemes, another counting consideration for servers and storage I/O is the International System of Units (SI). (See http://physics.nist.gov/cuu/Units/units.html to learn more about SI.)

There are SI standards for bits vs. bytes, as well as base 2 vs. base 10 bits and bytes.

Table 3.1 shows a simple example of counting using base 2, increasing by 8, 16, 32, 64, and 128, which could represent a number of bits for addressing or the size of something. For example, 8 bits can represent 256 colors of a single pixel, or 2 ^ 16 (i.e., 16 bits) represents 65,536 colors.

Table 3.1. Example of Scaling and Counting Bits and Bytes (Base 2)

Bits	Number
2 ^ 8	256
2 ^ 16	65,536
2 ^ 32	4,294,967,296
2 ^ 64	18,446,744,073,709,600,000
2 ^ 128	340,282,366,920,938,000,000,000,000,000,000,000,000

Another example is 2 ^ 32 (i.e., 32 bits), which represents about 4 GBytes of addressable data, whereas 2 ^ 64 enables significantly more data or memory to be addressed. Even though you might have a server that supports 64-bit processors, the software being used will determine how much memory can be accessed as well as other functions.

> *Tip:* If something looks like it is off by a factor of 8, the problem might be misinterpreting bits for bytes or vice versa. Similarly, if a storage or memory space capacity seems to be missing more than expected, check whether base 2 or decimal numbers are being used.

On relative data access and movement time, consider:

- CPU memories, including registers and cache (ps to ns)
- Local main memory (e.g., DRAM) access (ns to μs)
- Remote memory (RDMA) and local storage, network I/O access (μs to ms)
- Remote server, storage I/O network access (milliseconds to seconds to minutes to hours)
- Physical shipment of data between locations over distance (minutes to hours to days)

A big truck or a Boeing 747 air cargo freighter has large bandwidth (ability to move large amounts of data), but it will also have higher latency vs. a network. A network will have lower latency than a truck or a cargo airplane, but it might lack the bandwidth.

Time (duration and speed) is measured in years, months, weeks, days, hours, minutes, seconds, and fractions of a second, as shown in Table 3.2.

Table 3.2. Time Measurement—Fractions of a Second (Decimal)

Symbol	Name	Fraction of a Second	Example
ms	Millisecond Thousandth of a second	1/1,000 0.001	100 ms = 1/10th of a second
μs	Microsecond Millionth of a second	1/1,000,000 0.000 001	100 μs = 10 kHz, 1,000 μs = 1 ms
ns	Nanosecond Billionths of a second	1/1,000,000,000 0.000 000 001	1 ns = 1 GHz (processor cycle time, speed)
ps	Picosecond A trillionth of a second	1/1,000,000,000,000 0.000 000 000 001	A really small amount of time

Note that data infrastructure resources are usually measured in terms os data moved (e.g., bandwidth, throughput, and data transfer rates) in MBytes or GBytes per second, or speed of access such as latency. Watch out for and keep MBytes vs. Mbits in perspective.

There are also some additional metrics to be aware of. For example, processors run at different frequencies measured in hertz, such as MHz or GHz, which correspond to the amount of work done in a nanosecond. There are also baud rates vs. line rate vs. data transfer rates, revolutions per minute for rotating magnetic HDDs, along with refresh rates of dynamic memory and video displays. We will look at those and others in more detail in their respective chapters.

All programs (i.e., the code or algorithms and data structures) get defined and stored as bits and bytes. Those bits and bytes are organized in some way for the programs that use them. This means that even a program is organized in some ways when it is stored on disk or flash SSD as well as when mapped into memory.

There is a hierarchy of bits and bytes along with how they are organized and built upon with additional layers to create and define their functionality. These bits and bytes are addressed and accessed by the processor running or executing the program code (algorithm) to do various tasks. The more bits there are, the more addresses or ability to define things to do or represent different data and information are available.

Here are some bits and bytes basics:

- Data is represented in binary (0's and 1's, that is, base 2) format.
- A bit is on or off (binary) and can be grouped into bytes.

- 8 bits make a byte; a byte makes an 8-bit word.
- Words can be larger than 8 bits (16, 32, 64 bits).
- Be careful to avoid confusing bits and bytes: Big B for Bytes, little b for bits.
- Various character encoding schemes and bits are used.
- American Standard Code for Information Interchange (ASCII) includes 7-bit and 8-bit character encoding
- Unicode includes 8-bit UTF-8, which incorporates ASCII.
- Extended Binary Coded Decimal Interchange Code (EBCDIC) is an 8-bit character code used mainly on IBM mainframe operating systems.
- Bytes get grouped into pages, blocks, sectors, extents, and chunks of memory (storage) of various size ranging from 512 bytes to thousands of bytes in size.

The International System of Units (SI) is used for describing data storage and networking in base 2 format (see Table 3.3). For example, 1 GiBi (Gibibyte) = 2 ^ 30 bytes or 1,073,741,824 bytes.

Table 3.3. Storage Counting Numbering and Units of Measures

			Binary Base 2 Value				Decimal Base 10 Value
kibi	ki	2 ^ 10	1024	10 ^ 3	kilo	k,K	1000
mebi	Mi	2 ^ 20	1048576	10 ^ 6	mega	M	1000000
gibi	Gi	2 ^ 30	1073741824	10 ^ 9	giga	G	1000000000
tebi	Ti	2 ^ 40	1099511627776	10 ^ 12	tera	TB	1000000000000
pebi	Pi	2 ^ 50	1125899906842624	10 ^ 15	peta	P	1000000000000000
exbi	Ei	2 ^ 60	1152921504606846976	10 ^ 18	exa	E	1000000000000000000
zebi	Zi	2 ^ 70	1180591620717411303424	10 ^ 21	zetta	Z	1000000000000000000000
yobi	Yi	2 ^ 80	1208925819614629174706176	10 ^ 24	yotta	Y	1000000000000000000000000

Another common unit of measure used for data storage and servers as well as their operating systems is gigabytes (GBytes or GB) in base 10 (decimal) format. One GByte represents 10 ^ 9 or 1,000,000,000 bytes.

3.2.2. Where Are My Missing Bytes?

Computer memory is typically represented in base 2, with disk storage often being shown in both base 2 and base 10. For example, the SSD that I used in my laptop while writing this book is advertised as 512 GB. Before any operating system, RAID, erasure code, or other formatting and overhead, the SSD presents itself as 476.84 GB.

A common question is what happened to the missing 35.16 GBs? The answer is that they did not go anywhere because they were never there, depending on what numbering base you were using or assuming. This has led to most vendors including in their packaging along with documentation what the accessible capacity before environment overhead is subtracted.

The importance of the above is to understand that if you need a specific amount of data storage capacity, you need to get what you need. This means understanding the various ways and locations of where as well as how storage capacity is measured.

Table 3.4. Disk (HDD, NVM, or SSD) Storage Capacities

Base 10 (Decimal)	Base 2 (SI)	Base 10 (Decimal)	Base 2 (SI)
256 GB	238.42 GiB	2 TB	1862.65 GiB
400 GB	372.53 GiB	4 TB	3725.29 GiB
480 GB	447.03 GiB	6 TB	5587.94 GiB
500 GB	465.66 GiB	8 TB	7450.58 GiB
600 GB	558.79 GiB	100 TB	90.95 TiB
1 TB	931.32 GiB	1 PB	909.49 TiB
1.8 TB	1676.38 GiB		

Factor in the overhead of partitions, controllers, RAID (mirroring, replication, parity, and erasure codes), dispersal copies, spares, operating and file system, volume managers, and data protection such as snapshots or other reserved space as part of storage capacity. Table 3.4 shows some common storage device capacities in both base 10 (decimal) and base 2 (binary) in SI notation.

3.2.3. Server Memory and Storage Hierarchy

The importance of main memory and external storage is that virtual machines need memory to exist when active and a place on disk to reside when not in memory. Keep in mind that a virtual machine (VM) or cloud instance is a computer whose components are emulated via data structures stored and accessed via memory. This also applies to virtual storage, networks, and data centers (infrastructures) that reside in software-defined memory (i.e., data structures).

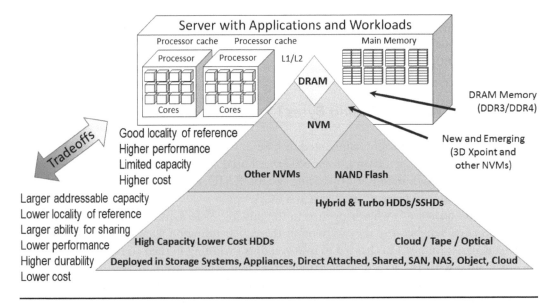

Figure 3.3   Server memory and storage hierarchy pyramid.

Server storage data hierarchy (Figure 3.3) extends from memory inside servers out to external shared storage including virtual, container, and cloud resources. Often discussions separate the server memory and data storage; however, memory is storage and storage is persistent memory. This separation has occurred in part due to focus by some on servers and memory, while for others it is about storage systems, software, or component devices. Servers and storage need I/O networking to communicate with other servers and storage, as well as users of information services both locally and remote.

An example of a storage or memory hierarchy is shown in Figure 3.3. This server memory storage hierarchy ranges from the fast processor core or L1 (level 1) and L2 (level 2) on-board processor memory to slow, low-cost, high-capacity removable storage. At the top of the pyramid is the fastest, lowest-latency, most expensive memory or storage. This storage is also less able to be shared without overhead with other processors or servers. At the bottom of the pyramid is the lowest-cost storage with the highest capacity while being portable and shareable.

Server processors L1 and L2 cache or onboard memories are very cost-effective for performance. However, while they are fast, they are also finite and relatively expensive compared to more expensive main memory (e.g., DRAM). Just as I/O to L1 and L2 is fast, yet finite capacity, DRAM is also faster than persistent storage.

However, such memories are also finite due to physical addressing and Dual Inline Memory Module (DIMM) slots along with cost. There are different tiers or types of persistent storage, from non-volatile memory (NVM) including NAND flash, Storage Class Memories (SCM), NVRAM, and others as used in SSD-based devices, as well as magnetic HDDs and tape, among others shown in Figure 3.3.

Figure 3.3 memory storage hierarchy PACE attributes include speed (Performance), persistence and durability (Availability), quantity and type (Capacity), along with energy and cost (Economics) to meet different environment needs. Memory and storage are like real estate: Location matters, not all properties are the same, and comparing on cost or size does not reflect their entire value.

As shown in Figure 3.3, L1 and L2 cache are very effective, yet there are capacity and cost limits. Similarly, persistent storage such as tape found on the lower end of Figure 3.3 has good cost-to-usable capacity, but for typical on-line applications, performance is not effective. This is again similar to real-estate, where there are high-value properties that have a good location for convenience, and also lower-cost, high-capacity warehouses for bulk storage of items that take longer to get to.

What this means is that, as with real estate, simply comparing something by its address or cost per square (or cubic) foot across various properties is comparing apples to oranges. For example, compare a retail storefront in a busy location that is small yet convenient at a high cost per square foot to a lower-cost-per-square-foot warehouse located miles or kilometers away. It is the same with storage memory: Lower cost per capacity may not be as effective as more expensive high-performance storage that has a lower cost per I/O or work being done for productivity.

Storage can be tiered with the applicable memory or storage technology and applied to the task at hand. At a lower cost (and slower) than RAM-based memory, disk storage, along with NVM-based memory devices, are also persistent. Note that in Figure 3.3, Cloud is listed as a tier of storage to be used as a tool to complement other technologies. In other words, it is important from both cost and performance perspectives to use the right tool for the task at hand to enable smarter and more effective information services delivery.

Why have all these different types of storage (memory); why not just one? The answer comes down to performance, availability, capacity economics (i.e., PACE) attributes for different applicators and data.

3.3. How Servers Organize, Access, and Use Storage

Over time, server, storage I/O access, organization, and management has evolved from very simple to more advanced. Over time, as resources (servers, storage, I/O networking and hardware, software and management tools) evolved, more basic data infrastructure functionality moved out of the applications. For example, functionality moved into databases, file systems, operating systems, containers, and hypervisors, as well as into storage systems or appliances.

Instead of an application having to handle most if not all of its data protection, the underlying data infrastructure is relied on to be resilient. On the left of Figure 3.4 we show how programs used to have to be highly optimized due to limited resources, including the performance of processors, storage, and networks, along with limited memory or storage capacity. In those environments, more of the actual optimization including actual data or storage placement was done by the programmers in a way that might resemble DevOps today, though with newer tools and resources.

As to when and where the actual software-defined environment began, besides from a software-defined marketing perspective, I will leave that up to you to decide. However, bits and bytes get defined into data structures that support programs to implement data infrastructures, which in turn support business or organizational applications.

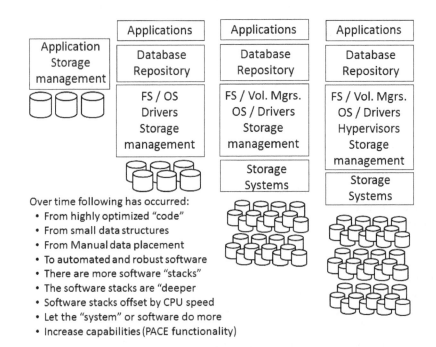

Figure 3.4 Data infrastructure hardware and software evolution.

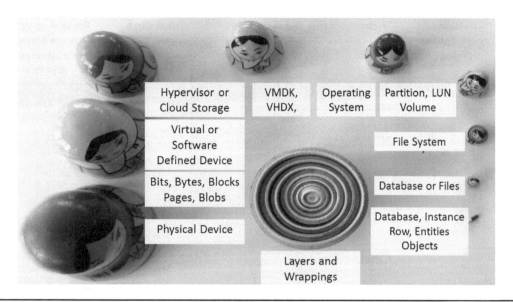

Figure 3.5 *Matryoshka* doll and data infrastructure items nested inside each other.

Moving data infrastructure functionality to other layers enables applications to focus on their core business functionality objectives. Moving toward the right in Figure 3.4 shows increased functionality leveraging databases, repositories, file systems, volume managers, hypervisors, and storage systems with increased intelligence and automation, as well as performance and capacity.

Figure 3.4 puts into context how things have evolved, including different layers of technologies, particularly software to provide abstraction, feature functionality, management, and other capabilities. The different software layers are stacked on top of each other, providing more advanced functionality as you go up or increase your altitude (i.e., get higher up from the hardware). However, keep in mind that there is plenty of software that is also resident in the hardware, from firmware to BIOS, among other items.

As shown in Figure 3.5, another way of viewing the layers or stacks is to think of different capabilities and data infrastructure items being encapsulated inside of another layer. This is similar to a *matryoshka* (a Russian doll inside another doll), where database rows, columns, entities, objects, logic, and rules are encapsulated in a database or file. That file resides in a file system, which in turn can be part of an operating system. The operating system can be bare metal on a physical machine, or as a guest (VM) on a hypervisor or a cloud instance.

In the case of a VM or cloud instance, the guest operating system is stored in a virtual disk such as a Microsoft VHDX or VMware VMDK, among others, which in turn is stored in a hypervisor file system or storage.

The hypervisor or bare-metal server instance can be stored on a partition, volume, or LUN as part of a virtual, software-defined, or legacy storage system including RAID virtual disks. All of the above along with other variations (and additional nested layers) reside on top of data infrastructure hardware. Figure 3.6 shows a generic logical example of how memory and storage get mapped, layered, nested, and containerized inside other layers.

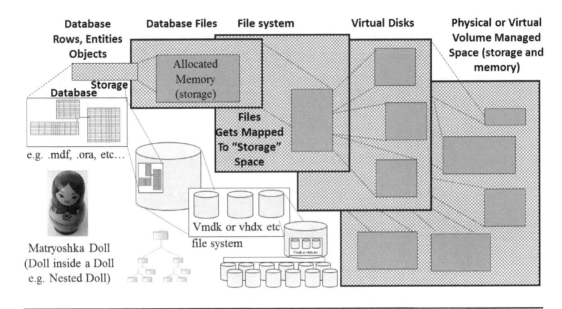

Figure 3.6 Memory and storage supporting data infrastructure applications.

A key concept in Figure 3.6 is to think of physical HDD and SSD based storage along with cloud and virtual as a large memory map that various data structures get defined and mapped into. Of course, that is oversimplifying, as there is much more going on that we will look at in more depth later.

As an example, Figure 3.7 shows a Windows operating system (top right) and its storage volumes running as a guest VM on a VMware hypervisor (middle right, management view).

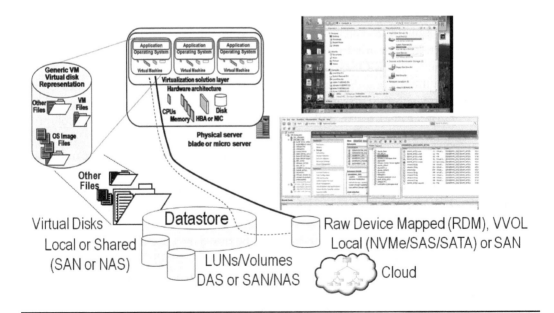

Figure 3.7 Virtual server bits and bytes being software-defined.

The top center shows the guest virtual machines and their virtual CPUs, memory, I/O, networking, and storage that run on the physical machine. The guest VMs are stored in their virtual disks (top left) such as a VMDK (or VHDX or Hyper-V, among others).

In Figure 3.7 the VMDK is stored in a data store that is part of a VMware VMFS file system or VVOL, which in turn is deployed on top of a physical storage device, virtual, or software-defined appliance or cloud storage.

Keep in mind that the operating system, hypervisor, file systems, partitioning and storage management tools, and drivers along with utilities are the algorithms of the programs that in turn define and use various data structures.

In the case of a hypervisor, the virtual CPU, memory, I/O networking, and storage along with other devices are simply data structures defined via software that reside in memory and are stored on persistent storage.

3.3.1. Resource Tenancy (Sharing vs. Dedicated)

Earlier I mentioned that memory and storage are similar to real estate in that location matters. There is another similarity, which is *tenancy*—meaning that, as with real-estate, a server, storage, or I/O resource may be dedicated or a single-occupant dwelling, or it may be a shared entity, like as an apartment building. For example, storage can be dedicated to a server along with associated I/O paths, or they can be shared. When they are shared, the resources can be in a common resource pool, or separated for privacy or the quality of service or experience.

For real estate, that can mean a shared apartment, condominium, townhome, duplex, or other structure with attached or detached storage and parking. For servers, storage, and I/O networking hardware and software, the tenancy can be isolated for privacy yet compete for performance with noisy neighbors. Servers, storage, and I/O resources can also be isolated while sharing performance—they can be isolated from somebody consuming too much without the neighbors being negatively bothered.

Data infrastructure resources tenancy includes:

- Shared nothing, shared something, shared everything
- Isolation for privacy or security, performance, and capacity
- Access control, monitoring, and metering for reporting
- Various granularities from data center to server, storage, and I/O

Various resource tenancy examples are shown in Table 3.5.

Tenancy may involve storage or data sharing. Storage sharing means being able to divide up a storage device, system, or appliance into two or more partitions, volumes, or LUNs that can be used for different things. The storage is shared, yet there is separation, with the granularity being the partition, LUN, or volume where different file systems can be formatted for use. Data sharing is more fine-grain and means that an individual file (or data) can be read and written with proper security by multiple users or computers.

Storage and data sharing may sound like the same thing, and the phrases are often used interchangeably; however, they are quite different and not at all interchangeable. For example, storage sharing means being able to have a disk drive or storage system accessible by two or

Table 3.5. Server, Storage, I/O, and Other Tenancy

Tenancy	Description	Considerations
I/O network SAN LAN MAN WAN	Virtual LANs (VLAN), SAN segments, virtual networks, virtual private networks (VPN), port blocks, LUN, volume mapping and masking	The granularity of I/O networking tenancy Time division multiplexing (TDM) vs. wave division multiplexing (WDM) Bandwidth and performance, along with availability and signal quality of service (QoS)
Server Hypervisor Operating System Containers Applications	Guest virtual machines (VMs), operating partitions, processes, threads, containers (e.g., Docker or LXC) micro services, hardware logical partitions (LPAR), shared hosting, virtual private servers (VPS), dedicated private servers (DPS), cloud instances, database instances, time sharing	Shared processor, memory, and I/O resources along with software such as operating systems, database, or other items. Some hypervisors protect against nosey and noisy neighbors for performance and availability as well as security (privacy), QoS. Understand the granularity of the multi-tenancy and its impact.
Storage	Device partitions, LUNs, volumes, virtual disks, storage pools, endpoints, namespaces, database, file systems, object repositories, mapping and masking, I/O QoS. Device vs. file, object or bite/byte-level sharing (granularity).	Share everything, nothing, or something. Access control isolation (tenancy) for performance or capacity. Virtual storage appliances, software-defined storage management, data protection, snapshots, backup and archives

more initiators or host computer servers. In Figure 3.8, each of the six servers shown has an internal dedicated direct attached storage (DAS) device (/dev/sdc). Those devices are not accessible to other servers unless they explicitly export the device to other servers using file or storage sharing software. Also shown on the lower left are three servers that each has a LUN or volume mapped from the shared storage system (e.g., Storage-A).

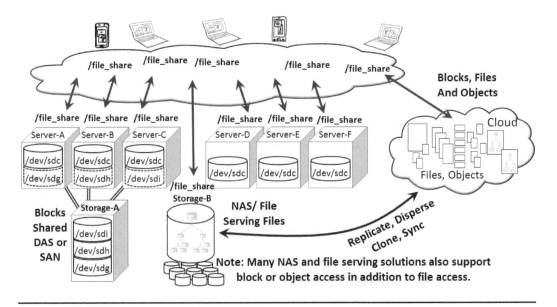

Figure 3.8 Storage and data sharing multi-tenancy.

For simplicity, Storage-A space capacity in Figure 3.8 is "carved up" or allocated as three separate LUNs, which in turn are accessed via Server-A (/dev/sdg), Server-B (/dev/sdh), and Server-C (/dev/sdi). Volume mapping and masking is used to prevent the different servers from seeing other than their own LUNs. This is an example of shared storage: Storage-A has storage shared, yet dedicated to a particular server.

Note that it is possible to share block storage across servers. However, some storage and data sharing software, such as a file system, is needed to provide data coherency and consistency. An example would be in a VMware environment where two or more physical machines can access and share the same LUN that is mapped into a VMFS datastore.

Also shown in Figure 3.8 in the lower center is a NAS storage system (Storage-B), which has exported a shared directory called /file share that is accessible from all the servers, plus by "edge" mobile devices such as laptops and tablets. Since file or data sharing software is used, individual files can be shared including, with proper configuration and authentication, reading and writing.

On the right-hand side of Figure 3.8, cloud storage is shown, including block (iSCSI) and NAS files (NFS, CIFS, HDFS), along with object accessible storage. Examples of NAS shared file access include Network File System (NFS), Hadoop Distributed File System (HDFS), Common Internet File System (CIFS, aka Samba, aka SMB, aka Windows file sharing). Note that some context is needed here, because SMB here means the server storage I/O protocol for local as well as remote access, as opposed to small/medium business or server message block, both of which are also commonly referred to as SMB.

3.3.2. Basic Storage Organization (Partitions, LUNs, and Volumes)

Block-based data access is still the lowest level of access and the fundamental building block for all storage. In the future, as object storage devices (OSD) evolve and the associated storage management software, drivers, operating systems, hypervisors, and other tools mature, some lower level activity can be expected to switch to that technology.

Block-based data access is relevant for cloud and virtualized storage as well as storage networks. Regardless of whether applications on servers (initiators) request data via an API, object, file, or block-based request from a file system, database, email, or other means, it ultimately gets resolved and handled at the lowest layers of blocks of data read from or written to disks or SSD storage.

Hardware and software defined storage devices can be

- Immutable, Write Once Read Many (WORM) or dynamic
- Opaque as to what is stored beyond bits and bytes
- Physical, virtual, cloud, internal or external, and encrypted
- Dedicated, shared, single or multiple drives
- Some number of bytes organized into blocks, pages, sectors, or objects
- Variable block sizes from 512 bytes to 4 KB, among others
- "Just a bunch of devices/disks (JBOD)" or managed
- Protected using RAID, erasure code, and other techniques

- Powered on, off, or in standby (sleep) mode
- Active, idle, or in any of various low-power modes
- Accessed via NVMe, SATA, SAS, USB, or another interface
- On-line, near-line, or off-line accessible
- Un-initialized, initialized, partitioned, or part of a pool or group or file system
- Mounted, dismounted, or in some maintenance mode
- Contents on disk data structure describing the device and partition
- A software-based signature block(s) or label describing the device
- Spare space for revectoring data from failed or failing locations

The storage blocks and their logical block addresses (LBA) or logical block numbers (LBN) can be thought of as a matrix or list of addresses that get mapped to the underlying storage media or storage system, depending where in the storage stack the volume is created.

Hidden from normal use or access, a device will usually have some spare space for automatically remapping or revectoring data when an SSD memory or HDD location goes or is detected as being bad. This space is not seen nor normally accessible by users or typical data infrastructure tools. When a device is initialized, some number of bytes (blocks) is used for data structures that describe the device, including metadata.

The metadata includes a software-defined signature, label, or name and ID or identifier of the device, format, partition tables, optional master boot record (MBR), as well as other information. Generally, the first several blocks of a device are used for partition and related information. However, some space at the end of a device may also be used.

In the past, it was possible to determine the starting LBN or LBA based on the physical disk platter or cylinder location. Today, however, HDDs and SSDs may not place data in a contiguous sequential manner (if fragmented, or space is full), instead leveraging firmware to optimize placement for performance, availability, and capacity. Some software or command settings that only erase the start of a storage device (e.g., to clean, remove, or clip a partition, label, or signature) may leave unremoved items at the end of the device.

For example, if you only do a "dd if=/dev/zero /of=/dev/deleteme bs=512 count=1024," you will erase the first 1024 × 512 bytes of the device. (*Warning:* Doing this could damage your drive and result in data loss!) However, there could be a spare or backup of those blocks at the end of the device. Tools such as Windows Diskpart, when told to erase, can clean both the start and the end of a device.

Some tools allow an entire drive to be deleted or erased, including writing zeros or ones everywhere—a process that is also known as bleaching or conditioning a device. *Another warning:* Make sure you have a good backup and follow the vendor's or other recommended procedures when doing a drive or device initialize, erase, or similar operation, as you will lose data!

Most open systems storage today is organized in fixed block architecture (FBA), meaning that all blocks, pages, sectors, and extents at the lower-level device, as well as up through the various software layers, have the same size. Note that "same size" here means that those units of storage allocated at a given layer all have the same allocation.

This means that at a device layer the blocks could be organized as 512 bytes, 520 bytes, or 4096 bytes (4 KB), among other options, and that a volume manager or file system could allocate at a larger size such as 8 KB. Exercise caution, however, to make sure that devices, partitions, volumes, volume managers, and file systems are aligned on proper boundaries to avoid extra overhead. The extra overhead of smaller allocations for larger data results in additional storage and performance being consumed to handle misaligned data, resulting in split or fragmented (extra) I/O operations.

Besides fixed block architecture, another option is variable block architecture (VBA), which has been used in IBM mainframe and some higher-level storage management software for allocating storage. In the case of open systems or other software that allocates storage such as objects in variable lengths, that data or those objects get mapped into FBA-based allocations.

The benefits of FBA are speed and simplicity of access, implementation, and commonality. The downside of FBA is that everything is the same vs. being able to adapt to a particular need as with VBA. The VBA benefit of flexibility, however, comes with increased complexity. Which choice is right depends on what you are trying to do, and that is why you may also see a hybrid of higher-level VBA with lower-level FBA.

For those who see or access data via a file system, database, document management system, SharePoint, email, or some other application, the block-based details have been abstracted for you. That abstraction occurs at many levels, beginning with the disk drive or SSD, the storage system, or controller, perhaps implementing RAID, erasure codes, disk pools, or forward error correction to which it is attached as a target, as well as via additional layers including storage virtualization, device drivers, file systems, volume managers, databases, and applications.

A drive (HDD, SSD, USB thumb drive) is the basic building block and can range from a single internal or external device in a server, workstation, laptop, tablet, DVR set-top box, or other appliance to tens, hundreds, or thousands of devices in a large storage system appliance (Figure 3.9). Regardless of whether there is a single drive, hundreds, or thousands, these devices are initialized via some software such as the operating system, hypervisor, or storage system (or at the factory).

Depending on how and where the storage is being used, that initialization may involve adding a file system to the device or adding the device to an existing file system. Another option is for the device to be added to a virtual array group or disk pool of a RAID card, storage system,

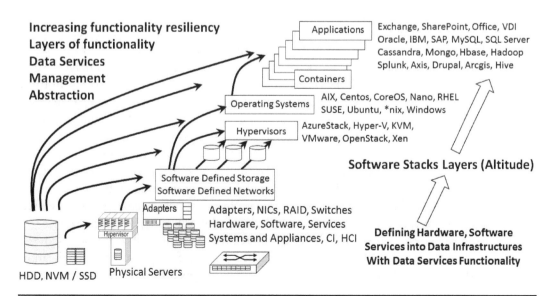

Figure 3.9   Software-defining storage bits and byte layers.

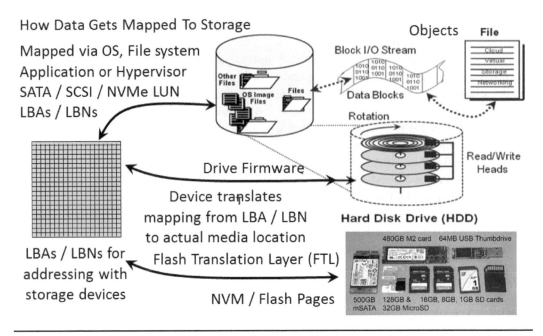

Figure 3.10   Device with partition and file or object storage access abstractions.

appliance, volume manager, or other storage management software. Moving from bottom left to top right in Figure 3.9, storage management software exists in the storage system, operating system, hypervisor, or third-party layered tools. That software can, in turn, create new virtual disks or software-defined storage with additional aggregating, and abstracting functionalities.

Various types of partitions can be used. An MBR uses a 32-bit address for LBN and supports up to 2-TB-sized devices. A globally unique identifier (GUID) partition table (GPT) uses 64 bits for LBAs, meaning larger volume sizes. GPT can be used on smaller devices, but it is generally found in larger volumes.

Working our way from the top down, the operating system, file system, or database as shown in Figure 3.10 is responsible for mapping the file system where the folder or directories are located to a specific disk drive, LUN, or volume on a storage system. A storage system that receives data from a server via file access as in the case of a NAS or file server is, in turn, responsible for mapping and tracking where blocks of data are written to on a storage device.

In Figure 3.10, an application creates a file (top right) and then saves it; for example, a document, video, or object is created and saved to disk (local, remote, physical, virtual, or cloud). An application works with the underlying operating system or file system to ensure that the data is safely written, with consistency, to the appropriate location on the specific storage system or disk drive. The operating system or file system is responsible for working with applications to maintain directories or folders where files are stored. Note that data and storage are accessed via flat linear address spaces that abstract the underlying physical storage media or service.

Data is accessed on the disk storage device (Figure 3.10) by a physical and logical address, sometimes known as physical block numbers and LBNs. The file system or an application performing direct (raw) I/O keeps track of what storage is mapped to which logical blocks on what storage volumes.

Within the storage controller, disk drive, or SSD, a mapping table is maintained to associate logical blocks with physical block locations on the disk or other media such as tape. When data is written to disk, regardless of whether it is an object, file, web database, or a video, the lowest common denominator is a block of storage (Figure 3.10).

Blocks of storage have been traditionally organized into 512 bytes, which aligns with memory page sizes. While 512-byte blocks and memory page sizes are still common, given larger-capacity disk drives as well as storage systems, 4 KB (8 × 512 bytes) block sizes now exist that also emulate 512 bytes (512E).

Note that the opacity of a block storage device is whatever can fit into the pages, blocks, or sectors as allowed by the file system, volume manager, storage management, database, or other software-defined data infrastructure management tools.

Figure 3.11 shows on the left a Windows Server example with a system disk (C:) partitioned and initialized, along with several other devices. The other Windows devices are initialized, partitioned, and configured as dynamic volumes using Windows Storage Manager into a multi drive parity RAID volume. The example shows the physical devices using the Windows "wmic" command, along with the "diskpart" command to list the disks, partitions, and volume information.

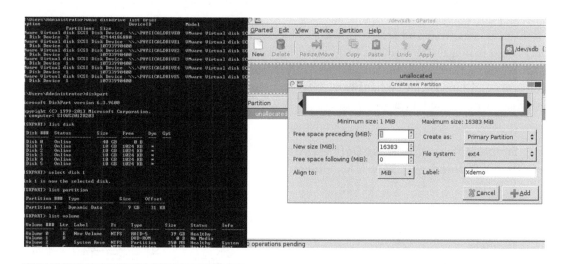

Figure 3.11   (Left) Windows partition and volumes. (Right) Linux Gpart partition.

On the right-hand side of Figure 3.11 is a Linux example using the Gpart tool to create a new partition and then define a file system on that device to make it ready for use. Note that in the Gpart configuration, parameters set alignment on MiB boundaries to avoid misalignment between files and upper-level applications, along with subsequent performance or space allocation issues.

Physical, virtual, and disk image formats include:

• Virtual machine disk (VMDK) and VVOLs for VMware
• Raw image formats (e.g., .img) and KVM QEMU, as well as qcow

- Virtual hard drive (VHD) and extended VHDX for Microsoft
- ZIP and self-extracting .EXE along with other formats.
- DMTF Open Virtualization Alliance (OVA) open virtualization format (OVF)
- Backup/restore save sets and container files (not to be confused with a docker container)
- International Standards Organization (ISO), CD/DVD image, files (.ISO) that can be copied to a USB or other device, or mounted with applicable software and a driver

3.3.3. How Data Gets Written to and Read from a Storage Device

Applications make I/O request for data via various software layers and utilities. An individual storage device (HDD or SSD) can be partitioned and formatted with a file system, or added to another existing file system for use. Another option is for a storage device to be added among others and aggregated as part of a disk or storage pool attached to a RAID adapter or storage system appliance (block, file, object, or database).

The actual I/O can be block, file, object, or API and application-based such as a read, write, get, put, or SQL query, among others. Applications can make requests to open, create, and write to a file, then close for later use, or open a file or storage device as raw, handling its own I/O operations.

Data is organized and stored in a file, object, or database or a repository system. Data access and management activities include listing data or metadata. Servers (or storage) initiate I/O requests (create, read, update, delete). Target storage (or servers) respond to I/O requests. Initiators and targets can be hardware, software, block, file, or object. Storage can be direct attached or networked, local or remote.

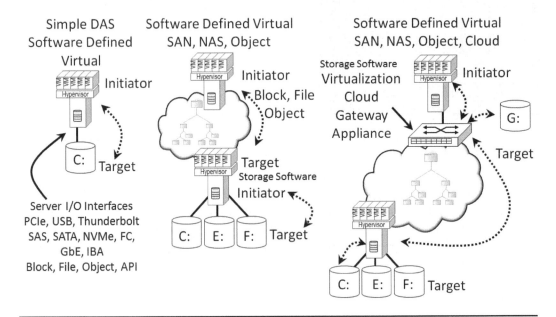

Figure 3.12 Initiator and target examples.

Initiators and Targets

A fundamental storage and I/O networking concept is that of the initiator (client or source) and targets (server or destination). There is as initiator and a target in all types of storage and access mechanisms across physical, virtual, and cloud technologies. The topologies, underlying implementations, and specific functionality will vary with vendor-specific products.

While initiators are typically servers, they can also be storage systems, appliances, and gateways that function as both a target and an initiator. A common example of a storage system acting as both a target and an initiator is local or remote replication. In this scenario, a server sends data to be written to the target device, which in turn initiates a write or update operation to another target

Servers or other initiators initiate I/O (Figure 3.12) requires (reads, writes, and status inquiries) of targets that then respond to the requests. Initiators are either taught (configured) targets or discover them at boot or startup. These targets can be block, file, object, or some other service-based destination providing and responding to storage I/O requests.

Initiators (upstream, master source, client, requester of work to be done)

- Can be physical or virtual, local or remote
- Are usually an application server, virtualization, or cloud access appliance
- Storage systems can initiate to other servers or storages and services
- Some host bus adapters (HBAs) can operate in target or initiator mode
- Can also be targets such as a gateway to other storage services

Targets (downstream, slave, destination, server, services the work to be done)

- Can be blocks, files, objects, or an API
- Can be local internal or external, dedicated or shared, as well as remote
- Can be a storage system, virtualization appliance, or cloud service
- Some can also re-initiated I/O to other storage or services

The same server can initiate I/O activity such as a file read or write request using HDFS, NFS, or SMB/CIFS over TCP/IP on an IP-based network to a NAS target. In the case of a file request, the initiator has a mount point that is used to direct read and write requests to the target, which responds to the requests.

Servers also initiate I/O requests to cloud targets or destinations, and clouds can access external media or other clouds. While the protocols can differ for the block, file, object, or specific API, the basic functionality of initiators and target exists even if referred to by different nomenclature.

Direct Attached Storage (DAS)

Similar to how block-accessible storage is the basic building block for other forms of storage access and deployment, direct attached storage (DAS) is also a foundation. DAS can be found

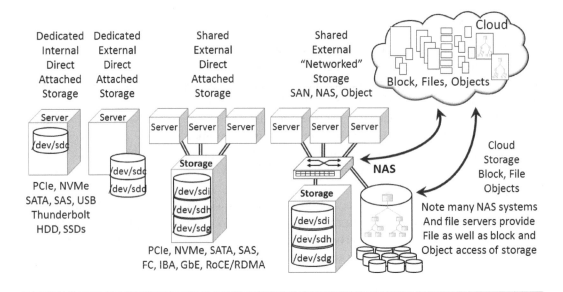

Figure 3.13 Server and storage I/O access models.

in small clusters or standalone deployments, workstations, laptops, and tablets. DAS is also found in larger systems including CI, HCI, and CiB solutions along with larger-scale clusters. Scale-out cluster nodes can be for compute, block, file, object, or database serving as well as private, public, and hybrid clouds or virtualization platforms. Many storage systems or appliances are servers with internal, external, or a mix of DAS that gets software-defined to be an SAN, NAS, object, database, or other storage or data server (Figure 3.13).

Figure 3.13 shows various storage access scenarios including dedicated internal DAS, dedicated external DAS, shared external DAS, shared external networked (SAN or NAS) storage, and cloud accessible storage. DAS is also called point-to-point, in that a server attaches directly to a storage system's adapter port using NVMe, InfiniBand, iSCSI, Fibre Channel, or SAS without a switch.

It is important to keep in mind that DAS does not have to mean dedicated internal storage; it can also mean external shared direct accessible storage using NVMe, SAS, iSCSI, Fibre Channel, FCoE, RoCE, or InfiniBand, among others. Note that a variation of the DAS example shown on the left of Figure 3.13 is CI, HCI, and CiB solutions that contain multiple servers and storage in an integrated packaged solution.

Networked Storage

Once again, context matters, in that sometimes people say or think SAN in terms of a shared storage system that can do block and file, or just block. Other people will think of SAN as being a shared storage system and switches. Some people will use NAS to mean just a file-sharing appliance or a particular vendor's product, or generically as a shared storage system that can do block, file, and object.

Keep the context in perspective to understand how and what SAN or NAS is being discussed: It might mean a shared storage system, or protocol, or an access method.

Networked storage can be block, also known as storage area network (SAN), file is known as network attached storage (NAS), and object, also known as content addressable storage (CAS), along with an application such as a database. SAN examples include Fibre Channel, FCoE, NVMe over Fabric, FC-NVMe, InfiniBand, SMB3 Direct, SRP, Ethernet-based iSCSI, NVMe, and SAS, while NAS includes NFS, CIFS, and HDFS, among others. Object includes AWS S3, Microsoft Azure, and OpenStack Swift, among many others.

What this all means is that block, file, and object storage can be accessed locally, or via a network including from cloud service providers. When you hear the term networked storage, understand the context if the discussion is about a type of storage system, appliance, service, protocol, or functionality.

3.4. Volumes and Volume Managers

Multiple storage devices (hardware, virtual, or software-defined) get aggregated, abstracted, and managed via volume managers and storage-defined storage management tools. These tools create and leverage disk or storage pools that in turn create virtual disk or virtual storage used by other higher-level applications or data infrastructure tools.

Volume and software-defined storage managers, in addition to aggregation and abstraction, also enable mirroring, stripe, bound, or spanned volumes along with erasure codes (EC), local reconstruction code (LRC), Reed Solomon (RS), and other RAID parity protection. Other features include replication, cache, tiering, snapshots, thin provision, and compression and dedupe along with expanding or shrinking of a volume.

Abstraction capabilities enable physical storage to be added or removed without disrupting other data infrastructure activities. Volume and software-defined storage managers are implemented in software and deployed as tin-wrapped appliances (storage systems), virtual storage appliances and gateways, or simply as software part of an operating system or hypervisor.

Examples include those based on an operating system, such as Windows Storage Spaces Direct (S2D), VMware VSAN, Oracle Automatic Storage Manager, as well as third-party tools from Datacore, Enmotus (FuzeDrive), Red Hat (Ceph), Starwind, and Nexenta, among others. Bundled solutions include those such as IBM SVC along with those from Dell EMC, HDS, HP, and many others. Some solutions are optimized for hypervisors or specific operating systems, while others, such as RexRay from EMCcode, are optimized for Docker containers.

Volume and software-defined storage manager characteristics include:

- Create virtual, dynamic, and other software-defined devices
- Software-defined storage to add abstraction and functionality
- Aggregate many smaller physical or virtual disks
- Can be located in various places (operating system, hypervisor)
- Transparent storage addition or deletion
- Enable snapshots, replication, transparent move, and migration
- Caching and other advanced storage management features
- Hybrid configurations leverage multiple layers of tools

Table 3.6 shows various types of common storage volumes and formatting.

**Table 3.6. Various Types of Volumes That May Be Used
in Combination with Each Other**

Type	Volume Type Description	Benefits	Caveats
Raw formatted	Raw with no file system or format; the application defines usage. File system (cooked) means that the device/volume has been defined for ease of use vs. raw (uncooked) bits and bytes.	Raw has no file system overhead, enabling fast direct access to storage. The file system adds a layer of abstraction, ease of use, agility, functionality, and interoperability. File systems manage the underlying (back-end) block or object storage.	Raw contents only know software that defines its contents and opacity. Applications may need more control. The file system defines the boundary limits, capabilities, and opacity of the volume.
Simple	A JBOD ("just a bunch of disks/drives") device with no protection or optimization.	Basic building that can be defined into pools or higher-level abstraction.	Limited to basic function and scale without being defined by software.
Bound	Concatenated or spanned volumes. Increase storage capacity without performance by combing two or more storage devices.	Easily expand a device or volume to have more storage space capacity. Implementation may support splitting bind set without total data loss.	Only provides space capacity boost, as data is not spread across multiple drives as with stripe or other volume configuration types.
Stripe RAID 0	Data is spread equally across two or more drives for performance; there is no data protection other than your backup!	Performance boost by spreading data across all drives. All I/O is chunked or shared into pieces stored on all the drives (assuming I/O is large enough).	No data protection in that loss of a single drive damages the aggregated volume, requiring the entire volume to be restored from a backup or other protection copy.
Mirror	Replicate or duplicate data copied to two or more devices in real-time (synchronous) or time-delayed (asynchronous) mode, aka eventual consistency.	All data is copied to two or more other devices, systems, or sites in real time or time-delayed, providing good data protection and possible enhanced read performance boost.	High overhead cost of additional storage capacity, for example, if two-way mirror, then 2:1, if three-way mirror, then 3:1 capacity and cost overhead incurred.
Parity erasure code, Reed-Solomon, LRC, SHEC	RAID 4, 5, 6, among others, forward error correction. Data is chunked and striped or spread across multiple drives along with recovery information to reduce space capacity overhead.	Lower capacity overhead cost than mirroring. Some implementations may enhance read performance, while others may have write penalties. Various options for balancing performance, availability, capacity, and economics.	Depending on specific implementation being used, potential exposure from long rebuild times with larger-capacity drives. Balance PACE attributes of the storage with those of your application.

3.5. Files and File Systems

File-based data access of unstructured data has seen rapid growth due to ease of use and flexibility for traditional environments as well as for virtual and cloud data infrastructures. File-based simplifies data access by abstracting the underlying block-based components, enabling information to be accessed via a filename.

File systems and files are also used as an underlying software layer to support objects in object storage systems such as OpenStack Swift, among others. With these and other types of object storage or database solutions, objects or data items are stored in files that are defined by the applications that use them.

File system software provides an abstraction of file access on a local or remote basis. Instead of knowing where the physical data is located on a disk or storage system, meaningful file-names are used, along with directory structures or folders, for organization purposes. With file sharing, the client or initiator makes requests to the filer (target or destination) to process I/O requests on a file basis. The file system software presents (serves) data from the storage device to other host servers via the network using a file sharing protocol while maintaining data coherency.

Many different types of file systems are implemented in different places with various focus and capabilities. Context matters when discussing file systems, as there are, for example, low-level file systems implemented inside storage devices. These file systems cannot be seen or accessed directly; rather, the device controller maps external requests to the internal file system.

For example, some NAND flash translation layer (FTL) controllers can implement a low-level file system for managing the various dies or chips. While it is a file system, the context is not to be confused with, for example, the file system on a server, workstation, or laptop. Other file systems get implemented behind the scenes at other layers and within the different software.

CI, HCI, and CiB converged solutions also tend to be built around some form of a distributed or clustered file system. Likewise, virtualization hypervisors such as VMware ESXi have their own VMFS file system in addition to Virtual Volumes (VVOLs). In addition to being local, file systems can be shared for remote access via a server or network attached storage (NAS) appliance. File virtualization or global name, gateways, and global file system software can be used to overlay several different file systems to create a federated or unified name space.

File system characteristics and attributes include:

- Architectures (32 bit, 64 bit, or larger, cluster, distributed, journal, parallel).
- User space such as FUSE-based or POSIX kernel mode implementations.
- Client may be embedded in the operating system, or as an optional plug-in.
- Industry standard and open for interoperability or proprietary.
- Local dedicated or shared along with remote.
- File system contains metadata about the file system and files within it.
- Files contain data, as well as some metadata about the files.
- Optimized for activity or bandwidth of big or small files.
- Root or anchor point, name, letter for mount points.
- Number of files and file size, number of directories or folders.
- Number of snapshots and replicas, granularity of locking.
- Access control lists (ACLs), permissions, encryption, quotas.
- Thin-provision, dedupe and compression, expand and shrink.
- Access (NFS, CIFS, HDFS), global namespace or federation.
- WORM, immutable and versioning, extended attributes for metadata.
- Filename size, character set, case sensitivity, symbolic links.
- Data consistency check and repair, symbolic links, linked clones.

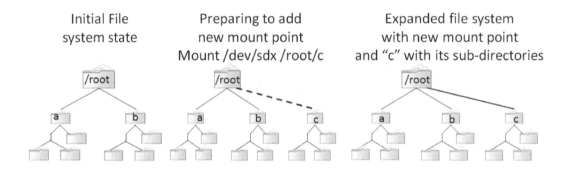

Figure 3.14   Adding a mount point and expanding a file system.

Depending on the file system, operation system, and drivers, different device or file systems can be "mapped" or mounted into a file system. For example, I have AWS S3 buckets mounted to Linux systems using tools such as S3fs. The S3 object storage buckets appear as a standard directory that I can copy or move data between local and remote locations using standard tools such as ep, copy, mv, rsync, among others.

Figure 3.14 shows an anchor point (i.e., /root) where additional directories and mount points exist. The left side of Figure 3.14 shows a file system in an initial state, before a storage device (or cloud endpoint) has been mounted into that file system (center). On the right side of Figure 3.14 is the expanded file system view with "a," "b," and "c" appearing.

For example, assume that some photos stored on a local NAS appliance are accessible as /media/photos and an S3 photo bucket is at /media/s3mnt/photos. Then a command such as "cp –uRvp /media/photos /media/s3mnt/photos" will create a copy in the S3 bucket. Note: Pay attention to your addressing and relative location offsets to avoid putting things in the wrong location.

Figure 3.15 shows on the left an Ubuntu Linux example of using the "df" command to display volumes and mounted file systems, along with a "ls –l" to display files in a directory. The

Figure 3.15   Linux and Windows partition and file system view examples.

Ubuntu system shown has both local devices initialized and configured with file systems that are mounted as part of its namespace. In addition to local devices, the Ubuntu system shown in Figure 3.15 also has shares from a NAS storage appliance endpoint mounted into the local namespace. Also shown in Figure 3.15 are an AWS S3 bucket endpoint mounted as a mount points using s3fs software so that it too appears in the local namespace.

The middle of Figure 3.15 shows Windows "wmic" and "diskpart" commands to show devices, partitions, and attributes. On the left of Figure 3.15 is a Windows "dir" command showing the contents of a directory (e.g., some chapters for this book).

File system activities include:

- Format, mount, unmount, rebuild, protect
- Snapshots, replicate, and mirror (read, read/write)
- Versioning, Rsync, and clone (full, partial, resumable)
- Consistent check and repair, data integrity
- Move, migrate, delete, resize, shrink, grow
- Report, status, listings, and attributes

3.5.1. Metadata Matters and Management

Volume managers, file systems, object storage repositories, and databases exist to manage data. However, part of management is knowing what you have. Organizations use IT applications to keep track of things, know what they have, where it is, and how it is being used or consumed or produced, among other things. IT, on the other hand, and specifically data infrastructures, rely on metadata for maneuvering resources, as well as managing data to be protected, preserved, and served.

Metadata exists for:

- Data center infrastructure management (DCIM), from facilities to software
- Application, database, repository, file system, and other settings
- Events, errors, activity, transaction, data protection, and other logs
- Files and objects describing their basic attributes or data being stored

Metadata matters include:

- File size and extents, create, modified, and accessed dates
- Application name or type, ACLs
- WORM and immutable attributes, policy and retention attributes
- Permissions (owner, group, world, read, write, execute, delete)
- Search and discovery for lookup or finding a document
- Policy and automated management including retention or disposition

Metadata is data about the data, which can mean attributes of a volume or file or file system or object. In addition, metadata also exists to keep track of how data infrastructure resources are configured and used from clustered storage and file systems to hypervisors and other software-defined items. In simple terms, metadata describes the actual data, when it was created, accessed, updated, by whom, associated applications, and other optional information.

3.5.2. File System and Storage Allocation

Figure 3.16 shows a file system formatted with 8 KB allocation (chunks, extents, clusters, pages), where some files are sparse and others are dense. In Figure 3.17, some files are contiguous while others are fragmented and aligned; file A has an extent named A1, files D and E show sparse usage (allocated but not used). Files D, E, and F are 8 KB or smaller. Also shown in Figure 3.17 is a contiguous (nonfragmented) file along with a fragmented file.

In Figure 3.17, a file system is shown allocated to a storage device with 512-byte sectors (blocks) on the left and with 4 KB-sized blocks on the right. In Figure 3.17, the larger block sizes enable more data to be managed or kept track of in the same footprint by requiring fewer pointers or directory entries.

For example, using a 4 KB block size, eight times the amount of data can be kept track of by operating systems or storage controllers in the same footprint. Another benefit is that with data access patterns changing along with larger I/O operations, 4 KB makes for more efficient operations than the equivalent 8 × 512-byte operations for the same amount of data to be moved.

In Figure 3.17 the 512-byte and 4 KB devices have the same amount of usable capacity in this example. However, the benefit is that as I/O size and space allocations increase, along with the size of devices getting larger, less overhead is required. Figure 3.17 also shows the importance of understanding the lower-level device sector size for HDD and SSD devices, as well as file system allocation (extent, cluster size) for optimizing performance and capacity.

NAS filers can have dedicated HDD or SSD storage, external third-party storage, or a combination of both. NAS systems that leverage or support attachment of third-party storage

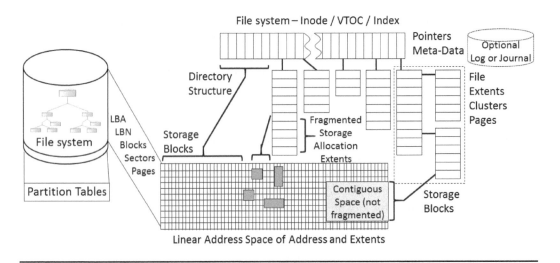

Figure 3.16   File system example.

File system formatted with 8KB allocation (chunks, extents, clusters, pages), some files are sparse others are dense. Some files are contiguous while others are fragmented and aligned, File A has an extent named A1, files D and E show sparse usage (allocated however not used). Files D, E and F are 8KB or smaller.

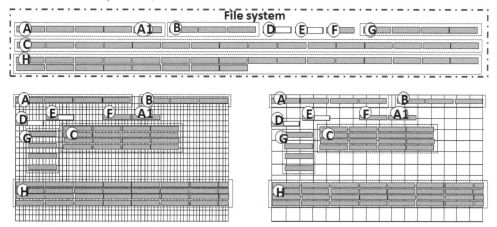

Physical or virtual device with 512 byte sectors Physical or virtual device with 4KB byte sectors

Figure 3.17   File system storage allocation and fragmentation.

from other vendors are known as gateways, NAS heads, routers, or virtual filers, among other marketing names. The advantage of using an NAS gateway is the ability to reuse and repurpose existing storage systems for investment protection or purchasing flexibility.

Some file systems (local and remote), along with access protocols and gateways, include, among others, Avere, AFP, AFS, Aufs, AWS EFS, Btrfs, Ceph (Redhat) and CephFS, CFXS (SGI), EMC, Dell, ext2-ext4, f2fs, fuse, Gluster (Red Hat), GPFS, HDFS, HTTP/Rest, HFS+, IBM, JFS, LTFS, Lustre (Seagate, DDN and others), Microsoft (FAT, FAT32, ExFAT, NTFS, ReFS, SOFS), KFS, NFS (NFSv3, NFSv4, NFSv4.1, NFSv4.2, pNFS), OpenStack Manila, Panzura, SAM QFS, SFS, Symantec/Veritas VxFS, S3fs, Union, XFS, ZFS (Oracle, Nexenta, CloudByte, SoftNAS and others), UFS, and VMware VMFS.

3.6. Cloud, Object, and API-Accessible Storage

Not all objects or object storage, systems, software, and servers are the same. As mentioned earlier, there are different meanings and context for the term *object*. Object storage can be accessed using object or API methods; the term can also refer to storage that stores data in back-end object architectures.

Object-based storage access, also previously known as CAS, continues building on the block and file storage access models. In other words, object storage is another layer of abstraction, flexibility, functionality, and scalability layered on blocks and files or jon other objects and application services. Object access is via an endpoint that provides a bit or byte stream to buckets, containers, objects, and blobs, among other constructs.

There are many different types of objects as well as usage of the term, so context matters. Likewise, containers can mean different things, thus context here is about object containers (e.g., buckets).

As discussed previously, file systems store files in a directory hierarchy structure mapped onto an underlying block storage device. Instead of a storage system reading and writing blocks of data as required by a file system, object-based storage works with objects. With block and file-based storage, applications coordinate with file systems where and how data is read and written from a storage device, with the data in blocks with little metadata attached to the stored information.

Compared to typical file systems, object storage repositories provide a larger flat name or address space for objects. This means more objects of various size can be managed without having to apply a global namespace or file system overlay or federation tool. More buckets or containers can exist in an object system namespace, which in turn contains objects or blobs.

The object repository software or system also contains metadata about the buckets; the buckets have metadata about the objects, and the objects contain metadata about themselves as well as the data in the object. Objects in an object repository also are configured for durability and can be tuned by policy to include optional replication, tiering, and other features.

Think of object storage repositories as a higher-level, additional level of abstraction above files to increase durability and scalability while housing more unstructured data in a cost-effective manner, similar to how databases house application-organized structured data.

Figure 3.18 shows at the top various applications, data protection tools, and other utilities that access buckets and objects using various APIs or file access gateways, among other methods. Buckets and containers along with their objects or blobs are owned by a project or

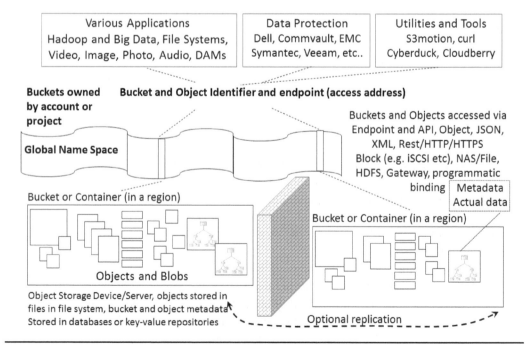

Figure 3.18 Object storage or service architecture example.

account, which may be global across different regions or implementations (e.g., public cloud servers, private or hybrid cloud deployments).

Note that buckets (or containers) exist in a specific region, as do their objects, which can optionally be replicated to another region via policy per bucket.

Object storage software, systems, solutions, and services usually have a flat namespace where the objects (or blobs) are stored. For compatibility and familiarity, many solutions also support buckets (or containers) along with their objects in a hierarchy similar to traditional file system folders, directories, and subdirectories. The buckets and objects are stored in a flat name structure, but via naming, you can enable items to appear in a hierarchy, for example, adding a slash similar to a directory name path.

Depending on the object storage implementation, buckets may have a unique name within the namespace, which namespace may be global across all regions or availability zones such as with AWS, Google, and others, or isolated to a specific site instance. Bucket names should conform to Domain Name Service (DNS) naming conventions, since they can be used with a CNAME redirect. A CNAME redirect is a DNS record that enables access addressing via a web Uniform Resource Locator (URL) to access a bucket or object. This capability enables a bucket or object to be accessed without revealing the cloud or object storage Uniform Resource Identifier (URI).

Objects get stored in a bucket; an object system can have many buckets, each with many objects, but an object belongs only to a single bucket. This means that while a bucket may require a unique name, the object name should have to be unique only within the bucket. Object names will also vary in length (e.g., a thousand byte characters or more) and types of characters.

Object names can include slashes to mimic the behavior and resemble traditional file system directory hierarchies. For example, a bucket can be something along the lines of "storageio/books/book4/chapter3.docx," which in this case would be a word document for Chapter 3 of this book. Figure 3.19 shows examples using command tools such as s3motion (left) for listing objects in an AWS S3 bucket, as well as via AWS S3 dashboard (right).

Figure 3.19 Listing objects using S3motion (left) and AWS S3 dashboard (right).

Objects contain data being stored along with metadata. Depending on the implementation, the size of objects will vary from 0 bytes to 5 TBytes (or more). Note that with some object storage, while supporting 5 TB-sized (or larger) objects, their APIs may require the objects be broken up into smaller chunks for ingestion and then reassembled once stored.

In general, objects and object storage (depending on the implementation) can be:

- Variable in size, objects from 0 bytes to 5 TB (or larger)
- Metadata per object from a few hundred to thousands or even megabytes
- Immutable, objects are written, then read, new versions created or appended
- Local via software on a server or storage system or via a service
- Support different access methods natively or via gateways
- Utilize access control lists and other security
- Accessed using Create, Put, Post, Get, Head, Delete
- Large or small in size, many numbers of objects in a bucket
- Actively read, or cold dormant such as an archive
- Opaque in that you define what is stored in the object
- Tend to be optimized for streaming (reads or writes)
- Can be good for logs, snapshots, storing large files, videos, photos, images
- May not be conducive or best fit for transaction workloads
- Optimized for sequential or stream writes and subsequent reads
- Some software or servers call objects blobs and buckets containers

How many objects and buckets, as well as the size of objects, will vary by the system or software implementation and in theory may be unlimited. However, if you are concerned, double-check to see what the vendor or service provider supports and has tested, besides what they support architecturally and in theory.

In object storage, instead of as a group of blocks of data, data is stored in an object that contains metadata as well as the information being stored. Objects are defined by an application or some other entity and organized in such a way that information about the data (metadata) is also attached to the data being stored—independently of a file system, database, or other organizational mechanism.

Objects get accessed and:

- Belong to an account or project that is part of a global namespace
- Are in a bucket or container that is owned by an account or object
- A bucket is in a specific region and availability zone (data center or system)
- Object access via an endpoint interface such as HTTP using S3, Swift, or other API
- Programmatic bindings, custom APIs, gateways, block, file, or object access
- Stored in flat namespace, however, can have a hierarchy of names
- Some solutions are using emerging key-value object drives
- Object storage can be dispersed for durability along with versioning
- Objects themselves are stored in multiple for protection
- Policies including retention, versioning, and replication
- Ouath and other security, including encryption options

Vendors and service providers with object-based storage solutions and tools include Amazon (S3 and Glacier), Caringo, Cloudberry, Commvault, Dell EMC, Google Cloud Storage (GCS), HDS, HP, IBM (Softlayer, Cleversafe), Microsoft, Nasuni, NetApp, OpenIO, OpenStack (swift), Oracle, Quantum, Rackspace, Red Hat/Ceph, Scality, Seagate, SwiftStack, VMware, and WD/Amplidata, among others.

3.7. Cloud Services (Public, Private, Hybrid)

In general, clouds (public, private, or hybrid) provide agility and flexibility, speed of deployment, resource abstraction and improved utilization, multitenancy for optimization and coexistence, resiliency, and durability, as well as elasticity and potential for cost savings or cash flow improvements vs. acquisition fees. Different types of clouds include applications as a service (AaaS) and software as a service (SaaS) (applications focused); platform as a service (PaaS) (development platforms); information as a service (IaaS) and information technology as a service (ITaaS) (information and data infrastructure); computing, networks, and storage; public (remote), private (on-premises), and hybrid (mixed) mode; solutions, services, and do it yourself (DiY).

There are many types of cloud computing options:

- Various tenancies, from shared to virtual and dedicated
- Application-centric or focused, including database services
- Containers and micro-servers for code or thread execution
- Shared hosting (e.g., websites or blogs for general populations)
- Virtual private and dedicated private server or instance; an instance is essentially a virtual machine running an OS image
- Image as a preconfigured or build-your-own OS installed with optional tools
- Various hardware or infrastructure abstraction tools that may exist in or under the OS
- Import and export or movement of virtual machines
- Access via standard IP tools including HTTP/Rest, RDP/XRDP, and SSH

Cloud network options include CDN, monitoring, diagnostic, data collection and analysis, load balancing, routing, firewall, direct-connect and high-speed access, traditional general-purpose Internet IP access, virtual private clouds and virtual private networks, NFV, and SDN. Cloud storage offerings include various focus, architecture, and service offerings; block, file, object, database, and repositories; enterprise, SMB, SOHO, and consumer services for photo, video, music, file sharing (Dropbox, Google Docs, Shutterfly); archiving, backup, replication and sync; bulk storage optimized for low-cost, high-capacity, durable (e.g., objects, general storage for block, file, or object access and external accessed storage and storage for internal access (from instances).

3.7.1. Cloud and Object Access

There is a common myth that all cloud storage is object-based and that all object storage has to be accessed via object APIs or an interface such as S3 or OpenStack Swift, among others. The reality is that there are different types of clouds and services for storage as well as computing and applications. In the case of storage, some cloud storage is object-based and accessible via S3 or Swift, among other interfaces.

However, object storage can also be accessed using gateways or other software via NAS files or block. Likewise, there is non-object-based storage in clouds, some of which are internal to a cloud, for example, AWS EBS (Elastic Block Storage) or EFS (Elastic File Storage) aka cloud NAS, as well as OpenStack Cinder (block) or Manila (Files), and Azure file storage, among others.

There are also queues, tables, and other relational database services such as those from AWS, Google, Microsoft Azure, and Oracle, among others, accessed via their programmatic bindings or database-centric bindings including SQL as well as ODBC, among others. Cloud resources can also be accessed via hardware, software, and virtual gateways as well as via hypervisors, operating systems, or utilities, along with storage systems that support cloud access.

Note that different tools provide various types of access to clouds, so be careful with a simple check box of cloud support or not. To avoid surprises, check to see what type of cloud access, cloud service, capabilities, and functionality are supported. Some example of cloud and object storage access interfaces include, among others, AWS Glacier and S3 (Simple Storage Service), Google (GCS), HTTP/HTTPS/Rest, Kinetic (Seagate and others), Microsoft Azure and ODX, OpenStack (Swift), SNIA (CDMI), and VMware VVOL, among others. Various others include API, programmatic bindings, HTTP/Rest-based, JSON, DICOM, IL7, ODBC, SOAP, UPnP, and XML-based, among others.

Various cloud storage and services can be accessed different ways:

- Natively from application, server, or storage system/appliance
- Using a gateway (hardware or software or virtual appliance)
- Via a utility program, tool, or application (essentially a gateway)
- Storage resources can be accessed as:
  - Block iSCSI
  - File (NAS), including CIFS/SMB/SAMBA, HDFS, NFS
  - Object and APIs
    - API, programmatic binding, Rest/HTTP
    - Torrent, JSON, XML, S3, Swift, DICOM, IL7, iRODS
    - Database or other requests (ODBC, SQL, and others)

Some cloud service providers and technologies include Amazon AWS, Cloud Stack, Dell EMC (Mozy, Virtustream), Google, IBM Softlayer, Microsoft Azure, OpenStack, and Rackspace. Other cloud services, solution, and technologies or providers include Adobe, ADP, Akamai, AT&T, Bluehost, Box, Crashplan/Code4, Datapipe, Dimension, Dropbox, Equinix, GoDaddy, Iron Mountain, Joyent, Jungledisk (Rackspace), Oracle, Salesforce, SAP, Verizon, VMware vCloud Air, and Xerox/ACS, among many others.

3.8. What's in Your Server Storage I/O Toolbox?

What's in my bits and bytes toolbox? AWS, Azure, GCS, Bluehost, Centurylink, and Rackspace, among other providers, for computing, I/O networking, storage, and application services. Tools such as s3motion, gsutil, curl, and Cloudberry, among others, for moving objects around, as well as s3fs for mounting AWS S3 buckets to local Linux file systems for access or data movement. Jungledisk for cloud and Retrospect for local backup (and file image-level duplication).

Acronis for image-level clone and partition resizing, Clonezilla for device or partition clone, along with dd. Gpart, grub, rufus, foglight, and seatools. JAM Treesize Pro, RVtools, Microsoft diskspd, Perfmon, wmic, and diskpart, in addition to Windows (7, 8, 10, 2012, and 2016), VMware tools, OpenStack. Various Linux tools including dd, htop, iostat, iotop, ssh,and iperf, and Dell Benchmark Factory as well as Spotlight on Windows (part of the toad tools), among others. Other tools include an S3 browser, S3cmd, Azure storage explorer and space sniffer, among others.

In addition to hardware and software tools, I also have accumulated many books (besides my own), some of which you can find on my bookshelf, www.storageio.com/bookshelf, that includes operating systems, hypervisors, system administration, programming, Unix/Linux, along with Windows-focused, database, Hadoop, cloud, virtualization, data protection, and many more. Besides books, there are also many different tutorials, labs, and additional content to expand your tradecraft at this book's companion site, www.storageio.com/book4.

3.9. Common Questions and Tips

Where does hardware stop and software start? This depends on your view of what hardware and software are. If you view the code that gets packaged and deployed as microcode in an Application Specific Integrated Circuit (ASIC), Field Programmable Gate Array (FPGA), or another processor as software, then software starts the low level. If you consider firmware on a card, drive, motherboard, router, or another device as software, then it starts at a low level. Same with BIOS or UEFI. Depending on your focus about altitude in the information and data infrastructure stack layers, hardware may be more software or less. After all, it's all bits and bytes.

Is block storage dead? why talk about it if the future is object storage? Block storage may have disappeared for some environments, or has been masked with layers of software. For other environments, block storage will remain, although it might be abstracted and defined by software.

Why do storage devices slow down as they fill up with data? Generally, it is not the actual storage device that slows down; rather, the device becomes more active doing busy work finding free space to store data. As a volume or file system fills up, finding contiguous free space takes longer. The result is that the drive is busier doing overhead work, which makes it appear slower.

When should I use block, file, or object? That will depend on what your applications need or prefer. If they can work natively with an object interface or API and their PACE attributes can be meet, then object is an option. On the other hand, if your application needs or wants a file interface along with associated file storage and its PACE attributes, that is an option. Some applications still want or prefer block or at least a local file system. It all comes down to how your application will or can access the storage, as well as the storage PACE capabilities. Additional comparisons are shown in Table 3.7.

Table 3.7. Comparing and Complementing Block, File, and Object

Block	File	Object
Block can be accessed raw, with the database or other application controlling and defining access. Some applications prefer to use block with a local file system for performance and another control such as databases, among others. Normally, block will have at least a file system. Block is good for lower-latency, frequently updated, and transactional items. Block may be used and abstracted by the file system.	General-purpose file and data sharing local or remote. While they are general-purpose, some file systems are optimized for parallel processing of large files and very big data, such as high-performance computing, energy, seismic, exploration, and similar applications. Some file systems are optimized for different things. Some file systems can perform direct raw I/O for databases, similar to block. File access is very common both for access to blocks as well as objects.	Sequential accessed read (Get) or write (Put) or streaming data. Large or small items, archives, backups, content (videos, audio, images), log files, telemetry, seismic, and similar big or small unstructured data. May not be well suited for update transactional activity. A hybrid approach is to use object for logs, retention, large files, or objects, with transactional and key-value repository on file or block.

Do virtual servers need virtual storage? While virtual servers can benefit from features found in many virtual storage systems, generally speaking, no, virtual servers do not need or require virtual storage.

Do IOPS only apply to block storage? It depends, but if you think of input/output operations per second as being either block, file, or object, then IOPs apply, although there will be different context and terminology.

3.10. Learning Experience

For this chapter's learning experience, practice with the following and see how you do. Granted, as with many server storage I/O fundamental topics, the actual answer may depend on the proper context. Hence, it's important to understand the context and comprehend question rather than memorize the answers.

- Look at your computer and determine how much total storage capacity is reported vs. free space (or in use) and how that compares to what you think you have or bought.
- Describe the difference between a partition, a LUN, and a volume vs. a file system.
- Explain why a 64-bit processor running a 32-bit operating systems is not able to access or use as much as memory as a 64-bit operating system.
- What is more important, having performance, availability, capacity, or low cost for server, storage, and I/O networking?
- Is S3 a product, service, protocol, software tool, or standard?
- Is tiering just for capacity, or can it be used for performance?
- How much raw usable capacity in gigabytes would an SSD advertised as 500 GB have?

3.11. Chapter Summary

There are many aspects to the server, storage, and I/O that span hardware, software, services, cloud, virtual, containers, and physical. Context matters with bits and bytes topics from objects to file systems and containers, among others. Bits and bytes get defined into blocks, partitions, LUNs, volumes, file systems, files, objects, and other data infrastructure items. Those data infrastructure items get defined by various software to support other tasks and application needs.

General action items include:

- Know the difference between base 2 (binary) and base 10 (decimal) counting.
- There is a server memory and storage hierarchy.
- Data infrastructure software includes many layers or stacks of functionality.
- I/Os get initiated by an initiator and resolved or survived by a target.
- Understand server, storage I/O tenancy, difference of storage vs. data sharing.
- Storage can be accessed as block, file, or object including various APIs.
- Bits and bytes get organize into blocks, partitions, volumes, and LUNs.
- Partitions, volumes, and LUNs get formatted with file systems to store files.
- Object storage repositories or databases get layered on top of file systems.
- Objects storage includes accounts or projects with buckets or containers.
- Buckets and containers contain objects or blobs along with their metadata.
- There are different types of objects, and context is important.
- There are various layers of data infrastructures software that get defined.

Bottom line: Hardware, software, and services get defined and converge at bits and bytes.

Chapter 4

Servers: Physical, Virtual, Cloud, and Containers

Applications define servers; software defines the applications that run on hardware.

What You Will Learn in This Chapter

- How servers provide computing for programs, processes, and threads
- Difference between bare-metal, virtual, cloud, containers, and converged solutions
- Server hardware, software, and services
- PCIe as a key enabler for server storage I/O and networking
- Engineered, COTS, white box, and commodity servers
- Physical and software-defined server fundamentals
- Servers as storage systems and storage as servers

In this chapter, we continue looking at the brains of data infrastructures, the server storage I/O building blocks, which are compute processors also known as computers or servers. What good is data if you cannot process it into information? That is the essence of servers, processors, or computers, whose purpose is to compute. Key themes, buzzwords, and trends addressed in this chapter include physical machines (PM) or bare-metal (BM) servers, Peripheral Component Interconnect (PCI) and PCI Express (PCIe), virtual server infrstructure (VSI) including virtual machines (VM), virtual desktop infrastructure (VDI), containers, cloud, converged infrastructure (CI), hyper-converged infrasture (HCI), cluster in a box (CiB), all in one (AiO), and cloud instances, among others.

4.1. Getting Started

Servers, also known as computers, run applications or programs to deliver information services. These programs are also responsible for generating as well as managing I/O data and

- Supports various application workloads
- Span different types of environments
- Compute node, network or storage platform
- Hardware and software defined functionality
- Commodity white box or custom engineered
- Standalone, Rack, Tower, Blade, Converged
- Compute, memory, I/O network and storage

- Servers also known as appliance, platforms, nodes or systems
- Single or Multiple Servers, Standalone or Clustered or Grid
- Some are disk less others have large number drives (HDD or SSD)
- Various expansion options exist depending on packaging
- Different options for defining how the server platform gets used

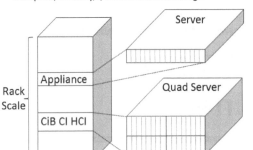

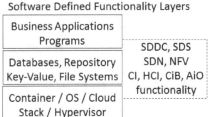

Software Defined Functionality Layers

Business Applications Programs	SDDC, SDS SDN, NFV CI, HCI, CiB, AiO functionality
Databases, Repository Key-Value, File Systems	
Container / OS / Cloud Stack / Hypervisor	

1U, 2U, 3U, 4U height, Single, Dual, Quad, Eight node
Single or multi-sockets, Various CPU, memory options
PCIe and other expansion slots, power, cooling

Figure 4.1   Software (applications) defining servers.

networking activity for data infrastructures. Another role that servers play is that of functioning as data or storage and networking appliances, performing tasks that in some cases were previously done by purpose-built systems. Figure 4.1 shows generic physical server hardware configured with various options and software-defined functionalities.

Hardware- and software-based servers get defined to perform different tasks supporting various applications as shown in Figure 4.1. On the left side of Figure 4.1 are various environments and package deployment scenarios. The top left of Figure 4.1 shows how a physical server can be dedicated to a specific function such as a storage system, network router, or application server.

Also shown across the top of Figure 4.1 is software defining a server to support general applications and operating systems, or a hypervisor, Docker containers, as well as cloud.

Servers are targeted for various markets, including small office/home office (SOHO), small/medium business (SMB), small/medium enterprise (SME) and ultra-large-scale or extreme scaling, including high-performance supercomputing. Servers are also positioned for different price bands and functionality scenarios.

General categories of servers and computers include:

- Tablets, mobile devices, laptops, desktops, and workstations
- Small floor-standing towers or rack-mounted 1U, 2U, and 5U, among others
- Medium-sized floor-standing towers or larger rack-mounted servers
- Blade centers and blade systems, twins, quads, and CiB
- Large-sized floor-standing models including mainframes
- Fault-tolerant, rugged, embedded processing or real-time
- Physical, virtual, container, and cloud-based
- Standalone, scale-out cluster, grid, RAIN (redundant array of independent nodes)
- Targeted special-purpose servers and application-specific appliances

- Converged, hyper-converged, CiB and AiO appliances
- Open Compute Project (www.opencompute.org)–based

Some examples of targeted special-purpose and application-specific usages include:

- CI, HCI, CiB, AiO for VSI, VDI, and other application workloads
- Database, Hadoop big data, analytics, and platform-as-a-service (PaaS) appliances
- Storage systems, storage appliances, networking routers or switches
- Purpose-built backup appliances (PBBA) or data protection appliances (DPA)
- Content distribution networks (CDN) or media content servers

4.1.1. Tradecraft and Context for the Chapter

Historically, there have been many different proprietary hardware as well as server software architectures, along with associated I/O networking and storage technologies. While there are still some proprietary hardware architectures, most servers have migrated to open technologies including those from AMD and Intel, among others. Likewise, most operating systems (OS) and hypervisors also now support various open hardware server, I/O networking, and storage technologies, and vice versa.

There are some hardware and software server combinations, such as IBM "Z" mainframe hardware and zOS operating system software, among others. IBM has a general-purpose server processor architecture call the PowerPC that is used in some of its proprietary servers.

Other vendors who have had proprietary servers and processors include HPE and Oracle. If you need to learn about HPE, IBM, Oracle, and other propriety hardware or servers, check out their websites along with associated white papers and tutorials, among other resources.

A context matter for servers includes a mainframe that can be used to describe a large physical server based on open or proprietary hardware and software. For many people, the term *mainframe* is associated with legacy (or existing) IBM zOS operating system (and its predecessors) and "Z" hardware servers.

The term *mainframe* is also used generically (mostly outside of IT) to refer to the main or central computer system, which may also be based on open systems. Hence, when you hear *mainframe,* understand the context, as it might be running legacy "Z" mainframe operating systems and applications, or it could be running Ubuntu, which includes Ceph, OpenStack, and other software typically found on open systems platforms.

Another server context matter is the term *server* itself. Servers have different names, from physical to virtual, cloud to logical. For example, a logical server includes the software that performs some function such as email, database, web, content, video, file and network, security, backup, or archiving, among others.

In the previous examples, what defines the type of server is software being used to deliver functionality. This can be confusing when looking at servers, because a server can support different types of workloads. Thus, it should be considered a server, storage, part of a network, or an application platform. Sometimes the term *appliance* is used to mean a server; this is indicative of the type of service the combined hardware and software solution are providing.

4.2. Server Fundamentals

Just as there are many different types of organizations, environments along with their applications having various PACE characteristics, there are many options when it comes to servers. Some are application-specific; others are general-purpose servers. There are physical hardware, virtual, and cloud servers as well as a container or micro services servers such as Docker, among others.

There are also functional servers such as a database, email, messaging, content, big data, Hadoop, file, or data serving that combine hardware and software. Also, there are various packaging options from do-it-yourself (DiY) to bundled, turnkey, converged, hyper-converged, cluster-in-box, and others. This prompts the question of why there are so many different types of servers and options. The answer is simple: Everything is not the same.

Servers include some type of:

- Command, control, communication, and compute (C4) capabilities
- Packaging of hardware and software (physical, virtual, cloud, container)
- Compute or processing capability
- Memory and storage for data
- I/O and networking for connectivity
- Management tools and interfaces

Servers vary in physical size, cost, performance, availability, capacity, and energy consumption, and they have specific features for different target markets or applications. The packaging also varies across different types of servers, ranging from small hand-held portable digital assistants (PDAs) to large frame or full cabinet-sized mainframe servers.

Another form of server packaging is a virtual server, where a hypervisor, such as Microsoft Hyper-V, VMware vSphere, Xen, and KVM, among others, is used to create virtual machines (VM) from physical machines (PM). Another option is containers, which are lightweight units of computing to support a fine-grained process or thread of work without the overhead of a full hypervisor or operating system.

Cloud-based computing resources can also leverage or require a VM. Figure 4.2 shows various types of servers, ranging from bare-metal PM that in turn wrap around software (e.g., tin-wrapped) to perform some function. Other examples include software wrapped around other software such as cloud or virtual to create and define virtual servers. Another example at the top right of Figure 4.2 is containers or micro-services such as Docker or Linux Containers (LXC), Windows, among others.

While not technically a type of server, some manufacturers use the term "tin-wrapped" software in an attempt not be classified as an appliance, server, or hardware vendor but still wanting their software to be positioned more as a turnkey solution. The idea is to avoid being perceived as a software-only solution that requires integration with hardware.

The solution is to use off-the-shelf commercially available general-purpose servers with the vendor's software technology pre-integrated and installed ready for use. Thus, tin-wrapped software is a turnkey software solution with some "tin," or hardware, wrapped around it.

A variation of the tin-wrapped software model is the software-wrapped appliance or a virtual appliance. With this model, vendors will use a virtual machine to host their software on

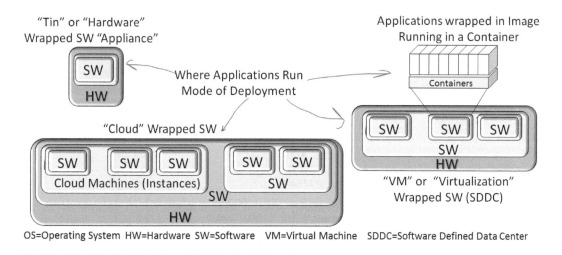

Figure 4.2   Different types of servers, hardware, and software.

the same physical server or appliance that is being used for other functions. For example, a database vendor or virtual tape library software vendor may install a solution into separate VM on a physical server with applications running in other VM or partitions. This approach can consolidate underutilized servers.

However, caution should be exercised to avoid overconsolidation and oversubscription of available physical hardware resources, particularly for time sensitive applications. Keep in mind that cloud, virtual, and tin-wrapped servers or software still need physical computing, memory, I/O, networking, and storage resources.

4.2.1. Server and I/O Architectures

Generally speaking, servers, regardless of whether they are physical, virtual, cloud, or a specific vendor implementation, have a common architecture. That architecture consists of a central processing unit (CPU) or processor, memory, internal buses or communication chips, and I/O ports for communicating with the outside world via networks or storage devices. Computers need to perform I/O to various devices, and at the heart of many I/O and networking connectivity solutions is the PCIe industry-standard interface shown in Figure 4.3.

Figure 4.3 shows a generic computer and I/O connectivity model that will vary depending on specific vendor packaging and market focus. For example, some computers will have more and faster processors (CPUs) and cores along with larger amounts of main memory as well as extensive connectivity or expansion options.

Other computers or servers will be physically smaller, lower in price, with fewer resources (CPU, memory, and I/O expansion capabilities) targeted at different needs. Note that while this is a physical server, virtual and cloud instance servers emulate the same basic architecture. With cloud and virtual servers, software defines data structures in memory to represent CPUs, memory, and storage devices, among other resources.

In Figure 4.3, the closest to the main processor has the fastest I/O connectivity; however, it will also be the most expensive, distance-limited, and require special components. Moving

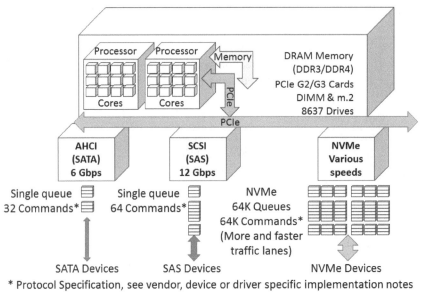

Figure 4.3   General computer and I/O connectivity model.

further away from the main processor, I/O remains fast, with distances measured in feet or meters instead of inches, but more flexible and cost-effective.

In general, the faster a processor or server, the more prone to a performance impact it will be when having to wait for slower I/O operations. Fast servers need lower latency and better performance I/O connectivity and networks.

Figure 4.4 shows common server resources in action using Spotlight on Windows (freeware version), which is part of my server storage I/O fundamentals toolbox. Resources being shown for the Windows server include, from left to right, network, CPU (four Intel Xeon cores at 2.2394 GHz each) running 93% busy with a large processor queue (work to be done and being done) with 89 processes and 2529 threads.

To the right of the CPU are memory metrics including physical and virtual free and in use, along with paging and cache activity. To the right of memory are paging file (top right) and disk storage space used and activity.

Note that on the top left of Figure 4.4 there are various icons that can be clicked on to drill deeper into specific process, CPU, memory, networking, and storage I/O activity metrics.

Figure 4.4 provides a visual representation of the primary server components regardless of operating system, PM bare metal, VM, or cloud instance. As mentioned, Figure 4.4 happens to be a Windows Server with a Microsoft SQL Server database server active.

Besides the SQL Server database, the workload running in Figure 4.4 is simulating users doing transactional workload. The workload is being generated by Benchmark Factory doing TPC-C transactions and simulates one hundred-plus virtual users.

Speaking of virtual users, this Windows Server is running as a guest VM on a VMware ESXi vSphere hypervisor on a Lenovo physical server that is a dual-socket Intel E5 2620 (12 cores with two threads per socket) for a total of 48 logical processors. Figure 4.4 would look similar if the Windows Server were on an AWS EC2, Microsoft Azure, or other cloud instance, although the quantity and speed of CPUs, memory, and other resources would vary.

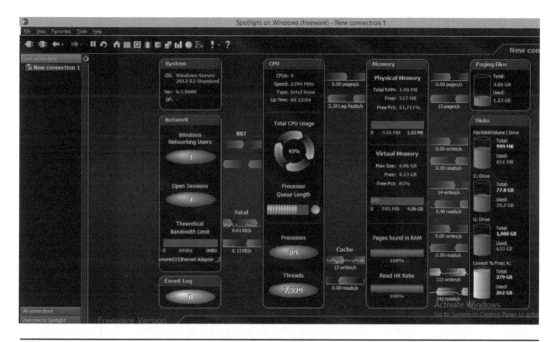

Figure 4.4   Server CPU, memory, and I/O in use.

Servers (physical, virtual, cloud instance) may be optimized for different applications:

- Compute-intensive with GPUs and lower-cost advanced RISC machine (ARM) processors
- Memory-intensive (quantity and speed of memory)
- I/O-intensive including on-board non-volatile memory or SSD
- Connectivity or expansion ports, high availability and resiliency

4.2.2. Applications PACE and Defining Your Server

Servers exist and get defined with hardware and software to support applications. Characteristics of servers to support different application PACE requirements include:

Performance   Speed (e.g., GHz) of processors (sockets, cores, threads) and memory (speed, type, registered, buffered, DDR3, DDR4, DIMM RAM and NVDIMM, SCM), storage and I/O network capabilities

Availability   Redundant components (power, cooling, controllers, drives, memory), error correcting code (ECC) and other reliability availability serviceability (RAS) features, Failover and configuration for availability, security including encryption and identity access management

Capacity   Number of cores, threads and amount of memory, quantity and type of I/O and networking expansion ports, internal storage space capacity

Economics   Up-front initial capital expenditures (CapEx), recurring operating expenditures (OpEx), value vs. low cost, energy power consumption, maintenance and software licenses along with other service fees

4.3. Bare-Metal, Physical Machines, and Hosts

Servers are used for hosting an operating system, hypervisor, or other software stack bundle. Physical machines (PM), bare-metal, virtual hosts, or hardware-based servers are the basis for where various software runs.

PM servers come in various sizes, shapes, and configurations at different prices that you can buy, rent, lease, or subscribe to. PM servers can be on-site at your premises, in a colocation facility (CoLo), with a managed service provider (MSP), or elsewhere.

In addition to offering cloud machine (instances) based on an underlying hypervisor (enabling virtual machines), many MSP and cloud providers have started offering instances as virtual private servers (VPS). Some also go a step further and offer dedicated private servers (DPS).

The lines blur in that some hosting and service providers offer plain bare-metal servers on which you install your own software, including hypervisors as well as operating systems. With DPS, the provider makes a dedicated physical server available to you that has a thin management layer from which you can then install your software; this is not a true bare-metal system in that you get some level of support.

Context matters when talking about hosting companies, cloud, and service providers, as these terms may refer generically to an instance or system as dedicated when in fact it is on a shared physical platform. By understanding what to ask and keeping things in the proper context, you can be informed about making smart data infrastructure decisions.

In general, servers have some form of power supply and cooling fans, along with a main or motherboard with memory and I/O capabilities. I/O capabilities include support for Universal Serial Bus (USB), HDMI, or VGA graphics, Ethernet LAN (or Wi-Fi) network connectivity, and various storage ports such as NVMe, SAS, SATA, Thunderbolt, and PCIe, among others.

Hardware defining the server or platform includes:

- Motherboards and add-in cards (AIC) including mezzanines, daughter cards, and risers
- Number and type of memory slots (e.g., DIMMs) per node or motherboard
- BIOS and UEFI capabilities, IPMI, VNC, PXE boot, and other management options
- Operating system and hypervisors or special software for bare-metal servers
- Software device drivers and supported configurations

An example of special software running on a server is storage systems that use industry-standard Intel or AMD x64 processors. These solutions, also known as storage servers or appliances, have their applications (e.g., storage functionality) bundled with an operating system or kernel that are combined and installed on the hardware platform. Examples include networking routers, load balancers, and others.

4.3.1. Fundamental Server Components

Common to all servers is some form of a main system board, which can range from a few square meters in micro towers to small Intel NUC (Next Unit of Compute) or Raspberry Pi-type desktop servers, to large motherboards or series of boards in a larger frame-based server.

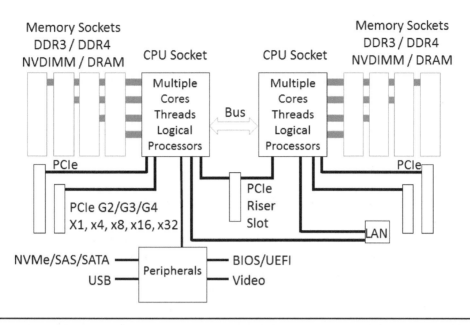

Figure 4.5   Generic dual-socket computer server hardware architecture.

For example, a blade server will have multiple server blades or modules, each with its own motherboard, which share a common backplane for connectivity. Another variation is a large server such as an IBM "Z" mainframe, Cray, or another supercomputer that consists of many specialized boards that function similar to a smaller-sized motherboard on a larger scale.

Some motherboards also have mezzanine or daughter boards for attachment of additional I/O networking or specialized devices. Figure 4.5 shows a generic example of a two-socket, with eight-memory-channel-type server architecture.

Figure 4.5 shows several PCIe, USB, SAS, SATA, 10 GbE LAN, and other I/O ports. Different servers will have various combinations of processor and Dual Inline Memory Module (DIMM) sockets along with other features. What will also vary are the type and some I/O and storage expansion ports, power and cooling, along with management tools or included software.

A server architecture such as shown in Figure 4.5 could be a standalone rack-mount or tower server, or part of a cluster in a box, where there are two, four, or eight (or more) servers (or blades) in a common enclosure.

What determines the type of packaging to use is what meets your needs, space or other functional requirements, preferences, or what a vendor provides.

Note that blades can be thin with new storage, or thick with storage and additional PCIe I/O expansion ports. CiB, CI, HCI, and other physical servers vary from two to four to eight or more nodes, each being an individual server that can have one or more sockets with multiple cores.

These types of systems are commonly found in consolidated CI or HCI solutions. We'll have more to say about CiB, CI, and HCI later in this chapter.

In general, servers have one of more of the following:

- Compute or central processing units (CPU) chips or sockets

- Cores per CPU socket or chips capable of single or multi-threading
- Main memory, commonly dynamic random access memory (DRAM)
- Optional sockets for expansion memory, extra CPUs, and I/O
- Attachments for keyboards, video, monitors (KVM—not to be confused with hypervisor)
- I/O connectivity for of peripherals including networks and storage
- I/O networking connectivity ports and expansion slots such as PCIe
- Optional internal disk storage and expansion slots for external storage
- Power supplies and cooling fans

4.3.2. Server Reliability, Availability, and Serviceability

Server reliability, availability, and serviceability (RAS) includes physical hardware components along with software and configuration settings. RAS components help to increase server availability or uptime by reducing downtime from device or component failure. Serviceability includes the ability to make physical or software repairs to a server rapidly, by swapping out failed components or making configuration changes or resets.

Hot-swappable or replacement components include field replacement units (FRU) such as power supplies and cooling fans, disk drives, and other items. Hot-swappable means the component can be replaced while the server is in use. To improve availability and resiliency, redundant components such as power supplies and cooling fans may exist. In an $N + 1$ model, for instance, there is some number (N) plus an extra component—for example, a single power supply plus one hot space, or seven disk drives and a hot spare.

Note that the hypervisor or operating system needs to support hot-swap, or hot-plug devices. Other RAS capabilities include mirrored memory along with error correction code for data integrity. In addition to having redundant power and cooling, along with storage devices, servers also support additional I/O and networking devices that can be configured for redundancy.

4.3.3. Processors: Sockets, Cores, and Threads

At the core (pun intended) of servers are the processors (virtual, logical, or physical) that execute the software that defines data structures and algorithms to perform various tasks. There are general-purpose, so-called open, and industry-standard general-purpose processors. Examples of general-purpose processors include those from Intel, AMD, IBM (Power series), and Broadcom.

In addition to physical general-purpose processors, there are also graphics processor units (GPU), reduced instruction set computing (RISC), lower-power advanced RISC machines (ARM), and other specialized processors. Besides physical processors, there are also virtual processors such as those found in hypervisors for virtual machines.

Processors are attached to a mother or main board installed into a socket. Depending on the type of server, there may be multiple sockets, each with multiple cores and threads. The cores are what do the actual work: Think of cores as multiple compute servers that exist within a single processor socket. The processors and cores installed in the sockets (Figure 4.6) can also share access to onboard memory, PCIe, and other devices.

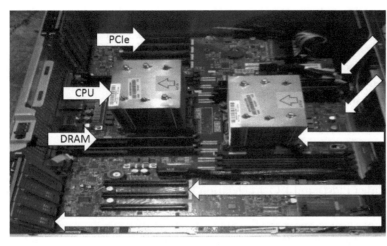

Figure 4.6   Server with dual socket processors.

In addition to a processor having cores that run at a given clock rate (GHz speed or instructions per second), cores can have one or more threads for doing work. Also, cores, processors, and sockets can run at variable speeds (faster for performance, slower to save power). If an application or software supports threads to spread work out concurrently and in parallel, then the cores and their threads can be leveraged.

Server processors and motherboards (Figure 4.6) scale up by having faster cores, while scaling out is having more sockets, cores as well as threads, in addition to being clustered (grid or ring) with other physical servers. For some applications that are not optimized for multiple cores and threads, faster cores may be better.

Software that can leverage more cores and resources may be a benefit, but faster cores are also important. Physical server compute and chassis vendors include AMD, Apple, Asus, Cisco, Dell EMC, Foxconn, Fujitsu, HPE, Huawei, IBM, Inspur, Intel, Lenovo, NEC, NVIDIA, Oracle, Quanta, Samsung, Seagate/Xyratex, and Supermicro, among others.

Different types of physical processors include:

- Uniprocessors or symmetrical multiprocessors (SMP)
- Proprietary processors such as IBM "Z" mainframes, "i" and "p" systems
- AMD and Intel x86 32-bit or x64 64-bit general-purpose processors
- GPU for graphic and other compute-intensive applications
- Reduced infrastructure set computing (RISC)
- Advanced RISC machines (ARM)
- ASIC and FPGAs for specialized processing
- Standalone server with one or more sockets and cores

Server compute attributes include:

- Server-side encryption and security features

- Local onboard cache (e.g., L1/L2)
- One or more sockets for processors
- Processors with multiple cores (e.g., logical processors) and threads
- Various GHz (or MHz) processing speeds
- Intelligent power management, low-power modes
- High-performance turbo modes (burst and sustained)
- I/O connectivity and expansion buses

Server compute-related features include:

- Trusted platform module (TPM) and I/O acceleration features
- Virtualization technology (VT) assist modes, extended page translation (EPT)
- Fast boot if OS and hypervisor support the feature
- Second-level address translation (SLAT) for memory nesting
- Physical address extension (PAE) memory access for 32-bit systems
- No execute (NX) bit if supported by the OS or the hypervisor secures memory locations

When you are looking at servers, pay attention to the processor capabilities, particularly if your software requires or supports special features such as virtualization technology or similar capabilities. Some software will also perform a processor-type check to see what features are supported, or the level of the processor regarding being able to support the specific software.

What this means is that even though most software may be plug-and-play across x86 or x64 processors today, there are still some that are plug-and-pray, or require specific hardware and software combinations including BIOS/UEFI or firmware updates.

4.3.4. Memory Matters

Application programs rely on memory in servers (and other devices) for storing and accessing the algorithms (code) that work with and process data structures to provide information services. Demand drivers for increased memory capabilities includes applications such as little data databases (SQL and NoSQL), big data Hadoop and SAP HANA or Oracle, among other analytics, video rendering and business analytics, modeling and simulations where large amounts of data are kept in memory for speed of access.

More memory is being installed into denser footprints in servers along with more sockets for processors. Similar to CPU processors, memory can be physical (e.g., main memory), as well as buffers inside processors and cores (e.g., L1/L2 cache), along with extended or virtual memory.

Some tradecraft context is required when discussing virtual memory, as there is the virtual memory associated with VSI hypervisors and their guest virtual machines (e.g., vMemory), as well as operating system- or application-based virtual memory.

The operating system and application virtual memory creates a larger available address space than what might otherwise be physically possible. The additional memory address pages get moved in and out of physical memory being stored on a page and/or swap file located on an HDD or SSD. The virtual memory found in a hypervisor (e.g., vMemory) functions similarly in that the available physical memory is allocated and shared, while additional vMemory may be paged in or out from an HDD or SSD page or swap file.

Server processors rely on a memory management unit (MMU) for managing and mapping real or physical memory to virtual memory, along with other functions. Many processors also have functionalities such as EPT (extended page translation) and SLAT (second-level address translation) for nesting and other advanced feature functionalities.

For security and safety, other memory and processor features include NX bit functionality to protect a certain address range from being manipulated, assuming the operating system or hypervisor software also supports the feature. Other common memory topics include error correcting code among other techniques to ensure data integrity of data in main memory.

With multiple processors, sockets, and cores being common on today's servers, a way of accessing a common memory map is needed to access regions of memory that are physically attached to different processors (e.g., on the same motherboard). A nonuniform memory access (NUMA) memory architecture processor has the ability for memory to be physically attached to different sockets and cores, yet accessed with low latency from a companion processor socket and its cores.

The processor sockets and cores communicate with each others' main memories via high-performance, low-latency interconnects such as AMD Opteron Hyper Transport and Intel Quick Path Interconnect (QPI). In addition to physical server NUMA, hypervisors also implement NUMA as well as some applications such as Microsoft SQL Server, enabling hardware-as well as software-based NUMA.

Moving out from the motherboard or main system board and beyond the limits of a QPI or Hyper Transport or even PCIe is what is known as remote direct memory access (RDMA). RDMA is a high-performance, low-latency technique for loosely or tightly coupled servers to access each others' main memory directly over a network connection. What makes RDMA different from other server-to-server communications techniques is leveraging faster, lower-latency physical network interconnects as well as a streamlined I/O network stack with less overhead.

Memory matters include:

- Quantity, capacity, speed, and type of server memory
- Packaging such as a dual inline memory module (DIMM)
- Speed and type, such as DDR3 or DDR4 or higher
- Server or laptop, workstation or mobile memories
- Buffered, unbuffered, and registered memories, and video RAM
- Non persistent dynamic random access memory (DRAM)
- Persistent NVM including NAND flash or 3D XPoint NVDIMM and SCM
- Hypervisor, operating system dynamic, virtual memory, and ballooning

Figure 4.7 shows an example of DDR4 DRAM memory and DIMM slots.

Hypervisors support memory ballooning to allocate and reclaim memory from guest virtual machines to improve effective memory use. Similar to operating system virtual memory, which leverages page and swap files, ballooning allows vMemory and the guest's memory to expand and contract as needed. The result is better use of available memory.

For example, in Figure 4.8 the VMware host shown has five active guests (virtual machines) that have a total memory configuration of 26 GB, yet the physical memory has only 16 GB of DRAM. Some of the guest VM are more active in using memory, while others can free up their allocations.

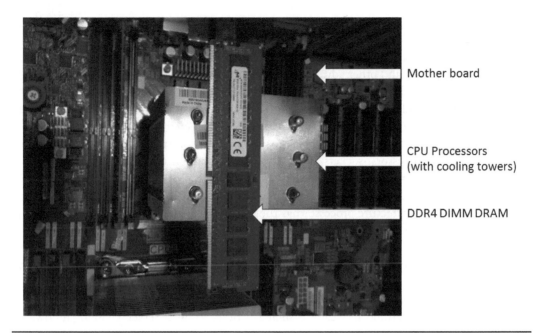

Figure 4.7   DDR4 DRAM module and server DIMM slots.

Non-volatile memory includes NAND flash and its variations [single-layer cell (SLC), multi-level cell (MLC), enterprise MLC (eMLC), triple-level cell (TLC), and 3D or vertical, among others]. Other NVM includes read-only memory (ROM), non-volatile RAM (NVRAM), NOR, phase-change memory (PCM), and 3D XPoint, among others available in various densities, speeds, and packaging from DIMM (e.g., NVDIMM to drives, cards, and PCIe AiC).

Figure 4.8   Physical and virtual memory allocation (VMware ESXi).

A tip or recommendation is to install as much DRAM and NVM as possible, both for main memory as well as secondary and paging memory storage. However, also focus on getting as much fast memory as possible, as fast servers need fast memory.

4.3.5. I/O and Expansion (Internal and External)

Various servers have different internal and external I/O and expansion or connectivity options. Depending on the physical size, target market, price band (cost), and other factors, how many expansion slots or connectivity ports will vary. What will vary includes the quantity, type, and speed of different ports or slots both internal and external. What will also vary is whether these ports or interfaces are exposed (external) or available for internal use to the server.

Server expansion ports vary by type of server, but they can include:

- CPU processor sockets, memory slots, PCIe, NVMe, and M.2 ports
- Mezzanine, daughter or RAID cards, SAS/SATA, USB, Thunderbolt
- Storage drive (HDD or SSD) and NVMe slots, LAN GbE and 10 GbE ports
- Keyboard, video, mouse

4.3.6. PCIe, Including Mini-PCIe, U.2, M.2, and GPU

At the heart of many server I/O and connectivity solutions is the PCIe industry-standard inter-face (see PCIsig.com). PCIe is used to communicate with CPUs and the outside world of I/O networking devices. The importance of a faster and more efficient PCIe bus is to support more data moving in and out of servers while accessing fast external networks and storage.

For example, a server with a 40 GbE NIC or adapter would have to have a PCIe port capa-ble of 5 GB per second. If multiple 40 GbE ports are attached to a server, you can see where the need for faster PCIe interfaces come into play.

As more VM are consolidated onto PM, as applications place more performance demand either regarding bandwidth or activity (IOPS, frames, or packets) per second, more 10 GbE adapters will be needed until the price of 40 GbE becomes affordable. It is not if, but rather when you will grow into the performance needs on either a bandwidth/throughput basis, or to support more activity and lower latency per interface.

PCIe is a serial interface specified for how servers communicate between CPUs, memory, and motherboard-mounted as well as AiC devices. This communication includes support attachment of onboard and host bus adapter (HBA) server storage I/O networking devices such as Ethernet, Fibre Channel, InfiniBand, RapidIO, NVMe (cards, drives, and fabrics), SAS, and SATA, among other interfaces.

In addition to supporting attachment of traditional LAN, SAN, MAN, and WAN devices, PCIe is also used for attaching GPU and video cards to servers. Traditionally, PCIe has been focused on being used inside of a given server chassis. Today, however, PCIe is being deployed on servers spanning nodes in dual, quad, or CiB, CI, and HCI deployments. Another variation of PCIe today is that multiple servers in the same rack or proximity can attach to shared devices such as storage via PCIe switches.

PCIe components (hardware and software) include:

- Hardware chipsets, cabling, connectors, endpoints, and adapters
- Root complex and switches, risers, extenders, retimers, and repeaters
- Software drivers, BIOS, and management tools
- HBAs, RAID, SSD, drives, GPU, and other AiC devices
- Mezzanine, mini-PCIe, M.2, NVMe U.2 (SFF 8639 drive form factor)

There are many different implementations of PCIe, corresponding to generations representing speed improvements as well as physical packing options. PCIe can be deployed in various topologies, including a traditional model where an AiC such as GbE or Fibre Channel HBA connects the server to a network or storage device.

Another variation is for a server to connect to a PCIe switch, or in a shared PCIe configuration between two or more servers. In addition to different generations and topologies, there are also various PCIe form factors and physical connectors (Figure 4.9), ranging from AiC of various length and height, as well as M.2 small-form-factor devices and U.2 (SFF 8639) drive form-factor device for NVMe, among others.

Note that the presence of M.2 does not guarantee PCIe NVMe, as it also supports SATA. Likewise, different NVMe devices run at various PCIe speeds based on the number of lanes. For example, in Figure 4.9 the U.2 device shown is a PCIe x4.

PCIe leverages multiple serial unidirectional point-to-point links, known as lanes, compared to traditional PCI, which used a parallel bus design. PCIe interfaces can have one (x1), four (x4), eight (x8), sixteen (x16), or thirty-two (x32) lanes for data movement. Those PCIe lanes can be full-duplex, meaning data being sent and received at the same time, providing improved effective performance.

PCIe cards are upward-compatible, meaning that an x4 can work in an x8, an x8 in an x16, and so forth. Note, however, that the cards will not perform any faster than their specified speed; an x4 in an x8 slot will only run at x8. PCIe cards can also have single, dual, or multiple external ports and interfaces. Also, note that there are still some motherboards with legacy PCI slots that are not interoperable with PCIe cards and vice versa.

Figure 4.9   PCIe devices NVMe U.2, M.2, and NVMe AiC.

Note that PCIe cards and slots can be mechanically x1, x4, x8, x16, or x32, yet electrically (or signal) wired to a slower speed, based on the type and capabilities of the processor sockets and corresponding chipsets being used. For example, you can have a PCIe x16 slot (mechanical) that is wired for x8, which means it will only run at x8 speed.

In addition to the differences between electrical and mechanical slots, also pay attention to what generation the PCIe slots are, such as Gen 2 or Gen 3 or higher. Also, some motherboards or servers will advertise multiple PCIe slots, but those are only active with a second or additional processor socket occupied by a CPU. For example, a PCIe card that has dual x4 external PCIe ports requiring full PCIe bandwidth will need at least PCIe x8 attachment in the server slot. In other words, for full performance, the external ports on a PCIe card or device need to match the external electrical and mechanical card type and vice versa.

Recall big "B" as in Bytes vs. little "b" as in bits; for example, a PCIe Gen 3 x4 electrical could provide up to 4 GB/s bandwidth (your mileage and performance will vary), which translates to 8 × 4 GB or 32 Gbits/s. In Table 4.1 there are a mix of Big "B" Bytes per second and small "b" bits per second.

Each generation of PCIe has improved on the previous one by increasing the effective speed of the links. Some of the speed improvements have come from faster clock rates while implementing lower overhead encoding (e.g., from 8b/10b to 128b/130b). For example, PCIe Gen 3 raw bit or line rate is 8 GT/s or 8 Gbps or about 2 GBps by using a 128b/130b encoding scheme that is very efficient compared to PCIe Gen 2 or Gen 1, which used an 8b/10b encoding scheme. With 8b/10b, there is a 20% overhead vs. a 1.5% overhead with 128b/130b (i.e., of 130 bits sent, 128 bits contain data and 2 bits are for overhead).

Table 4.1. PCIe Generation and Sample Lane Comparison

	PCIe Gen 1	PCIe Gen 2	PCIe Gen 3	PCIe Gen 4
Raw bit rate	2.5 GT/s	5 GT/s	8 GT/s	16 GT/s
Encoding	8b/10b	8b/10b	128b/130b	128b/130b
x1 Lane bandwidth	2 Gb/s	4 Gb/s	8 Gb/s	16 Gb/s
x1 Single lane (one-way)	~250 MB/s	~500 MB/s	~1 GB/s	~2 GB/s
x16 Full duplex (both ways)	~8 GB/s	~16 GB/s	~32 GB/s	~64 GB/s

By contrast, older generations of Fibre Channel and Ethernet also used 8b/10b, having switched over to 64b/66b encoding with 10 Gb and higher. PCIe, like other serial interfaces and protocols, can support full-duplex mode, meaning that data can be sent and received concurrently.

Bit Rate, Encoding, Giga Transfers, and Bandwidth

Let's clarify something about data transfer or movement both internal and external to a server. At the core of a server, there is data movement within the sockets of the processors and its cores, as well as between memory and other devices (internal and external). For example, the QPI bus is used for moving data between some Intel processors whose performance is specified in *giga transfers* (GT).

PCIe is used for moving data between processors, memory, and other devices, including internal- and external-facing devices. Devices include host bus adapters (HBAs), host channel adapters (HCAs), converged network adapters (CNAs), network interface cards (NICs) or RAID cards, and others. PCIe performance is specified in multiple ways, given that it has a server processor focus which involves GT for raw bit rate as well as effective bandwidth per lane.

Effective bandwidth per lane can be specified as half- or full-duplex (data moving in one or both directions for send and receive). Also, effective bandwidth can be specified as a single lane (x1), four lanes (x4), eight lanes (x8), sixteen lanes (x16), or 32 lanes (x32), as shown in Table 4.1. The difference in speed or bits moved per second between the raw bit or line rate and the effective bandwidth per lane in a single direction (i.e., half-duplex) is the encoding that is common to all serial data transmissions.

When data gets transmitted, the serializer/deserializer, or serdes, converts the bytes into a bit stream via encoding. There are different types of encoding, ranging from 8b/10b to 64b/66b and 128b/130b, shown in Table 4.2.

Table 4.2. Low-Level Serial Encoding Data Transmit Efficiency

Encoding Scheme	Overhead	Single 1542-Byte Frame			64 × 1542-Byte Frames		
		Data Bits	Encoding Bits	Bits Transmitted	Data Bits	Encoding Bits	Bits Transferred
8b/10b	20%	12,336	3,084	15,420	789,504	197,376	986,880
64b/66b	3%	12,336	386	12,738	789,504	24,672	814,176
128b/130b	1.5%	12,336	194	12,610	789,504	12,336	801,840

In these encoding schemes, the smaller number represents the amount of data being sent and the difference is the overhead. Note that this is different yet related to what occurs at a higher level with the various network protocols such as TCP/IP (IP). With IP, there is a data payload plus addressing and other integrity and management features in a given packet or frame.

The 8b/10b, 64b/66b, or 128b/130b encoding is at the lower physical layer. Thus, a small change there has a big impact and benefit when optimized. Table 4.2 shows comparisons of various encoding schemes using the example of moving a single 1542-byte packet or frame, as well as sending (or receiving) 64 packets or frames that are 1542 bytes in size.

Why 1542? That is a standard IP packet including data and protocol framing without using jumbo frames (MTUs or maximum transmission units).

What does this have to do with PCIe? GbE, 10 GbE, 40 GbE, and other physical interfaces that are used for moving TCP/IP packets and frames interface with servers via PCIe.

This encoding is important as part of server storage I/O tradecraft in terms of understanding the impact of performance and network or resource usage. It also means understanding why there are fewer bits per second of effective bandwidth (independent of compression or deduplication) vs. line rate in either half- or full-duplex mode.

Another item to note is that looking at encoding such as the example given in Table 4.2 shows how a relatively small change at a large scale can have a big effective impact benefit. If the bits and bytes encoding efficiency and effectiveness scenario in Table 4.2 do not make sense, then try imagining 13 MINI Cooper automobiles each with eight people in it (yes, that would be a tight fit) end to end on the same road. Now imagine a large bus that takes up much less length on the road than the 13 MINI Coopers. The bus holds 128 people, who would still be crowded but nowhere near as cramped as eight people in a MINI, plus 24 additional people can be carried on the bus. That is an example of applying basic 8b/10b encoding (the MINI) vs. applying 128b/130b encoding (the bus), and is also similar to PCIe G3 and G4, which use 128b/130b encoding for data movement.

PCIe Topologies

The basic PCIe topology configuration has one or more devices attached to the root complex (Figure 4.10) via an AiC or onboard device connector. Examples of AiC and motherboard-mounted devices that attach to PCIe root include LAN or SAN HBA, networking, RAID, GPU, NVM or SSD, among others. At system start-up, the server initializes the PCIe bus and enumerates the devices found with their addresses.

PCIe devices attach (Figure 4.10) to a bus that communicates with the root complex that connects with processor CPUs and memory. At the other end of a PCIe device is an end-point target, a PCIe switch that in turn has end-point targets attached. From a software standpoint, hypervisor or operating system device drivers communicate with the PCI devices that in turn send or receive data or perform other functions.

Note that in addition to PCIe AiC such as HBAs, GPU, and NVM SSD, among others that install into PCIe slots, servers also have converged storage or disk drive enclosures that support a mix of SAS, SATA, and PCIe. These enclosure backplanes have a connector that attaches to an SAS or SATA onboard port, or a RAID card, as well as to a PCIe riser card or motherboard connector. Depending on what type of drive is installed in the connector, either the SAS, SATA, or NVMe (AiC, U.2, M2) using PCIe communication paths are used.

In addition to traditional and switched PCIe, using PCIe switches as well as nontransparent bridging (NTB), various other configurations can be deployed. These include server to server for clustering, failover, or device sharing as well as fabrics. Note that this also means that

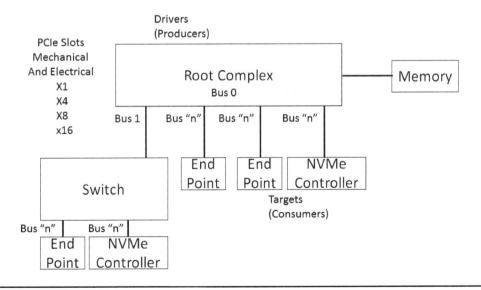

Figure 4.10 Basic PCIe root complex with a PCIe switch or expander.

although it is traditionally found inside a server, PCIe can today use an extender, retimer, and repeaters extended across servers within a rack or cabinet.

A nontransparent bridge (NTB) is a point-to-point connection between two PCIe-based systems that provides electrical isolation yet functions as a transport bridge between two different address domains. Hosts on either side of the NTB see their respective memory or I/O address space. The NTB presents an endpoint exposed to the local system where writes are mirrored to memory on the remote system to allow the systems to communicate and share devices using associated device drivers. For example, in Figure 4.11, two servers, each with a unique PCIe root complex, address, and memory map, are shown using NTB to any communication between the systems while maintaining data integrity.

PCIe considerations (slots and devices) include:

• Physical and software plug-and-play (good interoperability)
• Drivers (in-the-box, built into the OS, or add-in)
• BIOS, UEFI, and firmware being current versions
• Power draw per card or adapters
• Type of processor, socket, and support chip (if not an onboard processor)
• Electrical signal (lanes) and mechanical form factor per slot
• Nontransparent bridge and root port (RP)
• PCI multi-root (MR), single-root (SR), and hot plug
• PCIe expansion chassis (internal or external)
• External PCIe shared storage

Various operating system and hypervisor commands are available for viewing and managing PCIe devices. For example, on Linux, the "lspci" and "lshw–c pci" commands displays PCIe devices and associated information. On a VMware ESXi host, the "esxcli hardware pci

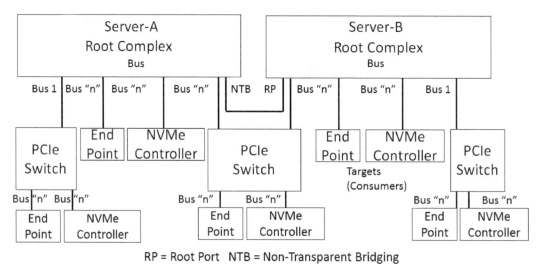

Figure 4.11   PCIe dual server example using NTB along with switches.

list" command will show various PCIe devices and information, while on Microsoft Windows systems, "device manager" (GUI) or "devcon" (command line) will show similar information.

4.3.7. LAN and Storage Ports and Internal Storage

Different types of servers will have various onboard internal and external ports as well as storage options. Some servers will have more PCIe expansion ports, while others may have more internal storage. Other servers may have more LAN networking, and others more memory, depending on their targeted deployment roles.

Various port and connector types include:

- USB (2 and 3) ports and VGA/HDMI video connectors
- PCIe, SAS, SATA, and converged NVMe U.2 (SFF 8639 drive) and M.2 slots
- Gigabit Ethernet (GbE) LAN-on-motherboard (LOM) ports
- 1 GbE and 10 GbE LOM or add-in-card ports
- Fibre Channel, InfiniBand, and converged network adapters (CNA)
- General-purpose and management networks
- Fixed ports, GBICS, SFP, QSFP transceivers
- 40 GbE and 100 GbE Ethernet adapters

Software and management tools include:

- BIOS and UEFI firmware or software
- Device drivers for hypervisor and operating system
- Performance, monitoring, and test tools such as iperf
- Configuration tools for enabling different features
- Remote direct memory access (RDMA)–capable devices
- RDMA-over-converged-Ethernet (RoCE) devices

While some servers do not need storage, most do. Even if the storage is in the form of NVRAM or a ROM, USB thumb drive, mSATA, or m.2 flash device, SD card, or NVDIMM, most servers will have some form of storage for booting the OS or hypervisor and associated software.

Some servers may be diskless, with no internal HDDs or SSDs and relying on network PXE boot or a metal-as-a-service (MaaS) server or other means of initial software load, or may boot from SAN or NAS. Storage performance, availability, and capacity capabilities vary depending on what the server is defined to do. Some servers have additional internal storage, or the ability to add drives, while others rely on external expansion.

Server storage options include, among others:

- Local internal or external dedicated direct attached storage (DAS)
- Attachment via PCIe using NVMe, SAS, SATA Thunderbolt, GbE, or USB

- Optional SAS, 10 GbE, InfiniBand, and Fibre Channel host bus adapters (HBAs)
- RAID adapters (PCIe or motherboard-attached) for internal or external storage
- Support for just-a-bunch-of-disks, SSD, or RAID/parity configured storage
- Access block SAN, file NAS or object, API, and cloud external networked storage
- Optional expansion storage enclosures for additional storage capacity
- Small-form-factor 1.8-in. and 2.5-in. drives (HDD, SSD, NVMe) and 3.5-in. drives
- Support for tape and optical drives, various software and device drivers

4.4. Server Packaging and Enclosures

Physical servers are available in various shapes and sizes with different options to meet diverse application needs. There are 19-in. (or 23-in.)-wide rack mount servers ranging from 1.75 in. (44.45 mm) or 1 Unit (U) tall to 2 U, 3 U, 4 U, 5 U, or many U tall, particularly with larger mainframes and chassis systems.

Some servers can be either rack-mounted horizontally or set up as a pedestal or tower vertically, using mounting brackets or rail kits. In addition to height (U) and width, servers also vary in depth. A single server with its processor sockets, memory, PCIe, and other I/O slots along with storage devices may occupy an entire server enclosure.

Another variation is having two separate servers share a physical enclosure with one on each side (e.g., half-rack wide), which can be powered up or down independently of the other. Some of these larger "two servers in a box," may be called "fat twins" due to their larger size.

An even denser configuration involves four separate servers (or nodes) in a single enclosure, also known as a cluster in a box. There are also denser solutions such as eight small blades, microservers, and high-density blade servers.

Although there are floor-standing pedestal, towers, and deskside or desktop servers, most servers are installed into racks and cabinets of various heights measured in Us. In addition to server base enclosures, there are also expansion chassis for housing additional storage devices or other peripherals.

Servers are typically housed in data centers, computer, telecommunications, or networking rooms and closets, among other habitats (places where they are kept). Refer to the Appendix for additional information on habitats for technologies and related topics.

Server packaging considerations include:

- Height, width, depth, and weight of a server enclosure
- Number of separate servers in an enclosure
- Processor sockets, memory, and I/O devices per server node in an enclosure
- Power, cooling, and device access (top, front, side) loading
- Bundled or converged, hyper-converged, or standalone system
- Servers may be standalone or bundled with software
- Cluster, converged, hyper-converged, or consolidated
- BIOS and UEFI capabilities along with management tools
- Hypervisor, operating system, or other bundled software
- Open Compute Initiative standards and packaging

4.4.1. Appliances: Converged, Hyper-Converged, and CiB

Servers can be obtained bare metal with no operating system, hypervisor, or other software provided. Some servers come with a hypervisor or operating systems already installed. A variation of traditional server is appliances that are preloaded with software to provide management, resource sharing, and hypervisor capabilities.

These solutions may be called appliances, solution bundles, or converged, among other names. In some cases, multiple servers are combined in a single 2U or taller enclosure such as a fat twin (two servers) or quad (four servers) and are called converged, hyper-converged, or cluster in a box.

Keep in mind that software requires hardware and hardware needs software, which in turn gets defined to become a solution. CI, HCI, CiB, hyper-scale, and scale-out solution approaches vary. Some CI, HCI, and CiB solutions are hardware packaged, while others are built on software that can runs on different hardware.

Remember that not all data centers or applications and environments are the same from a size, scale, scope, or functionality perspective, as shown in Figure 4.12. There are numerous types of CI, HCI, CiB, and solution bundles: Some focus on small SMB or ROBO and workgroups, whereas other CI, HCI, and CiB solutions are positioned for larger SME and small enterprise environments.

Converged solutions are rack-scale for the larger enterprise and web-scale or cloud providers. In some cases, the software and hardware solution might be targeted for a given market as shown Figure 4.12, or the software can be obtained and deployed to meet different needs. Two architectures are shown in Figure 4.13: disaggregated (left), where compute and storage resources are scaled independently; and aggregated (right), with resources scaling together.

Both aggregated and disaggregated architectures have their benefits and trade-offs, depending on different scenarios and workloads. Disaggregated architectures have been common with client server, SAN, NAS, and other scenarios and are used in some converged solutions.

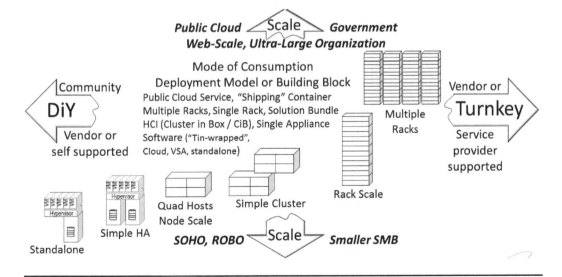

Figure 4.12   Server CI, HCI, CiB, and solution bundle approaches.

Disaggregated	Aggregated
• Similar to how servers and storage have scaled	• Simpler All-in-One (AiO) enables scale down
• Converged or traditional SAN, NAS or Object	• Can scale-up for larger environments
• Flexibility to size resources to different needs	• Reduces resource integration complexity
• Heterogeneous, interoperable across resources	• Ease of use and ease of acquisition
• Scales for larger or complex environments	• Usually homogenous across resources

Figure 4.13 Aggregated and disaggregated architectures.

Aggregated architectures have been more common on servers with dedicated DAS as well as some hyper-converged solutions.

An important theme is that one size and solution approach does not fit every environment. What works for a smaller environment may not scale without adding complexity. Likewise, a larger solution may not scale down without increasing complexity.

Figure 4.14 shows an example of a hardware-independent CI that can scale-out using Microsoft Windows Server combined with Scale Out File Service (SOFS) and other components. This solution, depending on the hardware packaging chosen, can be called a CiB or in some cases an HCI or simply a clustered server and storage platform. In Figure 4.14 there are multiple PM servers or hosts that serve as nodes in the cluster, grid, or stamp. In this figure there could be Windows 2012 R2 servers or 2016 storage spaces direct (S2D) implementation, among others. Storage Replica is a feature within Windows server for performing replication across servers including of underlying S2D devices.

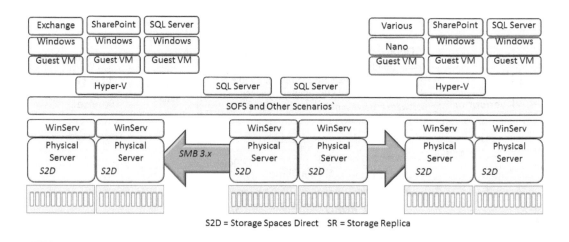

Figure 4.14 Physical and virtual servers.

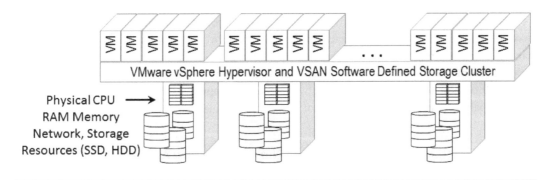

Figure 4.15   Physical and virtual servers.

Stamp is a term that describes a cluster, grid, or collection of nodes working together to provide a common service such as storage for cloud, virtual, object, or other functionality. For example, a storage stamp could be an SOFS or S2D cluster as shown in Figure 4.14 that is an instance of collective storage functionality. In a large environment, there could be multiple clusters, grids, rings, or stamps of resources.

Figure 4.15 shows another variation of a CI environment using VMware VSAN as an example of software that can be deployed on various hardware systems. In addition to deploying your own VSAN converged environment using existing hardware and software license that comply with the hardware compatibility list (HCL), VMware and their partners also offer EVO:RAIL, which is a combination of VMware ESXi, VSAN, and associated management tools along with hardware from various vendors (you can choose your options).

Stamp, CI, CiB, HCI, clustered, and bundled server storage considerations include:

- Bare metal, root, VM, or container-based access and management
- Aggregated or combined server compute and storage that scale together
- Disaggregated server compute and storage that scale independently
- Hardware bundled with software and management tools
- Software unbundled from hardware; use your own hardware
- Ability to use external DAS, SAN, NAS, or other storage
- Storage can be shared to servers outside the solution cluster
- Tightly or loosely coupled; shared everything, something, or nothing
- Solution optimized for applications such as VSI or VDI or data protection
- Options for high-performance computing, I/O, storage capacity, or memory
- Choice of hypervisors or other software management and data protection tools
- Solution optimized for small SMB/ROBO, or SMB/SME, or rack scale

4.5. Scaling Servers (Up, Down, and Out)

There are many different approaches to scaling with servers for performance, availability, capacity, and economics to meet different needs. Some software prefers fewer yet faster processors, while others want or need scale-out approaches for resiliency or other requirements.

Some applications lend themselves to one, while others are agnostic how the scaling is accomplished. Too few servers, for example, can create a single point of failure, while too many can cause complexity or introduce additional overhead. Table 4.3 shows various scaling options. All of the approaches shown, as well as their variations, have benefits and caveats. What will be best for your own scenario will depend on application PACE considerations, as well as capabilities of the hardware and software technologies being used.

Table 4.3. Server, Storage and Networking Scaling Approaches

Scaling	Description
Down	Opposite of scale-up, where smaller or fewer resources are used.
Out	Horizontal scaling. Many devices such as servers, storage, or network devices create a grid, cluster, RAIN, ring, fabric, network, or other topology to support scaling up of the workload from a performance, availability, and capacity standpoint.
Up	Vertical scaling. Larger, faster, the more capable server, storage, or network is used to consolidate from many smaller devices—going from many smaller servers to a few larger servers, from many smaller network switches to larger ones, from many storage devices or systems to fewer larger (or faster) devices or systems, or from slower cores to faster (and perhaps more) cores. Also known as scaling vertically.
Up and Out	Combines multiple larger, faster resources (scale-up) to scale-out beyond an individual shelve, rack, cabinet, or data center. This means the best of both worlds where it is applicable or possible.

Server scaling configuration options include, among others:

- Two or more pairs of servers for performance and availability
- Active/active for automatic failover and resume (or restart)
- Active/passive for standby with restart or roll-back, roll-forward
- Grids, clusters, rings, or combinations of servers
- RAIN or cluster

4.6. Management Tools and Security

Some tradecraft context is needed here, as we are discussing KVM as it pertains to keyboard/video/mouse. Note that KVM also refers to server virtualization software, which can be found in Linux operating systems and which we will discuss later.

With the proper context, KVM, as its name implies, refers to keyboard/video/mouse, spanning physical, virtual, and even cloud servers. Currently, the term *headless server* means one without a dedicated monitor or console, yet able to support attachment of a video device locally or remote. This also means the server's software, from the BIOS/UEFI or platform management to the hypervisor and operating system, support remote access using a command-line interfaces (CLI) including power shells, as well as graphical user interfaces (GUIs). There is a trend toward future servers (physical, virtual, cloud, and application) to be headless.

Moving forward, the headless trend takes another step by removing GUI and local login capabilities and substituting management via remote platforms. An example of this is Microsoft Windows Nano, which is a Windows server with all of the GUI and 32-bit (WOW) software removed. The benefit is a smaller, light-weight operating system able to do more in a given footprint (e.g., memory, CPU, and storage), including supporting more VMs or containers.

In the meantime, many other physical, virtual, and cloud servers will continue to have various things to manage and interact with. These include virtual keyboard video monitors (KVMs), something that most hypervisors can do by providing virtual consoles. This also means supporting tools such as SSH for remote command-line login, PowerShell, API, remote desktop protocol (RDP), XRDP, among others.

For example, in my lab, servers or other devices that need a KVM attach to a KVM switch, which also has a physical keyboard, mouse, and monitor. Using the KVM switch, I select which system I need physical access to, some of which are bare metal and others are hypervisor-based servers. In addition to using a KVM switch, I also have remote access using SSH, PowerShell, RDP for Windows and XRDP for Linux, among other Microsoft and VMware tools.

Having remote access tools is handy. For example, I can use SSH tools such as putty to access my VMware hosts or guests, as well as local bare-metal servers, or remote cloud instances. I can also use web browser tools to access hypervisors, guests, or cloud resources as well as local networking devices including switches, routers, printers, NAS, and access points.

Speaking of remote access, having a copy of your virtual toolbox at a cloud site comes in handy. For example, including common software, scripts, and keys safely parked in a secure cloud location saves time when you have to work with different remote sites.

Useful ports and interfaces on servers include:

- VGA for servers where you need to attach an older monitor
- HDMI, DisplayPort, mini-plug, and VGA and audio ports
- USB ports for KVM as well as attaching other devices
- GbE or LAN ports dedicated or shared for management
- Serial ports (yes, there is still a need for them)
- USB can also be used for monitoring UPS and temperature sensors

Other software and management tools include Intelligent Platform Management Interface (IPMI), Chef, Secure Shell (SSH) tools such as putty and bitvise, file transfer protocol (FTP), Virtual Network Computing (VNC) remote access, Intel Active Management Technology (AMT), RDP and XRDP, Microsoft Windows Management Interface (WMI) and System Center, along with various power shell CLIs for different platforms.

Additional tools include Chef, Puppet, JuJu, metal as a service (MaaS), with a trend toward smaller mobile devices—having the ability to use the above and other tools from a phone or tablet is evolving from novelty to a necessity. For example, on my phone I have an RDP client, Ping, SSH, and, for fun, a db meter, among other tools.

4.7. Operating Systems

Servers have some operating system that can range from a very simple microkernel or embed-ded kernel as part of an application, or a hypervisor, to a fuller feature function version. Some operating systems are lighter in weight than others, having basic command-line or shell inter-faces, while others have GUI and many applications.

For example, there is Windows 10 desktop as well as Windows Server 2016 (among other versions). Windows Server can be installed with or without the GUI tools to streamline its footprint.

Another version is the Windows Hyper-V server, which for an interface relies on PowerShell or a basic text menu of a few commands. More advanced management is handled via remote Windows Management Interface (WMI) tools. There is also a more streamlined version of Windows Server called Nano (a Windows container platform) that not only eliminates the GUI, it also removes all extra overhead including legacy 32-bit applications and code. Microsoft Windows is just one example of an operating system with many variants; the same applies to different *nix software.

Keep in mind that servers rely on storage, and storage systems often leverage the underlying servers and their operating system tools or technologies for storing persistent stateful data. Servers also use persistent storage as ephemeral storage for temporary cache, buffers, and workspace for stateless data. This includes physical, virtual, container, and cloud compute instances.

Even on cloud solutions and hypervisors, as well as containers with robust management interfaces or PowerShell, some server storage I/O tradecraft skills and understanding of basic operating system and system administration is a must. Some operating systems are virtual-aware or enlightened, including para-virtualized support. What this means is that instead of simply running as a guest or in emulation mode, the enlightened or para-virtualized software is optimized and hypervisor-aware to improve performance. Likewise, some hypervisors have integration and para-virtual or enlightened drivers and services. For example, Hyper-V has Linux Integration Services (LIS), which may be in the box with the operating system or supplied as an add-on.

Another variation is VMware ESXi, which has LSI Logic Parallel and LSI Logic SAS storage adapters, as well as the VMware Paravirtual virtual HBA. The VMware Paravirtual HBA is a more optimized, hypervisor-aware component that can improve performance while reducing CPU consumption. There are similar variants with network drivers among other devices.

Common operating systems include:

- Windows (desktop, server, and Nano)
- Various Unix and Linux (Centos, CoreOS, Oracle, RHEL, SUSE, Ubuntu)
- Oracle Solaris, HP HP-UX and OpenVMS, IBM zOS, AIX, OS400
- VxWorks real-time operating system kernel

Common hypervisors and data center operating systems tools include:

- Citrix and Xen, KVM/Qemu, Microsoft Hyper-V, VMware vSphere ESXi
- Azure Stack, OpenStack, Oracle, and IBM solutions
- Virtualbox, Vagrant, Mesos, and ScaleMP

Virtual resources include:

- CPU and memory, video and display devices
- Networking switches, routers, and storage adapters
- Storage devices (HDD, NVM SSD, SCM), CD and USB devices

4.8. Hypervisors and Virtual Server Infrastructures (VSI)

Physical servers support software-defined data centers, also known as virtual data centers, along with traditional applications. In addition to being software-defined to support virtual servers and containers, physical servers also get defined to be data and storage servers or appliances, network switches, routers, and caching devices, among other roles.

Virtualization in general and server functionality in particular includes:

- Abstraction—management transparency of physical resources
- Aggregation—consolidation of applications, operating systems, or servers
- Agility—flexibility to adapt dynamically for common data infrastructures tasks
- Elastic—able to expand or adjust to PACE requirements as needed
- Emulation—coexistence with existing technologies and procedures
- Multi-tenancy—segmentation, isolation of applications, users, or other entities
- Provisioning—rapid deployment of new servers using predefined templates

Server virtualization can be implemented in hardware or assisted by hardware, implemented as a standalone software running bare metal with no underlying software system required, as a component of an operating system, or as an application running on an existing operating system.

VMs are virtual entities represented by a series of data structures or objects in memory and stored on disk storage in the form of a file (Figure 4.16). Server virtualization infrastructure vendors use different formats for storing the virtual machines, including information about the VM itself, configuration, guest operating system, application, and associated data.

When a VM is created, in the case of VMware, a VMware virtual disk (VMDK) file or Microsoft Virtual Hard Disk (VHDX) or DMTF Open Virtualization Alliance (OVA) Open

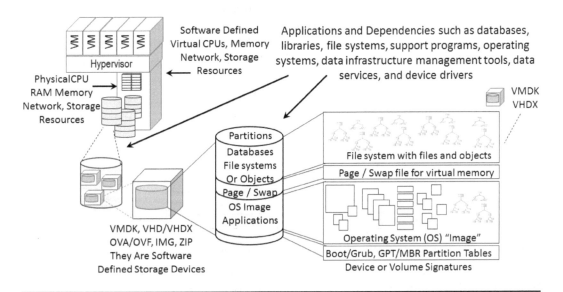

Figure 4.16 Virtual servers and virtual disk representation.

Virtual Format (OVF) is created that contains information about the VM, an image of the guest operating system, associated applications, and data.

A physical server operating system and guest applications, settings and data are converted to a VM via physical-to-virtual (P2V) migration. P2V involves creating a source image of the physical server's operating system installation and configuration, boot files, drivers, and other information along with installed applications and mapping it into a VMDK or VHDX.

There are some commercial as well as free tools available to assist with P2V along with virtual-to-virtual (V2V) migration between different hypervisors as well as virtual-to-physical (V2P) migration. I am sometimes asked why anybody would every need to do a V2P, as everything is going virtual and away from physical. My answer is simply that I have had images which were virtualized, only to have a need to put them back as bare metal on a PM for a particular task. Instead of doing a clean install, assuming I have the bits and binary code or ISOs, it can be easier to simply do a V2P.

With a hypervisor-based VM (Figure 4.16), the VM presents to the guest operating system what appears to be a CPU, memory, I/O capabilities including LAN networking and storage, and keyboard, video, and mouse devices. Hypervisors have virtual NICs, HBAs, and virtual LAN switches, all implemented in memory, creating virtual software-defined networks. The virtual LAN switch is used by the virtual NIC to enable the VM to communicate using IP via memory, instead of via a traditional physical NIC and LAN, when operating on the same physical server.

Depending on configuration, underlying hardware resources such as CPU processors, memory, disk or networking, and I/O adapters may be shared or dedicated to different partitions. Hypervisors also support a pass-through mode, where guest operating systems and their applications along with device drivers can access an actual physical device without going through the hypervisor software stack per se.

Figure 4.17 shows, from left to right, a traditional server, type-1 and type-2 hypervisors, and containers. The containers example in this figure could exist as a guest virtual machine as well.

Type-2 hypervisors shown in Figure 4.17 include Microsoft Hyper-V, which supports guest OS or virtual machines with Windows Server as the underlying host operating system. In addition to Hyper-V, other type-2 hypervisors include VMware Workstation (not to be confused with a virtual desktop), Oracle Virtual box, and others. The advantage of type-2 is the ability to leverage underlying operating systems while sharing resources.

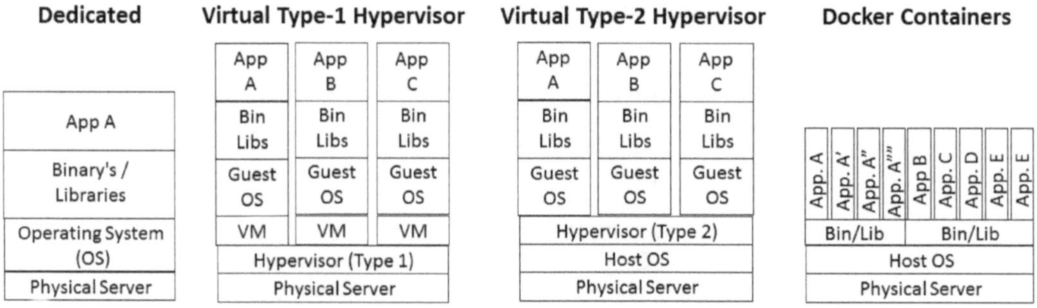

Figure 4.17   Bare metal, type-1 and type-2 hypervisors, and container services.

Also shown in Figure 4.17 are containers such as Docker, LXC, VMware VIC and Photon, and Microsoft Nano, among others. Not to be confused with physical servers densely packed into a 40-ft intermodal shipping or other container, Docker and other containers function as micro-servers. These containers take a step beyond sharing physical resources such as with a hypervisor by leveraging common software binary libraries and operating services. The result is a smaller footprint to run application software that does specific tasks, yet that does not need a full-blown operating system as with a type-1 or type-2 hypervisor.

Type-1 hypervisors (Figure 4.17) run bare metal on physical servers that in turn host virtual VM and their guest operating systems. Depending on underlying hardware processor support such as VT and available memory, type-1 hypervisors can also host other type-1 hypervisors in what is known as nesting.

For example, for testing or other purposes, VMware vSphere ESXi type-1 hypervisor is installed on a physical machine that in turn has several VM with different OS. One of those VM, properly configured, can have another version of VMware ESXi installed as a guest, which in turn could have several VM with different OS guests. How much can you nest? How much CPU, RAM, and I/O including networking do you have for the given physical machine?

Figure 4.18 shows an example of nested virtual servers being used to create a test, development, proof of concept, learning, or training environment. Using nesting, for example, a single server (or multiples in a cluster) can have a hypervisor such as VMware ESXi running on the bare-metal physical machine. As shown in Figure 4.18, on top of the hypervisor are four virtual machines, one (right) with yet another hypervisor installed and additional guest VM on top of that.

This approach can be useful, for example, to "stand-up" or set up a new test environment, such as to verify the functionality of software or management tools, or simply to try out a new version of a hypervisor or operating systems. Another use for nesting is to set up a smaller version of your production environment to practice common tasks such as backup/restore, replication, or other data protection tasks.

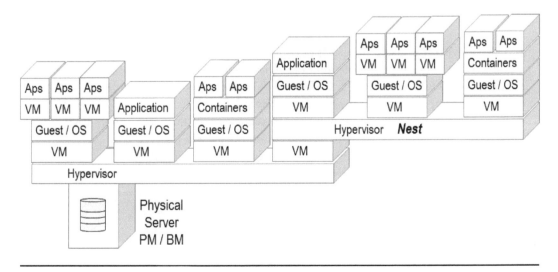

Figure 4.18   Example of nesting virtual servers, hypervisors, and cloud stacks.

Note that not all hypervisors support nesting or nesting of other hypervisors. Likewise, some cloud machine instances may not support nesting of other hypervisors. As an example, if I try to enable a Hyper-V role on a Windows server running on an AWS EC2 instance, an error message appears indicating that virtualization cannot occur. This is because the underlying hypervisors that EC2 uses do not allow nesting or at least nesting of Hyper-V on Windows servers (at least when this was written).

For those hypervisors that do support nesting, changes to configuration files and settings might be required to expose underlying virtualization technology, as well as promiscuous mode networking. Refer to your specific hypervisor as to what it supports for nesting, along with best practices, hints, tips, and tricks.

Besides the type-1 and type-2 hypervisor architecture models, there are also variations such as para-virtual (PV) and hardware virtual machines (HVM). HVM enable operating systems (e.g., guest images) to run natively without any special software or driver changes as though they were running on a BM system. HVM leverage underlying hardware capabilities in a pass-thru mode including VT and memory extensions, I/O, and GPU resources.

Note that PV applies not only to types of hypervisors, there are also PV devices and associated drivers such as virtual networking and I/O cards or adapters. Historically, PV was a preferred or recommended route for higher performance, lower latency due to optimizations. However, there may also be compatibility or other issues.

With faster PM servers, along with enhancements to virtualization and hypervisor software, some current-generation HVM-based instances can be as fast as or faster than traditional PV-based ones. Check your hypervisor or service provider's best practices, tips, and recommendations as well as various HCL to determine what's applicable for your environment.

In addition to free and premium versions along with various hypervisor, guest, network, and storage management tools, vendors are also providing tools for migrating VM to or from various cloud services. Likewise, many cloud providers are providing their own tools and plug-ins to hypervisor vendors' tools and frameworks, such as VMware for Amazon or Microsoft, among others.

Hypervisor and virtualization considerations include:

- Type of hypervisor (type-1, type-2) or containers and partitions
- How much multi-tenancy isolation of resources is needed
- Amount of RAM, video RAM, CPU, and I/O devices
- Software compatibility and support for different hypervisors
- Vendor service and support for hypervisors (some vendors may not support VM)
- In the box vs. add-on or third-party and community device drivers
- Para-virtual or enlightened drivers and devices and guest integration services
- Cross-platform and hypervisor management tools and plug-ins
- P2V, V2V, V2P, P2C, and V2C conversion and migration tools

4.8.1. Virtual Desktop Infrastructure

While not a server per say, virtual desktop and associated infrastructures (VDI) have a connection (pun intended) to VSI and hypervisors (Figure 4.19).

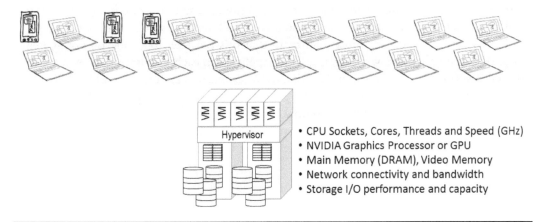

Figure 4.19 VDI and VSI with server, storage, and I/O networking considerations.

VDI is a popular use case deployment for VSI as part of desktop, application, and other consolidation or transformation initiatives. The server (as well as I/O networking and storage) demands will vary depending on the type of implementation, such as zero-client or applications being used, among others. VSI is where some VDI applications and their associated data infrastructures exist.

Some VDI deployments leverage traditional disaggregated VSI compute and storage models, while others use variations of converged solutions, including aggregated hyper-converged configurations. Besides server CPU socket speed, some cores, and threads, other server considerations include main memory, video memory, GPU or graphics accelerators, as well as network and storage I/O performance as well as space capacity.

4.9. Containers and Microservices

Containers are another form of computing service that are part of a trend of containerizing applications for portability, ease of development, deployment, and use. Often containers use the metaphor of an intermodal shipping container like those seen on railroad trains and large ships.

The metaphor centers around the commonality of shipping containers deployed across different shipping mediums relying on a standard set of interfaces and packaging. What's inside those shipping containers can vary, some being cooled for fresh fruits, fish, meat, poultry, vegetables, or other perishable items, while others are not cooled or even heated for some bulk commodities.

Likewise, application containers can run the different process tasks using Windows or Linux services software defined by their creator, similar to what goes into a shipping container.

There are also some context matters regarding containers, as the term can be used to refer to micro-services such as Docker, Linux, VMware (integrated containers), and Windows, as well as cloud-based lightweight compute or application processing instances.

The term *containers* can refer to an intermodal physical shipping container-like structure in which servers, network, and storage hardware can be pre-installed for at-scale deployments.

There are also database containers, object and cloud storage containers also known as buckets, and data protection archive and backup containers, also known as save sets, among others.

Micro-services is another context matter, in that this can refer to a collection or aggregation of different containers working as a larger application, each performing its own specific task, process, function, or subsystem activity, or in response to additional workload demand. Another variation is that some simply call containers micro-services in general. Thus the context of the conversation and topics is essential.

For our purposes here, the focus will be on micro-services such as Docker, Rocket, Windows, and Linux LXC, among others. Instead of describing what micro-services containers are, let's start with what they are not.

Container benefits include:

- Speed up development, test, operations deployment, and management (DevOps)
- Streamline application and resulting image life-cycle management
- Simplify application packaging, distribution, and installation process
- Build, ship, and deploy container images locally or via cloud service
- Complement other orchestration and data infrastructure tools and technologies
- Provide some isolation and tenancy for applications vs. a shared environment
- Smaller, with less overhead than using a virtual machine with a full operating system
- Take up less storage space and memory, and start faster due to lower overhead

Container characteristics include:

- Registries exist for storing and managing container image life cycles.
- Public and private registries exist for flexibility.
- Lightweight simple tenancy (isolation) for image portability.
- The image is an application, process, or task that runs on the container.
- Container images can be standalone or part of an application fabric.
- Small footprint accomplished via packing only what is needed.
- Dependency libraries exist via the container server where image runs.
- Stateless, where data is not preserved, or stateful (persistent) data.
- Can leverage host and external storage for stateful data.
- Various drivers including for external persistent storage access.
- Built by adding layers to base or existing images; add only what you need.
- Leverage copy on write of shared read data to reduce overhead.
- Linux- and Windows-based images run on Linux and Windows servers.
- Development and management via different operating environments.

Containers are lightweight, streamlined compute instances that have some additional isolation vs. a regular process running among tens, hundreds, or thousands of others on the same host operating system. Containers, in general, do not have the same tenancy (isolation) as a virtual machine guest or as the sole occupant of a PM or bare-metal host server and operating system. However, there are micro-services (container) implementations that provide increased security and tenancy isolation from other running containers, their associated images, and resource usage.

Dedicated	Virtual Type-1 Hypervisor	Virtual Type-2 Hypervisor	Containers
App's run as is using their binaries (Bins) and Libraries (Libs) on an operating system (OS) that has a dedicated server. Very good isolation.	App's run as is using their Bins and Libraries in guest VM. OS runs as is in VM. VM runs on guest host OS sharing underlying resources. VM, may be "enlightened", para virtualized (optimized). Good multi-tenant isolation.	App's run as is using their Bins and Libraries in guest VM. VM is similar to a dedicated server however sharing underlying resources. VM, may be "enlightened" or para virtualized (optimized).	App's are written to use containers. Containers "light weight" and isolated in a multi-tenant environment sharing OS and where applicable binary's, libraries. By default the container storage can be stateless or ephemeral (temporary), as well as statefull (persistent) based on configuration.

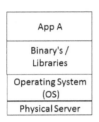

Figure showing the four stacks:

Dedicated: App A / Binary's / Libraries / Operating System (OS) / Physical Server

Virtual Type-1 Hypervisor: App A, App B, App C / Bin Libs (x3) / Guest OS (x3) / VM (x3) / Hypervisor (Type 1) / Physical Server — zVM, VMware ESXi, Xen

Virtual Type-2 Hypervisor: App A, App B, App C / Bin Libs (x3) / Guest OS (x3) / Hypervisor (Type 2) / Host OS / Physical Server — VMware Workstation, Hyper-V

Containers: App A, App A', App A'', App A''', App B, App C, App D, App E, App E / Bin/Lib, Bin/Lib / Host OS / Physical Server — Container

Figure 4.20 PM and BM, hypervisors and containers comparisons.

Containers are not hypervisors or full-blown heavy operating systems, nor are they intended to run several concurrent processors or applications (Figure 4.20). In some ways, micro-services containers will remind some of previous-generation timesharing, while others may have a *déjà vu* moment to Solaris containers or HP-UX partitions, among others, including some databases that also have similar isolation and sharing.

Keep in mind that all applications (software) require some compute resource to run on, which can be a cloud or virtual machine that in term runs on a physical processor along with associated memory, storage, I/O, and networking. In addition to the virtual or physical hardware, there is also an operating system along with associated libraries, among other dependencies.

The application can have an entire PM or BM dedicated to it, or may coexist and share those resources. Another option is that the application can have a dedicated (or shared) virtual machine along with its guest operating system, libraries, and associated dependencies.

The VM provides tenancy and isolation from nosy (security and privacy concerns) as well as noisy (impacts performance) neighbors (other applications) compared to existing on a shared server. Multiple VM can coexist to share a physical resource, improving the hardware efficiency and tenancy isolation and thereby aiding application effectiveness.

VM have their own guest operating system and application libraries as well as other dependencies installed, all of which add up to additional overhead for smaller, simpler applications.

Containers such as Docker, Rocket, VMware (integrated), and Windows Nano use an underlying operating system, libraries, and other dependencies. This results in being able to have an image that contains only the application and its direct dependencies, so it is much smaller, simpler, and easier to deploy across different runtime environments. By being a self-contained package (container) with the application, image means the container can be moved around (shipped) and then run in different places (location independence).

While VM and their associated guest operating systems, libraries, and applications can have portability, they tend to be rather large, measured in hundreds to thousands of MB (GB) in size. Containerized images (applications or processes) have less bulk because they contain only what they need. This is like packing for a trip where you do not know what you will need so you put everything that might be required into a large suitcase, trunk, or container.

If you have traveled, you know what it is like to carry all of that extra baggage around, not to mention the cost, as opposed to packing as light as possible and using what's available where you go. Containers do not have the same bulk size associated with a VM, so they are usually a fraction of the size and hence much more portable.

There are common management interfaces that span from Linux to Windows and other environments for interfacing with micro-server containers such as Docker. However, similar to the binaries or executable programs and their associated support libraries for Linux, Windows, and other operating systems, there are also similar (at least as of this writing) constraints with containers and their image deployments today.

For example, the images and container themselves, at least as of this writing, are tied (have an affinity) to a particular underlying host operating system, or hypervisor. For example, those from Linux run on Linux, Windows on Windows, or those on Linux on a hypervisor such as Hyper-V on Hyper-V, and so forth. Note that container images can be developed on different platforms and then built and deployed on others, including those hosted by a PM, BM or DPS, a VM or VPS, as well as cloud instance or micro-service such as those provided by AWS, among others.

Container tools include those from, or part of, Aufs, AWS, Btrfs, CoreOS, Dell EMCcode such as Rexray, Docker, Google, Kubernetes, Mesos, Microsoft (Including Windows Server, Hyper-V, and Visual Studio), OpenStack, Pivotal, Swarm, Union, and VMware, among others. These include development, configuration, orchestration, life-cycle management including tracking, versioning, rollback, and rolling upgrades, among other activities.

We will revisit containers in other chapters from a storage I/O networking perspectives.

4.10. Cloud Machines and Instances

Cloud services that provide compute capabilities such as AWS Elastic Cloud Compute (EC2), Microsoft Azure, Google, IBM Softlayer, HPE, and many others present instances or what are essentially virtual machines. Service providers may refer to the compute capability as an instance, CM, or VPS and DPS, among other names.

Think of cloud machines (CM) or instances as VM that run in a cloud service. Note that most CM cloud instances are in fact virtual machines, the exception being DPS or application-as-a-service (AaaS) and software-as-a-service (SaaS) solutions along with some container services.

Different cloud services use various underlying data center operating system stacks such as AzureStack, Photon, OpenStack, CloudStack, Mesos, Microsoft, or VMware, among others. Likewise, the underlying hypervisors support the cloud compute or virtual machine on the services being offered, with some using native or modified versions of KVM, Xen, Docker, as well as Microsoft Hyper-V and VMware ESXi using different operating systems.

Common operating systems include Centos and CoreOS among other Linux including RHEL, SUSE, and Ubuntu. Similar to an on-premise VSI, the CM or instances are configured in different ways; for example, some will be set up for low cost, low performance, and small amounts of memory, I/O, and storage.

Similar to a VSI or shared hosting environment, something to watch out for with these types of CM or instances is noisy neighbors or other guests that are consuming resources. CM can be configured with larger amounts of CPU, RAM, I/O, networking as well as different operating systems (images). In some cases you may even have a DPS or an instance with a dedicated physical server. Some cloud service providers are physically separating some servers in a different part of their data center or on isolated networks for security.

Cloud server and compute considerations include:

- Dedicated or shared underlying resources (multi-tenancy)
- Amount and type of compute, I/O, networking, memory, and storage
- Specialized instances with GPUs or other capabilities
- Type of operating system images provided—Can you bring your own?
- Ability to migrate your existing VM to or from the cloud service
- Management tools, interfaces, dashboards, and plug-ins
- Granularity of billing (microseconds for containers, hours or days for servers)
- Additional fees including for IP addresses, storage I/O, and other activity
- Network configuration and security firewall options or requirements
- Availability, accessibility, durability, access security reporting, SLO and SLAs
- Reserved instances, on-demand or spot pricing options
- Can you bring your own software licenses, as well as build your own images

4.11. Putting the Server Together

Let's start to wrap up this chapter by putting various server-related topics and themes together that combine with storage, I/O networking, and other resources to create data infrastructures. Figure 4.21 is a variation as well as a combination of different things we have discussed so far in this chapter, as well as a preview of things to be covered elsewhere in this book.

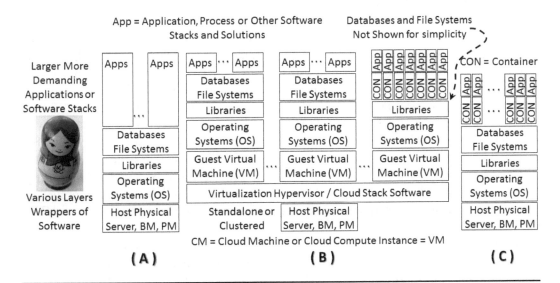

Figure 4.21   Comparison of various server options and deployments.

From left to right, you should recall the Matryoshka doll from Chapter 3 (bits, bytes, blobs, and software-defined building blocks). The Matryoshka doll is used as a metaphor to represent different layers and wrappers of software that can sit inside of, or on top of, another layer. This includes nesting, virtualization hypervisors, and cloud instances, as well as containers.

In Figure 4.21 on the left (A) is a PM or BM server with an operating system, libraries to support applications and other software dependencies, databases, repositories and file systems, as well as applications that require more resources for their given PACE requirements and workload needs.

In the center (B) of Figure 4.21, a hypervisor is layered atop the BM, PM, or physical host server to create isolation (tenancy) for the various guest operating systems with virtual machines. VM enable workloads that do not need a full dedicated PM to be aggregated and share resources. Keep in mind, however, that aggregation (consolidation) can cause resource aggravation (bottlenecks), which we discuss in following chapters.

The applications can be heavy or light in terms of resource demand, but each VM has isolation from nosy (snooping) as well as noisy (loud or resource disrupting) neighbors. Also shown in (B) is a VM running containers to provide limited tenancy to smaller lightweight applications, processes, or tasks. The containers are used to support those applications or functions that do not need the overhead of a full VM. On the right of Figure 4.21 (C), an OS with containers is shown representing hundreds if not thousands (depending on resources and workload demands) of light application images working independently or as part of an application fabric based on micro-services.

Note that the examples in Figure 4.21 can be based on standalone or clustered traditional or converged solutions and packaging. Likewise, they can be used in different combinations to create hybrid deployment scenarios to meet specific needs and application PACE requirements. Which is best for your data infrastructure will depend on your environment and application PACE requirements.

4.12. What's in Your Server Toolbox?

Various physical and software tools should be in your server (or virtual and cloud server) toolbox. In addition to various tools, there are various tradecraft skills and tricks you should also have as part of your toolbox, such as knowing that CTRL-ALT-F2 on a Ubuntu Linux server stuck on the GUI login screen takes you to the command prompt (you still need to login).

Or that instead of sending a VMware vSphere hypervisor guest virtual machine a CTRL-ALT-DEL, send it a CTRL-ALT-INSERT. Alternatively, to access the VMware ESXi shell command line involves an ALT-F1 (ALT-F2) to return to the Direct Console User Interface (DCUI). Accessing various servers' (or storage systems') BIOS or UEFI and system setup will vary.

If needed, keep a handy "cheat sheet" note near your keyboard or KVM switch to help you remember what the server setup, boot menu options are. Also, have an SSH tool such as putty, which I use for accessing various physical, virtual, and cloud servers, along with VNC and RDP tools.

My toolbox also includes various types and sizes of servers, operating systems, hypervisors, OpenStack, and different applications as well as databases and Hadoop along with various virtual storage appliances. Software tools include workload generation and benchmarking tools such as Benchmark Factory, Docker, Hammerdb, and Hadoop/HDFS (dfsio and teragen).

Others include Iometer, Benchmark Factory, fio, Iorate, Microsoft Diskspd, PCmark, Sysbench, and Vdbench. Monitoring, reporting, and configuration tools include Acronis image, Clonezilla, Cloudberry, Cyberduck, Grub, Linux (dd, lshw, lspci, iotop, htop, fdisk, gdisk, gpart), Microsoft (diskpart, wmic, perfmon, powershell), Rufus, Seatools, putty (ssh), RDP and XRDP, Spotlight on Windows, and Visualesxtop, iperf, ntttcp, among others. In addition to hardware, software, and services tools, I also have some items on my physical and virtual bookshelves; some examples of these can be found in the Appendix.

4.13. Common Questions and Tips

Does having 64 vs. 32 bits mean faster data movement and performance? It depends on your software—whether it can leverage the extra bits for addressing and ability to do more work in a given amount of time.

What is a white box server? A white box server is a generic server that does not have a big-name-brand logo on it. Big-name-brand logos include Cisco, Dell EMC, Fujitsu, HPE, Huawei, IBM, Lenovo, and Oracle, among others. Some white box servers include those from Foxconn, Intel, and Supermicro, among others. Other terms for the white box are generic industry standard, commodity, or commercial off-the-shelf (COTS) server. The term *white box* has gained popularity in the industry in part because some vendors feel that *white box* has more appeal than *commodity*.

What is an engineered server or system? An engineered system (server, network, storage, service, or solution) is one that has been customized or tailored to meet specific environment, workload, and application PACE requirements. Instead of a one-size-fits-all, use-what's-available approach, an engineered solution or server can be custom manufactured with a specific number of processor sockets and processor, memory, I/O expansion, storage devices, power supplies, and cooling fans, among other items.

An engineered server can also be as simple as one configured to fit specific needs vs. whatever is the cheapest to use and deploy. There are also engineered VM and CM tailored to meet different needs. The key is that some engineering thought has gone into the configuration of the server vs. simply using what's available.

Can a container be used to replace virtual machines? In some situations, yes, depending on what your application or functionality needs are. If you simply need a lightweight resource to run a simple image, process, or task without the overhead of a virtual machine, then container-based images can be a good fit. However, just because you can do something with a container does not mean it is always the best solution.

Is all software plug-and-play today across different servers? Some software will do processor checks to ensure that a certain level of resources, functionality, or performance are available. Likewise, some operating systems include in-the-box drivers for NVMe and other emerging technologies, while others are plug-and-pray (plug the device or driver in and pray that it works).

If the cloud is the future, why spend time learning about servers? Clouds still use physical servers that get abstracted, masked, and software virtualized or defined. Even when virtualized or software defined, you are still working with virtual servers, so understanding what's underneath the hood is important for knowing how to configure and make other decisions.

If containers are the future, why learn about hypervisors, operating systems, and servers? Not all environments will move everything to containers, just like not all environments will virtualize 100% of their applications or move everything to the cloud. Most environments will for at least the next decade have some hybrid combination of physical, virtual, cloud, containers, and other options for supporting their data infrastructures and application needs.

Will SCM eliminate the need for DRAM and storage? Maybe someday, or in the not-so-distant future for some early adopters and corner cases. You may find some servers using persistent-storage class memories such as 3D XPoint, among others, to replace DRAM, as well as NAND flash-based storage. Likewise, you may see servers that can directly access persistent-storage class memories without having to use PCIe or other interfaces. In the near term, however, look for SCM to be used in traditional roles while physical interfaces, processors, adapters, and of course software drives can leverage them.

Why not add more memory? If your servers and budget can accommodate more memory to meet your performance needs, add as much as you need to get the effective benefit.

Does a data pond, pool, or lake require a block, file, or object storage? The type of data and your applications will determine what your data reservoir or repository should be. For example, if you are only going to be accessing objects via S3 or another API, that will determine your storage need. On the other hand, if you need a mix of files including HDFS, NFS, and S3 objects, then that needs to be considered. Look beyond the interface and access method, paying attention to application PACE needs and capabilities of the underlying storage.

Does software-defined virtual and cloud eliminate hardware? Generally speaking, no, in that software requires hardware; however, where that hardware is located, how much of it you need, as well as how it gets used does change. You could eliminate some local hardware if you keep in mind that you will still need networking equipment. Likewise, you may also find value in having a local cache or buffer or a backup copy of your remote or cloud data. Virtual systems require hardware. However, you may be able to reduce the amount of hardware or where it is located.

With virtualization, can I get down to having only a single server? In theory, yes you could, but that would also introduce a single point of failure (SPOF). Depending on your organization's need and tolerance for disruption, you will want at least two servers, although one can be smaller than the other for resiliency.

What is tin-wrapped vs. software vs. shrink-wrapped software? Shrink-wrapped refers to the software you buy (or get for free) that comes in a box or other packaging, including via download, that you can then install on your server (physical or virtual). The benefit is that you get to decide how and where the software is installed; the caveat is that you provide the hardware or virtual appliance to run the software. Tin-wrapped refers to buying or obtaining software that is packaged with hardware as part of a solution bundle or turnkey appliance. The benefit is ease of use, and for some there may be accounting or purchasing advantages; the caveat is that you are dependent on the solution provider to choose the hardware for you.

Can you clarify PM, BM, CM, cloud instance, VM, container, DPS, and VPS? A physical machine (PM) is a bare-metal (BM) server on which some operating system, hypervisor, or other software is installed on. A cloud machine (CM) is a virtual machine instance (guest) running on a PM or BM (host) with a particular operating system image similar to a virtual machine (VM). A cloud instance is a CM or cloud-based VM, similar to a virtual private server (VPS). A DPS is a dedicated private server, similar to a PM with some management software installed (think of it as your own private server with a hypervisor such as KVM or Xen installed). Containers in the context of Docker or micro-services are highlight instances for running your application or process without the overhead of a hypervisor or full operating system.

4.14. Learning Experience

Here are some simple exercises to explore your current servers, workstations, or even laptops to learn more about their CPU, memory, I/O, and other capabilities.

On a Linux system, use the commands *lshw –c cpu or lscpu* or *cat /proc/cpuinfo* or, to determine whether processor virtualization technology exists, *egrep '(vmx|svm)' /proc/cpuinfo*. To display memory information, use *lshw –c memory*, and for PCIe devices, *lshw –c pci* or *lspci* combined with your favorite grep or other script output filtering techniques.

For Windows, from a command line, you can use the *wmic* command that displays information ranging from the motherboard and BIOS to network, CPU, disk, memory, and more—for example, to find CPU, *wmic CPU list brief*, and memory, *wmic memory chip list brief* or *wmic memphysical list brief* and *wmic disk drive list brief*. You can also download the Microsoft devcon tool, which is a command-line alternative to the GUI device manager. Also, Windows defines environment variables that you can access via *echo*, for example,

```
C:\>echo %PROCESSOR_ARCHITECTURE% %PROCESSOR_IDENTIFIER%
x86 x86 Family 6 Model 37 Stepping 2, GenuineIntel
```

4.15. Chapter Summary

Fast servers need fast applications, databases, operating systems, hypervisors, adapters, memory, adapters, I/O networks, and storage. Likewise, applications have various PACE attributes and corresponding server resource needs. Just as there are different types of environments and application workloads, there are many server options.

Some software and applications get deployed on appliances or tin-wrapped software, while others are deployed on or in virtual machines or on cloud instances. All servers have common tenants, including a processor for compute, memory, and storage, along with I/O networking and other ports.

General action items include:

- Understand the difference between a PM, VM, CM, and container instance.
- Know the similarities and commonalties across physical, virtual, and cloud.
- PCIe is a key enabler for server storage I/O networking hardware and software.
- Servers can support various applications including storage and networking.
- Hardware and software get defined and managed to support different applications.
- Servers are the platform basis for many storage systems, appliances, or solutions.
- There are different generations of physical, virtual, and cloud instance servers.
- Servers are where hardware and software converge, being defined for different tasks.
- Inside, under the hood between the outside wrappers of a CI, HCI, CiB, and AiO solution, are hardware such as servers, I/O networks, storage, and software, along with management tools.

Bottom line: Applications need servers of some type, the software requires hardware somewhere, where and how the software gets packaged and deployed will vary. The best server will be the one that meets your application PACE requirements.

Chapter 5

Server I/O and Networking

The best I/O is the one you never have to do; the second best is the one with least overhead.

What You Will Learn in This Chapter

- Servers communicate with people and devices by performing I/O operations
- Server I/O fundamentals for accessing storage and other resources
- Inter- and Intra-system server storage I/O networking
- Physical and virtual server I/O

This chapter looks at general server input/output (I/O) fundamentals as well as logical I/O and networking. In this chapter, the focus starts with server I/O (also known simply as IO) fundamentals from inside physical and virtual servers as well as external resource access—in other words, server and storage-defined networks for performing I/O to move data, as well as to support management-related tasks for data infrastructures. The chapter expands from server I/O fundamentals to logical and physical network basics. Logical and physical network hardware and software from a local and long-distance perspective are covered in Chapter 6. Key themes, buzzwords, and trends addressed in this chapter include logical and physical networks, server I/O performance, ULP, ports, TCP, SCSI, NVMe, and related topics.

5.1. Getting Started

Everything is not the same in most IT data centers and data infrastructures or environments across different applications. Although there are similarities, given the diversity of applications and usage needs as well as types of servers and storage devices, there are many different protocols and network transports. Servers are used as platforms for running various applications, from business information to those focused on data infrastructure including network and storage appliances, among others.

Servers perform I/O to support data infrastructure and application functionality including:

- Accessing, sharing, synchronizing, and moving data
- Block, file, object, and application storage service
- High-availability (HA) clustering and workload balancing
- Data protection (availability, durability, backup/restore, replication, business resiliency/ business continuity/disaster recovery (BR/BC/DR)
- Web and other client access, including PDAs, mobile, IoT, and terminals
- Voice and video applications, including Voice over IP (VoIP)
- Management, command, control, and monitoring functions

Besides running applications that require computing processing resources, servers also perform various I/O operations. I/O includes operations internal and external to a server, such as accessing memory, NVM persistent storage (SSD and HDDs), and access to other servers, storage, and network and client devices as well as cloud services.

Server I/O and networking connectivity have similar characteristics to memory and storage. The closest operation to the main processor should have the fastest I/O connectivity; however, it will also be the most expensive, distance-limited, and require special components.

Moving farther away from the main processor, I/O remains fast with distance measured in feet or meters instead of inches, but more flexible and cost-effective. An example is a PCIe bus and I/O interconnect which is slower than processor-to-memory interconnects but able to support attachment of various device adapters with very good performance in a cost-effective manner.

Generally speaking, as an I/O occurs farther from the core (processor) and main memory, the slower it becomes, meaning a longer response time. Figure 5.1 shows an application (top right) that resides in a software-defined virtual server (top left). The application makes server

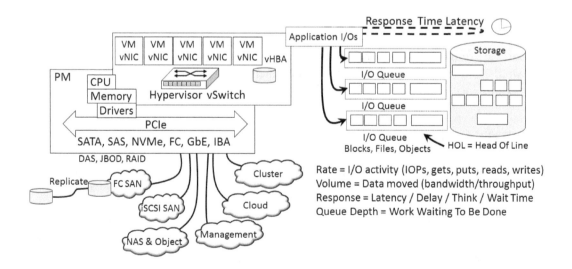

Figure 5.1   Server I/O fundamentals.

I/O requests, doing reads, writes, gets, puts, updates, lists, among other operations, on various data accessed by the server (bottom of Figure 5.1).

In Figure 5.1 the server (top right) hosts various applications that do I/O operations such as database transactions, or reading and writing files and blobs or objects, among other activities.

Also, the server I/O activity can be random or sequential, large or small, as discussed in Chapter 2, depending on the application PACE (performance, availability, capacity, and economics) characteristics. Server I/O rates are the number of operations being done in a given amount of time, which will vary in size or type (read, write, status). The server I/O bandwidth is the amount of data being moved (throughput), which can be measured or calculated as I/O rate × I/O size for a given amount of time.

5.1.1. Tradecraft and Context for the Chapter

Data infrastructures made up of servers, storage, I/O networking hardware, software, and services exist to support applications and their data as defined by various data structures. Also, keep in mind that server I/O exists to support different applications needs for access and managing data from physical, software-defined virtual, container, and cloud resources. Likewise, a software-defined data center is one that has various resources defined as data structures residing in memory, with which different applications (programs) can interact and manage.

Context matters for the server I/O include ports, which can be physical (PHYs) connector ports of various types in different locations (motherboard, PCIe expansion cards, and network and storage devices). Ports can also be logical, such as a TCP/IP port for common services such as 8080 for HTTP among others. There are also virtual ports in virtual server networks, as well as software-defined environments.

Another context term for server I/O tradecraft is *fabrics,* which may refer to internal and external I/O and network topologies, but may also be used in reference to data access or management activities.

Virtual SAN (VSAN) is another term that requires some context to understand. In the past it has been used to refer to a virtual subnetwork segment in a SAN, similar to a virtual LAN (VLAN). Note that VSAN refers to a VMware software-defined storage product, as well as generically to a software-defined virtual SAN. Thus, when you hear VSAN, keep context in perspective.

NLM is another server I/O term that can have multiple meanings, including network lock management or network loadable module. Similarly, VSA can mean virtual storage array, virtual storage appliance, or virtual storage application, among others.

Watch for other context issues regarding server I/O throughout this chapter.

Data consistency and integrity are fundamental terms that involve server I/O as well as other areas and topics covered in this book. As their names imply, data is considered consistent when it is in the form, format, and representation as intended when stored or updated. For example, consistent data means that if "wOrd" is stored, it does not change into "w0rd" (change from O to zero). Consistency and data integrity come into play as data is moved in and around servers, over networks or stored directly, as well as indirectly when cached. We'll have more to say about data consistency and integrity later.

Server I/O and network management activity can be done:

- In band—Management activity occurs over the same network as I/O or other traffic.
- Out of band—Management activity occurs over a separate network (out of the way of data).

Servers perform I/O via internal and external networks that may be either dedicated to a specific functional use or general-purpose. Similar to how there are different layers of applications, software, and device drivers, there are various types and layers of server I/O. For example, there may be one network on which all server to server, client, or users to the server, as well as server to storage traffic, occurs. Alternatively, separate networks may be used for accessing the servers from users or clients, from server to server, as well as from server to storage along with a management network.

Machine-to-machine (M2M) server I/O includes:

- Data, control, and management I/O (replication, configuration, logging)
- Server to client, mobile, Internet of devices (IoD), Internet of things (IoT), and other devices (general access of servers)
- Server to server or cloud (access cloud or other server resources and storage)
- Server to storage systems or servers (access primary or protection storage and data)
- Storage to storage or cloud services (access local and remote cloud storage services)
- Cloud to cloud or another service (hybrid cloud and application resource access)
- Cloister, grid, and scale-out heartbeat and management

These server I/O operations can occur in different directions for various purposes. For example, applications on servers (or clients) as well as storage systems (e.g., servers running storage application software) access or are accessed via north/southbound activity (Figure 5.2).

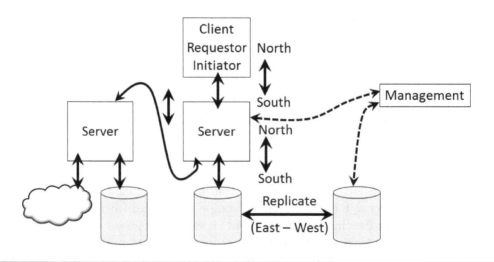

Figure 5.2 North/southbound (access) and east/westbound (management) traffic.

Northbound, for a server, means facing the southbound access of a user client, server, or storage system requesting services. For a storage system, northbound refers to the front-end access ports that other servers or storage systems use for accessing data or other services. Southbound, for a server, means facing some other server, storage, or resource's northbound interface, requesting data or other services. For a storage system, southbound refers to the back-end access ports where NVM flash SSD, HDD, tape, cloud, or other resources are attached.

For a server or storage system, east/west means the internal I/O fabric, network, or back-plane for moving data and performing management tasks, as well as the external management fabric or control plane between nodes. Management activity includes monitoring, configuration, system health status checks, cluster heartbeat, data migration, and copy or replication services, among other activities.

5.1.2. Server I/O PACE and Performance Fundamentals

Different applications will have various PACE attributes that determine their server I/O pattern. Server I/O patterns range from I/O-intensive to light or minimal, from reads to writes, random to sequential, many small or a few large-sized, and various other combinations or permutations.

Server I/O PACE characteristics include:

- Performance—I/O activity rates (IOPs, gets/puts, reads/writes, messages, packets, frames, objects, commands, files) sent, received, or listed; data moved (bandwidth, throughput, Kbytes sent and received); latency or response time (how long it takes to do the work); queues (server, device, network), and resource usage (CPU time including system over-head and productive users as well as memory). Note that latency is associated with longer distances and slower storage devices. However, latency can also occur locally with fast devices, depending on I/O network congestion or other issues.
- Availability—Resiliency and data accessibility including multiple paths for resiliency and high availability. Data consistency and integrity to prevent dropped, lost, or corrupted data that is also safe, secure, and kept private. Data replication and copies for protection and durability.
- Capacity—Available I/O resources including physical connectivity ports, slots, connectors, bandwidth internal to a server, as well as on a local or remote basis. Limits on upload or download speeds. Low-latency, high-availability networks or interfaces vs. low-cost, low-quality, best-effort I/O paths and services. Effective usable including protected, overhead for protection, formatting, file system, RAID/parity (including erasure codes), allocated vs. used.
- Economics—Cost for resources, including lost productivity due to errors or retransmission resulting in rework or lost productivity. Other costs include server I/O resources from hardware to software to services.

Server I/O metrics and related topics include:

- Average, peak, minimum, sustained, percentiles, standard deviation of metrics
- Bandwidth of data moved (how many bytes read/written)

- Latency and response time (application, system, device)
- Frames, messages, I/Os, objects, files, frames, or packets per second
- Availability and quality of service (errors, timeout, retransmits, dropped or blocked)
- Noisy top talkers or traffic between source and destinations

Response time or latency is how long it takes for an I/O to occur, which can be measured end to end (i.e., from the application to the I/O device) or at a given point such as an I/O device (e.g., storage SSD or HDD).

Another server I/O topic is queues that exist in many locations as buffers to hold work until it can be processed. Longer queues can be observed with busy systems and can be an indicator of productivity activity if response times are low or normal with high activity rates. On the other hand, long queues with low rates, low bandwidth, and poor response time can indicate a performance bottleneck, congestion, or other problem.

Figure 5.3 shows an application at the top left running on a server that issues I/O to some device with the associated activity flowing from the top left down to the lower middle and lower right. The purpose of Figure 5.3 is to highlight the main components involved with and that impact server I/O on a local as well as remote basis.

In general, the faster a processor or server is, the more it will be prone to a performance impact when it has to wait for slower I/O operations—resulting in lost productivity. Consequently, faster servers need better-performing I/O connectivity and networks. "Better-performing" here means lower latency, more activity (IOPS, messages, packets, files, an object get or puts), and improved bandwidth to meet various application profiles and operations.

In later chapters, we will look more closely at storage devices and storage systems with regard to performance, capacity planning, and metrics, along with benchmarking. For now, however, let's look at a couple of simple server I/O examples including a transactional database workload, file processing, and general block I/O.

Figure 5.4 shows a sample transactional database workload with different numbers of users and server I/O being done to various types of HDD configurations. The server I/O performance metrics are transactions per second (TPS) activity represented in the solid bars for

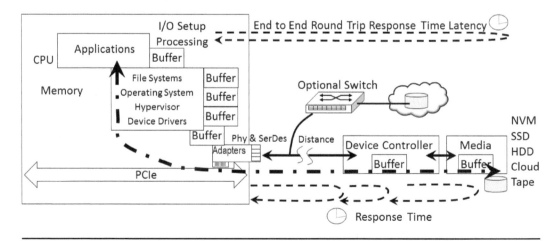

Figure 5.3 Server I/O layers and response time.

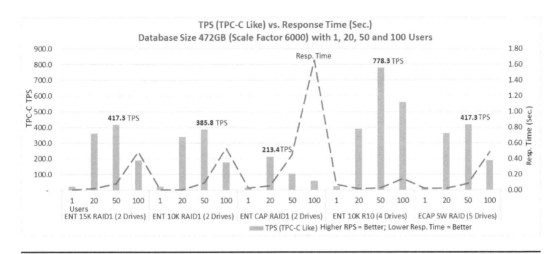

Figure 5.4   Various users doing database server I/O to different HDDs.

various numbers of users. The dashed line represents the response time or latency at different workloads (numbers of users) as well as using different types of storage devices (HDDs).

In Figure 5.4, the workload is a Transaction Processing Council (TPC) on-line transaction processing (OLTP) TPC-C. There is one server acting as an application client running Benchmark Factory to generate the TPC-C workload, simulate the users, and record the resulting metrics. The client is doing I/O over a 40 GbE connection to a database server with various types of HDDs in different configurations, each with a 472-GB database running for 90 minutes.

Another server I/O workload example is shown in Table 5.1, where a large number of big files (5 GBytes) are created and then randomly accessed over a 10-hour period. The I/O profile or pattern is 90% reads, randomly accessing the files, with an I/O size of 128 Kbytes with 64 threads or concurrent workers (users). The resulting metrics are shown in Table 5.1, including file read and write rate, response time, CPU usage percentage, as well as bandwidth across different types of HDD configurations including RAID 1 (R1) and RAID 10 (R10).

Instead of looking at the large server I/O to big files, Table 5.2 shows metrics similar to those in Table 5.1, but with 1,638,464 smaller 16-Kbyte files spread across 64 directories. The files are accessed over a 10-hour time via 64 threads or workers doing 90% reads with a 1-Kbyte I/O.

Table 5.1. Performance Summary for Large File Access Operations (90% Read)

	Avg. File Read Rate	Avg. Read Resp. Time (s)	Avg. File Write Rate	Avg. Write Resp. Time (s)	Avg. CPU % Total	Avg. CPU % System	Avg. MBps Read	Avg. MBps Write
ENT 15K R1	580.7	107.9	64.5	19.7	52.2	35.5	72.6	8.1
ENT 10K R1	455.4	135.5	50.6	44.6	34.0	22.7	56.9	6.3
ENT CAP R1	285.5	221.9	31.8	19.0	43.9	28.3	37.7	4.0
ENT 10K R10	690.9	87.21	76.8	48.6	35.0	21.8	86.4	9.6

Table 5.2. Performance Summary for Small-Size (16-KB)
File Access Operations (90% Read)

	Avg. File Read Rate	Avg. Read Resp. Time (s)	Avg. File Write Rate	Avg. Write Resp. Time (s)	Avg. CPU % Total	Avg. CPU % System	Avg. MBps Read	Avg. MBps Write
ENT 15K R1	3415.7	1.5	379.4	132.2	24.9	19.5	3.3	0.4
ENT 10K R1	2203.4	2.9	244.7	172.8	24.7	19.3	2.2	0.2
ENT CAP R1	1063.1	12.7	118.1	303.3	24.6	19.2	1.1	0.1
ENT 10K R10	4590.5	0.7	509.9	101.7	27.7	22.1	4.5	0.5

Keep in mind that server I/O requires some amount of CPU for processing the various application codes, as well as the different software drivers in the I/O stack. This means memory used for buffers as well as system (CPU) overhead for moving data between buffers.

When looking at server I/O performance, in addition to looking at activity such as IOPs or TPS as well as response time, bandwidth, and queues, also pay attention to how much CPU is used for doing the IOs. The CPU used will be a mix of overall as well as system overhead. By reducing the impact of server I/O, such as using less CPU per transaction or IOP or packet sent and received, more work can be done with the same server.

Likewise, if a server is spending time in wait modes that can result in lost productivity, by finding and removing the barriers, more work can be done on a given server, perhaps even delaying a server upgrade. Techniques to remove bottlenecks include using faster server I/O interfaces such as PCIe Gen 3 and NVMe along with NVM including flash SSD as well as other fast storage and networking interfaces and devices.

Another server I/O example is shown in Figure 5.5, where a 6-Gbps SATA NVM flash SSD (left) is shown with an NVMe x4 U.2 (8639) drive that is attached directly to a server. The workload is 8-Kbyte random writes with 128 threads (workers) showing results for IOPs (solid bar) along with response time (dashed line). Not surprisingly, the NVMe device has a

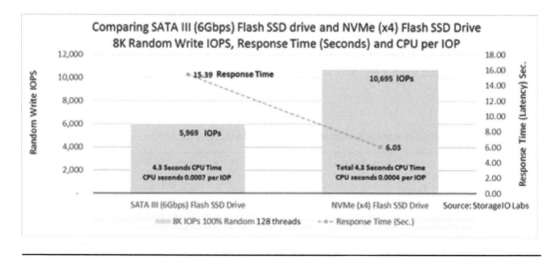

Figure 5.5   A 6-Gbps SATA NVM flash SSD vs. NVMe flash SSD.

lower response time and a higher number of IOPs. However, also note how the amount of CPU time used per IOP is lower on the right with the NVMe drive.

To address various application challenges and demands, the server I/O and networks need to support good performance, flexibility, and reliability in an affordable manner. There are various types of servers and storage to support different applications and environments and also many options for server I/O networking connectivity. These different options vary by performance, availability, capacity, and capability, as well as economics, functionality, and distance.

5.1.3. Server I/O and Networking Fundamentals

The server I/O is focused on moving bits, bytes, blocks, and blobs from one place to another, local or remote, across physical, virtual, and logical resources, software, and device drivers. In addition to being characterized by PACE application attributes, the server I/O also involves some amount of distance, CPU, and memory consumption. Distances vary from the motherboard, chassis, cabinet, data center, and campus or across continents.

The server I/O involves physical (along with logical components) or software-based such as virtual residing on a virtual server or cloud environment on a public, private, or hybrid platform. These networks can be configured for low latency or high bandwidth, highly resilient or low-cost, among other attributes.

Types of server I/O networks include:

- Internal—Inside the server, including CPU to memory and storage
- External—Outside a physical (or virtual) server to storage and networks
- DAS—Direct attached storage (internal or external, dedicated or shared)
- SAN–Storage area network block-accessed storage
- LAN—Local area network including, wired or wireless
- MAN—Metropolitan area network and campus
- WAN—Wide area networks including remote site or cloud

Figure 5.6 shows various server I/O networks, interfaces, protocols, and transports. This figure shows both physical and software-defined virtual networks (inside virtual machines and hypervisor devices and software device drivers. Also shown are logical networks including upper-level protocols (ULPs) for doing local I/O inside a server, between servers and storage, and communicating with clouds.

ULPs will be covered in more detail later. In brief, they are logical network protocols such as TCP/IP and SCSI command set, among others, that set the rules of the road on how data and I/O move from place to place.

In general, the farther an I/O has to go or be serviced away from the processor where the application is running, the longer it will take. This is where context may come into play to put things in perspective.

In Chapter 3, various time fractions and measurements were shown in Table 3.2. For those who are focused on the server I/O close to the processor or in memory, time measurements are in microseconds, nanoseconds (billionth of a second), and picoseconds. Moving farther out,

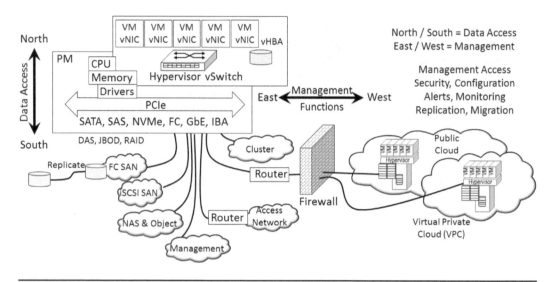

Figure 5.6   Server I/O access (server to storage, server to server, storage to storage).

NVM and storage-focused I/Os tend to be focused in the millisecond (one thousandth of a second) or microsecond (one millionth of a second) range.

Moving farther out over LAN, MAN, and WAN as well as to a cloud server, I/O access can be measured in seconds or milliseconds. To put this in perspective, a network ping from a server in Minneapolis to another on the U.S. West Coast might be measured in tens of milliseconds (or less), while a local LAN ping might take a fraction of a millisecond. These times might seem fast for some networks, but they are slow compared to microsecond or faster speeds.

While it would be nice to have all server I/O requests be handled in microseconds or less, there is also the reality of how much memory, storage, and other servers can be accessed locally on a server and at what cost. If you have a large enough budget, you can build very large and scalable server I/O solutions today. On the other hand, what do your applications require for server I/O based on their PACE attributes?

Also, note that distance, which may be measured in centimeters or inches, feet or meters, or miles and kilometers, is only one factor in determining server I/O performance. In addition to physical distance, other factors include the speed of interfaces and their type, such as copper electrical; there are also protocols, as well as software stack layers.

The higher up (above other software layers and stacks) an application that is doing I/O sits, the more layers of software (databases, file systems, operating systems, drivers, hypervisors and their drivers) as well as devices have to be traveled. Thus, you may have a fast physical server I/O path, but protocols and other software along with their configuration can impact performance. Thus, fast applications need fast servers with fast software and hardware to accomplish fast I/Os with low latency.

5.1.4. Inter- vs. Intra-Server I/O Activity

You are probably familiar with the terms Internet and intranet. An intranet is a private network such as within your organization, while the Internet is widely accessible across organizations.

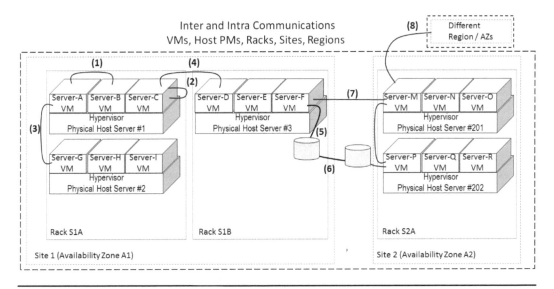

Figure 5.7   Inter- and intra-server and I/O network communications.

The same concept applies to servers and storage I/O, management and other activities that can occur inter- (across or among different systems, servers, devices, cabinets, datacenters, zones, or regions) as well as intra- (within). For example, inter-system, Internet, inter-cabinet, inter-cluster, and inter-node refer to spanning different systems, networks, cabinets, clusters, or nodes, respectively. On the other hand, intra-system, intranet, intra-cabinet, intra-cluster, and intra-node refer to work or communications occurring within a system, network, cabinet, cluster, or node.

Figure 5.7 shows, on the left, (1) two virtual machines communicating intra-physical server, (2) application communicating in a VM (intra-VM), (3) a VM communicating inter-physical servers with another VM, (4) two VM communicating between two different physical servers (inter-) in two different racks (inter-) in the same (intra-) availability zone (AZ).

The middle of Figure 5.7 shows (5) a VM communicating with its local storage (intra-), which (6) in turn is replicated to another storage system at a different location (inter-). Also shown is (7) a VM communicating with another VM at a different site (intra-) but in the same region (intra-). Finally, the right side of Figure 5.7 shows a VM communicating with another VM or storage at a different region (inter-).

5.2. Server I/O Basic Building Blocks

Server I/O building blocks include physical hardware devices, cables, connectors, and transceivers, along with diagnostics tools. Besides physical hardware, there are also software components from applications, file systems, operating systems, hypervisors, and device drivers. There are also protocols and command sets that are software-implemented in the various devices that define how data is transmitted.

Software drivers are defined to implement different layers of I/O protocols including flow-control, multipath failover for redundancy or load balancing, security, and data integrity, among other functionalities. To help get the server I/O basic building block discussion underway, Table 5.3 shows some common terms along with applicable content and description.

Table 5.3. Simplex, Half-duplex, Full-duplex

Term	Description
Simplex	Sends data in only one direction.
Half-duplex	Uses two simplex data paths for sending data in opposite directions (send and receive). Even though there are separate transmit and receive paths, only one side can talk at a time, with each side taking turns talking and listening.
Full-duplex	Uses two simplex data paths for sending data in opposite direction (send and receive). Data can be sent in both directions concurrently, with each side able to talk and listen. The benefit is potentially double the performance of half-duplex.
Congestion	Occurs due to too much traffic on a given I/O or network path. Think of a busy freeway during rush hour.
Blocking	Occurs when some traffic prevents other I/Os from completing promptly. Head-of-line blocking is where traffic at the front of the line or queue is slow and blocks other server I/O traffic from occurring. Think of driving behind a very slow driver on a busy freeway.
Oversubscribed	Refers to a resource being overallocated or having more users, consumers, requesters of the service, device, or I/O path than what can be supported. In a normal mode where there is light traffic, a server I/O path or network can accommodate many users who do not consume many resources. However, if everybody accesses the network or interface at once and that network or I/O path is shared, then congestion or blockage can occur. An alternative is to have dedicated ports, paths, links, and connections, but that will usually add to cost.
MAC	Media Access Control address is a unique identifier for network and some devices in both hardware as well as software-defined and virtual networks. Ethernet, for example, uses a 48-bit MAC.
Links	Physical or logical connections between devices and components or software.
Tunnel	In addition to referring to a physical path through which a cable runs, network tunnels provide a means of encapsulating other network ULPs or traffic onto a common network.
Buffers	Queues or cache for holding data in various locations through the I/O stack.
Failover	Failover, failback, MPIO, load balance, and others
WWN	World Wide Name is a unique name similar to a MAC address used for addressing SAS, SATA, and Fibre Channel devices such as HDDs and SSDs.

Servers perform I/O to various devices including other servers using copper electrical or fiber-optic cables, along with wireless [WiFi, WiMax, free space, and other radio-frequency (RF) technologies]. Data can be transmitted in parallel, where, for example, a byte (or larger) has each of its bits sent side by side on a separate wire. Additional wires may be used for command, control, and management.

The benefit is that each bit gets its lane or wire; the challenge is that more physical space and wires are needed. Some of you may remember or recall thick heavy cables, or flat ribbon cables such as for IDE, or ATA, parallel SCSI, block mux bus, and tag, among others. These were parallel and relatively easy to implement, but they required much physical space and had distance limitations.

Figure 5.8 shows an application that communicates with hardware devices via various layers of software. Here, for example, host bus adapters (HBA) have devices attached or are attached to networks and switches. I/O protocols are implemented in software drivers as well as in the hardware devices. Note that the speed of the I/O is tied to how fast the various software layers and drivers, along with protocols, can move data. Even though, for example, a fiber

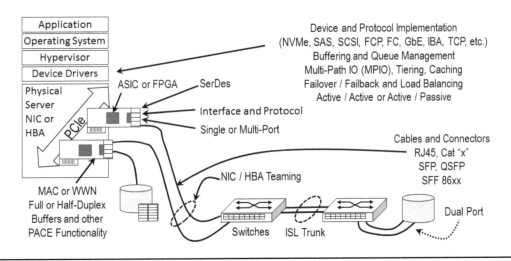

Figure 5.8   Server I/O hardware and software.

optic cable might be used to connect a server to another server or storage system, data is not necessarily moving at the speed of light. While the fiber optic cable uses light waves to transmit data serially, the effective data rate is determined by the serial-deserializer (SerDes, pronounced Sir-Deez) transmit rate, encoding, and all associated protocols, device drivers, and devices.

Today, most communications are sent serially, that is, bits are sent one after the other in a given direction. The benefit is that fewer wires and smaller connectors can be used while supporting distance over copper, fiber, or wireless to the limits of a given interface and protocol. To keep up with performance, serial transmission of data uses a faster clock speed to move data faster, as well as leveraging newer encoding schemes that have less overhead, such as 64b/66b or 128b/130b.

At the physical or Phy layer, data gets converted to serial or back to parallel using a SerDes. The SerDes can be implemented in custom chips such as ASICs or FPGA that perform I/O functions. SerDes range in speed that corresponds to the server I/O interface and networks such as 1 Gb and 10 Gb Ethernet (or faster), 12 Gb SAS, 40 Gb or 100 Gb InfiniBand, among others.

Some server I/O activity is for higher concurrent or parallel bandwidth. For example, many data items or I/Os can be done at the same time, or broken up and sent concurrently as parallel streams and then reassembled. Other applications may be serialized where they need bandwidth, however the data cannot be broken up or chunked into smaller pieces.

Some context is needed here in that a low-level physical cable may be running in serial mode, but upper-layer networks may be able to do parallel or concurrent data transfers. Likewise, upper-level networks and protocols or device drivers and applications may be able to use multiple independent serial networks and transmit data in parallel over them. The key point is that some applications can benefit from having say 10 × 1 GbE network connections in a server where there are many smaller data transfers or server I/Os. Other applications may need lower latency or aggregate bandwidth for fewer, yet later I/Os that benefit from a single 10 GbE (or faster) interface. Then there are scenarios where 10 × 10 GbE is needed (or wanted).

To protect data from accidental changes during transmission, enabling data integrity, cyclic redundancy check (CRC) error detection codes are commonly used for server I/O and other data or storage communications. For some number of bytes or blocks being transmitted, a check value is calculated and sent with the data. The receiving end can use the check value to detect any data corruption and take appropriate action to maintain data integrity.

5.2.1. Cache and Locality of Reference

The best I/O is the one that you do not have to do. However, servers need to do I/O activity to communicate with the outside world, including interacting with people, or machine to machine (M2M). M2M activity includes server to server, server to storage, storage to storage. as well as to other devices that access servers and their services—for example, various mobile, desktop, and other IoT along with IoD that need to access server applications, services, or resources, including applications as a service (AaaS), storage as a service (SaaS), backup as a service (BaaS), and disaster recovery as a service (DRaaS), among others.

Realizing that servers need to do some I/O, the second-best I/O is the one with the least impact on performance or cost to the application. An I/O that can be resolved as close to the application as possible (locality of reference), such as via memory or cache, should be more effective than having to go longer distances, even if locally inside a server.

While latency can increase even on a local basis, it is also often associated with distance, including physical geographic distances from meters to kilometers or farther. Another aspect of distance, however, is how many layers of software ("altitude") an I/O traverses from application to intended destination or system being communicated with. Also, the number of systems and applications that must be traversed will also have an impact on server I/O performance as well as network configuration or usage and quality-of-service (QoS) policies. Caches help to minimize the impact of slower server I/O by keeping local data local or close to the applications.

With server virtualization and consolidation, many disparate systems and their applications are aggregated onto a server, which results in more aggregate I/O activity (Figure 5.9).

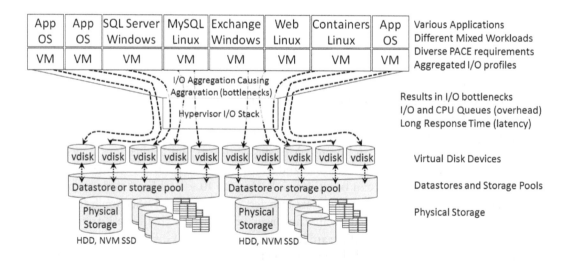

Figure 5.9   Server I/O aggregation causing aggravation.

This aggregation results in the aggravation of multiple different I/Os converging on common resources, all mixed up together in what some call an I/O blender effect.

Read cache such as write-through (synchronous) enable data to be written to a target device (no cache on the write, put, or update) for strong consistency; however, subsequent reads are cached. Note that cache implementations need to have data integrity to prevent against stale data (i.e., data is changed, but the cache copy is not current).

Besides avoiding stale reads (or writes), cache implementations also needs to be refreshed or "warmed" (loaded with data) after a system restarts before it can effective. Also, to read cache, there are also reads and writes such as write-back (asynchronous), where writes or updates are also cached for eventual commitment to storage.

A variation of the cache is journals and logs used for applications, file systems, operating systems, and snapshots, along with metadata management among other things. The logs or journals maintain a persistent copy of data until it can be safely verified as being written with integrity to its destination (e.g., another server, cloud server, or storage). Logs and journals thus provide data consistency and crash coherency in the case of a system failure, so that data is not lost beyond its recovery point objective (RPO).

Caching includes read-ahead (prefetch) as well as read cache (keep a copy of written data local), also known as write-through (the write is kept for subsequent reads) and write-back. With write-back, data is kept in a buffer (preferably persistent or non-volatile for data consistency and integrity), so that the application does not have to wait for the I/O to complete. With write-through, applications have to wait for the I/O to complete. Think of write-back as asynchronous or eventual consistency, while write-through is synchronous or data consistent.

Caches and micro-tiering exist in many different locations:

- Metadata cache locally, with actual data stored elsewhere
- Application-centric, including as databases or key-value repositories
- Application buffers and caches using server RAM or local SSD
- Storage and network protocol (blocks, files, objects)
- File system and volume manager cache using RAM and/or SSD
- Operating system and hypervisor-based caches using RAM and SSD
- Network-based cache using a combination of RAM and SSD
- Storage system or storage appliance-based RAM and SSD cache
- Cloud gateway or access appliance using RAM, SSD, or HDD as a cache

The importance of understanding locality of reference and caching is to facilitate knowing where to place different applications to meet their PACE requirements. Likewise, understanding PACE application requirements helps to know what resources should be configured for performance, availability, capacity, or to meet cost objectives. For example, assigning different application workloads and their data to the applicable server, storage or I/O networking tier, and class or category of service helps to meet service requirements.

Knowing about the application's data access patterns and activity helps to determine when and where to use local server cache, in-memory buffers or RAM disks, NVM storage such as NAND flash, NVMe and SSD for indices, metadata, pointers, as well as temp and scratch areas. From a server I/O or networking standpoint, this means configuring to avoid congestion or traffic bottlenecks on some paths and underutilization on others.

5.3. Access Methods and Personalities

Similar to servers and storage, there is not currently a single type of network or I/O interface that addresses all usage scenarios across performance, availability, capacity, and economics. However, there are some common ones. For example, from the core (pun intended), there is PCIe, which has traditionally been kept inside physical servers, but is now being extended inter-server within racks or cabinets.

PCIe supports various protocols including NVMe among others. Various I/O cards and adapters extend the server's reach from in the cabinet to long distances to meet different application PACE requirements.

Access methods determine how the servers access other servers or storage resources such as block, file, object, or application services, for example, via a NBD, NVMe, SMB3 direct, SCSI or SATA block command using PCIe, iSCSI, Fibre Channel, InfiniBand, NVMe over Fabric, FC-NVMe or SAS; HDFS, NFS, CIFS/SMB, or pNFS file via IP or UDP using GbE; or HTTP/Rest, iROD, JSON, SOAP, ODBC, S3, Swift, or VVOL, among other objects and application APIs over an IP-based network.

Personalities determine how the service or resource function performs or is optimized. For example, a storage service may have a NAS NFS file access method using TCP/IP via a 10 GbE interface. However, that storage service personality may be optimized for data footprint reduction (DFR) including dedupe, compression, and archiving of items.

The solutions personality may be optimized for sequential streaming of data rather than random updates of smaller files. On the other hand, a general-purpose NAS that supports NFS (as well as CIFS/SMB, HDFS, and others) may be optimized for many concurrent and parallel operations of different sizes, including large or small reads and writes. As another example, some object access storage and cloud resources have personalities optimized for large write once, read many of large files, objects, or backup and archive save sets, while others are more flexible.

The key point here is to understand that there is more involved than the physical and logical access method, which in turn is defined by personalities and how they function. In other words, avoid judging a server (or storage) I/O resource simply by its cover, or connector or access method; look under the cover at what's inside, for what matters to meet application PACE needs.

5.3.1. Logical Server and Storage I/O

As we continue our journey from the core of the server on various paths (physical, logical, and virtual) via hardware and software that an I/O can take, let's have a closer look at logical and physical networks. To level-set and provide a frame of reference, Figure 5.10 shows an example of the Open Systems Interconnection (OSI) model for network physical and logical stacks.

Figure 5.10 shows a comparison of various networking and storage I/O interfaces and protocols to support different application needs. Transmission control protocol (TCP) and Internet Protocol (IP) are core protocols for supporting local- and wide-area application access and data movement. Also shown in Figure 5.10 is how other networks such as Fibre Channel map or correspond to the OSI and IP models.

OSI Layer	Description	Data Organization	Applications Characteristics	Fibre Channel
7	Application		FTP, HTTP, Telnet, SNMP, SSH	SCSI/FCP, FICON, ULPs
6	Presentation		ASCII, EBCDIC, Unicode, XML, HTML	
5	Session		Logical Ports, Services, SQL, NFS, RPC	
4	Transport	Segment	Datagrams, Flow control, TCP/UDP	FC-4
3	Network	Packet	Address, Headers, IP/ICMP	FC-3
2	Datalink	Frame	MAC Address, PPP/SLIP	FC-2
				FC-1
1	Physical	bit	Data Encoding (e.g. 64/66b)	FC-0

Figure 5.10 OSI stack and network positioning.

In Figure 5.10, the lowest level (low altitude) represents the physical layer including cabling and transceivers, along with encoding schemes. Moving up in altitude from the lower levels of the OSI model are bits organized into frames, packets, or segments as part of performing data movement or access.

The importance of Figure 5.10 is being able to put in the proper context how bits and bytes get organized and defined to move over various networks when a server does an I/O on behalf of an application. Bits get grouped into bytes, and then frames, packets, and segments. When a server does I/O, the bytes are grouped based on the specific protocol by the device driver and other software into one or more operations. Those I/O operations might be a single read, write, get, put, list, or other function that in turn results in one or more segments, packets, or frames.

Different protocols have various frame sizes or maximum transmission units (MTU). For example, TCP/IP has standard frames, as well as large or jumbo frames where more data can be transferred with less overhead. When discussing segments, packets, frames, blocks, or transfers, keep in perspective the context of where in the server I/O stack the conversation is focused.

Table 5.4 shows the vertical access of various physical server I/O networks along with horizontal different logical or upper-level protocols and command sets. These and other physical, as well as logical and ULPs, exist to support different server I/O application needs. The NVMe

Table 5.4. Examples of Physical Networks (Vertical) and ULP Command Sets

PHY/ULP	SCSI	AHCI/SATA	NVMe	FCoE	FICON	IP/UDP
PCIe		SATA Express	Yes			VIA RDMA
SAS	Yes	Yes				
SATA		Yes				
Ethernet	Via IP (iSCSI)		RoCE/fabric NVMeoF	Yes		Yes
Fibre Channel	SCSI_FCP aka FCP		RoCE/fabric FC-NVMe		FC-SB2	EOL
InfiniBand	SRP		RoCE/fabric			IPoIB

command set is implemented on PCIe including using cards, M.2, and 8639 drives as well as over fabrics using RDMA over Converged Ethernet (RoCE, pronounced Rocky). NVMe is also supported with an underlying FC network (i.e.,FC-NVMe), where NVMe is a ULP similar to SCSI (i.e., SCSI_FCP). The SCSI command set has been implemented on SAS, Fibre Channel (FC), and InfiniBand (SRP) and IP in the form of iSCSI.

LAN networking has evolved from various vendor-specific network protocols and interfaces to standardize around TCP/IP and Ethernet. Even propriety interfaces and protocols such as IBM Mainframe FICON, which evolved from ESCON, on propriety fiber optic cabling, coexist in protocol intermix mode on Fibre Channel with open-system SCSI_FCP.

Context clarifies that SCSI_FCP, also known simply as FCP, is what many people refer to as Fibre Channel. Some simply refer to FCP and Fibre Channel as Fibre Channel. However, FCP is just a ULP on the actual Fibre Channel transport. Likewise, Ethernet commonly supports TCP/IP-based networks, and TCP/IP-based networks are routing implemented with Ethernet. Thus, the two are sometimes referred to as one and the same. Context matters!

5.3.2. Namespace and Endpoints

Data infrastructure resources including servers, storage, and networking and associated services are organized under and managed via namespaces. Context is important with the namespace in that there are program- or application-specific namespaces, file system, and object storage access names spaces, along with database, network, and other namespaces.

What namespaces have in common is organizing unique names and versions of resources (blocks, files, objects, devices) along with any associated address for access. For example, a local or remote filesystem has a namespace that identifies folders and directory names as well as files.

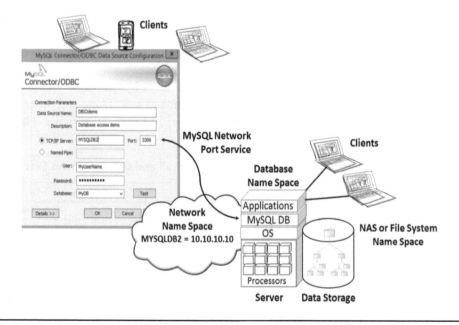

Figure 5.11   Various namespaces, including databases.

Object-accessed storage has namespaces holding the names of buckets, containers, objects, and blobs that in turn are used for locating and accessing those resources. Figure 5.11 shows various namespace examples, including DNS (e.g. MYSQLDB2), database name DBIOdemo along with storage devices, NAS file systems, and network database server services (e.g., port 3306).

Another example is key-value or other repositories that maintain namespace of database entities such as databases, tables, columns, or other objects. Networks also leverage namespaces for managing and mapping names to addresses. DNS have servers to manage and resolve namespaces, which can be the Internet, or an internal intranet or subnet.

Namespaces include worldwide names such as with Fibre Channel, SAS, and SATA devices, IP DNS, NAS file, and object access, among others. For example, http://storageioblog.com is an Internet domain name that resolves to an IP address on my server (where that happens to be currently located). Within my intranet, there is a different domain name such as onsite. storageio.com or some variation that is managed by my domain controllers and DNS systems.

For networks, routers handle routing and address translation between different namespaces and the devices or addresses within them. There are different types of routing for various networks and protocols. Some routing-related terms include network address translation (NAT), SNAT (secure NAT), and DNAT, as well as virtual extensible LAN (VXLAN), VRF (virtual routing forwarding), and generic routing encapsulation (GRE), among others.

Storage systems, appliances, software, and services, as well as servers offering applications or data services, present some endpoint from or into a namespace to attach to. The endpoint can be a physical, virtual, or cloud service providing block, file, and object or application services such as databases that are attached. As already mentioned, there are the different physical server I/O interfaces and network transports along with various protocols as seen in Figure 5.12.

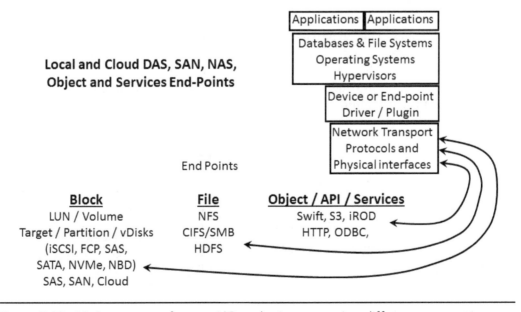

Figure 5.12 Various types of server I/O endpoints accessing different namespaces.

Remember that there are many different types of object access or services. However, S3, Swift, and other buckets or containers, as well as objects and blobs, can be accessed using HTTP/REST with a Uniform Resource Locator (URL).

For object access using HTTP/REST, error codes include 404, Not Found; 408, Request Timeout; 500, Internal Server Error; 502, Bad Gateway; 503, Service Unavailable; and 504, Gateway Timeout; among others. Some of these should be familiar from when you have tried to access websites or other resources via the Internet using HTTP (i.e., through a browser).

Note that a CNAME DNS record can be created to allow object access using your preferred domain but masking the bucket endpoint. For example, using Google Cloud Storage (GCS), you can set up a CNAME DNS entry similar to

my.sample.com CNAME c.storage.googleapis.com

which then allows "storage.googleapis.com/mybucket/Image1.jpg" to be accessed using a URL such as "http://my.sample.com/Image1.jpg."

File endpoints are provided via an access path to the filesystem, file server, gateway, or another service. There are local files where a filesystem exists on a local DAS or networked storage device. Files can also be accessed via file servers using NFS, CIFS/SMB, and HDFS, among other access methods. One resource access name format is Universal Name Convention (UNC) of \\servername\fileshare for Windows and //server name/file share for Unix/Linux. Notice the difference between the backslash "\" used for Windows and the forward slash "/" used for Unix/Linux.

In the case of block server I/O, a LUN, volume, virtual disk, or partition is presented from a local SATA, SAS, NVMe, or USB device. A networked storage device presents a LUN, volume, or virtual disk via iSCSI, shared SAS, NVMe, Fibre Channel (FCP for open systems, FICON for zOS mainframe), InfiniBand, SRP, or NBD, among others.

A hypervisor, operating system, software-defined storage or volume manager or file system mounts a block device, and it is the corresponding file system unless raw (i.e., non-file I/O) is being done to it. For example, some databases and other applications may bypass a file system, instead doing their storage and I/O management of block devices. Some examples of block endpoints are shown in Table 5.5.

Other types of server I/O endpoints include database sockets and named pipes as well as ODBC access. Other endpoints include VMware VVOL (virtual volumes).

Table 5.5. Various Block Endpoints

Block Endpoint	Comment
iqn.1998-01.com.vmware:myvm23-18b12413	iSCSI initiator
iqn.2012-07.com.lenovoemc:storage.MYNAS2.MYNAS2iSCSI3	iSCSI target
naa.5000c5003011467b	VMware storage device
PhysicalDisk0	Windows PowerShell Get-PhysicalDisk
\\.\PHYSICALDRIVE0	Windows wmic disk drive list brief
/dev/sda1	Ubuntu Linux device
sudo mount /dev/nvme0n1 /FASTDISK	Ubuntu mount NVMe device

5.3.3. Connection-Oriented and Connectionless Transport Modes

Two different modes of communication and transmission of data across networks are connection-oriented and connectionless. These two modes characterize how different networks and protocols handle data transmission.

A connection-oriented transport mode is the opposite of a connectionless transport that sends data in a continuous stream from one source to another, for example, from the server to another, or from a server to a storage device. The transport guarantees that all data will be delivered to the other end in the same order as it was sent (in-order delivery), without data loss or duplication of data. The data transmission occurs through a defined process including connection establishment, data transfer, and connection release. A common example of a connection-oriented transport is the Transmission Control Protocol (TCP).

In a connectionless transport mode, data is transferred between the source and destination with no previous setup (connection establishment, transmission, session close). Individual data packets can take different routes between the source and destination. Connection-oriented-mode data takes the same route unless there is a disruption to the route. Each packet is individually routed to its destination in a packet-switching (connectionless) transport mode using the most expedient route determined by some routing algorithm. The packets are reassembled at the destination into their appropriate sequence. An example of a connectionless transport mode is User Datagram Protocol (UDP). Connectionless transport modes are also called packet-switching networks.

5.3.4. RDMA (Remote Direct Memory Access)

Remote Direct Memory Access (RDMA), as its name implies, enables I/O and data transfer direct from different servers' (remote) memory via a low submicro second latency network and a set of software device drivers. RDMA enables fast data access or movement with lower latency compared to the traditional server I/O over legacy general-purpose networks.

RDMA speed is achieved by enabling data to be moved between the memory of one server and another server (or storage device) with CPU off-load, bypassing some common I/O software stacks.

Various networks support RDMA, including 10 GbE (and higher speeds) and InfiniBand for server-to-server I/O. Various hypervisors including Microsoft Hyper-V and VMware vSphere ESXi support RDMA, as do many operating systems and upper-level applications including databases and storage software. An example is Microsoft SMB 3 Direct, which enables low latency, high-performance storage access and server-to-server I/O such as with a server storage I/O software bus. Some context is needed here, in that Microsoft SMB supports general-purpose file sharing with CIFS; however, with SMB 3 Direct, there is also a low-latency capability with performance characteristics similar to those of traditional block server I/O.

Another example is iSCSI Extension for RDMA (iSER), which uses TCP/IP with RDMA services (iWARP) for low-latency, high-performance block storage access. Another example of RDMA use has RoCE-enabled networks that support low-latency server I/O protocols such as NVMe over Fabric. The benefit of using RDMA is lower latency with higher performance, although not all hardware or software support it. Thus, there is a trade-off involving balancing lower cost with commonality across hardware and software vs. higher cost, lower-latency, higher-performing RDMA-enabled technology.

5.3.5. RoCE (RDMA over Converged Ethernet)

RDMA over Converged Ethernet (RoCE, pronounced "Rocky"), as its name implies, leverages converged 10 GbE (or faster) to support RDMA over longer distances using standard networking technologies. This capability enables low-latency protocols such as NVMe over fabric to be deployed using RoCE as the underlying network transport. The benefit is being able to use NVMe as a low-latency block storage access protocol of shared data center storage as an alternative to SCSI_FCP (Fibre Channel) or iSCSI, among others. There are different versions of RoCE, which continues to evolve, some leveraging higher-performance networks while others can use general Ethernet and L3 routing. The industry trade group supporting RoCE activities can be found at www.roceinitiave.org as well as at the Open Fabric Alliance.

5.3.6. SATA, SCSI, and NVMe Protocols

Common block storage protocols for server I/O include SATA, SCSI, and NVMe, which are implemented on various physical, logical, and software-defined interfaces. For example, the SCSI command set is implemented on IP for iSCSI and SAS, on Fibre Channel for SCSI_FCP (i.e., FCP) or what many refer to simply as Fibre Channel, as well as on SCSI RDMA Protocol (SRP), found on InfiniBand.

SATA devices can coexist with SAS and NVMe (U.2 or 8639) devices on the same physical interfaces as well as via SATA Express on M.2 or PCIe. NVMe can be implemented via PCIe add-in cards, or M.2 slots, as well as U.2 (SFF 8639) drives in addition to RoCE for NVMe over the fabric, FC-NVMe, and NVMeoF. Other block access protocols are Network Block Device (NBD), found on some Linux systems, Ceph RADOS Block Device (RBD), and Dell EMC ScaleIO. For zOS mainframes, block protocols are implemented on Fibre Channel using FICON, which is a ULP that can coexist in protocol intermix mode with SCSI_FCP traffic.

In general, the block storage protocols are implemented via a server source side driver that functions as an initiator, and a target destination controller that handles I/O requires such as reads, writes, and status, among others. Both the initiator and the target can be implemented in hardware via an adapter or chipset, as well as via software in the hypervisor, operating system, or add-in software. A server that is also functioning as a storage system or appliance can be both a target and an initiator.

5.3.7. TCP/IP and TCP/UDP

Transmission Control Protocol (TCP), including Internet Protocol (IP) and User Datagram Protocol (UDP), are common network protocols users for server I/O, among other functions. TCP/IP (i.e., IP) and TCP/UDP (i.e., UDP) are ubiquitous in their use and in turn support various upper-level applications or server I/O functions. IP and UDP have evolved from Version 4 (IPv4) to Version 6 (IPv6), expanding from a 32-bit address (2 ^ 32 or 4,294,967,296) to

128 bits (2 ^ 128 or 340,282,366,920,938,463,463,374,607,431,768,211,456), resulting in a more available address.

IP-enabled servers and storage devices include an IP address such as dotted quad, for example, 10.10.10.10, and a network mask such as 255.255.255.0 (or other variation), primary and secondary DNS such as 8.8.8.8 (Google DNS), and a gateway address such as 10.0.0.1. Besides dotted quad, addresses may include Classless Inter-Domain Routing (CIDR) with a /24 following an IP address.

For example, a system with the address of 10.10.10.10/16 is in subnet 10.10.0.0/16. Subnets are useful for segmenting and organizing traffic and in conjunction with VLAN for traffic isolation (tenancy).

Addresses can be static assigned, or dynamic via DHCP, or via Automatic Private IP Addressing (APIPA), such as 169.254.0.0 to 169.254.255.255. Using Network Address Translation (NAT), local intranet or subnet addressed devices can communicate via routers to other devices on the Internet. Routing enables network traffic to flow between a source and destination(s). Tools such as "route –n" or "netstat –rn" can be used to display routing information on a system.

VLANs provide tenancy or isolation of traffic on the same network, where systems in a given VLAN do not see the traffic from other VLANs and vice versa. VLAN IDs range from numbers 1 to 4095; for example, a system in VLAN 2 will have a VLAN ID of 2.

Ports on a switch are associated with different VLANs, and VLAN traffic is tagged to facilitate routing and enable isolation or tenancy on a shared network. Trunk ports on switches enable VLAN traffic to flow between switches and systems in a network (virtual or physical).

IPv4 headers are 20 bytes whereas IPv6 are 40 bytes to handle the larger source and destination address along with other information. Payload or data being sent in a standard IPv4 transmission is 1500 bytes plus header and other overhead bytes. These overhead bytes include a preamble, the interpacket gap (IPG), frame header, frame check sequence (FCS), IPv4 (or IPv6) header, and TCP header.

For example, a maximum transmission unit (MTU) of 1500 bytes results in 1538 bytes being sent, which is 94.93% efficient. There are also jumbo frames of various sizes; for example, an IPv4 9000 MTU results in 9038 bytes being transmitted (99.14% efficient). Depending on the device driver and other software configuration as well as IP version, larger MTUs can also be configured, as well as smaller ones.

"Percent efficient" refers to the number of payload or data bytes vs. overhead. This overhead ties back to the discussion in Chapter 4 about low-level encoding schemes (e.g., 8b/10b, 64b/66b, and 128b/130b). For larger data transfers or server I/O, a larger MTU means less overhead. However, if the MTU is too large and there are smaller transfers, then there is the extra overhead of moving those bytes.

What this means is that there are various levels of efficiency that add up and can have an impact depending on what type of server I/O activity is occurring. Knowing your applications PACE characteristics including size of I/O operations regardless of whether block, file, or object is important to understand, along with underlying software and hardware protocol configurations.

TCP (IP and/or UDP) are commonly deployed on physical, virtual, and software-defined network transports. TCP is also used for both general access of applications and server-to-server communication, including cluster heartbeat and data movement. TCP is also used for data storage access and movement, including block, file, and object as well as databases, among other services such as replication. TCP supports ports (context is logical vs. physical) numbered from 1 to 65535 that are well defined or known for different functions. There are also sockets that provide a stream interface supporting various APIs.

Protocols and applications utilizing IP or UDP include Domain Name Service (DNS), Dynamic Host Configuration Protocol (DHCP), and various NAS including CIFS/SMB/ SAMBA, HDFS, NFS, and pNFS. Additional protocols and applications that use IP include Netty, Remote Desktop Protocol (RDP), XRDP, and PCoIP, and SSH, PCoIP, SSH, FTP, and HTTP-based REST object access, including S3 and Swift. Others include Address Resolution Protocol (ARP), Challenge Handshake Authenticate Protocol (CHAP), Internet Protocol security (IPsec), Multi-Path I/O (MPIO), CDMI, iSCSI, iSNS and NBD (Network Block Device), ICMP (used by Ping), WebDAV, ODBC, SOAP, NTP (Network Time Protocol), VXLAN, XML, and JSON.

5.3.8. Protocol Droop and Deterministic Behavior

Deterministic behavior occurs when a server I/O interface or network and protocol performs in a predictable manner. Traditional server storage I/O interfaces, protocols, and networks such as SAS, SCSI, SATA, and Fibre Channel are deterministic in that their performance is predictable. Some context is needed here, in that "performance being predictable" refers to how the network or interface itself behaves, as opposed to the application workload and storage device.

Nondeterministic behavior means that the network, protocol, or interface has variable performance independent of the application workload or destination (target device). Traditionally, Ethernet has been nondeterministic; with enhancements, it can also operate in a deterministic manner more suitable for some time-sensitive server storage I/O operations.

Performance over distance can be affected by the type of fiber optic cabling being used, the type and power of optics being used, configuration including some splices and connectors, latency from a distance being covered, and the type of network being used on the fiber optic cabling. The latter is what is called protocol droop and may not be specific to fiber optics. However, I am including it here as it is appropriate before we go to the next chapter to talk about metropolitan and wide-area storage networks.

Droop occurs with some protocols over distance, resulting in a loss of bandwidth and increased latency. While the type of fiber being used (MMF or SMF) and the transceivers will determine physical transmission limits, it is the networking protocol among other factors that regulates the actual speed and distance. The clocking rate and encoding scheme of the lower-level physical networking protocol also contribute to speed, whereas the flow control mechanisms and buffers also affect distance.

Interfaces such as Ethernet Data Center Bridging (DCB) and Fibre Channel, which provide deterministic or predictable behavior rely on some flow control mechanism to ensure timely delivery of information. Fibre Channel uses buffer credit systems similar in concept to your checking account in that, in principle, you can only write checks (send data) for the amount of funds (buffers) you have available. When you run out of funds (buffers), you have to (or you should) stop writing checks (sending data) until it is safe to resume. From a performance standpoint, if you have enough buffers (buffer credit count), you can compensate for propagation delays by keeping the channel or "pipe" or network full while sending over long distances.

Networking interfaces such as Gigabit Ethernet and IP are not as susceptible to the impact of droop, as their applications tend to be nondeterministic in that storage applications rely on deterministic behavior channels. However, IP-based WAN and MAN interfaces are typically built on a SONET or another network that can interject latency.

5.4. Physical Server and Storage I/O

Looking at Table 5.6, one might question why so many different networks and transports are needed: Why not just move everything to Ethernet or TCP/IP?

The simple answer is that the different interfaces and transports are used to meet different needs. What this also means is being able to utilize the most applicable tool or technology for the task at hand. An ongoing trend, however, has been protocols, transports, and networking cabling convergence, as shown in Table 5.6. In essence, the number of protocols has essentially converged to one for open-systems block I/O using the SCSI command set [SAS, Fibre Channel, and Fibre Channel over Ethernet (FCoE)], iSCSI or SRP on InfiniBand (native SCSI on InfiniBand vs. iSCSI on IP on InfiniBand).

Table 5.6. Example Server I/O Transports and Comments

Transport	Acronym	Comments and Characteristics
Peripheral Component Interconnect Express	PCIe	Various form factors and packaging including AiC, M.2, U.2 (8639), among others; different speeds G1, G2, G3, G4
Serial attached SCSI	SAS	1.5 Gbps, 3 Gbps, 6 Gbps, 12 Gbps, and faster
InfiniBand	IBA	SDR, DDR, QDR, FDR
Fibre Channel	FC	1 Gb, 2 Gb, 4 Gb, 8 Gb, 10 Gb, 16 Gb, 32 Gb, 128 Gb (4 × 32Gb)
Ethernet	GbE	1 Gb, 2.5 Gb, 5 Gb, 10 Gb, 40 Gb, 50 Gb, 100 Gb, 400 Gb
Universal Serial Bus	USB	USB 2.0 (480 Mbps), USB 3.0 (5 Gbps), USB 3.1 (10 Gbps)

5.5. Server I/O Topologies

Server I/O occurs inside physical, virtual, and cloud servers as well as with external resources and services. Similarly, various physical and logical software-defined topology configurations address different requirements. Some topologies are established inside a virtual or physical server, while others go within a cabinet or rack; still others are for within (intra) or between (inter) data centers or service providers.

Following are some basic server I/O topologies, from server to storage and associated networking hardware and software as well as services.

- **Point to point.** Direct connection between a server and another device (server or storage system). The benefit is simple connection to a device without the need for a switch or other data infrastructure components. The caveat is that a single point-to-point connection creates a single point of failure (SPOF). Another caveat is how many device ports are availability for connectivity or lack of device sharing. Point to point can be done using

Ethernet (including iSCSI), Fibre Channel, InfiniBand, NVMe, PCIe, SAS/SATA, or USB, among another server I/O devices.

- **Hub.** Switches have, for the most part, replaced hubs. However, some hubs are still used, a common example being USB hubs. The benefit is expanded connectivity or number of ports sharing performance and access across those ports. The caveat is performance and lack of isolation or tenancy vs. using a switch.
- **Switched.** Also known as a directors or expanders, switches connect multiple servers, storage, networking or other devices together. For example, while switches are commonly associated with Ethernet and IP networks, or Fibre Channel and FICON SANs, they also exist for InfiniBand, SAS, and PCIe. A variation of switches are the expanders found inside storage enclosures where many SSD or HDD are located. The expander (a variation of a switch) enables shared access with faster speeds than with a hub.

  The difference from a hub is that a switched port can operate at full line speed (depending on implementation) between source and destination. Note that some switch or director ports aggregate multiple physical access ports into a shared internal connection (fan-in, fan-out) within the switch, while others provide a full-line-rate, nonblocking implementation. The benefit is increasing scale, connectivity, performance, and resiliency (when redundant paths are implemented) to meet specific needs.

  Note some context is important here, in that the presence of a switch in a server storage I/O and data infrastructure conversation does not mean part of a general IP or management network. Context is important to clarify that, for example, a PCIe switch in a storage system or part of a rack-scale solution can be a private network if you prefer a server SAN, regular SAN, processor area network, among other names, however, separate from regular networks. Just because something has a switch does not mean its part of regular general-purpose networking.
- **Networked.** Also known as a fabric, comprising multiple switches and or directors among other devices configured in various topologies.
- **VLAN.** A virtual LAN is a logical segment that provides isolation or tenancy for certain traffic on a shared physical (or virtual) network. There are also virtual SAN segments, not to be confused with virtual SAN storage appliances or software-defined and virtual storage.

The size and complexity of an environment, as well as applications, determine the scale and type of network topology, as well as the amount of resiliency, multiple-paths, separate physical networks for north–south data movement, as well as east–west traffic management.

Additional VLAN can be defined as implementing networking multitenancy, including QoS.

Figure 5.13 shows a server I/O and network topology with servers across the top; below that is an access layer above an aggregation layer, with a top-of-row (TOR) layer of switches (managed or unmanaged) below that. The aggregation layer functions as a backbone or trunk for both north–south and east–west traffic management. Additional servers or storage systems attach to the TOR, which may be in a cabinet or rack.

Also shown in Figure 5.13 are interswitch links (ISL), as well as logical and physical links that are teamed together to attach to another server (top right). The server at the top right hosts a virtual server that in turn has a software-defined virtual network topology with virtual adapters, switches, and other components.

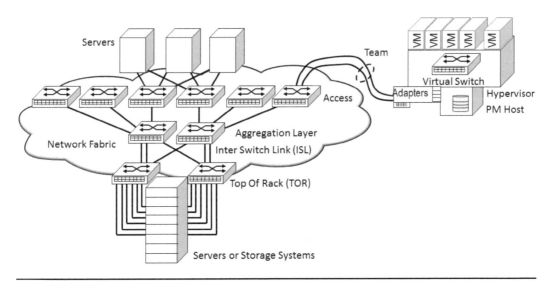

Figure 5.13   Server I/O with physical and virtual network topologies.

An access point (AP) is an endpoint—logical or physical, wired or wireless—where an I/O connection can be made to a server, or from a server to another resource, destination, peer, or service. Examples include Wi-Fi access points, which may be public or private. Another type of access point is a security gateway or firewall appliance providing virtual private network (VPN) or virtual private cloud (VPC) access capability. An access point can also be a cloud or managed service gateway implemented as part of some other software, service, system, or solution, or as a standalone appliance or software.

An access *layer* refers to where a server, client, storage system, or another device connects to a network to provide access to various data infrastructure services. Aggregation layers connect access with resources (servers, storage, networking devices, or services, including cloud).

Resources may, in turn, have internal networks for communicating among servers or storage devices in a rack, cabinet, or chassis. Similarly, depending on the size of the environment, there may be separate access networks, management, and resource networks in addition to redundant paths for resiliency, high availability, and performance load balancing.

The size, scope, and complexity of an environment will in part determine the scale of the access layer, aggregation, and other layers. The simpler the environment, the flatter will be the network topology. Likewise, more complex (not necessarily large) environments may affect the scale or size of the server I/O and corresponding data infrastructure I/O networks.

Table 5.7 shows some basic server I/O topologies that can be deployed in various ways.

A core switch sits at the core of a network as a single device or with other switches attached in various topologies. Often a core switch will be a physically large device with many network connections, trunks, or ISLs attached to it, and edge or access switches along with storage or other target devices. In some scenarios, a core switch may be a physically smaller device functioning as a convergence point between the large edge or access switches and directors and other switches that provide access to storage or other targets.

Edge or access switches, also known as top of rack (TOR), middle of row (MOR), or end of row (EOR), vary in size but provide an aggregation point for many servers to fan-in to storage

Table 5.7. Server I/O and Networking Topology Examples

Topology	Example	Description	Benefit and Caveat
Cascade daisy chain		Two or more switches connected with at least one ISL between them. Switches at the end of the cascade only connect inward to other switches. A ring is created by connecting the outermost switches to each other.	The benefit is increased capabilities beyond a switch. Multiple ISLs or trunks can be used to increase bandwidth between switches. The caveat is that a loss of connectivity between switches can result in a broken or segmented network.
Fan-in, fan-out		Many-into-one or aggregation is not as much a topology as an access option. The inverse of fan-in is fan-out or one-to-many. Note that fan-in, fan-out can occur from devices accessing a server, or a server accessing many devices.	The benefit is aggregating many devices to a common connection(s) point or vice versa. The caveat is potential for oversubscribed or blocking congestion performance, depending on configuration implementation.
Mesh fabric		Multiple switches connected in various ways, either directly or via an access and aggregation layer.	The benefit is scaling beyond the limit of a single switch (scale-out or horizontal); however, ports are used for scaling.
Ring		A group of switches all connected. For resiliency, multiple ISLs can be used between switches, or dual rings can be deployed.	The benefit is simple connectivity. However, loss of a switch results in a broken or segmented network if resiliency is not implemented.

systems or other devices connected to the core of a network. These switches can be Ethernet, Fibre Channel, or InfiniBand among others.

Historically, edge or access switches have been physically smaller, as a means to reduce cost and aggregate multiple slower servers onto a higher-speed ISL or trunk attached to a larger core switch or director. For example, ten standalone servers may each only need about 1 GbE of bandwidth or equivalent IOPS, packets, or frames-per-second performance.

Server I/O fabrics and network topologies can be flat or hierarchal with many layers from access and edge to aggregation, core, and top of rack, among other variations. A server I/O network can be as simple as a point-to-point connection between one or more servers attached to and sharing one or more storage devices. A server I/O network can also be as complex as multiple subnets (segments or regions) spanning local, metropolitan, or global sites and using multiple topologies and technologies. All these, and others, can be combined in various permutations to fit your application, environment, and data infrastructure needs.

5.6. Server I/O Management Topics

Networks and I/O connectivity are important and, with continued convergence, the lines between storage interfaces and networks are becoming blurred. Networks and I/O require networking, server, storage, and data professionals to coordinate the various interdependencies across the different technology domains. Various tasks and activities are needed to keep I/O and network capabilities in good order to support virtual data centers.

Common server I/O management tasks include:

- Planning, analysis, forecasting, reporting
- Acquisition, testing, and deployment decision making
- Monitoring of performance, capacity, use, and errors
- Configuration changes and resource provisioning
- Troubleshooting, diagnostic, testing, and validation resources
- Security and data protection

Separate I/O paths and networks for management are sometimes referred to as management networks or subnets, VLAN or segments, or as a control and management plane (which plays off the notion of a backplane for a server), also known as a management or control fabric. Management traffic is also referred to as east–west activity, including getting or reporting on status, configuration, software or firmware updates, data movement or migration, replication, and event or activity logging.

Management networks, fabrics, or control planes that support intradevice east–west traffic should have performance equal to the amount of data that will be moved over them. If your east–west traffic includes supporting event or other activity logging, plus virtual machine movement and migration, as well as data replication or copy functions, you need to have sufficient I/O network bandwidth to support those functions.

Security and data protection for server I/O networks include encryption of data in-flight (and at rest), along with identity access authorization controls and certificates. Server and I/O network management tools include ones for hypervisors as well as operating systems. Also, various third-party software tools and hardware appliances support server and I/O networking management activities. Another network technology is the OpenFlow protocol for implementing software-defined networks via the Open Network Foundation.

5.7. What's in Your Server Storage I/O Toolbox?

Tools can be big or small, physical, software, or virtual, or cloud-based. Also, tools can be subsets or components of what are needed to support larger activities. In the context of enabling and supporting a data infrastructure, there are various building-block components such as servers, storage, I/O networking hardware, software, and services that themselves can be considered tools. However, there are also tools and techniques for enabling building block components into a resilient, flexible, scalable, and productive data infrastructure.

For example, what's the best tool, technology, or technique for enabling server I/O in the cloud, software-defined virtual, CI, HCI, or physical environment? Depending on what you have in your toolbox, you can apply different technologies in various ways to address server I/O demands.

Of course, if all you have is a hammer or all you know how to use is a hammer, then every server I/O problem, challenge, or scenario is going to look like a nail. Different applications will have different server I/O PACE characteristics and workload demands. This chapter has looked at various technologies, techniques, and tools to address and enable server I/O across multiple environments.

Testing and diagnostic equipment can be utilized for design and development of components as well as testing of components individually and as part of an overall storage network. More examples include network and physical cable monitoring and diagnostic devices and workload generators. Some of these devices are protocol-based, while others work at the networking interface layer and still others at the physical cable level.

Various tools can be integrated and complemented by additional management software tools as part of an overall storage and storage networking management solution set. For example, a component may work fine by itself and with other devices under certain conditions, but when additional devices and workload are added, things may change.

Unit and component testing for verification and validation and for stress testing of an entire system under various workloads may uncover issues not seen in normal testing, such as impact on memory or discovery of memory and network leaks. The combination of various equipment, operating systems, patches, and device drivers can add to the complexity of testing and diagnostics.

Some tools to include in your server I/O physical, virtual, and cloud toolbox include esxtop and visual esxtop for VMware hypervisors, Linux ethtool, fio, dd, htop, ifconfig, iperf, ntttcp, iostat, iotop, *lshw –class network* or *lspci –k | grep net, netperf*, Windows arp, diskspd, ipconfig, ntttcp, and perfmon. Other tools include hcibench, iometer, netstat, nslookup, ping, spotlight on Windows and UNIX, sysbench, and vdbench, among many others.

Another toolbox item is to have a list or know where to find vendor hardware, software, and services compatibility lists. Most vendors maintain compatibility lists for various drivers, management tools, hardware, software, and services for different configuration permutations, along with caveats or release notes. In addition to hardware, software, and services tools, your tool box should include skills or tradecraft on how to use these and other technologies.

5.8. Common Questions and Tips

How can you measure or determine IOPs or bandwidth if you know one but not the other? If you know how many I/O operations (activity) are occurring, say, 120,215 with an average of 8 KB, your bandwidth is 120,215 × 8 KB = 984 MBps. If you do not know the activity or IOPs. but you do know the amount of data transferred (KB read or written) and I/O size, you can determine IOPs. If you know bandwidth and IOPs, you can determine I/O size.

Are IOPs only for block storage? IOPs tend to be associated with read or write block I/Os, but there are corresponding I/O operations for file I/Os (file reads, writes, creates, deletes, etc.) and object I/Os (gets, puts, head, lists, etc.), as well as transactions per second (TPS) for databases, among other metrics. The key point to remember is that IOPs refer to the amount of server I/O activity or rate; its companion is bandwidth or amount of data moved (volume or flow), along with latency or response time and queue depth.

What's the best server I/O protocol and network for storage? It depends! The decision as to what type of server and storage I/O interface, protocol, and topology is often based on cost, familiarity with technologies, their capabilities, and, in some cases, personal or organizational preferences.

Why not just run everything over TCP/IP? Some day, perhaps, IP, in general, may be the path to a single network for the server I/O. However, that will not happen overnight due to the support of legacy processes and practices and other reasons. For now, the stepping stone is Ethernet together with increased use of IP in addition to other technologies such as PCIe and NVMe for server I/O.

When should redundant paths be used in a network or fabric? We will discuss redundancy and resiliency in later chapters, but, in general, they should be used wherever you require availability or high availability. This includes from a server to a local device regardless of whether you use PCIe, SAS, SATA or Ethernet, Fibre Channel, InfiniBand, or NVMe, among others.

Also, redundant paths should usually be used on wide-area networks, depending on application PACE requirements. Keep in mind, though, that having multiple paths to a common network provides only local fault containment or fault isolation, as the common network will be a potential single point of failure (depending on its resiliencies. Also, path resiliency enables performance as well as load-balancing opportunities.

Can you explain time to first byte and its impact? This simply means how long (access, response time, or latency) before the first byte, block, blob, or object is accessed. Depending on where the data is located, there can be a slight delay for the first read to complete. However, subsequent reads or access can be faster, depending on cache, buffers, read-ahead, and pre-fetch, and on updates or writes, or write-back caching. We'll have more to say on this topic in Chapter 6.

5.9. Learning Experience

As a learning experience, use a tool such as ping to ping some known address. Use a tool such as ipconfig (Windows) or ifconfig (Linux) to display networking information including IP addresses that can then be ping. Review the Appendix for other learning materials that apply both to this topic and Chapter 6. Also, visit www.storageio.com/performance, where there is an evolving list of tools, technologies, techniques, and tips about server storage I/O networking performance, benchmarking, and related topics. Speaking of Chapter 6, continue learning there, as it is the companion to this chapter.

5.10. Chapter Summary

The good news is that I/O and networking are becoming faster, more reliable, and able to support more data movement over longer distances in a shorter time frame at a lower cost. As with server and storage technology improvements, the increase in networking and I/O capabilities is being challenged by continued demand to move or access more data over longer distances in even less time and at an even lower cost with reliability.

General action items include:

- Minimize the impact of I/O to applications, servers, storage, and networks.
- Do more with less, including improved utilization and performance.

- Consider latency, effective bandwidth, and availability in addition to cost.
- Apply the appropriate type and tier of server I/O and networking to the task at hand.
- Virtualize server I/O operations and connectivity to simplify management.
- Convergence of networking transports and protocols continues to evolve.
- Fast applications and storage need fast server I/O (bandwidth and low latency).

Bottom line: The best server I/O is the one you do not have to do; second best is the one that has the least overhead impact on your applications.

Chapter 6

Servers and Storage-Defined Networking

Are your servers and storage well connected?

What You Will Learn in This Chapter

- Server and storage I/O networking devices and components
- Direct attached, local, metropolitan, wide-area, and cloud access
- Hardware and software management tools for I/O networking
- Server and storage I/O access over distances

This chapter is a companion to the I/O generalities, software, and protocols covered in Chapter 5. In this chapter, the focus moves out further, looking at physical, virtual, and software-defined networks spanning local, metropolitan, wide-area, and cloud. Key themes and trends addressed in this chapter include data infrastructure physical networks and devices that support DAS, SAN, LAN, NAS, MAN, and WAN, among other capabilities. Some of the buzzwords include converged, unified, SDN, NFV, ULP, SAN, LAN, GbE, Fibre Channel, NVMe, NVMe over Fabric using RDMA along with NVMe over Fibre Channel (FC-NVMe), SAS, TCP/IP, InfiniBand, and related topics.

6.1. Getting Started

In Chapter 5 the focus was on general server input/output (I/O), including performance, protocols, and other software-related items. This chapter expands the conversation to look at server I/O hardware along with networks for local, metropolitan, and wide-area access. In addition to moving out further from the server, this chapter also extends from logical networks, including upper-level protocols, to cover physical networks and devices along with services. Given that everything is not the same in the data center, including different applications as well as other considerations, keep in mind that there are many options for servers to communicate with storage and other resources.

General server I/O characteristics include:

- Location—Physical, virtual, and cloud-based, including access
- Distance—Ranging from meters with PCIe to thousands of kilometers with WAN
- Functionality—Types of applications, protocols, quality of service (QoS), and features
- Interoperability—Media, types of transports and devices working together
- Reliability—No or as few errors as possible while being resilient to failures
- Management—Configuration, remediation, reporting, provisioning, policy enforcement
- Cost—Upfront and recurring capital as well as operating expenses
- Accessibility—What resources or services are available and where

Figure 6.1 shows common server storage I/O networking hardware and software components for both physical as well as software-defined cloud and virtual implementations. These components include device drivers and other software management tools in the physical and virtual servers. Physical and virtual adapters, along with switches, gateways, load balancers, and firewalls, as well as various servers (active directory [AD], DNS, DHCP, management, web, and database, among others) and storage are shown.

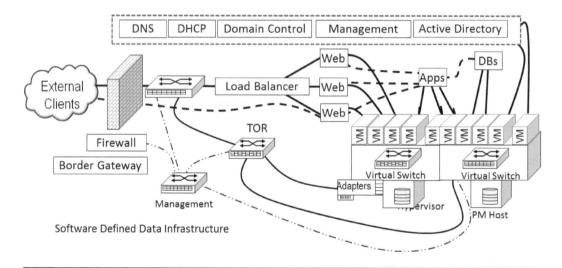

Figure 6.1   Server storage I/O networking and software-defined data infrastructures.

6.1.1. Tradecraft and Context for the Chapter

Network hardware needs software, software needs hardware including networking components as well as services. Networking services are made up of hardware and software defined to provide management, control, and data movement local and over distance. Similar to servers and storage, different types of I/O networking technologies, tools, and techniques support various needs because, of course, everything is not the same. Servers perform I/O operations with storage and other devices.

Context and tradecraft terms applicable to this chapter include the following:

- **Head-of-line blocking and blockage:** Something at the head of a line (HOL) or queue that blocks or slows down whatever is behind it—for example, a slow transaction, I/O request, get, or put that is taking a long time, thus slowing down faster traffic. The result of HOL blocking is blockage or congestion similar to what you would experience on an automobile highway that converges from many to a single lane during rush-hour traffic, or when there are slow-moving vehicles blocking or causing traffic delays.

  In the case of an I/O or network path, a slow transaction, IOP, read, write, get, or put, either local or remote in a serial sequence of activity, could cause whatever is behind it to run slower. The quality of service (QoS) and other management control techniques along with protocol implementations can help alleviate bottlenecks or their resulting impact.

- **Hose or pipe:** A physical network cable (copper electrical or fiber optic), wireless, or other connection including connection services from network bandwidth providers through which data flows. The term *pipe* relates to the notion that data flows like water through a pipe, that the network pipe is the cable or medium on a local or wide-area basis that data moves on, over or through— Hence filling the network or I/O pipe (bandwidth) with data (i.e., fully utilizing it).

  Another term for pipe is hose, as in a flexible garden hose that can mean a smaller, less rigid network or I/O pipe. You might hear the phrase "open the spigot, or hose, or fire hydrant" and data flowing, which means a large amount of data is moving fast. Note that with virtual and cloud servers, a network pipe can be implemented in memory as a virtual network connection, such as with a virtual switch.

- **Isochronous:** Recall from previous chapters that *asynchronous* refers to time-delayed or random communications where an application does not have to wait for notice that data is written (i.e., consistent), which is also known as eventual consistency. *Synchronous* refers to data being sent and waiting for receipt or consistency acknowledgment. Isochronous is another variation of communications in which data is sent in a steady stream such as from a video camera or other data collection device that is time-sensitive. Some examples of interfaces that support isochronous modes include I2C and SMbus on server motherboards and component devices for management (as opposed to data movement). Other examples include USB and Thunderbolt devices that can stream data from cameras or other data acquisition devices and instruments.

- **Links:** Links can be either physical, such as a connection between servers or to storage and networking devices, or logical. A logical link is established via a command protocol sequence to enable communication between applications or software on different systems via logical ports or sockets. A physical link connects via physical ports and sockets. Data flows over these physical and virtual links at different speeds or width, depending on underlying hardware and the software protocols and configuration being used.

- **Speed of light (photons) vs. speed of data (protocols):** With fiber optic communications it is common to hear that data moves at the speed of light. While the underlying physics and low-level technology enable photons of light to move at the speed of light, actual data movement is limited to the speed of the serial-deserializer (SerDes) and optical transceivers, interface, and protocols, as well as configurations. For example, a 1 GbE network using fiber optic cable operates at only 1 Gbps, whereas the speed of light is 299,792,458 meters per second.

While fiber optic cable and light photons can move faster than 1 Gbps, the communication equipment (switches, routers, gateways, servers, encoding), software, and protocol configuration limit the effective optical speed to 1 Gbps (or less). With faster communication equipment, servers, applications software, and configuration, faster speeds are possible using optical technology.

- **Time to first byte:** Often with I/O and networking performance for servers and storage, the focus is on how many IOPs, gets, puts, lists, reads, writes, updates, deletes, files, packets, frames, objects processed per second, or amount of bandwidth and latency. These are usually looked at in terms of average, peak, or minimum; burst or sustained; or standard deviation. However, another consideration is what is the time to the first or the last byte of data.

   Note that instead of first byte, the time can also be to the first block, blob, object, record, file, video, or other item. For some types of processing, the focus is on when the first byte or usable amount of data is ready or can be processed, while for other scenarios, the entire dataset must be presented to start processing. Also, many metrics look at the steady or bursty state of moving data and may not reflect ramp-up time to get to a sustained data or activity rate.

   Once read-ahead or pre-fetching starts to populate buffers, caches can become more effective on subsequent reads. Likewise, for writes or updates, write-behind, write gathering, or grouping, among other techniques, can accelerate these operations. For time-sensitive applications, keep in mind whether the focus is on sustained or burst rate, as well as time to first byte or last byte.

6.2. Server and Storage I/O Networking Devices

I/O and general-purpose data networks continue to converge to enable simplified management, reduce complexity, and provide increased flexibility of IT resource usage. Converged networks and virtualized I/O are taking place at both the server level internally with PCIe enhancements, as well as externally with Ethernet, Fibre Channel, and InfiniBand.

Various network and I/O transports and protocols have been on a course of convergence for several decades. These are used for implementing storage-defined networks including block, file, and object usage. The evolution has been from vendor-proprietary interfaces and protocols to open industry standards such as Ethernet, SAS, Fibre Channel, InfiniBand, and NVMe, among others.

Characteristics of server I/O devices includes:

- Physical port connector types
- Interface, transport, and protocols

- Number of ports and port types
- Bandwidth and performance
- Availability and path failover
- Distance enablement and interoperability

Server and storage I/O networking devices and software include:

- Device drivers, management, and monitoring software tools
- Unmanaged and managed switches and directors
- Bridges, routers, gateways, load balancers, firewalls, and gateways
- Distance enablement and bandwidth optimization
- Adapters, network cache, and acceleration devices
- Physical cabling, connectors, and conveyance systems
- Probes, analyzers, sniffers, test, and diagnostics equipment
- Wireless and Wi-Fi access points

6.2.1. Cabling and Connectors

Tying I/O and networks together, in addition to adapters, switches, bridges, routers, and gateways, are cabling in the form of copper or fiber optic and associated transceivers. There are many different types of copper and fiber optic cables, which have various price points and support for distance and performance. Likewise, there are different types of transceivers to match cable speeds and distances.

Transceivers are the technologies that attach to copper or fiber optic network cables and interface with adapters, NICs, switches, or other devices. You may recognize transceivers by different names such as RJ45, Cat "x," CFP2, MPO, GBIC, SFP, SFP+ and QSFP, QSFP+, QSFP-DD, MQSFP, FlexE, among others for attachment to Twinax (copper) or SMF and MMF (optics). Some transceivers are fixed, while others are removable—enabling switch or adapter ports to be reconfigured for different types of media. An example of a fixed transceiver is an Ethernet LAN port on your laptop or server.

Wireless networking continues to gain in popularity; however, physical cabling using copper electrical and fiber optic cabling continues to be used. With the increased density of servers, storage, and networking devices, more cabling is being required to fit into a given footprint. To help enable management and configuration of networking and I/O connectivity, networking devices, including switches, are often integrated or added to server and storage cabinets.

For example, a top-of-rack or bottom-of-rack or embedded network switch aggregates the network and I/O connections within a server cabinet to simplify connectivity to an end-of-row or area group of switches. Lower-performing servers or storage can use lower-cost, lower-performance network interfaces to connect to a local switch, then a higher-speed link or trunk, also known as an uplink, to a core or area switch.

Cable management systems, including patch panels, trunk, and fan-in, fan-out cabling for overhead and under-floor applications, are useful for organizing cabling. Cable management tools include diagnostics to verify signal quality and dB loss for optical cabling, cleaning and repair for connectors, as well as asset management and tracking systems.

A relatively low-tech cable management system includes physically labeling cable endpoints to track how the cable is being used. Software for tracking and managing cabling can be as simple as an Excel spreadsheet or as sophisticated as a configuration management database (CMDB) with an intelligent fiber optic management system. An intelligent fiber system includes mechanisms attached to the cabling to facilitate tracking and identifying cabling.

Figure 6.2 shows some examples of things that are in my server storage I/O networking toolbox, including various cables, connectors, interposers, gender changers, software tools, and utilities (on CD or USB thumb drives). Some of the cables and connectors shown include Ethernet RJ45, QSFP for InfiniBand or Ethernet, SFF-8470 for SAS or InfiniBand, USB-to-SATA converter cable, and SFF-8643 (Internal) and SFF-8644 (External) SAS Mini-HD. The 8643 connector is also used for PCIe x4 to NVMe 8639 and for SAS.

Also shown in Figure 6.2 are SFF-8687 (mini-SAS internal); not shown is the external SFF-8688 connector. The SAS interposer shown at the lower right enables an SAS device to connect to an SATA backplane connector that is wired or connected to an SAS controller.

Note that the SAS-to-SATA interposer works *only* with physical SATA backplanes or connectors that *are attached* to SAS controllers; because of the nature of the protocol, SAS cannot connect to SATA, the connection works only the other way. Also shown are physical tools such as screwdrivers, magnifying glass, and cable tie straps with labels.

Figure 6.2 shows various tools, cables, connectors, USB thumb drives, and CDs, among other items that are part of my server storage I/O networking toolbox. Not shown are various software tools, utilities, and drivers for configuration, monitoring, reporting, and testing.

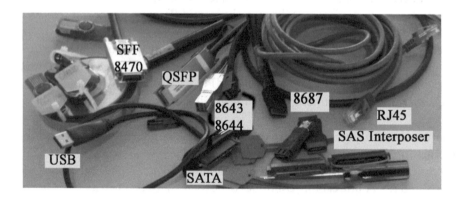

Figure 6.2   What's in my server storage I/O networking toolbox.

6.2.2. Host Bus Adapters and Network Interface Cards

Host adapters provide connectivity between a server and some I/O network such as Ethernet, Fibre Channel, InfiniBand, SAS, or SATA, among others. Also known as host bus adapters (HBAs), host channel adapters (HCAs) and network interface cards (NICs) usually attach to a server's PCIe bus.

Servers are not the only resources that use HBAs in some shape or form. Some storage systems simply use off-the-shelf adapters with drivers interfaced to their storage operating systems. Other storage systems have dedicated fabric attachment (FA) or host-side ports on director or controller cards that are used for attaching to servers or SANs and back-end ports for attaching to storage devices.

Adapter attachment to the PCIe or processor bus may be via an add-in-card (AiC), or via a server motherboard surface-mount attachment point, or via a support chip. Also, to supporting various protocols and interfaces, some adapters support more than one I/O network interface and protocol or can be reprogrammed in the field. For example, some converged network adapters (CNA) can be put into Converged Ethernet or InfiniBand, while others can vary between Ethernet and Fibre Channel.

Adapters have a unique lower-level (physical or virtual) address such as MAC, as well as upper-level protocol identifiers. For example, Ethernet, NICs or adapters have a 48-bit MAC address, while Fibre Channel, SAS, SATA, and other devices have unique worldwide names (WWN).

Note that, historically, MAC addresses were fixed into devices' hardware or firmware, but today some devices allow changing the MAC addresses. Similarly, virtual adapters can also have their unique generated addresses changed to meet different needs.

An adapter configured as target functions as a northbound front end responding to storage or other resource requests. Servers can also be configured with both front-end northbound-facing adapters as targets as well as back-end southbound-facing adapters as initiators to target NVM SSD and HDDs, among other storage devices.

For example, a server with adapters that are configured for front-end target and back-end initiator defined with various storage software would become a storage system or appliance. Storage can be internal and dedicated to a server, or external and shared. Servers can also share storage with two or more other servers in a cluster with either DAS or networked storage. Various types of adapters are shown in Figure 6.3.

Some adapter ports are surface-mounted on the motherboard, including LAN-on-motherboard (LOM) along with SAS, USB, and SATA, among others. Given the many different types of server packaging, there are also various types of physical adapters ranging in height, length, and form factor.

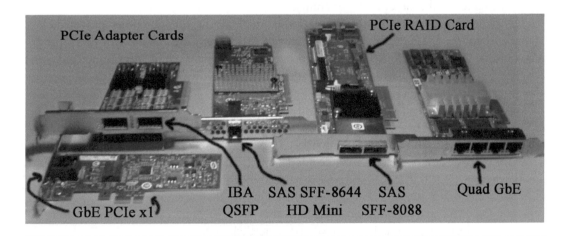

Figure 6.3 Various types of PCIe adapters.

Adapters such as for blade servers are mezzanine cards that plug into an upper-level slot or rise. In addition to CNAs, some adapters are referred to as universal target adapters, while others can be put into either target or initiator mode.

Some vendors ship hardware adapter or RAID cards already installed in servers that are partially or fully deactivated. When needed, you can acquire a software (or physical) key that enables additional functionality or simply turns the card on. What functionality and the cost to enable it will vary by server vendor and card or adapter manufacturer. For example, some server motherboards may have SATA ports enabled by default. However, there may be an SAS controller that is activated via a separate physical or software configuration key. Another example is a RAID adapter that may have basic RAID levels enabled, while advanced functions, including erasure code or rapid rebuild, caching, or other functions, may require an upgrade key.

Mezzanine and daughter cards are technologies that enable access to external I/O and networking resources. Physical servers or physical machines utilize physical adapters for performing I/O and networking functions, including access to other devices on a local or remote basis. Virtual servers or virtual machines typically access a virtual adapter that in turn uses a physical adapter or NIC via a Hypervisor or abstraction layer.

Adapter cards vary in terms of the number of ports they have, port protocol, and physical connector type, narrow or wide ports, as well as PCIe slot requirements (physical, mechanical, and electrical). Some adapters have multiple ports that use a single controller on the card to increase connectivity, although this configuration may become a performance bottleneck. Other adapters have multiple ports but also two or more controller chipsets and wide PCIe connectivity to maintain performance. Speaking of protocols, adapter cards may be dedicated to a particular protocol or interface such as Fibre Channel or GbE or SAS and SATA, or support multiple protocols as a converged and reprogrammable adapter.

Note that the number of ports times their port and protocol speed may not determine performance—it's what's on or inside the card that does. Performance and functionality will also vary by the type of card as well as the amount of cache (and or battery backup), speed, device hot-swap, pass-through modes, and server boot, among other options, and software support (device drivers functionality, hypervisors, operating systems, or cloud stack) including multipathing for load balance and availability.

The physical packaging of adapters varies as much as the functionalities provided. For example, there are different physical form factor PCIe adapter cards, along with ports, protocols, buffers, off-load, and other functionalties. Some adapters, such as mezzanine cards, are designed for the small dense footprint of blade servers, while others support multiple ports to maximize PCIe expansion port capabilities.

Historically, adapters have been single-purpose, such that an InfiniBand HCA could not be a Fibre Channel HBA, or a Fibre Channel HBA could not be an Ethernet NIC or an Ethernet NIC could not be an SAS adapter, though they all plugged into a PCIe bus. What is changing with CNAs is the ability to have a single adapter with multiple ports that can support traditional Ethernet LAN traffic including TCP/IP on the same physical port as FCoE using DCB as peers. Some CNA adapters also support the ability to change a Fibre Channel port to an Ethernet and FCoE port for redeployment purposes. This flexibility enables an adapter to be purchased and then repurposed or provisioned as needed.

Some adapters can be removed from PCIe expansion slots or as mezzanine cards in blade servers, while others are fixed onto mother or main circuit boards on servers or storage system

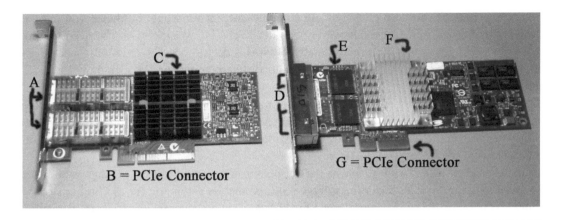

Figure 6.4 Common adapter card components.

controllers. A marketing term used by some vendors is LAN-on-motherboard (LOM). Note that some adapter cards also include hardware RAID with various functionalities along with caching and I/O acceleration.

In Figure 6.4 on the left is a dual port (QSFP) QDR (quad data rate) InfiniBand (56Gb) card (A). This card can be configured either as a dual-port InfiniBand device or as a 40GbE Converged Ethernet device via software settings. Also shown is a PCIe Gen 3 x8 card connector for the InfiniBand, Converged Ethernet card (B). The controller (C) for the card on the left resides under the large heat sink that looks like small cooling towers. The SerDes for the QSFP ports sits between the physical connection and the controller on the card.

On the right in Figure 6.4, the quad-port GbE card operates only in Ethernet mode. SerDes and port chipsets (E) sit between the physical port connectors (D) and the controller (F) in this card.

From a performance standpoint, pay attention to what the adapter card is capable of doing as well as potential bottlenecks or points of contention. Are there more ports of a given type and speed than what the controller chip (ASIC or FPGA) can handle with the PCIe interface, creating bottlenecks, or are the resources properly balanced? For example, if there is a pair of 40GbE ports on an adapter card, yet the controller chip only has the internal performance of a single 40GbE path, that will cause a bottleneck if both ports are active at the same time.

Another bottleneck example is a fast physical port (or more) on a card with a fast controller chip that interfaces with the server motherboard via a slow PCIe connection or configuration. In Figure 6.4 the card on the left also supports CPU offload for protocol handling including RDMA along with other acceleration, and jumbo (i.e., MTU) frames, among other features.

For applications that do not need the performance of all ports at their maximum speed, oversubscribed configurations can provide cost-effective connectivity options. However, be careful not to use a single card with multiple ports, thinking that those ports give you full redundancy even if attached to different devices, switches, or networks. While you will have path resiliency, a single adapter card will also be a single point of failure. Granted, you could set up two or more servers in a cluster, so that loss of an adapter on a server would cause a failover to another cluster node.

You might ask whether it is more cost- and service-effective to have dual adapters and eliminate the need to failover. On the other hand, what about using the single adapter with a failover cluster (if you have one)? For some people, the answer would be that if you have a cluster, why not failover? Others, however, would ask why failover if you do not need to and can isolate the fault locally with redundancy. What this decision comes down to is what level of availability and performance you need for your different applications, balancing cost, risk, and business impact or benefit.

Unless there is no need for resiliency, dual or multiple HBAs should be used and attached to redundant fabrics. Also, to provide redundancy and resiliency, multiple adapters and I/O paths can also improve availability by supporting failover during maintenance of different components.

Path manager software, available as part of operating systems, volume managers, and file systems or via third parties including storage vendors, enables automatic failover for high availability on adapters. In addition to failover for HA, path managers can provide load balancing across the ports and metrics for management purposes. Additional functionality can be added to path managers, including encryption, compression, or cloud access.

In addition to the physical adapters, there are also device drivers whose functionality will vary depending on a given operating systems or hypervisor. Hypervisors also support virtual adapters that attach to virtual and physical devices. There are also paravirtual device drivers and adapters that are further optimized for performance with a given hypervisor.

Note that some hypervisors for their virtual storage software may require that RAID adapters be placed into a pass-through mode (i.e., functioning like an HBA), treating the storage as JBOD ("just a bunch of disks"), with the software doing RAID and other data protection.

Why use a RAID adapter if it is going to be used as a JBOD or HBA? For some operating systems or hypervisors and virtual storage or software-defined storage, the answer may be to get some buffering or additional device queues to boost performance for SAS and SATA devices.

In addition to device drivers, software configuration, operating system, hypervisor, and target or destination device compatibility, another performance and functionality consideration for adapters is their firmware version. Verify your adapter's device driver, firmware, and other support functionalities with the manufacturer, as well as operating systems, hypervisors, and third-party vendors' hardware compatibility lists (HCLs).

6.2.3. Switches and Directors

In addition to adapters for servers and access ports on storage systems (or other access points or targets), other networking devices include switches and directors. Switches (which includes directors) are connected via cabling (copper or optical) to servers, storage, and other networking devices.

Similarly, virtual switches are connected via logical or in-memory connections to a physical adapter that in turn connects via a physical connection to a physical switch. Switches can be used simply to move data between ports, or managed to provide additional configuration, monitoring, and other capabilities. Fundamentally, switches forward frames or packets from a source to a destination. The destination can be singular or plural, as in broadcast. Virtual switches such as in a Hyper-V environment can be configured to access external networks, sharing a path with other applications on the underlying host server. Virtual switches can also be internal for communication between VM, or private, among other configuration options.

There are also different layer switches, including for physical layer 1, frame layer 2, and protocol routing layer 3. Like adapters and other components, switches can be physical with proprietary software packaged into ASIC and FPGA or using general-purpose commodity hardware as well as virtual or software-defined. For example, hypervisors such as VMware vSphere ESXi and Microsoft Hyper-V have virtual switches that exist in memory defined by software. There are also virtualized switches that use software deployed on commodity server hardware with multiple physical network ports.

Switches and directories can be dedicated to a specific protocol with different types of physical media types, or can support multiple protocols and interfaces. Some examples of common data infrastructure switches include Ethernet, Fibre Channel, InfiniBand, PCIe, and SAS. Note that in the case of SAS, another name for a switch is an expander—or another name for an expander is a switch. For example, some storage enclosures or drive shelves (including servers) for SAS HDD and SSD have expanders as part of their backplane.

Expanders are found in storage server systems as well as storage enclosure backplanes as well as on some adapter cards. Modular switches and directors support different cards or plug-in modules to support various features including multiprotocol, routing, bridging, or security, among others. Switches and directors also have support for different physical media including fiber optic and copper electrical transceiver types and speeds.

A switch (or director) is a device that performs layer 2 switching functions between storage networking devices. Unlike a hub concentrator with shared bandwidth between ports, a switch provides dedicated bandwidth between ports. Switches and directors can be used as standalone or single-device fabrics, in pairs to create multiple single-device SAN islands, or connected to create fabric.

Fabrics or networks are used to increase the number of ports beyond what is physically available on fabric devices and to support different topologies for various applications and environments. Switches and director ports can also isolate local traffic to particular segments, much like traditional network switches isolate LAN traffic.

Storage networking switching devices can range from simple four-ports to large multi-protocol devices with hundreds of ports. A Fibre Channel switch provides the same function as a standard network switch in that it provides scalable bandwidth for various subnets or segments. Converged switches support two or more different types of networks such as Fibre Channel and Ethernet, or Ethernet and InfiniBand.

A networking director is a highly scalable, fully redundant switch supporting blades with multiple interfaces and protocols, including Fibre Channel, Ethernet, and Fibre Channel over Ethernet, FICON and wide area, bridging, or other access functionality. Directors vary in port size and can replace or supplement fabrics of smaller switches. Fabric devices, including directors and switches, can be networked into various topologies to create very large resilient storage networks of hundreds to thousands of ports.

Interswitch links (ISLs) are used to connect switches and directors to create a fabric or SAN. An ISL has a port on each switch or director to which the ISL is attached. For redundancy, ISLs should be configured in pairs. For physical switches and directors, there are also software-defined and virtual switches that are part of hypervisors and cloud stacks such as OpenStack among others.

A core switch sits at the core of a network as a single device or with other switches attached in various topologies. Often a core switch will be a physically large device with many network connections, trunks, or ISLs attached to it and edge or access switches along with storage or other target devices. In some scenarios, a core switch may be a physically smaller device functioning as a convergence point between the large edge or access switches and directors and other switches that provide access to storage or other targets.

Edge or access switches, also known as top-of-rack (TOR), middle-of-row (MOR) or end-of-row (EOR) switches, vary in size and provide an aggregation point for many servers to fan-in to storage systems or other devices connected to the core of a network.

Historically, edge or access switches have been physically smaller as a means to reduce cost and aggregate multiple slower servers onto a higher-speed ISL or trunk attached to a larger core switch or director. For example, ten standalone servers may each need only about 1 GbE of bandwidth or equivalent IOPS, packets, or frames per second performance.

6.2.4. Bridges and Gateways

Other networking devices include gateways, bridges, routers, and appliances. These devices enable bridging between different types of networks handling interface, protocol, data translation, and other functionalities on a local- or wide-area basis.

Gateways, bridges, routers, and appliances include:

- Physical medium and logical protocol conversion
- SAN, LAN, MAN, and WAN distance enablement devices
- Cache and performance acceleration devices
- Bandwidth and protocol or application optimization
- Protocol conversion (SAS to FC, FC to FCoE, USB to Ethernet)
- Security, including firewalls and encryption devices
- Cloud point-of-presence appliances, gateways, and software
- Storage services and virtualization platforms

6.3. Local Server I/O and Storage Access

Servers access local storage including dedicated internal and external direct attached storage (DAS) as well as networked block, file, or object storage. Storage is accessed using ports on motherboards, mezzanines, and PCIe AiC host and RAID adapters. Adapters can be initiators that perform southbound access of storage resources. In addition to being configured as initiators, some servers' adapters can also be configured as targets.

Local server storage access includes:

- PCIe (internal AiC) and NVMe (AiC, M.2, and U.2 or SFF-8639 drives)
- SAS, SATA, mSATA and M.2, Thunderbolt, and USB
- Ethernet, Wi-Fi, and Bluetooth

Note that the above, in addition to being used for local or direct attached server and storage I/O connectivity, can also be used in broader configurations options. For example, while PCIe

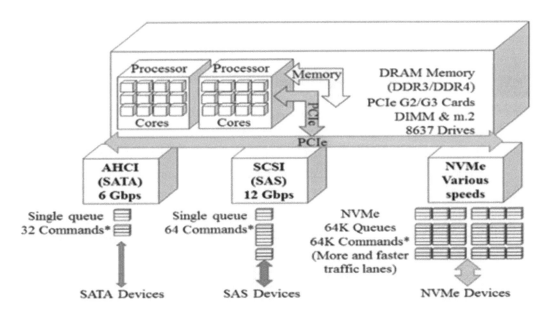

Figure 6.5 Various local server storage interfaces (PCIe, SATA, SAS, and NVMe).

has been common for use inside servers, today it is also being extended to interserver intra-cabinet configurations using NVMe.

SAS is also used both for local direct attached dedicated storage as well as shared including SAS SAN, such as with Microsoft SOFS among other scenarios. SAS, SATA, and NVMe are also used for front-end connectivity, as well as back-end such as in storage systems or servers for accessing storage devices.

Figure 6.5 shows common server I/O connectivity including PCIe, SAS, SATA, and NVMe.

Table 6.1 shows relative server I/O performance of some NVM flash SSD devices across various workloads. As with any performance comparison, take these data and the following with a grain of salt, as your speed will vary. The results shown in Table 6.1 compare the number of IOPs (activity rate) for reads and writes, random and sequential, across small 8 KB and large 1 MB I/Os.

Also shown in Table 6.1 are bandwidth or throughput (amount of data moved), response time, and the amount of CPU used per IOP. Note how NVMe can achieve higher IOPs with a lower CPU per IOP, or, using a similar amount of CPU, do more work at a lower latency.

The workload results in Table 6.1 were generated using a vdbench script running on a Windows 2012 R2 server and are meant to be a relative indicator of different protocols and interfaces; your performance will vary. Refer to www.storageio.com/performance as well as the Appendix for more details on vdbench and other related tools.

SSD has been used for decades to help alleviate CPU bottlenecks or defer server upgrades by reducing I/O wait times and CPU consumption (wait or lost time). With NVMe, the capabilities of underlying NVM and storage memories are further realized. Devices used include a PCIe x4 NVMe AiC SSD, 12 Gb SAS SSD, and 6 Gb SATA SSD. What determines which technology to use includes distance, protocol, and device supported, required functionality,

Table 6.1. Relative Performance of Various Protocols and Interfaces

		8 KB I/O Size				1 MB I/O Size			
NAND flash SSD		100% Seq. Read	100% Seq. Write	100% Ran. Read	100% Ran. Write	100% Seq. Read	100% Seq. Write	100% Ran. Read	100% Ran. Write
NVMe	IOPs	41,829.19	33,349.36	112,353.6	28,520.82	1,437.26	889.36	1,336.94	496.74
PCIe	Bandwidth	326.79	260.54	877.76	222.82	1,437.26	889.36	1,336.94	496.74
AiC	Resp.	3.23	3.90	1.30	4.56	178.11	287.83	191.27	515.17
	CPU/IOP	0.001571	0.002003	0.000689	0.002342	0.007793	0.011244	0.009798	0.015098
12 Gb	IOPs	34,792.91	34,863.42	29,373.5	27,069.56	427.19	439.42	416.68	385.9
SAS	Bandwidth	271.82	272.37	229.48	211.48	427.19	429.42	416.68	385.9
	Resp.	3.76	3.77	4.56	5.71	599.26	582.66	614.22	663.21
	CPU/IOP	0.001857	0.00189	0.002267	0.00229	0.011236	0.011834	0.01416	0.015548
6 Gb	IOPs	33,861.29	9,228.49	28,677.12	6,974.32	363.25	65.58	356.06	55.86
SATA	Bandwidth	264.54	72.1	224.04	54.49	363.25	65.58	356.06	55.86
	Resp.	4.05	26.34	4.67	35.65	704.70	3,838.59	718.81	4,535.63
	CPU/IOP	0.001899	0.002546	0.002298	0.003269	0.012113	0.032022	0.015166	0.046545

connectivity, and performance, as well as the cost for a given application. Another consideration is what a given server or storage system supports as connectivity options both physically as well as logically (i.e., software protocols and drivers).

Various NVM flash SSD devices are shown in Figure 6.6. At the left in Figure 6.6 is a NAND flash NVMe PCIe AiC, at top center is a USB thumb drive that has been opened up to show a NAND die (chip). In the middle center is an mSATA card, bottom center is an M.2 card. Next on the right is a 2.5-in. 6 Gb SATA device, and on the far fright is a 12 Gb SAS device. Note that an M.2 card can be either a SATA or NVMe device, depending on its internal controller, which determines which host or server protocol device driver to use.

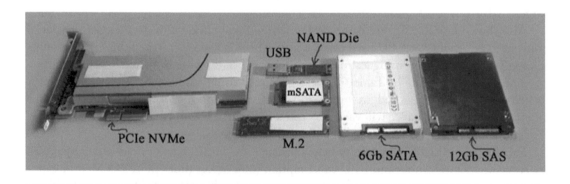

Figure 6.6   Various NVM flash SSDs.

6.3.1. PCIe, M.2, and U.2 (SFF-8639) Drive Connectors

The role of PCIe has evolved over the years, as has its performance and packaging form factors. Building off what was covered in previous chapters pertaining to PCIe, in addition to its role for attaching adapters and other devices via AiC slots, PCIe form factors also include the M.2 small form factor that replaces legacy mini-PCIe cards. M.2 (also known as next-generation form factor or NGFF) is a relatively new form factor. NGFF can be used to support SATA as well as NVMe, depending on the card device installed and host server driver support.

There are various M.2 NGFF form factors, including 2230, 2242, 2260, and 2280. M.2-to-regular physical SATA converter or adapter cards are also available that enable M.2 devices to attach to legacy SAS/SATA RAID adapters or HBA.

Another PCIe form factor and connector is the U.2, also known as an SFF-8639, shown in Figure 6.7. On the left of Figure 6.7 is a view toward the backplane of a storage enclosure in a server that supports SAS, SATA, and NVMe U.2. (SFF 8639). On the right is the connector end of an 8639 NVM SSD showing addition pin connectors compared to an SAS or SATA device. Those extra pins provide PCIe x4 connectivity to the NVMe devices. The U.2 (8639) drive connectors enable a device such as an NVM or NAND flash SSD to share a common physical storage enclosure with SAS and SATA devices, including optional dual pathing.

A bit of tradecraft here is to be careful about judging a device or component by its physical packaging or interface connection. In Figure 6.6 the device has SAS/SATA along with PCIe physical connections, yet it's what's inside (i.e., its controller) that determines whether it is a SAS-, SATA-, or NVMe-enabled device. This also applies to HDD and PCIe AiC devices, as well as I/O networking cards and adapters that may use common physical connectors yet implement different protocols. For example, the SFF-8643 HD-mini SAS internal connector is used for 12 Gbps SAS attachment as well as PCIe to devices such as 8630.

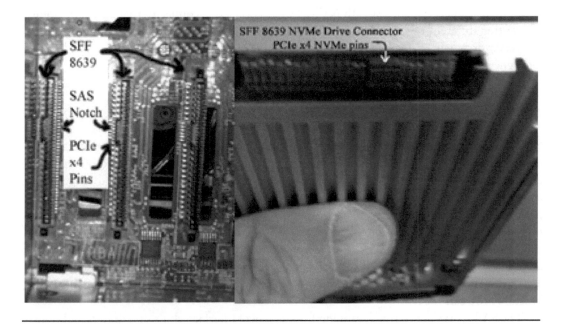

Figure 6.7   PCIe NVMe U.2 (SFF 8639) drive.

Depending on the type of device inserted, access can be via NVMe over PCIe x4, SAS (12 Gb or 6 Gb), or SATA. The U.2 connector enclosures have a physical connection from their backplanes to the individual drive connectors, as well as to PCIe, SAS, and SATA cards or connectors on the server motherboard or via PCIe riser slots.

While PCIe devices including AiC slot-based, M.2, or U.2 can have common physical interfaces and lower-level signaling, it's the protocols, controllers, and drivers that determine how software is defined and used. Keep in mind that it's not just the physical connector or interface that determines what a device is or how it is used, it's also the protocol, command set, and the controller and device drivers.

6.3.2. Serial ATA (SATA)

Serial ATA (SATA) is also known as Advanced Host Controller Interface (AHCI) and is the successor to parallel ATA (PATA). SATA is a block server I/O protocol and connector type for accessing storage devices including HDD and NVM NAND flash SSD. SATA implementations include inside servers, external (eSATA), as well as small-form-factor cards for mobile devices (mSATA). However, mSATA systems are giving way to M.2 solutions.

Note that for physical connection, a SATA device can attach to a SAS port, but not the other way around.

There are also virtual SATA and AHCI along with extended AHCI devices and drivers, for example, for the VMware vSphere hypervisor. Some context is needed in that there are SSD and HDD that have SATA interfaces and lower-cost, lower-performing internal functionality. However, there are also higher-performing and more durable HDD and SSD with SATA interfaces. Again, be careful about judging a device simply by its interface; instead, understand what the device itself is capable of doing, or how it performs.

Note that SATA can coexist with NVMe, such as via a U.2 connector that allows flexibility of inserting different types of devices (NVMe, SAS, SATA) into a common storage enclosure in a server or storage system. In the near term, SATA HDD and SSD will continue to be used for higher-capacity and cost-sensitive deployments, coexisting with NVMe and SAS.

6.3.3. Serial Attached SCSI (SAS)

Serial attached SCSI (SAS) is as an interface for connecting HDD to servers and storage systems; it is also widely used for attaching storage systems to servers. SAS based on the SCSI command set continues to evolve, with support for faster, 12 Gbps speed and longer cable lengths with good interoperability and price-to-performance ratio. The combination of price, performance, shared connectivity, and distances is well suited for converged, clustered, and high-density blade or hyper-converged server environments as a stepping stone to NVMe.

SAS is being used as a means of attaching storage devices to servers as well as a means of attaching HDD to storage systems and their controllers. Various SAS (and SATA) components are shown in Figure 6.8. Note that when using the U.2 connector, SAS, SATA, and PCIe-based NVMe can coexist.

The SAS ecosystem is broad, ranging from protocol interface chips, interposers and multi-plexers to enable dual port of SATA disks to SAS interfaces, SAS expanders, and SAS host PCIe

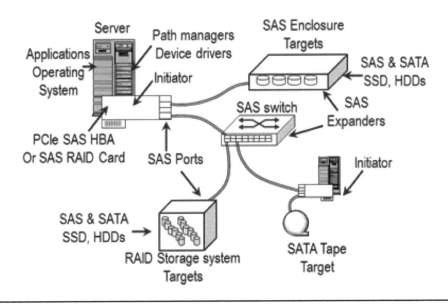

Figure 6.8 SAS and SATA components.

as well as internal and external RAID controllers, 3.5-in. and 2.5-in. SAS SSD and HDD and other components. SAS HDD are being deployed inside storage systems across different price bands and market segments for both block-based and NAS file-based storage solutions.

For high-density scale-up and scale-out environments, storage is moving closer to servers in the form of shared and switched SAS. While cloud service and the servers providing the functionality may be located some distance from the servers accessing them, a good cost-effective back-end storage solution is being used increasingly on shared or switched SAS. In the past, there was a gap in terms of connectivity or number of servers that could be attached to a typical shared SAS or DAS storage system. This has changed with the increase of native 12 Gbps ports and using the SAS switch to increase the fan-out (from storage to server) or fan-in (servers to storage) number of attached servers.

Another example of storage getting closer to servers is virtualization that leverages industry-standard processes with external storage. Last, but not least, is consolidation, which results in servers and storage coming closer together.

Figure 6.9 shows configurations of shared and switched 12 Gbps SAS storage supporting different application or environment needs. On the left side of Figure 6.9 are six single attached servers with a dual attached NAS gateway or storage appliance. A growing industry trend is the use of NAS file serving appliances that can attach to shared or switched storage, such as SAS, for hosting unstructured data including virtual servers. The six servers are configured for non-high availability (HA), while the NAS storage is configured for HA.

In the middle of Figure 6.9 are three servers dual-attached in an HA configuration with a NAS gateway. The NAS devices that are shown in Figure 6.9 leverage the shared direct attached or switched SAS storage, supporting traditional and virtual servers. For example, two of the servers, the physical machines in the middle of Figure 6.9, are configured as a cluster hosting virtual machines for hypervisors or PM. On the right side of Figure 6.9, the configuration is

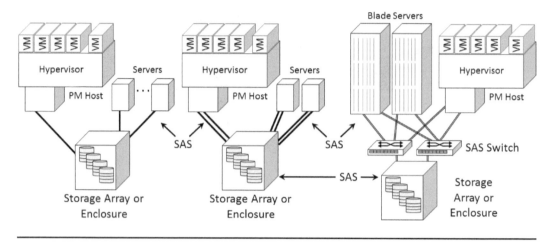

Figure 6.9   Shared and switched SAS.

enhanced by adding a pair of 12 Gbps 16-port SAS switches and expanders and a high-density blade system. Other examples of shared SAS SAN are Microsoft SOFS (Scale-Out File Server) or Storage Spaces Direct (S2D) in Windows Server.

6.3.4. NVM Express (NVMe)

NVMe is a new protocol alternative to AHCI/SATA and the SCSI protocol used by SAS. NVMe has been designed from the ground up for accessing fast storage, including flash SSD, leveraging PCI Express (PCIe). The benefits include lower latency, improved concurrency, increased performance, and the ability to unleash a lot more of the potential of modern multi-core processors.

NVMe, leveraging PCIe, enables modern applications to reach their potential. NVMe is one of those rare, generational protocol upgrades that comes around every couple of decades to help unlock the full performance value of servers and storage. NVMe does require new drivers, but once it is in place, it plugs and plays seamlessly with existing tools, software, and user experiences.

While SATA and SAS provided sufficient bandwidth for HDD and some SSD uses, more performance is needed. Near term, NVMe will not replace SAS or SATA—they can and will coexist for years to come, enabling different tiers of server storage I/O performance.

NVMe unlocks the potential of flash-based storage by allowing up to 65,536 (64K) queues each with 64K commands per queue. SATA allows for only one command queue capable of holding 32 commands per queue, and SAS supports a queue with 64K command entries. As a result, the storage IO capabilities of flash can now be fed across PCIe much faster, to enable modern multicore processors to complete more useful work in less time.

These and other improvements with NVMe enable concurrency while reducing latency to remove server storage I/O traffic congestion. The result is that applications demanding more concurrent I/O activity along with lower latency will gravitate toward NVMe to access fast storage.

Initially, NVMe is being deployed inside servers as back-end, fast, low-latency storage using PCIe add-in-cards (AIC) and flash drives. Similar to SAS, NVM SSD, and HDD that support

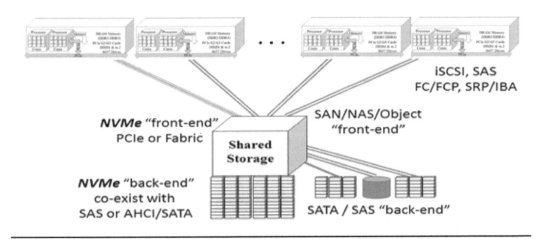

Figure 6.10   Various configuration options using NVMe.

dual-paths, NVMe has a primary path and an alternate path. If one path fails, traffic keeps flowing without causing slowdowns. This feature is an advantage to those already familiar with the dual-path capabilities of SAS, enabling them to design and configure resilient solutions.

NVMe devices, including NVM AIC flash, will also find their way into storage systems (Figure 6.10) and appliances as back-end storage, coexisting with SAS or SATA devices. Another emerging deployment configuration scenario is shared NVMe direct attached storage with multiple server access via PCIe external storage with dual paths for resiliency.

Even though NVMe is a new protocol, it leverages existing skill sets. Anyone familiar with SAS/SCSI and AHCI/SATA storage devices will require little or no training to implement and manage NVMe. Since NVMe-enabled storage appears to a host server or storage appliance as a LUN or volume, existing Windows, Linux, and other OS or hypervisor tools can be utilized.

On Windows, for example, other than going into the device manager to see what the device is and what controller it is attached to, it is no different from installing and using any other storage device. The experience on Linux is similar, particularly when using in-the-box drivers that ship with the OS. One minor Linux difference of note is that instead of seeing a /dev/sda device as an example, you might see a device name like /dev/nvme0n1 or /dev/nvme0n1p1 (with a partition).

Keep in mind that NVMe, like SAS, can be used as a back-end access from servers (or storage systems) to a storage device or system (Figure 6.10), for example, JBOD SSD drives (U.2), PCIe AiC, or M.2 devices. Like SAS, NVMe can also be used as a front end on storage systems or appliances in place of, or in addition to, other access such as GbE-based iSCSI, Fibre Channel, FCoE, InfiniBand, NAS, or object. Note NVMe can also run on Fibre Channel (FC-NVMe) as well as Ethernet and InfiniBand using RDMA.

What this means is that NVMe can be implemented in a storage system or appliance on both the front end—that is, the server or host side—as well as on the back end—the device or drive side, which is similar to SAS. Another similarity to SAS is NVMe dual pathing of devices, permitting system architects to design resiliency into their solutions (Figure 6.11).

Also shown in Figure 6.11 are SAS and NVMe devices with a primary (solid line) and a second, dual path (dotted line) for resiliency. If the primary path fails, access to the storage device can be maintained with failover so that fast I/O operations can continue while using SAS and NVMe. The website for the industry trade group for NVM Express is www.nvmexpress.org.

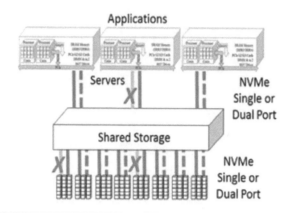

Figure 6.11   NVMe with dual-path resiliency.

6.3.5. Ethernet (IEEE 802)

Ethernet, also known by 802.x (where x is various extensions) is part of an IEEE standard along with IETF as a layer 2 (L2) data link. Ethernet is a popular industry standard network-ing interface and transport used in consumer as well as enterprise environments for a variety of different applications.

The popularity of Ethernet can be traced to its interoperability and affordability, which have helped make it a ubiquitous networking standard that continues to evolve. Ethernet has emerged and replaced various proprietary networking schemes over the past several decades while boosting speeds and increasing interoperability.

Ethernet is deployed on various physical media including fiber optic and copper electrical cabling as well as wireless (Wi-Fi). Ethernet also is available via various Telco's or network ser-vice providers on a metropolitan basis called Metro-Ethernet, among other marketing names. This means that Ethernet service is provided directly to your premise without the need to have a MPLS, SONET/SDH, WDM, DSL, ATM, or another gateway, modem, dialer, or router.

Ethernet speeds include legacy 10/100 Mbps, 1000 Mbps or 1 GbE, 2.5 and 5 GbE (NBASE-T uses Cat 5e and Cat 6 cabling for distances up to 100 meters), 10, 25, 40, 50, 100, and 400 GbE. Depending on the implementation, a 40 GbE network can be made up of 4 × 10 GbE or 2 × 25 GbE links; a 100 GbE can be made up of 10 × 10 GbE links aggregated or 4 × 25 GbE links. A 50 GbE can be made up of 2 × 25 GbE links.

In addition to various speeds, cabling media, and topologies, Ethernet has several other capabilities, including link aggregation, flow control, quality of service, virtual LAN (VLAN), and security. Other capabilities include support for power over Ethernet (PoE) to provide elec-trical power to low-powered devices, simplifying cabling and management. No longer asso-ciated with just LAN access of computers and applications, Ethernet is also being used for metropolitan and wide-area services and supporting storage applications.

Ethernet has come a long way, running on both copper and optical networks at speeds up to 100 GbE (400 GbE in development) for trunks, as well as being used for Wi-Fi networks. With IEEE data center bridging (DCB) enhancements including 802.1Qbb (priority flow control, PFC), 802.1Qaz enhanced transmission selection, ETS), DCB capabilities exchange protocol (DCBX), and others, the platform for enabling additional storage and LAN consoli-dation is developing.

With DCB and enhanced Ethernet, marketed as Data Center Ethernet (DCE) or Converged Enhanced Ethernet (CEE), Ethernet adds traditional storage interface characteristics to address concerns about deterministic performance and lossless data transmission. Today's Ethernet is a far cry from what was used or even available a decade or more ago.

The Spanning Tree Protocol (STP), IEE 802.1D, enables a loop-free logical topology when physical paths for redundancy exist. STP has enabled Ethernet to have redundant links that are not used for active data movement until a failover, when they are automatically enabled.

A challenge with STP is that, unlike Fibre Channel, which can leverage redundant links to improve performance using Fabric Shortest Path First (FSPF), standby links are nonutilized resources. To improve resource usage or maximize ROI on networking expenditures as part of a converged networking environment, IETF RFC 5556 or Transparent Interconnect of Lots of Links (TRILL) and many other Ethernet features exists. Note RDMA over Converged Ethernet (RoCE) enables NVMe support.

Note that some servers have their adapters set to "no promiscuous mode" to avoid having to see lower-level network (L2) traffic not destined for them (to offload performance). For servers with virtual machines, putting adapters (physical or virtual) into promiscuous mode allows those VM to see traffic and determine whether it is for them. For example, a server host adapter that also has a VM guest that is not in promiscuous mode might reject a packet not destined for the host, yet destined for the guest VM.

Refer to your operating system and hypervisor provider's best practices and network configuration guides for specifics about when and where to enable or disable promiscuous mode, as well as security considerations. Note that with nested hypervisors you might also have to enable promiscuous mode on the virtual adapters; again, refer to your specific operating system and hypervisor vendor's best practices.

6.3.6. Fibre Channel over Ethernet (FCoE)

Ethernet is a popular option for general-purpose networking. Getting Fibre Channel mapped onto a common Ethernet-based converged or unified network is a compromise between different storage and networking interfaces, commodity networks, experience and skill sets, and performance or deterministic behavior.

The key point with FCoE is that it is Fibre Channel mapped directly onto Ethernet, as opposed to being mapped onto IP or UDP. This means the entire Fibre Channel frame is mapped natively onto Ethernet, making it a peer to other traffic such as IP or UDP, as opposed to being encapsulated to ride on top of IP or UDP.

For example, with iSCSI, the SCSI command set is mapped onto IP, which in turn gets mapped onto Ethernet, similar to how SCSI_FCP implements the SCSI command set mapped onto Fibre Channel. Likewise for distance enablement, Fibre Channel frames get mapped onto IP with FCIP whereas, with FCoE, Fibre Channel is mapped onto Ethernet without the need for IP (i.e., layers are removed)

6.3.7. iSCSI (SCSI on TCP/IP)

iSCSI (SCSI mapped onto TCP/IP), which is the SCSI command set mapped onto IP, is a means of supporting block-based storage access over Ethernet LAN and WAN using existing hardware.

While iSCSI adapters with TCP offload engines (TOE) that improve performance by offloading host servers of TCP/IP and iSCSI protocol processing overhead exist, many deployments leverage software-based initiators and drivers with standard onboard NICs, or converged network adapters (CNAs). Various network topologies are supported by iSCSI, including direct connect or point-to-point as well as switched fabrics both local and over distance.

The benefit of iSCSI has been the low cost of using built-in 1 GbE network interface cards/chips (NICs) and standard Ethernet switches combined with iSCSI initiator software. Other benefits include ease of use and scalability. In addition to hardware iSCSI initiator (NICs), there are also software iSCSI adapters such as those found in hypervisors including VMware vSphere ESXi or various operating systems and third-party add-in software tools.

Likewise, there are both hardware and software iSCSI targets available for different operating systems. A challenge of iSCSI is less performance compared to faster dedicated I/O connectivity, and when a shared Ethernet network is used, increased traffic can impact the performance of other applications. iSCSI can operate over 10 GbE and faster networks. However, this approach requires expensive adapter cards, new cabling, and optic transceivers and switch ports that increase the cost of a shared storage solution requiring high performance.

For smaller environments, iSCSI traffic might exist on the same network as server-to-server, client-to-server, and management traffic. For larger environments or where applications need more server I/O performance, separate networks should be implemented.

Other iSCSI considerations and configuration options include NIC teaming, binding and bonding for performance, persistent reservations, IPsec and Chap security, MTU, and Multi-Path I/O (MPIO) for load balancing and availability. Another component is the Internet Storage Name Service (iSNS) software service, which can exist on different operating systems to provide iSCSI name and address resolution.

6.3.8. Fibre Channel (FC)

The value proposition, or benefit, of Fibre Channel (FC) has been the ability to scale performance, availability, capacity, or connectivity over longer distances with speeds of 32 Gbps or faster. A challenge of Fibre Channel has been the cost and complexity. Larger environments can absorb the cost and complexity as part of scaling, but it remains a challenge for smaller environments.

Most FC deployments utilize fiber optics with electrical connections used mainly in the backplane of servers, storage, and networking devices. Fibre Channel distances can range from a few meters to over 100 km, depending on distance enablement capabilities including optics and flow control buffers, adapters, switches, and cabling.

FF-BB-6 includes support for VN2VN connectivity. The VN2VN functionality enables the establishment of direct point-to-point virtual links between nodes in a FCoE network. This enables simpler configurations for smaller environments. Zoning may not be needed in these network designs, resulting in lower complexity and cost.

Regarding support for domain ID scalability, due to architectural limitations, Fibre Channel fabrics cannot have more than 239 switches, where each switch consumes a domain ID. In real-world deployments, this limit is lower, creating a problem when designing a largely converged network SAN. To solve this problem, FC-BB-6 fully defines how to support the deployment over Ethernet of the Distributed Switch Architecture. FC supports multiple concurrent upper-level protocols (ULPs) for open systems and mainframe server-to-storage, storage-to-storage, and, in some cases, server-to-server I/O operations. FC ULPs include FC-SB2, more commonly known as FICON, along with SCSI Fibre Channel Protocol (FCP), which is commonly referred to simply as Fibre Channel.

Topologies supported include point to point (no switches involved), core edge with access switches at the edge attached to larger switches, or direct at the core. Core edge can also be thought of as fan-in or fan-out where multiple servers converge and fan-in to a shared storage system or, if viewed from the storage system, fan-out. Fan-out from a server perspective would involve a group of servers with attachment to many storage devices. Various other topologies are possible, including a switch to switch on a local or remote basis.

N_Port_ID virtualization (NPIV) uses an ANSI T11 Fibre Channel standard to enable a physical HBA and switch to support multiple logical World Wide Node Names (WWNN) and World Wide Port Names (WWPN) per adapter for shared access purposes. Fibre Channel adapters can be shared in virtual server environments across the various VM. However, the various VM share a common WWNN and WWPN address of the physical HBA. The issue with shared WWNN and WWPN across multiple VM is that, for security and data integrity, volume or LUN mapping and masking need to be implemented.

By using NPIV supported by a target operating system or virtual server environment and associated HBAs and switches, fine-grained allocation and addressing can be performed. With NPIV, each VM is assigned a unique WWNN and WWPN independent of the underlying physical HBA. By having a unique WWNN and WWPN, VM can be moved to different physical servers without having to make changes for addressing of different physical HBAs or changes to Fibre Channel zoning on switches.

Also, NPIV enables fine-grained LUN mapping and masking to enable a specific VM or group of VM to have exclusive access to a particular LUN when using a shared physical HBA. Fibre Channel is promoted by the Fibre Channel Industry Association (FCIA) (http://fibrechannel.org) with protocols and standards involving ANSI T11 among other committees.

Fibre Channel over IP (FCIP) is a technology for interconnecting FC-based storage networks over distances using IP. FCIP implements tunneling utilizing IP as a transport to create a virtual Fibre Channel tunnel through the IP network to carry Fibre Channel frames in IP packets. This is similar to writing a letter in one language and putting it in an envelope to be sent through the postal system to some other destination in the world. At the receiving end, the envelope is opened and the contents of the letter are read. Along the route, those handling the letter do not have to understand the contents of the envelope, only where it came from and its intended destination.

FCIP takes Fibre Channel frames regardless of what the frame is for (FCP, FICON, or IP, among others) and places them into IP frames (envelopes) for transmission to the receiving destination. At the receiving destination, the envelopes are opened, and the contents are placed back on the Fibre Channel network to continue their trip. Another FC ULP is NVMe (FC-MVMe).

6.3.9. InfiniBand (IBA)

InfiniBand (IBA) is a unified or converged interconnect that can be used for storage and networking I/O as well as interprocess communications. IBA can be used to connect servers to storage devices, storage to LAN, and servers to servers, primarily for within-the-data-center applications. IBA enables low-latency remote memory-to-memory (RDMA) transfers to occur with fewer overheads to improve storage, networking, and other activity.

Given its high performance, low latency, and relatively low cost, IBA is popular in large-scale clusters of servers. This includes high-performance computing (HPC), also known as super-computing, as well as other scale-out, compute, application, and data environments that need high-performance, low-latency connectivity. InfiniBand is also used in some storage systems as a back-end or internal interconnect running TCP/IP and RDMA or some other protocol.

As a unified interconnect, IBA can be used as a single adapter capable of functioning as multiple logical adapters. Some IBA adapter cards that are converged can be put into InfiniBand or Ethernet nodes.

InfiniBand components include:

- Software device drivers and subnet managers
- Configuration and diagnostic tools including ibping
- Various cables and cable connectors
- Converged adapters and mezzanine cards
- Switches and gateways

InfiniBand ULPs include, among others:

- IPoIB—IP over InfiniBand
- NVMe—Using RoCE for NVMe over RDMA fabric
- RDMA—RDMA over InfiniBand
- SRP—SCSI on InfiniBand

IBA supports two modes of operation for IPoIB, including datagram mode (DM) and connected mode (CM). DM, the unreliable datagram (UD) transport, is used with the interface MTU—4 bytes (i.e., the IPoIB encapsulation header). With CM the Reliable Connected (RC) transport is used, allowing MTU up to the maximal IP packet size and reducing the number of IP packets sent for handling large data transfers or movements.

Note that different driver implementations may only support DM with certain adapters on various operating systems or hypervisors.

InfiniBand is commonly deployed for server-to-storage access, as well as server to server, including as a cluster or grid node interconnect. In addition to the application or compute server-to-server low-latency cluster node interconnect, IBA is also found as the backbone, back-end fabric or internal fabric for many clustered storage systems. For example, vendors including Dell EMC (Isilon, XtremIO), IBM (XIV), and Oracle, among others, use InfiniBand for east–west intracluster I/O traffic between server nodes. The server cluster nodes communicate northbound to application servers via Ethernet and Fibre Channel using various ULPs. Southbound I/O traffic from the server cluster nodes can be via PCIe, NVMe, or SAS, among others. In addition to being used for intracluster or east–west server I/O traffic, InfiniBand is also used for application-to-storage (north–south) traffic using IPoIB, SRP, or NVMe with RoCE, among others.

InfiniBand-based RDMA solutions also include software-defined storage such as Microsoft Scale-Out File Server (SOFS) and Storage Spaces leveraging SMB 3 direct. Some operating systems and hypervisors have InfiniBand drivers in the box, while optional drivers are available via device manufacturers such as Intel and Mellanox, among others.

IBA supports various topologies including point to point between two adapters, ring or daisy chain with dual-port adapters (one port connects to each adjacent server port), as well as switched fabric or mesh. IBA has a subnet manager (OpenSM) that assigns local identifiers (LIDs) to connected ports on an InfiniBand fabric and also maintains a routing table of LIDs. Subnet managers (SM) can be implemented in hardware via a switch or software on a server.

Various generations and speeds for InfiniBand are shown in Table 6.2.

Table 6.2. InfiniBand Generations and Performance Characteristics

		Per Lane Gbps	Per Port 4 Lanes Gbps	Per Port 12 Lanes Gbps	Encoding	Latency Microseconds
SDR	Single data rate	2.5	10	30	8b/10b	5
DDR	Double data rate	5	20	60	8b/10b	2.5
QDR	Quad data rate	10	40	120	8b/10b	1.3
FDR	Fourteen data rate	14.0625	56.25	168.75	64b/66b	0.7
EDR	Enhanced data rate	25	100	300	64b/66b	0.7
HDR	High data rate	50	200	600	64b/66b	0.5

Some Intel server core chipsets support a native integrated PCIe InfiniBand port that is physically extended to an external connection. However, most deployments rely on external PCIe adapter cards. A host adapter for IBA called a host channel adapter (HCA) connects a server to a switch for attachment to storage devices or other servers.

Gateways and others enable IBA networks to communicate with Ethernet and Fibre Channel networks and devices. IBA has found success in high-performance compute or extreme-compute scaling environments, where large numbers of servers require high-performance, low-latency communication, including for cluster and grid applications. InfiniBand is promoted by the InfiniBand Trade Association (http://infinibandta.org) and is part of the Open Fabric Alliance (OFA) (https://openfabrics.org).

6.4. Metropolitan and Wide-Area Networking

In addition to local-area networks and storage-centric SAN, there are also metropolitan-area networks (MAN) and wide-area networks (WAN). As their names imply, LAN are local in a building, data center, or perhaps a small campus. MAN includes larger campus or metropolitan areas, while WAN span larger geographies from hundreds to thousands of kilometers.

Metropolitan and wide-area networking are used for:

- Remote data protection (archiving, backup/restore, BC, BR, DR, HA)
- Remote mirroring and replication
- Connecting remote sites and branch offices with other locations
- Accessing public (private and hybrid) clouds, MSPs, and Co-Lo sites
- Providing fiber and optical services (dedicated and shared)

The good news about MAN and WAN is that they enable distance to support BC, DR, BR, and HA protection against on-site, local, or regional incidents or threat risks. Also, MAN and WAN enable general access with other locations and data centers, including cloud.

Distance technologies enable access to colocation, managed service providers, along with public clouds or for creating hybrid clouds. If the distance is the good news to protect against various threats or enabling access to remote services, the bad news is the associated performance as well as accessibility, cost, and complexity that can accompany MAN and WAN networking.

Remember that not all environments, applications, and technologies or services are the same, having different PACE attributes or capabilities. Distance choices depend on application performance and availability requirements, distance, budget, and availability of services locally and remotely. The key to survivability and business continuance is distance, whether across the computer room, across the campus, across town, in the region, or globally.

General MAN and WAN server I/O and networking considerations include:

- Do all of your applications need to send or receive data synchronously?
- How far away do your different applications need to be from a primary source?
- Does all of your data change on a regular basis, or is some fairly static?
- Are all of your RTO and RPO the same, or are there differences?
- What applications and data need better protection and access than others?
- Privacy and security of applications, data, and interfaces
- What are your PACE application requirements and characteristics?

6.4.1. Enabling Server I/O and Storage over Distance

Similar to the increases in computer performance and storage capacity, network bandwidth has increased along with the ability to support more interface types over longer distances. The cost economics of long-haul networking technologies has improved in most cases, with the exception being that some older, slower interfaces have seen instances of price increases.

When comparing network and bandwidth services, it is important to look beyond stated line rate or spec sheet numbers, instead learning what the effective performance and corresponding latency will be. For example, a network service provider may offer a lower rate for a shared high-speed bandwidth service; however, that may be based on upload or download speeds and be independent of latency.

Another example is a higher-cost bandwidth service that may appear to have the same amount of bandwidth but on closer investigation reveals that there is a better effective (usable) bandwidth with lower latency and higher-availability service.

Understanding what layers are involved in the network is important, because each layer adds complexity, cost, and latency. Latency is important to understand and to minimize for storage and data movement where consistency and transaction integrity are important for real-time data movement. The specific amount of latency will vary from negligible, almost nonexistent, to noticeable, depending on the types of technologies involved and their implementations.

Depending on how the I/O traffic is moved over a network, additional layers and technologies can be involved. The trade-off in performance and latency is the ability to span greater distances using variable-cost networking and bandwidth services to meet specific business requirements or application service-level objectives.

Distance is often assumed to be the enemy of synchronous or real-time data movement, particularly because latency increases with distance. However, latency is the real enemy, because even over short distances, if high latency or congestion exists, synchronous data transmissions can be negatively impacted.

Virtual and physical data centers rely on various wide-area networking technologies to enable access to applications and data movement over distance. Technologies and bandwidth services to support wide-area data and application access include Asynchronous Transfer Mode (ATM), Digital Subscriber Line (DSL), Wave Division Multiplexing (WDM) along with Dense WDM (DWDM), Metro Ethernet Microwave, Multiprotocol Label Switching (MPLS), Satellite, SONET/SDH, 4G Cellular, WiFi and WiMax, T1, T3, optical carrier (OC) networking using OC3 (3 × 51.84Mbps), OC12 (622.08 Mbps), OC48 (2.488 Gbps), OC96 (4.976 Gbps), OC192 (9.953 Gbps), or even an OC768 (39.813 Gbps), along with wavelength and bandwidth services.

General considerations about wide-area networking include:

- Distance—How far away do applications and data need to be located?
- Bandwidth—How much data can be moved and in what time frame?
- Latency—What are application performance and response-time requirements?
- Security—What level of protection is required for data in-flight or at remote access?
- Availability—What are the uptime commitments for the given level of service?
- Cost—What will the service or capability cost initially and over time?
- Management—Who and how will the network service be managed?
- Type of service—Dedicated or shared optic or another form of bandwidth service?

Metropolitan and wide-area networks are designed and implanted in a series of layers. An implementation may include multiple networking interfaces (SONET/SDH, ATM, Frame Relay, Ethernet, wave, and time-division multiplexing) and different physical media (copper and optical) to support IP and other ULPs.

With a network, the various characteristics of different technologies can be extended and leveraged to meet different service requirements. Incumbent Local Exchange Carrier (ILEC) and Common Local Exchange Carrier (CLEC) are examples of some telecommunications sources of bandwidth. MAN can also be implemented using SONET/SDH Optical Carrier (OC-x) network interfaces provided by a service provider, or implemented on your private optical transport network (OTN).

From a distance perspective, MAN using SONET/SDH can span hundreds to thousands of kilometers. However, for practical purposes, a MAN covers distances of 100–150 km. MAN can support both asynchronous and synchronous applications. The caveat is that storage and server-based applications are typically sensitive to latencies involving distances of greater than 100 km when in synchronous mode. Greater distances are possible in asynchronous and semi-synchronous modes, as well as with vendor-specific variations of multistage copies and mirroring using MAN and WAN interfaces. Additional content about LAN, MAN, and WAN, networking, including DWDM and optical or wavelength services, is provided in the Appendix.

6.5. Software-Defined Networks (SDN) and Network Function Virtualization (NFV)

Server I/O and networking involve hardware, software, and services that are configured and defined for various deployment scenarios. Recall from previous chapters that hardware needs software, the software requires hardware, and how they are configured via algorithms and data structures (programs) defines what they can do.

In the late 1980s and early 1990s, some networking switches, routers, and gateways were based on what was then called commercial off-the-shelf (COTS) and is today known as commodity hardware and server operating systems. Some of these early systems were proprietary servers and operating systems, as the proliferation of Intel (or AMD) x86-type or similar servers was still relatively new. Eventually, those early switches, directors, routers, bridges, and gateways gave way to custom processors, ASICs and FPGAs, along with proprietary operating systems or network operating systems and their networking applications, which are what we have today.

What we also have today is a return to using COTS or commodity hardware and operating systems. This time around, they are based on Intel or AMD x86- or x64-type architectures or Linux-based kernels with various network application software for implementing software-defined networks (SDN) and network function virtualization (NFV) along with network virtualization (NV).

Remember that virtualization means more than simply consolidation, as that is a use case. In general, virtualization means abstraction as well as enabling emulation that in turn combine to support various use cases (such as consolidation, among others). In general, the focus is moving beyond virtual switches and adapters such as those found in hypervisors and cloud stacks such as Microsoft Hyper-V, OpenStack, and VMware, among others. For example, additional features and functionality can include virtual networking services such as intrusion detection and prevention, identity management and authentication, load balancers and firewalls, along with routing. Using the popular "as a service" names, there are firewall as a service (FWaaS), load balancer as a service (LBaaS), VPN as a service (VPNaaS), among many others.

Benefits and value propositions for SDN, NFV, NV, and other variations of virtual and software-defined networking for servers, storage, IoT, and IoD, among other uses, include:

- Leveraging open technology, standards, and commodity hardware or services.
- Providing centralized control and insight of distributed resources, including those hosted in the cloud or managed service providers.
- Management tools and interfaces defined via software tools to adapt to various environments as well as defining how underlying hardware or services are configured.
- Reduced management bottlenecks associated with complexity and scaling of network environments.

Other benefits, value propositions and why SDN, NFV, and NV include:

- Facilitating scaling with stability while reducing or masking complexities.
- Flexibility and agility to adapt to various needs or application PACE requirements along with policy and QoS-based management.

- Enhanced interoperability with the public, private, and hybrid cloud as well as legacy environments and data infrastructure resources while reducing vendor dependencies and points of potential technology or service lock-in.

Note, however, that while the focus and marketing messaging may be to remove hardware vendor lock-in, beware of potential vendor software or services lock-in, if lock-in is a concern.

Common themes about SDN, NFV, NV, and other virtual networking include:

- Present resources to applications and other data infrastructure resources or services
- Orchestration of resources, configuration, and service delivery
- Abstract and aggregate physical resources similar to server virtualization
- Increase intelligence programmable by defining software for network control
- Admin has visibility into the entire data flow
- Applications see a logical switch
- Helps to mask or abstract fragmented networks

Some context and tradecraft reminders are useful here, including about the data plane and control plane. The data plane (or fabric or network) is where data moves from source to destination; it is also known as north–south traffic. Note that some management traffic such as moving virtual machines or virtual disks as well as replications may occur on the data plane.

The control plane (or fabric or network) provides command, control, and insight (metrics, events, logging, notification) and is also known as east–west traffic. Table 6.3 shows examples of various forms of software-defined and virtual networks along with characteristics.

What all of the approaches in Table 6.3, among others and variations, have in common is that they rely on software and underlying hardware (or services) being defined by other software to enable resilient, flexible, scalable, cost-effective server and storage I/O networking, locally and over distance, for different applications.

Figure 6.12 shows an example of a Hyper-V virtual switch (left) and its assignment to a virtual machine (right).

Figure 6.13 shows as an example on the left a VMware vSphere ESX virtual switch and NIC assigned to support VM on a host, and various host mappings for server and storage I/O virtual networking on the right.

Even though software-defined networking—or software-defined anything for that matter—has a focus on software rather than hardware, the reality is that the various software-defined and network virtualization approaches that rely on software also need to leverage faster COTS or commodity hardware resources (as well as cloud services).

In general, software-defined solutions require faster processors, more interoperability and functionality, and technology maturity. More intelligence will move into the network and chips, such as deep frame and packet inspection accelerations, to support network and I/O QoS, traffic shaping and routing, and security, including encryption, compression, and de-duplication.

Some examples of SDN, SDNM, NFV, and NV, among other approaches, include solutions from providers such as Amazon, Arista, Brocade, Cisco ACI, Google, Microsoft Hyper-V, OpenStack, Plexxi, VMware NSX, as well as others.

Table 6.3. Virtual and Software-Defined Networks

Name	Abbreviation	Description and Characteristics
Software-defined network	SDN	As its name implies, a programmable network infrastructure where software is defined to control and abstract underlying network-related resources while being transparent to applications and other data infrastructure entities. The term SDN is used within the industry rather broadly and can mean different things, so context is important.
Software-defined network management	SDNM	SDNM has a focus and emphasis on enabling the management of networks. It enables network administrators (or others) to manage and control networks with a layer of abstraction around resources (hardware, software, services). The abstraction layer enables higher-level functionality while masking underlying resources or their complexity, similar to server, storage, or other data infrastructure software-defined virtualization techniques. SDNM (and SDN) can leverage separate data and control planes (or fabrics). Note that SDN and SDNM need some communication between resources on the management control plane and data control plane.
Network function virtualization	NFV	Can be as simple as a virtual appliance handling network-related functions or services such as firewall, monitoring, load balancing, or access across local or cloud. NFV can function independent of SDN and vice versa, but, like many things, can be better together.
Network virtualization	NV	Network resources such as adapters, switches, bridges, routers, firewalls, load balancers, and gateways are implemented in software, or their software is implemented in memory via software, as opposed to running on dedicated hardware. This is similar to a virtual server implemented (or emulated) in memory via software instead of using dedicated hardware. A virtual network is a variation of NV, and NV is a virtual network (VN).
Virtual LAN	VLAN	Similar to multitenancy virtual servers, VLAN provide a virtual private network from other users within established quality-of-service parameters.
Virtual private network	VPN	Security and multitenancy technique for enabling remote access to various users, systems, or clouds. To the user, system, or others accessing the VPN or equivalent capabilities, traffic flows over a shared network yet is secured from other tenants.
Virtual private cloud	VPC	A hybrid cloud that uses public resources configured for private needs, including virtual networks, VPN, or other security and access to enable the use of the VPC.

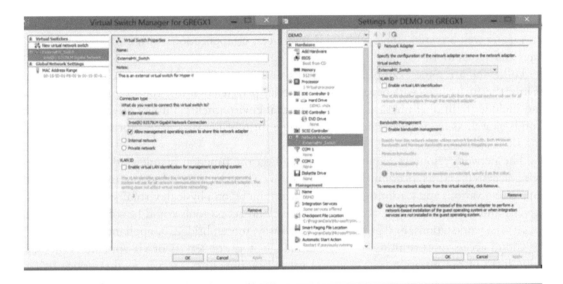

Figure 6.12 Microsoft Hyper-V virtual switch example.

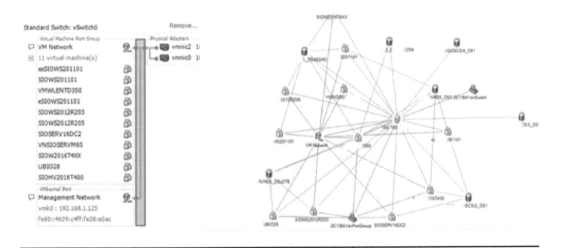

Figure 6.13 VMware vSphere ESX virtual switch example.

6.6. Accessing Cloud Storage and Services

Chapter 5 started out at high level and worked down from applications and server I/O to logical networks and protocols. In this chapter we began with the server I/O and networking devices from hardware and protocols on a local to wide-area basis, spanning physical, local, and virtual and software-defined solutions. Let's now move up from the lower-level physical and logical networks including various protocols to using those resources collectively for server and applications I/O. We'll use cloud and remote sites (as well as local ones) to show how to pull these various topics, technologies, tools, techniques, and tradecraft together in a useful way, using web browsers, tools with GUI interfaces, or applications with built-in cloud gateways and access tools.

Following are some examples of accessing and interacting with various cloud resources. While the focus here will be on clouds (public, private, hybrid), the examples extend to and include traditional physical as well as virtual environments. Public, private, and hybrid clouds as well as managed services rely on some form of network access for servers, storage, and their applications.

The network resources required will depend on the function or role of the cloud service being accessed and its location. Figure 6.14 shows a generic data infrastructure environment supporting various applications—local as well as remote, on physical, virtual, cloud, container servers, and including mobile access. Here, the cloud access tools could be software installed on servers, workstations, or laptops, or via a gateway, router, bridge, or appliance.

Cloud access tools running on workstations, laptops, or servers or via appliances include some form of optimization, security, and other functionalities specific to the service being accessed. In addition to public or private clouds, hybrid and virtual private clouds also provide tenancy across cloud resources.

TCP is a common denominator for accessing cloud services on a local or wide-area basis, spanning block, file, object, and API or application service access methods. Depending on the service or cloud access tool, various protocols and access methods may be used, for example, iSCSI, NFS, SMB/CIFS, or HDFS for NAS file-serving sharing, or S3, Swift, ODBC, CDMI, XML, JSON, HTTP, or REST along with other industry and vendor-specific APIs.

Figure 6.15 shows a mix of dedicated and private networks via public shared standard Internet or web access. These dedicated or private networks enable servers and applications, together with storage and the data they support, to communicate with public or remote private

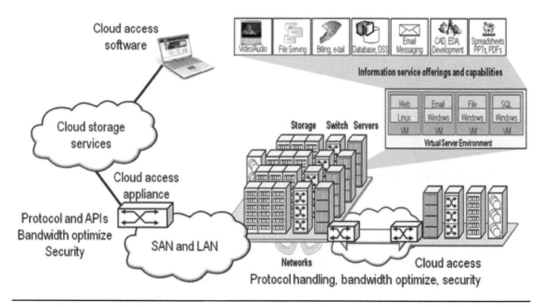

Figure 6.14   Local, physical, virtual, and cloud data infrastructure resources.

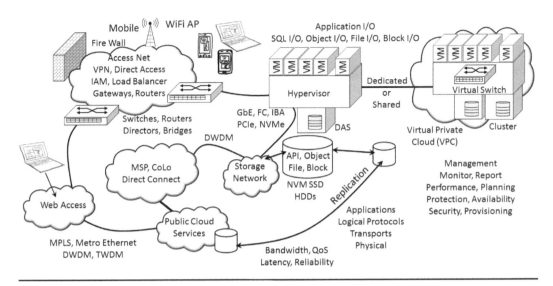

Figure 6.15   Overview of applications, protocols, and virtual and physical networks.

clouds, co-lo's, MSPs, hosting, or off-premises private data centers. Also shown in Figure 6.15 are physical, virtual, and cloud servers along with shared networked storage (object, file, and block) along with direct attached local storage including NVM SSD as well as HDD.

In Figure 6.15 the underlying complexity, as well as components inside the cloud-shaped figures, are masked. The cloud-shaped figures represent local storage or general-purpose networks, wide-area networks, the Internet, as well as public and private clouds. What is not shown in Figure 6.15 includes many items discussed in this chapter, such as switches, directors, adapters, firewalls, load balancers, bridges, gateways, routers, cabling, specific server or storage devices, local and wide-area protocols, technologies, and services.

Other things that are shown in Figure 6.15 are services such as AWS Direct Connect, which is a high-speed, low-latency connection from their hosting and co-location partners into the AWS environment. Note that normally, when you access a cloud service such as AWS, it is via general-purpose (Internet or web) networks and services. Services such as Direct Connect provide a special higher-performance access than traditional web access. Check with your cloud provider or partner to see what types of specialized and higher-performance or secure connectivity options they can provide.

Other cloud access tools and gateways include AWS CLI or Storage Gateway, Azure Storage Explorer, Commvault Simpana, Cloudberry, Cyberduck, Dell EMCcode S3motion (among others), Google gsutil, Nasuni, S3 Browser, and S3FS, among many others.

Many storage systems, as well as backup or data protection solutions and appliances, also include some cloud access or gateway that works with various public services, private as well as hybrid. Check with your preferred supplier to see what cloud services they support, along with features, functionalities, costs, and capabilities.

The following are some example of using various tools to interact with the local, public, and hybrid cloud as well as other data infrastructure services. For example, the Google storage tool (gsutil) can be used to display Google cloud storage information. Simply download the free tool, supply endpoint information, and it displays useful information.

```
gsutil ls gs://<some_endpoint>
```

Examples of accessing clouds include listing endpoints along with access keys (fictional) across different providers (public or private), in this case, using the Dell EMCcode S3motion tool.

```
C:\Users\greg\s3motion --listClients
{ clients:
   [ { name: 'azdemo@4',
       accessKeyId: 'AnAzdem@4',
       secretAccessKey: 'This<>mykey1@&yTewz%P@o==',
       endpoint: 'https://winazdemo1.blob.core.windows.net/' },
     { name: 'Googleit3',
       accessKeyId: 'This<>@id;)1Time$oU@Me&(*(1@&yTewz%P@o',
       secretAccessKey: 'This<>mykey1@&yTewz%P@o==',
       endpoint: 'storage.googleapis.com' },
     { name: '1ofmybuckets',
       accessKeyId: 'This<>@id;)$oU@Me&(*(1@&yTewz%P@o',
       secretAccessKey: 'This<>mykey1@&yTewz%P@o==' }
   {} ] }
```

Here is an example of using S3FS to present an S3 bucket endpoint as a mount point on a Linux system, similar to accessing a local NAS or another share.

```
# Excerpts from Linux a /etc/fstab configuration file
#
# Enable S3 bucket endpoint as a Linux mount point

sudo s3fs -o passwd_file=<file_with_endpoint_credentials> -o
use_sse nonempty -o use_cache=/tmp/cache_siodemo /demo/s3mnt

# Mount CIFS/SMB share NASshare from NASserver as Linux mount
point

//NASserver/NASshare /demo/NASshare cifs credentials=<file_with_
endpoint_credentials>,iocharset=utf8,sec=ntlm 0 0
```

From a Linux command prompt, you can then perform various operations such as *ls, cp, rm, mkdir, rsync,* among other activities. This includes moving items within the NAS namespace of a specific NAS server, as well as namespace for a given Linux system to AWS namespace. For example, rsync the NAS share to a subfolder NASshare in the S3 bucket, with the –av indicating to preserve attributes (ownership, permissions, dates) with verbose output.

```
Rsync -av -exclude-from='<exclusion_file>' "/demo/NASshare"
"/demo/s3mnt/NASshare"
```

Here is an example of using AWS CLI to synchronize the contents of two different buckets.

```
aws s3 sync "s3://oneofmybuckets/vms/" "s3://anotherbucekt/vms"
```

Moving further up in altitude (i.e., higher in the technology stack) and getting closer to applications, Figure 6.16 shows using Microsoft SQL Server Management Studio (SSMS) on a local system to access a remote SQL server located in the Microsoft Azure public cloud. On the left of Figure 6.16, using SSMS, a right-click on a database opens Properties and then Tasks to display Deploy Database to Microsoft Azure SQL Database (if it does not already exist). Also shown on the left of Figure 6.16 are specifying endpoint and security authentication credentials to access the database. Not shown are configuration settings via Azure portal to the specific firewall and other parameters to allow or gain access to the Azure SQL Database being set up or used.

On the right side of Figure 6.16 is a view of SSMS including local databases as well as the Azure database with results from a query to the cloud database.

What this has to do with the server I/O and networking is that common, standard networks, physical, logical, virtual, and various protocols, are used to support the above access. These include TCP (IP and UDP) as well as Ethernet (wired or Wi-Fi) along with other technologies, firewall, router, and related settings.

Needless to say, there is much more involved, and many other examples of the server I/O and networking with your servers, storage, and other resources will be explored as we continue in other chapters.

Figure 6.16   Access Azure SQL server from local SQL Server Studio.

6.7. What's in Your Server Storage I/O Networking Toolbox?

In addition to the material in this chapter as well as what is in the Appendix, other toolbox items include adding a TCP/IP and networking fundamentals book or two or three to your bookshelf. Depending on your area of focus, also add some Microsoft, OpenStack, and VMware networking as well as general server storage I/O reading to your toolbox. This includes not only hardware and software networking, but also applications and tools such as databases, among other things that will be covered in later chapters. Also, review the list of recommended reading in the Appendix, which includes various books, blogs, and other resources.

Software tools to have in your toolbox include those previously mentioned in the bits and bytes, processor, and general I/O chapters, for example, Linux ethtool, lspci, htparam, ping, traceroute, iperf, diskspd, vdbench, Benchmark Factory, cosbench (object access), NTttcp, caspa, wireshark, netscout, sysbench, perfmon, iotop, fio, htop, and visual esxtop, among others.

For hardware, handy toolbox items include various cable connectors, interposers, and gender changers, as well as physical tools such as screwdrivers, pliers, and plastic ties and labels for basic cable management.

6.8. Common Questions and Tips

Why not use wireless everywhere? If wireless is fast enough both for bandwidth and latency, as well as reliable with no timeouts or dropped transmissions and can avoid interference from other wireless traffic, is safe and secure from intrusion, interfaces and protocols exist for your server and storage applications, then why not? However, there are usually some of the above and other barriers that prevent widespread wireless use for all server I/O activity, at least for now.

What is oversubscription? When there are more consumers or users of a service than resources, it is called oversubscription and can result in congestion, blockage, or delays. Oversubscription can occur by design or by accident. Oversubscription by design can be used to balance the cost of additional resources with normal usage patterns. For example, a network can have adequate physical GbE or Fibre Channel ports for connectivity that converges into a single back-end, trunk, or other backbone connection. Under normal traffic workload, there is enough bandwidth to move data or allow data access with adequate latency. However, under heavy usage conditions, such as when all of the physical ports become active, a bottleneck can occur. Another example is if you are at an event and can get a wireless or cellphone connection, yet are not able to move data because the network is busy. The opposite of oversubscription is having enough back-end bandwidth or performance to support all of the peak activity, which can result in underutilization at times. However, the choice to use oversubscription designs vs. alternative approaches will depend on your applications I/O PACE requirements.

Does the presence of fiber optics guarantee performance? Not necessarily. You can be presented or sold or told about a fiber service only to find out that it is bandwidth or speed limited. For example, the speed of the fiber service might be 1 GbE, which may be or seem fast but is only a fraction of the speed at which the actual fiber network could run.

Keep in mind that it is the combination of the protocols, service configuration, routers, gateways, bridges, switches, and distance that determine what your effective performance will be. Remember also that when somebody says data moves at the speed of light on fiber networks, your performance moves at the speed of the protocols and configuration of that fiber network or service.

Is latency only a distance issue? While it is commonly associated with distance, latency can occur locally even on fast networks, due to the configuration, routing, blockage, or other issues. While bandwidth is often a focus for networks in particular when accessing the Internet or the web and remote applications or services, also pay attention to latency and errors or retransmits.

Ping is a handy tool available on most operating systems and even for iPhones or other mobile devices to check the latency of a network.

What's the best server I/O protocol and network for storage? It depends! The decision about what type of server and storage I/O interface and topology is often based on cost, familiarity with available technologies, their capabilities, and, in some cases, personal or organizational preferences. The right interface for you is what is applicable for your business needs, recovery and survivability requirements, and budget constraints. Be sure that your wide-area and other service providers are using unique paths so that you have no single points of failure in your network design and configuration.

Are networks less reliable today than in the past? In terms of capability, accessibility, and affordability, networks should be more flexible, scalable, and resilient than in the past. As is the case with many technologies, it is how they are deployed, configured, and managed to particular service levels, often under specific financial considerations, that cause networks to have different perceived or actual availability. Like servers and storage, networks should and can be more resilient today, if best practices are applied even in cost-sensitive environments.

How is a flat network different from how Fibre Channel SAN are designed? If you are familiar with designing SAN, other than some terminology differences, you should be very comfortable designing flat layer 2 networks.

Is FCoE a temporal or temporary technology? Depending on the time frame, all technologies can be seen as temporary. Given that FCoE probably has at least a ten-year temporal timeline, I would say that, in technology terms, it has a relatively long life for supporting coexistence on the continuing road to convergence. Note that server and storage I/O and networking infrastructures take time to deploy in larger environments while protecting investments in hardware, cabling, people skill sets, and management software tools. Some organizations will move to different technologies faster than others.

What comes next after Ethernet? For at least another decade or two, I see Ethernet or the various forms of 802.x continuing to evolve, with faster speeds, more efficiently encoding, better QoS, and other capabilities. That is not to say that some other interface will not appear and become viable, rather that Ethernet continues to gain momentum and traction from consumer PDAs to enterprise servers and storage.

With a dependence on networks (local and remote) for cloud access, what happens if someone decides to turn off the network? Look for ways to prevent your network or access to public resources being a single point of failure. Networks can and do go down for periods of time, so design to eliminate points of failure, including having alternative connections or access points. With public clouds that depend on public Internet access, loss of web access can be a disaster, so talk with providers about alternative options for accessing their services. For example, can you call a backup or archive service provider and, with appropriate security authentication, arrange to have your data returned via some physical medium if needed?

Should we use two separate service providers for a redundant remote network connection? That depends on whether the different carriers or service providers use unique paths or simply share common network paths. Some will say not to use a single carrier or service provider for redundant paths, as they are a single point of failure. Although that may be so, if the carrier or service provider has unique, diverse networks along with associated management, detection, and other capabilities, this may not be an issue. On the other hand, if you use two different service providers, make sure that their paths are in fact unique and not using the same network somewhere along the path between your site and the destination.

6.9. Learning Experience

The following are some learning exercises and experiences to practice and apply what has been covered in Chapter 6 as well as some of the items in Chapter 5. In a later chapter there is additional coverage of performance, benchmarking, workload simulation, and metrics that matter for server and storage I/O. Try pinging a web or blog site using site tools such as http://tools.pingdom.com/fpt.

Use the ping command on your server or system to test access and latency with another device:

```
Ping some.device.ip.address
```

Use iperf to test network performance between different servers. For example, set up one system to function as an iperf server responding to other systems acting as iperf clients. Multiple iperf clients can access the iperf server.

The following is an example of an iperf client command to perform I/O, specifying TCP window and buffer sizes with results shown in gigabytes for 300 iterations and displaying results every 2 seconds. Download iperf from www.iperf.fr.

```
iperf -c ip.of.server -m -fG -d -t 300 -i 2 -w 64k -l 128k
```

An alternative to iperf is the Microsoft NTttcp testing tool. Some other commands include using ethtool, iplink, *lspci | grep –i eth, netstat –I, or ifconfig –a*, among others.

6.10. Chapter Summary

General server I/O considerations include available networking services such as performance bandwidth and latency, availability, capacity or connectivity, as well as economics and compatibility. Another consideration is what applications require what local and remote services, as well as security, among other PACE attributes. With a focus on applications, does all data need to move between locations, and if so, at the same or different time frames?

General action items include:

- Context is important when working with server I/O and networks.
- Everything is not the same in the data center or across applications.

- Various tools and techniques exist to support diverse server I/O needs.
- Latency results from distance, but latency can also have local impact.
- Understand the relationships among applications, hardware, and software.
- Know how the various server and I/O components interact.

Bottom line: The best I/O is the one that an application and server do not have to do. However, the second best is the one that has the least impact on performance while being handled securely over local and wide areas using various techniques and technologies.

Part Three

Storage Deep Dive and Data Services

In Part Three, including Chapters 7 through 12, our deep dive continues as you learn about various data storage technologies as well as data services. Data services functionality enables data infrastructures and the applications they support to manage and optimize resources. This includes access, performance, availability, capacity, and management in cost-effective and efficient resource optimization as well to protect, preserve, secure, and serve applications, data, and resulting information services.

Buzzword terms, trends, technologies, and techniques include archive, backup/restore, copy data management (CDM), BC, BR and DR, clustering, durability, erasure codes, Flash, HDD, multi-tenancy, NVM (Flash, SCM, SSD, 3D XPoint, NVDIMM), RAID, replication, and erasure codes, RPO, RTO and SLO, SEcurity and access, snapshots, checkpoints, consistency and CDP, storage efficiency, and data footprint reduction (DFR) including compression, de-dupe, Thin provisioning, and virtual storage, among others.

Chapter 7

Storage Mediums and Component Devices

Memory is storage, and storage is persistent memory.

What You Will Learn in This Chapter

- Data storage mediums provide persistent storage for data and applications.
- Why there are different types of storage mediums for various applications.
- Storage media and medium fundamentals and PACE characteristics.
- Comparing storage mediums beyond cost per capacity.
- Storage space capacity vs. I/O consolidation.
- Storage device metrics that matter.
- Looking beyond the interface.

We are near the halfway point in this book, having covered the basic bits, bytes, blocks, and objects, file systems, access, servers, and I/O networking fundamentals. Let us now move deeper into the topic of storage, which is an extension of memory, which is storage; storage is persistent memory, memory is storage.

This chapter looks at the mediums and basic component building blocks that get hardware and software defined into solutions or service. In subsequent chapters, we will look at how the storage components discussed here, along with server and network resources discussed earlier, get combined with data services and delivered as solutions. The storage components discussed are the key themes, buzzwords, and trends addressed in this chapter, including non-volatile memory (NVM), storage class memory (SCM), hard disk drives (HDD), solid-state devices (SSD), hybrid, and NAND flash, among other related technologies and topics for enabling data infrastructures spanning physical, software-defined, virtual, container, and cloud environments.

7.1. Getting Started

A recurring theme in this book is that everything is not the same in data centers and data infrastructures. With different sizes and types of environments along with diverse applications, there is a need for various types of data infrastructure resources, including servers, input/output networking, and connectivity along with storage devices and mediums.

Unlike a one-size-fits-all approach, the various types of data infrastructure resources provide the ability to align different options to your specific environment and application performance, availability, capacity, and economics (PACE) requirements. Storage exists in the largest of the cloud hosting and managed service providers, from enterprise, SME data centers to smaller SMB and ROBO, along with SOHO and consumer environments, among others.

Storage gets deployed and used in:

- Internal, external, and portable formats
- Storage systems and appliances
- Servers (standalone, cluster, converged-infrastructure [CI], hyper-converged [HCI])
- Workstations, PCs, laptops, tablets, and mobile devices
- Media/TV, gaming, cameras, and entertainment devices

Similar to having different types of servers and I/O networks, there are various types of data storage resources. These data storage resources come in various packaging options including physical hardware, virtual, cloud, and software-defined, along with different PACE attributes.

Storage components include:

- Storage systems, appliances, and servers
- Storage controllers and storage processors
- Enclosures, drive shelves, bays, and slots
- Storage devices and mediums
- Software-defined, virtual, cloud, and management tools

Memory and storage can be

- Volatile or non-persistent, where data is lost when power is removed
- Persistent or non-volatile memory (NVM) such as NAND flash
- Semiconductor and storage class memory (SCM)
- Modifiable, supporting both reads and writes (R/W) or updates
- Write once, read many (WORM) or read-only
- Fixed or removable, hot-swappable and portable
- Magnetic (HDD or tape) and optical

Volatile storage or memory includes semiconductor-based dynamic random access memory (DRAM). Semiconductor-based NVM includes NAND flash along with other SCM such as Phase-Change Memory (PCM) and 3D XPoint, among others. Other persistent storage includes magnetic media such as HDD and tape, as well as optical media (CD, DVD, Blu-Ray).

In addition to drives with integrated media, there are also storage mediums such as magnetic tape, optical, and flash, where the media is packaged in a cartridge (a physical container or carrier) that works in conjunction with a drive or reader.

The context here is that a drive is often thought to be an HDD or SSD. However, a drive can also be an optical reader or a tape drive. In the case of an optical or tape drive, the media (optical or tape) are inserted, used, ejected, removed, and used for portability or data exchange. Tape and optical drives range from simple desktop (or internal) to large-scale robotic libraries.

Note that there are also HDD and SSD solutions where a device can exist in a special cartridge or carrier for portability. Portability may be needed to move bulk or large quantities of data between locations or off-site for data protection.

Most cloud services offer cloud-seeding capabilities ranging from individual HDD or SSD (as well as some for tape) to large-scale migration appliances—for example, moving data between data centers, or cloud-seeding (e.g., planting data seeds to grow in cloud) such as sending data to a cloud provider in a bulk move (e.g., shipping data). However, we are getting a bit (pun intended) ahead of ourselves; for now, let's get back to the basic building blocks.

Storage devices such as HDD, SSD, tape and optical media can be standalone for internal use or external to a server, workstation, laptop, or another system. These devices are combined with others components into a storage solution (enclosures, arrays, appliances).

Storage solutions range from just-a-bunch-of-disks (JBOD) in an enclosure to increased functionality with RAID or parity protection and other data services. Storage solutions with multiple devices are referred to as storage arrays, libraries, and appliances. Storage systems and solutions, as well as data services, are discussed in subsequent chapters.

Figure 7.1 shows various storage media and enclosures.

Common storage (and memory) considerations and attributes include:

- Heat (or cold), humidity, and altitude
- Noise and vibration tolerance
- Interface type, including dual-path
- Protocol and personality, including abstraction
- Form factor (height, width, depth)
- Packaging (internal, external, mobile)
- Power consumption and weight
- Performance, availability, capacity, economics

Figure 7.1   DRAM, HDD, SSD, and enclosures.

7.1.1. Tradecraft and Context for the Chapter

Recall that memory is storage, and storage is a persistent memory, as well as that everything is not the same in data centers or information factories across different applications, environments, and organizations. Storage gets accessed by servers via I/O paths (local internal, external, remote) to support applications as being part of data infrastructures.

Servers run the algorithms (programs) that access data structures (data), networks move data locally and remote; it is storage where the bits, bytes, blocks, blobs, buckets, and objects reside. This includes application programs along with their settings and metadata, as well as actual data, all of which gets protected, preserved, and served from some form of storage (Figure 7.2).

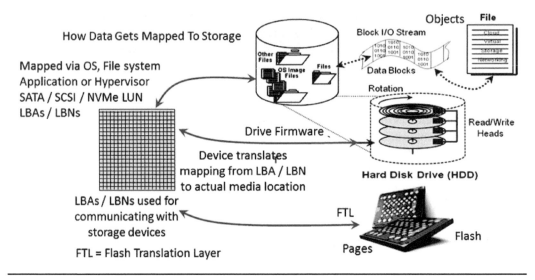

Figure 7.2   Data storage media fundamentals (bits, bytes, blocks, buckets, and blobs).

In Chapters 3, 4, and 5 we discussed how data is organized, from low-level encoding (e.g., 8b/10b, 64/66b, and 128b/130b) to bits, bytes, blocks, blobs, and objects. This included how the bits, bytes, and blocks are organized into partitions, volumes, file systems, object repositories, and databases, among other data structures.

In Chapter 3 we looked at how those data structures are accessed as block, file, object, and application services via different physical as well as logical protocols. Software defines the data structures of objects, blobs, buckets, containers, databases, files, volumes, blocks, bytes, and bits which are stored as zeros and ones (i.e., in binary). It is the underlying physical medium that determines how those items are stored in conjunction with upper-level software and device drivers.

Following are some storage tradecraft context terms.

- **Mount:** A device can be physically mounted, installed, connected to, or inside a server, storage system, appliance, or another system. A storage device also gets software or is logically mounted to a hypervisor, operating system, file system, or application using the device. A logical mount means that the software gains insight into the device, opening and accessing any on-disk structures (ODS) and metadata. This includes partition, file-system, volume, or other metadata to prepare the device or volume for use. In other words, the logical mount makes the device (or service) and its endpoint known and accessible, ready for use from a software perspective.

- **Bind:** Another term that requires context is bind, which can be similar to a logical mount. However, another context topic is *binding* a device with affinity to a particular system, controller, or port. A variation of bind is a bound volume or device, where two or more devices are made to look like a single device or namespace. The context is important, to avoid confusing a bound or concatenated (appended) device with a striped (evenly spread data) device. With a stripe, the data is spread evenly across the device. With bound, concatenated, or spanned volumes or devices, data fills one device or another, with new storage added to the existing space. Context also comes into play with bind or bound in that a device can be dedicated (has an affinity) to a specific physical or logical server, operating system, or hypervisor. Similarly, a device may be able to failover or be shared among multiple controllers, adapters, servers, or operating systems.
- **CSV:** Can refer to a *comma-separated value* file such as for use with Excel spreadsheets, as well as a *clustered shared volume*.

In addition to the context that matters for different storage device terms, there are also various ways storage devices get used. Storage can be used in different ways, including as persistent space or ephemeral (short temporal duration). This is where context comes into play, as a persistent storage medium and device will not lose data when powered off. However, how that device gets configured or defined can determine how the data is retained. For example, a HDD or SSD is persistent storage vs. DRAM, yet the SSD or HDD can be configured as an ephemeral, transient, temporal, nonpersistent storage tier.

Note that while a storage medium or device may support reads and writes, it can be configured and defined by software (or firmware) to be WORM. Likewise, applications may use persistent storage as temporal or nonpersistent storage—meaning the data exists on the device, but it is not backed up or in any other way protected against loss. Examples include temporary work or scratch space. Context is important in that the type of medium as well as how it is defined for use determines the availability, accessibility, and durability of data.

7.2. Common Storage Device and Media Matters

While data storage media have some unique attributes which will be covered later in this chapter, there are also many common characteristics besides those already mentioned. Besides performance, availability, capacity and economics (PACE) attribute, there are also many common metrics that matter across storage devices, mediums, and media.

Additional common storage medium and media matters include:

- Compatibility, packaging, device interface, and interoperability (logical and physical)
- Device or medium organization and architecture: how data is stored
- Environmental power, cooling, heat, and vibration considerations
- Management, including software device drivers and firmware

General device compatibility considerations include:

- Mechanical (physical mounting, space footprint, cooling, fixed, removable, hot-swappable)
- Electrical and interface (power, dual-port, dual-path, interface, and protocol)
- Software (device driver, OS/hypervisor, file system, utilities, format, interface)
- Usage—such as concurrent, shared, dedicated, serial access vs. concurrent and parallel
- Connectivity interface protocol (physical and logical)

Management considerations include the software implemented as firmware for device interfaces with servers, storage systems, controllers, or other appliances for configuration, as well as command, control, data movement, integrity and consistency checks, status and performance reporting, error notification, and implementing various features or functionalities. Device firmware can be updated via different tools; in larger storage systems or appliances, this is usually handled as part of the system's operating software.

For standalone or other devices, vendors usually provide a software tool that enables storage device firmware to be upgraded if needed. Many storage devices also support Self-Monitoring Analysis Reporting Technology (SMART) capability. Operating systems, hypervisors, controllers, or other SMART-enabled utilities are then able to obtain and report on the health, status, and other information about a device.

Environmental considerations include electrical power consumption in various modes of operation such as idle or inactive, sleep or low-power, low-performance mode, turbo or accelerated performance mode, start-up surge or burst activity, and sustained steady state.

A by-product of storage devices such as SSD, HDD, and tape drives consuming electricity to do work is heat generated, usually measured in British thermal units (BTUs). Heat needs to be removed via cooling. However, another device consideration is its tolerance to external heat, humidity, and altitude, along with noise and vibration.

Device or medium architecture considerations include fixed block architecture (FBA) or variable block architecture (VBA) using logical block address (LBA) and logical block number (LBN) or key value (recall that we discussed these and other related items in Chapter 3). FBA devices have sector sizes including 512 bytes and 4 Kbytes (also known as advanced format or AF). There are also some devices that support variations such as 520 bytes among others, along with Data Integrity Format (DIF), which is a special end-to-end (E2E) feature.

DIF is implemented in some storage device media and when also implemented in the application, database, operating system, device drivers, and storage system, provide complete end-to-end data integrity consistency checks. Note that 4K or AF devices can also be put into a 512-byte emulated (512E) mode for backward compatibility with older controllers, adapters, hypervisors, device drivers, and operating systems. Context comes into play here, as you need to keep in mind that we are referring to the sector size of the physical device as opposed to the logical partition, file system, virtual, or other allocation.

Tip: To prevent performance issues, pay attention to allocation and storage alignment configuration such that upper-level storage allocation aligns with lower-level storage boundaries. This is important with hypervisors such as Microsoft Hyper-V and VMware vSphere ESXi among others, as well as across different applications, including databases and others.

Other features or characters found on storage devices include some form of data footprint reduction (DFR) such as compression or compaction, device-level de-duplication, and security including drive-level encryption. In the next several chapters we will look more closely at data services implemented in servers, storage systems, appliances, and applications beyond the basic functionalities discussed here.

7.2.1. Storage Device Media PACE and Metrics that Matter

Different applications and workloads have various PACE attributes and associated resource needs. Storage devices, like servers and I/O networks, have different PACE capabilities to meet various application needs. Some applications need more, while others require less performance to support activity, along with different availability, capacity, and economic considerations.

Metrics may include the peak, minimum, average, and standard deviation, counts, rates, time, space, or percentages. There are simple metrics such as IOP and capacity, as well as compound metrics—for example, IOP or transaction per watt of energy consumed, or watts of energy per capacity along with cost per capacity or cost per transaction, IOP, or work activity.

Performance

Performance includes activity such as IOP (reads, writes, status, or other commands) of various size, which can be random or sequential and streaming. Media such as DRAM, NVM SSD, and HDD are optimized for mixed random and sequential, reads as well as writes, from concurrent or parallel workloads.

Some devices are optimized (their personality) for high performance measured in IOP or objects processed, while others are designed for throughput bandwidth or bytes transferred, along with latency or response time. Another performance consideration is the amount of CPU time consumed per I/O operation, which is a function of the protocol and device driver software.

Activity such as IOP, transactions, gets, puts, reads, writes, status, or other commands per second (or time interval), bandwidth or bytes transferred, I/O size, reads, writes, random or sequential, can be measured. Another indicator is time to first byte, block, or a significant amount of data transferred before stabilized and sustained performance.

Other performance metrics include seek and rotational latency time for HDD, as well as spiral or sustained transfer rates. While not necessarily a storage device or media metric, others include fragmented, split, or misaligned I/O requests. Metrics including small 4 KB and 8 KB as well as larger 64 B, 128 KB, and 256 KB I/O sizes as well as MB- or GB-sized data transfers is an indicator of performance.

Tip: For larger I/O sizes that result in lower IOP, the bandwidth or throughput should be higher; latency may also increase. Likewise, for smaller I/O sizes with lots of activity, you should see more IOP, with bandwidth equal to the IOP × I/O size, as well as lower latency (and CPU per IOP).

Tip: For IOP, besides the I/O size, also pay attention to reads or writes, random and sequential, and latency as well as queue depth (number of workers). Sequential reads should show better performance vs. sequential writes, which can be better than random reads or writes depending on the technology.

> *Tip:* Keep in mind that SSD or HDD drive-level read-ahead or other caching, I/O size, and reads vs. writes will have an impact on actual device and application performance. Look at multiple metrics to get the big picture of what a device is capable of to meet application PACE needs.

Availability

Availability for storage devices, components, and media include reliability, availability, and serviceability (RAS), uptime, durability, background media, data integrity and consistency, error detection and correction, as well as security. Security features commonly found in storage devices and media include drive-level encryption such as that supporting AES, TCG, FIPS, and OPAL, among others (see Chapter 10).

Storage devices have the ability to detect read and write errors as well as automatically revert data from bad blocks of storage on SSD and HDD to spare good blocks or sectors. Availability and data protection along with resiliency capabilities are provided by data services implemented in servers, storage systems, or appliances above the device layer.

Availability metrics include SMART, Annual Failure Rate (AFR), and durability, such as terabytes written (TBW), or full disk writes per day (DWPD). Other metrics include errors, retransmissions, bad block replacement, and re-vectoring (moving data to good locations, marking bad blocks as out of service). Nonrecoverable read errors per bits read, or bit error rate (BER), is also known as recoverable bit error rates (RBER) and unrecoverable BER (UBER).

BER numbers are represented such as, for example, $10 \wedge 17$ for enterprise-class HDD and SSD, which indicates on average the statistical probability that an error could occur for that number of bits read (or written). AFR is a percentage indicating on average for a given sample size the expected failure rate. This metric replaced the older legacy metric of or mean time between failures MTBF. The lower the AFR number, the better, while the higher BER number specified is better.

Capacity and Economics

Raw and usable capacity, uncompressed and compressed, are made up of some number of sectors of a specific size. Sector sizes include standard 512-byte native (512N); 512-byte emulated (512E), and 4 KB (4096 bytes) native (4KN), as well as 520, 524, 528, 4160, 4192, or 4224 bytes per logical block (sector). Note that bigger is usually better with sector sizes. However, care must be taken to ensure that the database, hypervisor, file systems, and other software can use the larger sector size, as well as that storage allocation will be properly aligned. Also, some software requires 512E or 512N format for backward compatibility, while other software may support 512E or 512N as well as 4KN, though not in a mixed environment.

While it should be abstracted, some software checks for and allocates space based on low-level physical sector size. Another capacity metric is the areal density or number of bits stored in a given amount of physical space, such as on an HDD platter, tape surface, or NAND flash die.

Capacities can be increased by stacking media in a device, such as more platters or flash die layers. In addition to usable storage space, other capacities include DRAM data buffers, read cache, and non-volatile write-back cache. Cost considerations include purchase, as well as any maintenance and recurring fees along with power, cooling, and enclosures.

Besides the list price for a given device and its capacity or performance, another consideration is the standard warranty from the manufacturer, as well as what is provided by the OEM or system vendor. Some cost and capacity metrics include cost per capacity per watt of energy used, or cost per IOP and cost per capacity and IOP or bandwidth.

7.2.2. Packaging, Form Factors, and Interfaces

Storage devices come in different types of packaging, with form factors and various interface and PACE attributes for diverse application needs. Some common storage packaging is internal, intended for use inside a server, storage system, appliance, workstation, laptop, or other device. Another common packaging is in external devices that attach to a server, storage system, or other device as expansion storage. Storage can also be fixed or removable as well as portable.

Figure 7.3 shows various HDD and SSD as well as PCIe add-in card (AiC) storage form factors, packaging, and interfaces. Note the different height of tall vs. low-profile drives with the same form factor.

In addition to PCIe AiC such as SSD or RAID cards with extra cache memory, other storage form factors (Table 7.1) include widths of 1.8 in., 2.5 in., 3.5 in., and 5.25 in. (i.e., bays for optical media such as CD/DVDs).

Other packaging options for storage include NVM on DRAM DIMM memory sticks, NVM DIMM (NVDIMM) sticks, PCIe AIC, SD and MicroSD cards on mobile devices, or as boot and software loading devices on higher-end systems. USB as an interface, and Firewire, Thunderbolt, Wi-Fi, and NFC are also used as storage attachment and storage devices.

Note that there are also interface and packaging enclosures as well as transposers that allow different packaging to exist in other configurations, such as those in Table 7.2. For example, I have some 5.25-in. (media bay) enclosures that support up to four 2.5-in. 12 Gbps SAS or 6 GBps SATA HDD and SSD. Using these 5.25-in. enclosures, I can use space on a server or workstation to add additional hot-swap storage attaching the HDD and SSD to internal controllers or adapters.

> *Tip:* Keep in mind power requirements (amps and volts) as well as vibration cooling in addition to mechanical or physical mounting, as along with an interface for servers or storage systems to access the storage devices. Other interface considerations besides physical connector and logical protocols, as well as single or dual path, include affinity or bound to a single controller, adapter, host server, and hypervisors or operating systems.

Figure 7.3   Storage devices (HDDs, SSDs and HDDs, PCIe AiC NVMe).

Table 7.1. Storage Device Form Factors

Form Factor	Characteristics
1.8 in.	Smaller form factor used by some SSD and formerly by some HDD
2.5 in.	Small form factor (SFF) that comes in various height profiles—standard, tall, low-profile, thin, and ultrathin, such as 7 mm and 9 mm, among other height options. Height varies with platters for HDD or stacks of NVM memory dies and chips of increasing capacity. Similalrly, fewer platters or layers of dies and chips in NVM flash SSD can enable thinner devices.
3.25 in.	Enterprise class includes high capacity as well as desktop or client class at low cost. Low-cost medium capacity for nonphysical space sensitive applications such as small workstations and set-top boxes (DVRs). Low profile, 7 mm, 9 mm, among other height options.
5.25 in.	This used to be an HDD and SSD device form factor a decade or more ago. Used today for optical media such as CDs and DVDs. Note, however, that there are SAS and SATA enclosures that can fit up to four 2.5-in. hot-swap SAS or SATA HDD and SSD into this form factor.
M2	Next-generation form factor (NGFF) cards that plug into slots on a server motherboard for SSD and other devices. M2 devices can be SATA or NVMe, depending on their controller. Current M2 SSD range from hundreds of GB to TB of space capacity.
mSATA	A small form factor card that fits into some laptops and other small workstations or devices as an SSD. Small form factor predecessor to NGFF that can support hundreds of GB of SSD storage.
PCIe AiC	Various form factors and interfaces (mechanical and electrical), as discussed in Chapters 4 and 5. Key things to note are whether they are tall or low-profile cards, as well as whether they require an x1 x4 x8 or x16 slot (also pay attention to mechanicals, e.g., slot size, as well as electrical, e.g., signal requirements for card and slot).
SD and microSD	Small removable NVM memory cards (SSD) used in cameras, cell phones, drones, laptops, and other mobile or IoT devices.
USB	Various thumb drives, SSD.

Another consideration if the device is shared is whether it is active/passive or active/active—that is, can it be shared actively by two or more hosts (physical or logical) vs. static failover. Various SFF interface specifications and related information can be found at http://www.sffcommittee.org/ie/Specifications.html.

7.2.3. Enclosures

Just as everything else is not the same in most data centers, with various applications, servers, I/O networks, and storage media, there are different types of enclosures. Besides being known as enclosures, other names include drive bays or slots, disk drawers, JBOD, and expansion shelves or trays.

Enclosures can be as simple as an expansion for adding SSD and HDD or other storage devices to an existing server or storage system and appliances, or they can also include server

Table 7.2. Storage Interfaces and Packaging Options

Interface Form Factor	Description
Ethernet IP	iSCSI or object key value (e.g., Seagate Kinetic), various NAS, and other forms of storage also use IP-based access. Note that the presence of a GbE or 10 GbE port does not mean a device is iSCSI or NAS or object; check the manufacturer's specifications to see what the device supports.
Fibre Channel (SCSI_FCP and FICON)	Fibre Channel drives (HDD and SSD) still exist, but they are being replaced by SAS, SATA, and NVMe devices. While Fibre Channel drives are being replaced, servers or hosts still attach to storage systems via Fibre Channel. In addition to standard SCSI command over Fibre Channel (SCSI_FCP), some IBM mainframes also communicate with storage systems via FICON, which is another ULP that runs on Fibre Channel.
NVMe (protocol, M2, U.2 (8639), PCIe AiC devices)	NVMe is a protocol that is supported via various physical interfaces including M2, U.2 (SFF 8639), PCIe AIC, and NVMe over fabric.
SAS and SATA	Many higher-end enclosures support both SAS and SATA devices at various speeds, while lower-end enclosures may only support SATA drives. There are also some lower-end enclosures that support 12 Gbps SAS as well as 6 Gbps SATA 2.5-in. and 3.5-in. devices. In addition to supporting various SAS and SATA drives, these enclosures support SAS, SATA, as well as other host attachment. Common connectors include SAS mini and mini-HD.
SFF 8639 U.2	PCIe x4 NVMe that also supports SAS and SATA devices.
Thunderbolt	Thunderbolt 1 and 2 use a Mini Display Port (MDP) connector. Thunderbolt 3 uses a USB Type C connector with performance up to 40 Gbps for short distances (a few meters) and various topologies.

processing capabilities. The external enclosures can be direct-attached to a host or server (or storage system and appliance) or connected via a switch, hub, or expander. Some configurations, in addition to dedicated direct connection, also include daisy chains or loops with single or dual paths.

Server enclosures are measured in rack unit (RU) height, where 1.75 in. is 2U 3.5 in. is 1U, and so on. An enclosure 2U tall would be a very tight fit for a 3.5-in. drive inserted vertically, whereas twelve 3.5-in. drives placed in three rows of four horizontally will fit. On the other hand, some enclosures support 16–24 2.5-in. drives (HDD or SSD) in vertical mounting (i.e., on their sides).

Enclosures come in different physical dimensions, from slimline 1U tall, to 2U, 3U, and 4U (or taller) by standard rackmount width and varying in depth and number of devices supported. For example, there are 2U 24- and 48-drive enclosures supporting 2.5-in. devices. In addition to rackmount enclosures, there are also various type of tower, pedestal, and desktop enclosures. The number of drives supported in an enclosure varies from a single drive (e.g., a desktop shoebox) to those that hold 60, 72, 80, 90, or more in 4U or taller form factors, including top-load, slide-out.

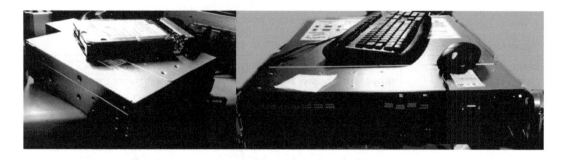

Figure 7.4   (Left) 5.25-in. × 2.5-in. hot swap. (Right) 16 × 2.5-in. enclosure.

In addition to having different types and sizes of drive slots or bays supporting various physical interfaces and protocols, enclosures also have one or more backplanes. These backplanes provide paths for data and commands as well as optional management traffic to move between drives and controllers or host adapters.

The backplanes can have expanders to increase the connectivity from a single host interface to multiple drives in an enclosure. While expanders are transparent, they can affect the speed of an enclosure or backplane. For example, if an enclosure has a 12 Gbps host port and 12 Gbps SAS drives, yet the backplane and expanders (or other components such as controllers) in the data path only support 6 Gbps, that will be the speed limit.

In Figure 7.4, on the left is a four-drive, 2.5-in., hot-swappable 12-Gbps SAS enclosure that fits into a 5.25-in. media bay of a server for storage expansion. On the right is a server with 16 × 2.5-in., hot-swappable 12 Gbps SAS (or 6 Gbps SATA) drive slots. Note that on the right of the figure there are four drives that are blue (or darker) in color. These are U.2 (8639) drive slots that support PCIe x4 NVMe, 12 Gbps SAS or 6 Gbps SATA devices.

Some enclosures are external while others are internal to servers and workstations and ranging from a few to many of different form factor sizes as well as interface types. There are enclosures that have servers (one or more) integrated to support traditional, converged, hyper-converged, cluster-in-a-box as well as storage system- or appliance-specific applications. Similarly, there are servers that have dozens or more of storage device slots or bays for adding SSD and HDD among other devices. Common host interfaces include SAS (12 Gbps), SATA (6 Gbps), PCIe (NVMe), U.2 (8639) and M.2 NVMe, and GbE (or 10 GbE), among others. Some enclosures have fixed interfaces and connector types, while others have flexible or modular I/O controllers.

Note that from a management standpoint, different hypervisors, operating systems, and utilities have varying amounts of capability for interfacing with enclosures. Likewise, some enclosures support more management capabilities than others along with their API or interfaces to server, storage, or other software.

Other common enclosure attributes include:

- Number of devices and form factor (3.5 in., 2.5 in., or 1.8 in., low or tall profile)
- Number and type of host interface, internal expanders or switches
- Front, rear, and slide-out shelf for top-loading

- Dual-port devices, active/active or active/passive I/O controllers
- Hot-swap drives, I/O controllers, cooling fans, and power supplies
- Redundant 80 Plus or 95% efficient power supplies and cooling fans
- Built-in RAID card, I/O controller, or server motherboard options
- Management interfaces and enclosure services

7.3. Aligning Technology to Application Needs

One size or technology approach does not fit all needs, criteria, or constraints. Consequently, there are many different types of storage media for different application requirements and various usage scenarios (Figure 7.5). Understand the various options to make informed decisions for various situations; don't decide just on the basis of cost per space capacity.

Storage systems, appliances, and servers implement tiered storage media, as do cloud and other services. Some context is appropriate here to clarify the difference between tiered storage (different types of storage) and tiering (moving data between tiers). Here the discussion is about tiered storage or the types of media; in Chapter 8 we will go deeper into storage tiering among other data services, and Chapter 12 looks at various storage (and server) solutions.

Note that most storage device components, including SSD and HDD, implement a variation of tiering or tiered media on a smaller scale. For example, many HDD have a small amount of DRAM as a data buffer, with some also having a NAND flash or NVM storage class memory as read and/or write cache in addition to the primary magnetic medium.

Common storage medium mistakes or missteps to watch out for include judging a device by its interface or packaging as opposed to its functionality, performance, availability, capacity, and overall economics tied to application PACE needs.

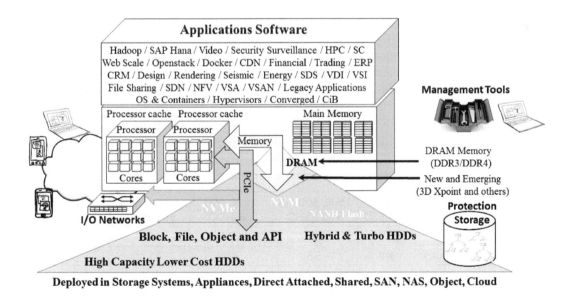

Figure 7.5 Types and tiers of storage mediums.

Table 7.3. General Application and Workload Storage Alignment

	Performance	Availability	Capacity	Economics	Type of Storage
Structured Databases Key value Transactional	Mixed I/O size, read and write workloads	HA, durable, protection copies	Varies	Cost per activity and work don	DRAM, SCM, NVM, SSD, fast 15 K and 10 K HDD with cache
Unstructured Big data, file serving	Small to large I/O size, mixed workloads	Durable and protection copies	Large capacity	Cost per capacity	Big data, file serving, videos, images
Protection copies Backups Snapshots	Larger size Sequential reads and writes	Durable copies of data and of metadata	Large capacity with data reduction	Lower cost per capacity and bandwidth	High-capacity low-cost SSD, high-capacity HDD and tape
Primary active data	Mixed random, sequential; reads, writes of various size	HA, durable, protection copies	Primary and protection copies	Cost per activity and IOP or bandwidth	Durable SSD, fast 15 K and 10 K HDD with cache
Bulk, content and reference data, active archives	Moderate, /-write sequential, larger I/O	Durable copies	Large capacity	Low cost per capacity	High-capacity HDD and SSD
Static archives and cold data	Mainly low WORM throughput	Durable copies	Large capacity	Low cost per capacity	Large capacity HDD, tape, optical
Videos Security Surveillance Images	Large I/O sequential read & write IOP for metadata	Durable copies	Large capacity	Mix of cost per capacity performance	High-capacity SSD, HDD, tape Metadata on SSD and fast HDD

Tip: Another item to watch out for is a focus on storage space capacity consolidation and efficiency optimization that just results in performance bottlenecks or other problems. Keep your applications and their various PACE attributes in mind; recall Chapter 2, where we discussed various application characteristics and associated criteria to avoid treating everything the same. Look at the functionality of the application or work being done.

Examples of online or active applications are those where the data is being worked with—read and written—such as file systems, home directories, databases, and email. Examples

Table 7.4. Storage Media Performance for Microsoft Exchange JetStress

	400 GB SAS SSD	15 K, 600 GB SAS SSHD	15 K, 600 GB HDD	6 TB, 7.2 K SAS HDD	3 TB, 7.2 K SATA HDD
Avg. total IOP/sec	2,791	334	252	174	95
Avg. reads resp. (msec)	1.4	6.6	8.2	11.6	17.4
Avg. writes resp. (msec)	1.3	4.7	33.9	33.0	49.2
Avg. IOP (read)/sec	1,486	181	139	89	52
Avg. IOP (write)/sec	1,305	154	113	85	42

of near-line or applications with idle data include reference material or repositories, backup, and archives.

Even though everything is not the same across different applications and data centers, there are similarities. Table 7.3 shows some general and common PACE characteristics together with types of storage and other considerations.

Table 7.4 shows as an example of how various storage media perform for a given workload, in this case, using Microsoft Exchange JetStress to simulate email processing.

7.4. Volatile Memory (DRAM) and Non-Persistent Storage

Memory is storage, and that storage is persistent memory. However, not all memory is persistent. The non-persistent or volatile memory includes DRAM, commonly found as the main memory (RAM) in servers, workstations, and laptops, as well as in storage systems, appliances, and networking devices. DRAM sits near the top of the server I/O storage (memory) hierarchy, below processor-based onboard cache (i.e., L0, L1, L2) and local memories. DRAM is fast for reads and writes, and relatively cost-effective. However, many systems have a finite number of physical slots for holding memory sticks (DIMM) that can be addressed by their processors.

There have been only a handful of memory generation classes from 1947 to the present day, although there have been many variations of each generation. These included volatile random access memory (RAM) in 1947, non-volatile programmable read-only memory (PROM) in 1956, static RAM (SRAM) in 1961, volatile dynamic RAM (DRAM) in 1966, non-volatile erasable programmable read only memory (EPROM) in 1971, non-volatile NOR flash in 1984, non-volatile NAND flash in 1989, and non-volatile 3D Xpoint in 2015. Note that there are many variations of these different generations, including implementations and densities, among other characteristics.

In addition to being used as main memory for holding application programs (algorithms and data structures) as well as data, DRAM is also used for holding the operating system and data infrastructure applications, device drivers, and buffers. To speed up application processors, some applications maintain large in-memory buffers as cache, including using an in-memory database, in-memory file systems, mem-cache, as well as in-memory RAM disks to hold data. For persistence, data is backed (saved or maintained) on some other form of persistent storage such as NVM or HDD.

While it is fast, DRAM is also volatile, meaning that in the absence of some form of a battery-backed (or capacitor) standby power system, loss of power means loss of data. This is where persistent and non-volatile memory (NVM) such as NAND or NOR flash, NVRAM, phase-change memory (PCM), 3D XPoint, and other emerging storage class memories (SCM) come into play. These NVM and SCM are available in NVDIMM and PCIe AiC drives as well as other packaging formats.

7.5. Non-Volatile Memory (NVM) and SSD

Solid-state devices (SSD) have existed in various packaging form factors, semiconductor media types, interfaces, and functionalities for decades. These date back to DRAM-based versions with proprietary interfaces that evolved to open systems. Functionality has also evolved, along with increases in capacity, reliability, and performance along with a reduction in physical form factor as well as cost. While IOP performance is usually mentioned as the value proposition or metric for SSD, there are also lower latency as well as increased bandwidth benefits.

The primary value proposition of SSD has been to boost the performance speed of applications, reducing wait-time overhead of servers and enabling non-volatile storage memory for harsh conditions. Benefits include more work being done in a given amount of time, reducing the time needed for a job, task, or application to occur, including response time.

A by-product of using some amount of SSD is the ability to make servers more productive, maximizing their usefulness or delaying upgrades. Besides maximizing the value of a server, SSD are also useful for getting more value out of software licenses, including databases among others, while eliminating data center performance bottlenecks.

Over the past decade or two, traditional DRAM-based SSD have evolved from having battery-backed data protection, expanding to include HDD to stage and destage persistent data, or to act as a "SCRAM" drive—a persistent memory for DRAM-based data that in the sudden case of a power loss, or shutdown command, can be quickly "scrammed" or moved to the HDD-based media, both powered by a battery or other standby power source.

In the early 2000s, with the increased capacity and lower cost of NAND flash improvements, vendors started introducing NVM as a persistent backing store for DRAM-based SSD, replacing HDD. Later in the decade, with a continued decrease in costs, increases in capacities, as well as need for durability, the proliferation of NAND flash-based SSD continued as components and as storage systems. Today, NVM, particularly NAND flash, are being used as replacements for HDD, or as a cache to complement their capacity in hybrid solutions.

Figure 7.6 shows various NAND flash SSD spanning different generations, physical packaging, and interfaces, as well as capacities. Larger capacities are shown on the left, including a 500-GB mSATA card, a 480-GB M2 card, along with 128-GB and 32-GB MicroSD cards.

Figure 7.6 Various NAND flash densities, capacities, and packaging.

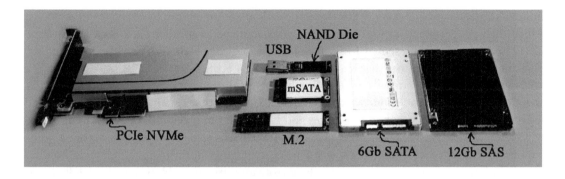

Figure 7.7 Various NVM including NAND flash SSD devices.

Note that the MicroSD cards are shown sitting on top of a MicroSD-to-SD card adapter. On the right are various SD cards ranging from 16 GB to 8 GB to 1 GB and, above that, a 64-MB (that is not a typo; it is MB, not GB) USB thumb drive with its flash chipset visible.

In addition to different physical packaging form factors and internal capacities, SSD also have various interfaces and protocols, including USB, Thunderbolt and mSATA, M2, U.2, and other NVMe, SATA, SAS, and PCIe, among others.

Figure 7.7 shows SSD devices commonly found in servers, storage systems, and other devices that vary in form factor packaging, interfaces, performance, capacity, functionality, and cost. Figures 7.6 and 7.7 illustrate how the physical form factors reduce the dies (chips) while capacities increase over time, along with different types of SSD for various use scenarios.

7.5.1. NVM, Flash, and SSD Fundamentals

Today the most common implementation of NVM for SSD in servers, storage, and other data infrastructure resources is NAND flash (commonly just flash). With flash, data bits are stored in cells, either single-level cells (SLC), dual- to multiple-level cells (MLC), vertical, or triple-level cells (TLC), or variations such as vertical and 3D, among others. 3D NAND refers to three-dimension or higher densities with more bits in a given density to increase capacity while reducing costs.

Each cell has a semiconductor floating gate that opens or closes when it is either reset (cleared) or programmed (data bit value stored). Each program/erase (P/E) cycle causes wear on the floating gate, which over time simply wears out. The number of P/E cycles, also known as durability, varies with the generation and type of NAND flash media.

Some older, lower-cost cells used in inexpensive thumb drive devices were in the 10,000 P/E cycle range, while current-generation, improved cells are in the 100,000 P/E range and higher. Keep in mind that it is only the writes (program and erase) that cause wear or decreased durability of NAND flash; you can do an unlimited number of reads. Also, write endurance measured in P/E, terabytes written (TBW), or drive writes per day (DWPD) can be managed via onboard (on or in the device) or external flash management, wear-leveling, and other software.

Context is needed to clarify that even though a file in a file system on a Linux or Windows system is deleted, its previously allocated blocks of data remain intact on the storage device unless a secure delete was done. With flash, the cells are grouped into bytes, pages, and blocks, which then get presented to hosts (servers, storage systems, controllers, adapters, laptops) for use. The flash storage presented to hosts appears as logical block address (LBA) and logical block number (LBN) sectors that may be 512 bytes or 4 KB in size.

Note that in addition to cells representing bits being allocated to the host, some cell bits and bytes are also reserved for error correcting code (ECC) and other data integrity purposes. Since low-level flash cells are grouped into pages and blocks that may be 64 KB, 128 KB, or larger, when there is a change, that entire "chunk" of storage needs to be changed. This is not a problem if the controller, adapter, or host software is managing, optimizing the changes holding data in buffers to group the writes and reduce write amplification. However, if a 4-KB chunk of data changes, additional writes will occur over time to that group of cells as data changes over time, a process known as write amplification (i.e., extra writes occur), which adds to wear and tear on the flash cells.

Flash SSD capacity and performance is achieved by using one or more dies (chips) of a given physical size, density measured in nanometers (nm), and capacity in bits or bytes. Multiple dies are grouped together to provide channels for bits to move in or out of the dies. Similar to HDD, flash and other NVM have an areal density which is the native bits per square inch (or other surface measurement).

As of this writing, state-of-the-art densities are in the range of 20–12 nm, having decreased from the range of 40 nm+ in the early 2000s. The result has been and continues to be physically smaller packaging that increases the effective usable capacity while reducing cost per capacity.

7.5.2. Flash SSD TRIM and UNMAP Garbage Collection

A by-product of flash P/E is that, over time, garbage collection needs to be performed to optimize performance. SATA SSD devices implement TRIM, while SCSI and SAS devices support UNMAP, both of which release and reset flash level blocks, pages, and cells. In addition to devices supporting TRIM and UNMAP, operating systems, hypervisors, or software-defined storage stacks, controllers, and systems also need to implement a similar feature.

In a flash SSD, the operating system, hypervisor, storage system, or other software can delete or change an item, but the old low-level blocks, pages, and cells also need to be cleared or reset as by TRIM or UNMAP. Many operating systems support TRIM and UNMAP. However, the feature may be turned on (or off) by default.

For Windows, if the following command returns "0," TRIM (or UNMAP) is enabled:

```
fsutil behavior query disabledeletenotify
```

If Windows TRIM or UNMAP are disabled, you can enable it using:

```
fsutil behavior set disabledeletenotify 0
```

For Linux systems such as Ubuntu you can do the command:

```
sudo fstrim -v <some_mount_point>
```

There's good news and bad news regarding TRIM and UNMAP: First, performance can be improved by releasing and having the SSD do a bulk clean-up to free up resources. The caveat is that by enabling this function automatically, the coordination between different software and utilities may be adversely affected from a performance perspective. The reason for this possible performance issue is that there a large number of changes have occurred, or are about to, and a large number of TRIM commands need to be processed, so the device could become very busy, causing brief delays. Check with your device and operating system or software vendor as to their specific recommendations for TRIM and UNMAP.

Areal density is the bits per square (or pie-shaped area) and an indicator of basic storage capacity. What this means is that one HDD with three platters can have the same usable capacity as an HDD with only two platters of the same form factor. The HDD with three platters has a lower areal density, making up for that with extra platters, read/write heads, and packaging.

7.5.3. Different Types and Classes of SSD

There are different types of NVM, including flash that is optimized for various needs. Some are optimized for frequent updates, combining DRAM or other technologies to enhance endurance and minimize the impact of wear. Others are optimized for WORM or infrequent writes while maximizing capacity at lowest possible cost.

Storage-class memories (SCM) are effectively a hybrid, or best of two worlds, that combine read and write endurance with performance closer to that of DRAM than NAND flash. However, unlike DRAM, SCM have the persistence of NVM with price points per given capacity between DRAM and NAND flash.

Does this mean at some point DRAM could be replaced by SCM? Perhaps, as there are many different types of SCM and next-generation memories in various stages of development. Some examples of SCM include 3D XPoint and PCM, and various NVDIMM, among others.

For the time being, look for SCM to take on more work where both read and write performance as well as endurance combined with persistence are needed. These include in-memory databases and large buffers or caches that need to maintain crash or controlled shutdown consistency. A benefit is being able to have data preserved and consistent at system start-up without having to cause fast applications to wait while caches are rewarmed (refilled and populated).

7.5.4. NVM and SSD Considerations

Like any storage media, NVM and SSD are safe for storing data as long as you follow best practices, including making sure they are protected with a clone, copy, sync, replica, or backup.

> *Tip:* Also, pay attention to context as well as big "B" as in Bytes along with little "b" as in bits when looking at as well as comparing SSD speeds, feeds, and associated specifications.

While often positioned as consuming lower power vs. HDD—some creative sales or marketers have even said that SSD require no power—the reality is that they do. The amount of power may be less than for some HDDs but some larger SSD may consume more power than some lower-performance HDD of different capacities. SSD do consume power; they do get warm, as anything that consumes power generates heat (BTUs). The key is how much work is being done. In a given density and footprint, how much work is being done per capacity? Pay attention to PCIe AiC and their power requirements, as well as other devices.

There is a finite number of times that a NAND flash cell can be programmed and erased (P/E), which is how writes are done. The number of P/E cycles in today's devices is high tens if not hundreds of thousands, differing with the type of media, as some SSD are optimized for higher capacity, fewer writes, and lower cost.

More cells stacked means more bits equals more capacity in a given footprint, which can reduce cost. However, the cost savings for more capacity also mean more impact when doing writes. Note that it is only writes that cause flash SSD devices to wear out; reads are unlimited.

Other considerations include the type of flash translation layer (FTL) management software along with durable RAM and other technologies that combine to increase the durability of the device. In this context, "device" means a USB thumb drive, MicroSD, SD, M2, mSATA, PCIe AIC, drive form factor, custom card or board, among other NAND flash SSD.

Another layer of flash SSD endurance and durability management exists with server operating systems, hypervisors, device drivers and other software, and adapters, along with storage systems or appliances.

> *Tip:* Look beyond the cost per capacity to see what the vendor shows for TBW, as this is an indicator of how many times a SSD device/solution can be written (i.e., its durability). For example, on a 10-GB SSD, 1 TBW means the device can be completely written and rewritten up to 10 GB × 100 times; 10 TBW for the same example means 10 GB × 1000 times, and so forth.

Keep in mind that a little bit of SSD in the right location for a given application scenario can have a large benefit. The trick is knowing or figuring out where that location is, what type of NVM SSD to use, how to use it, as well as how to implement to reduce bottlenecks without introducing management complexities.

7.6. Magnetic Hard Disk Drives

The magnetic hard disk drive (HDD) first appeared in 1956, about ten years later than memory such as RAM, which appeared in 1947. Over the subsequent 60-plus years it has gone through numerous changes, from physical size to usable storage space capacity, price, and performance, to availability and accessibility, among other changes. In the past, HDD were relied on for both performance as well as capacity and for availability or protection of data.

Over time, more of the performance workload has shifted toward RAM, NVM, and SCM storage mediums, with HDD being used for capacity, as well as a hybrid mix of capacity and performance. HDD are also used for both primary storage as well as for protection copies, including archives, backups, BC/DR, and other uses.

HDD attributes include interface type and protocol along with single- or dual-port support, the number of protocol command queues, self-diagnostics and power-on self-tests (POST), sector size, usable and raw capacity, and revolutions per minute (RPM) as part of its performance, read, and or read/write cache.

Many HDD have some amount of DRAM data buffers, which may range from a few megabytes to hundreds of megabytes. Also, many HDD are seeing additional NVM being added as a persistent read cache along with applicable performance-optimized algorithms in their firmware.

HDD are packaged in various form factors, the primary ones today being 2.5 in. and 3.5 in. While there are some performance models, most 3.5-in. disks are capacity-oriented, some optimized for lower power, lower performance, and colder data while others are for active workloads.

For example, there are low-cost desktop or client 6-TB (or larger) 3.5-in. HDD with SATA interfaces, as well as enterprise-class models of the same capacity, but with SATA as well as 12 Gbps SAS. The lower-cost drives, while having 6 Gbps SATA interfaces, may lack performance and other features including BER and AFR metrics compared to enterprise-class models.

Note that HDD, in addition to being SFF 2.5-in. or large form factor 3.5-in. size, can be standard, tall, low-profile, or thin thickness (7 mm or less). The number of platters stacked inside a HDD, along with their associated electronics, determines the device thickness. Capacity can be improved by adding more platters of a given areal density, but doing so will increase the height.

What differentiates the different drives is what's inside—for example, the number or type and processing capabilities of the internal processors (chips), DRAM data buffer size, NVM read cache size, NVM write buffers, as well as algorithms for performance and other optimization. A common mistake is judging an HDD simply by its interface and capacity as opposed to what's inside. The smaller-form-factor 2.5-in. drives are available in both enterprise as well as client desktop models optimized for different PACE attributes.

Externally, besides physical form factor and mounting screw locations, HDD have different interfaces for access, such as SAS and SATA, among others. Internally (see Figure 7.2), there are some number of platters (disks) attached to a motor (spindle) that spins the disks past a read/write head at a given RPM. The read/write heads are fixed to the end of an actuator that swings or flies out and over the disk platter surface.

Data is organized on HDD platters based on circular tracks made up of sectors that separate data bits. Multiple platter tracks are referred to as cylinders and sometimes as cylinder sector heads or CSH addressing. While there is some FUD (fear, uncertainty, and doubt) that HDDs still use CSH addressing, the reality is that, like NVM SSD and other forms of physical and virtual storage, HDD utilize LBA and LBN addressing.

One technique that was used in the past for boosting HDD performance was short stroking. Part of an HDD's space capacity was given up in exchange for restricting the HDD to using only certain tracks and sectors. A similar approach that was also used before the popularity of SSD was simply to install more HDD for their performance and leave a large portion of the space capacity underutilized. The idea of short stroking is to keep most of the I/Os close to the center of the HDD, where performance should be faster. HDD use techniques such as zone bit recording (ZBR) to balance the geometry of inner and outer track circumstances.

Areal density is the bits per square (or pie-shaped area) and an indicator of basic storage capacity. What this means is that one HDD with three platters can have the same usable capacity as an HDD with only two platters of the same form factor. The HDD with three platters has a lower areal density, making up for that with extra platters, read/write heads, and packaging.

Other HDD topics include spiral transfer data rate for moving large sequential data with sustained performance, as opposed to smaller random IOP and associated seek time. Read-ahead is a technique to pre-fetch data at the drive level to speed up subsequent I/O operations. Note that read-ahead is also done in some adapters, raid controllers, storage systems, and servers.

Historically, RPM was a primary focus of or factor for determining speed performance. RPM still plays a significant role in performance. However, there are other considerations,

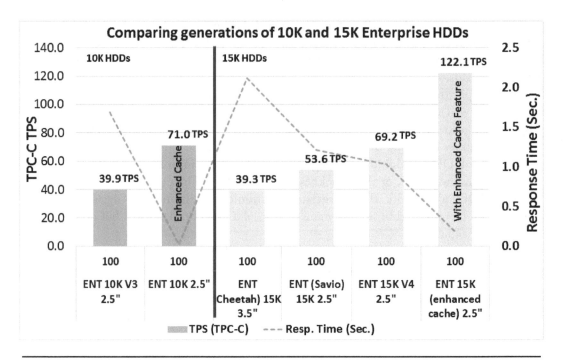

Figure 7.8   Comparing performance of HDD over time.

including cache, read-ahead, I/O grouping, and other optimizations. There are various RPM for different types of drives, such as 5400 (5.4K), 5.9K, 7.2K, 10K, and 15K, with lower speed for lower-cost, capacity-centric workloads and higher speed for faster, lower-latency loads.

Some newer lower-RPM HDD can perform as fast or faster than their predecessors of the same or larger form factor. This is due to internal algorithms and firmware optimization, among other techniques. For example, there are some 2.5-in. 10K RPM devices that perform faster than previous-generation 3.5-in. 15K RPM devices. However, your performance may vary.

> *Tip:* The point is to look beyond just the RPM and interface; look more closely at the drive and what it is capable of for your environment. Figure 7.8 shows how various generations of 10K and 15K RPM drives have improved on performance over time. The performance workload is for database OLTP transactions per second.

7.6.1. Types and Classes of HDD

There are different types of HDD besides the type of interface, form factor, and capacity, which also include enterprise and desktop classes. There are also performance-optimized as well as capacity-optimized models, including some with firmware to support video surveillance storage, NAS, or general storage.

Figure 7.9 shows an OLTP database workload transactions per second (TPS) for various numbers of users with different types of enterprise-class 2.5-in. SAS HDD. The TPS (dashed line) workload is scaled over various numbers of users (1, 20, 50, and 100), with peak TPS per HDD shown. Also shown is the space capacity used (bottom of stacked bar, dark color), with total raw storage capacity in the crosshatched area (top of stacked bar). The drives include

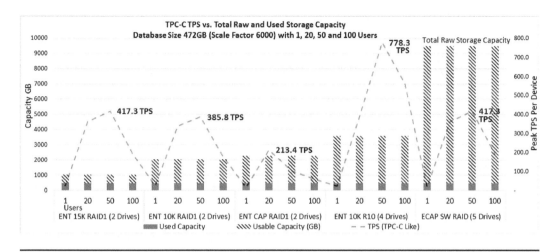

Figure 7.9   TPS with cost per TPS and storage space capacity used.

12 Gbps SAS, 600-GB, 15K RPM, and 1.8-TB, 10K RPM with built-in read/write cache acceleration, along with 7.2K RPM 2-TB 12 Gbps SAS HDD, all of which have 2.5-in. form factors.

In Figure 7.9, the 2.5-in. 1.8 TB SAS 10K HDD with performance-enhanced cache while not as fast as the 15K, provides a good balance of performance, space capacity, and cost effectiveness. A good use for 10K drives is where some amount of performance is needed as well as a large amount of storage space capacity for less frequently accessed content.

A low-cost, low-performance option would be the 2 TB 2.5-in.-capacity HDD, which have good cost per capacity, although lower performance than a 15K or 10K drive. A four-drive RAID 10 along with a five-drive software volume (Microsoft Windows Server) are also shown. For apples-to-apples comparison, look at costs vs. capacity, including number of drives needed for a given level of performance.

Figure 7.10 is a variation of Figure 7.9 showing TPC-C TPS (solid bar) and response time (dashed line) scaling across 1, 20, 50, and 100 users. The 15K HDD has good performance in

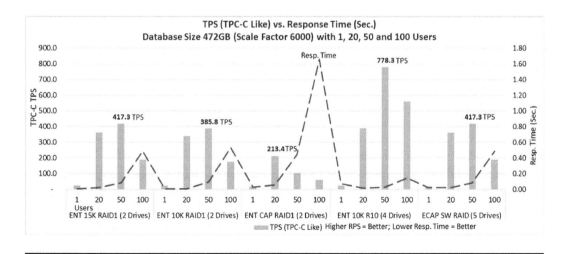

Figure 7.10   Database TPS scaling with response time.

Table 7.5. Performance, Capacity, and Cost Analysis for Small File Processing

	Average File Reads per Second (RPS)	Single Drive Cost per RPS	Multiple Drive Cost per RPS	Single Drive Cost per GB of Capacity	Cost per GB Usable (Protected) Capacity	Drive Cost (Multiple Drives)	Protection Overhead (Space Capacity for RAID)	Cost per Usable GB per RPS
15K R1	3415.7	$0.17	$0.35	$0.99	$0.99	$1190	100%	$0.35
10K R1	2203.4	0.40	0.79	0.49	0.49	1750	100%	0.79
7.2K CAP R1	1063.1	0.38	0.75	0.20	0.20	798	100%	0.75
10K R10	4590.5	0.19	0.76	0.49	0.97	3500	100%	0.76

apples-to-apples RAID 1 comparison. Note the performance with four-drive RAID 10 using 10K RPM HDD that provides good speed but requires more drives and costs.

In terms of cost per space capacity, the capacity drives have a good cost per gigabyte. A hybrid solution for an environment that does not need ultrahigh performance would be to pair a small amount of flash SSD (drives or PCIe cards) with 10K and 15K performance-enhanced drives with the capacity HDD along with cache or tiering software.

Table 7.5 is an example of looking at multiple metrics to make informed decisions as to which HDD will be best suited to your specific needs. For example, RAID 10 using four 10K drives provides good performance and protection along with large usable space. However, it also comes at a budget cost (higher price).

Table 7.5 shows that the 15K HDD on an apples-to-apples basis (i.e., same RAID level and a number of drives) provides the best performance. Also factor in space capacity, performance, different RAID levels, or other protection schemes along with costs.

Capacity 2-TB HDD have a low cost per capacity, but they do not have the performance of other options, assuming your applications need more performance. The right HDD for one application may not be the best one for a different scenario; multiple metrics as shown in Table 7.3 need to be included in an informed storage decision-making process.

7.6.2. HDD Solid-State Hybrid Disk Considerations and Trends

Several times over the past decades people have declared the end of the HDD due to running into a brick wall—the super parametric barrier wall. Each time it has looked like the end of the HDD might finally occur. However, some new improvement in recording has always occurred, such as perpendicular to replace horizontal, new surface coatings to increase aerial density, and other techniques. Historically, an issue with magnetic recording device has been the super parametric barrier, or how close data bits can be placed together safely without interfering with other adjacent data bits.

Different HDD use various encoding schemes for how the data is organized physically on a drive, such as perpendicular recording (also used with tape). In perpendicular recording, bits are stored perpendicular to the platter surface, thus taking up less space than laying them out horizontally. Perpendicular recording enables more bits to be placed in a given amount of space (aerial density) vs. legacy horizontal recording.

Other low-level techniques for increasing aerial density and consequently capacity include shingled magnetic recording (SMR), in which tracks of data are organized closer together. For SMR, think about shingles on a roof that overlap each other, or rows of corn in a corn field. Think about how many rows of corn you could plant in an acre or given the amount of space if the width of the rows or the distance between the corn plants could be reduced, without impacting the corn plant itself.

There are various implementations of SMR, some of which are plug-and-play and transparent to a given operating system, controller, or hypervisor, while others rely on software and device drivers to provide additional benefits. The advantage of SMR is that more data can be stored in a given amount of space, with reads having similar performance to a normal drive. Writes can be slower, given the extra time needed to group the data and then write it, making SMR better suited to archival or static reference data. On the other hand, put some fast cache such as RAM or NVM in front of SMR and you can have an interesting solution for some applications.

Another type of HDD low-level recording or data encoding technology is heat-assisted magnetic recording (HAMR). The benefit of HAMR is also to increase the amount of data that can be safely saved in a given aerial density footprint. Traditionally, some amount of clean air exists inside a HDD. However, newer drives are appearing that have helium gas inserted into the drives——not to make them lighter, but rather to improve on capacity density in a given physically smaller footprint.

Some HDD also have additional NVM (non-flash) for persistent write cache, enabling them to function in a turbo, enhanced cache, or hybrid solid-state HDD (SSHD) to boost performance. These enhanced-cache drives boost performance in large-capacity drives to narrow the processor–performance–capacity gap. Note that some early hybrid HDD (HHDD) or SSHD supported only basic read optimization. Today's HHDD, SSHD, or enhanced-cache and turbo drives optimize both reads and writes.

Another type of HDD is the emerging Ethernet attached key value or object drives. These drives, such as the Seagate Kinetic, use a GbE interface instead of SAS or SATA. However, the GbE does not support iSCSI, NAS, or other common IP-based storage access protocols. Instead, these drives have an API that gets implemented by operating systems, hypervisors, file systems, bulk object, cloud, and other storage software for storage access using a key value. Unlike traditional drives that have software protocol and management support built into operating systems, file systems, hypervisors, and other applications, these new devices need more time for their new software-defined infrastructure to be rolled out.

7.7. Magnetic Tape, Optical, and Removable Media

In addition to semiconductor-based storage class memories including NVM such as NAND flash, 3D XPoint, and PCM among other SCM, as well as magnetic HDD, other types of storage media include magnetic tape and optical media.

Despite being declared dead (like HDD) for decades, tape (magnetic) is still being enhanced as well as being used for various applications that leverage its off-line, removable, and relatively low cost per capacity. Similar to a powered-off HDD or SSD, tape consumes no power when it is not in use. The difference is that HDD and SSD are usually powered on to leverage their storage capabilities, while tape media tends to spend more time powered off. In other words, tape is well suited for cold or inactive data vs. hot, warm, or active data. Granted, the tape can be front-ended with various solutions that combine NAS or a virtual tape library or other variations to make tape appear virtually on-line or near-line.

Optical storage media include CD, DVD, and Blu-Ray, among others. Other forms of removable media besides tape and CD include HDD (or SSD) packaged in hot-swap specialized container packaging for mobile and portable use. Some of these removable devices have proprietary interface and docking or packaging interfaces, while others utilize USB, Firewire, Thunderbolt, and RDX, eSATA, or SATA.

Removable media require a drive into which a tape cartridge or optical media can be inserted. Optionally, for HDD solutions, the HDD exists inside a housing or portable enclosure that may have a proprietary docking interface to a docking device, or a USB, Thunderbolt, eSATA, or SATA interface. Tape and optical drives can be standalone or installed in a stacker, autoloader, or larger robotic library.

Tape libraries or robotic libraries can support multiple tape drives and tens to hundreds or thousands of tape cartridges. These libraries also feature import/export areas where individual tapes can be ejected and removed to be sent to an on-site or off-site storage facility or vault for safekeeping.

Optical attributes and use cases include:

- Relative low cost per capacity
- For CD and DVD, broad interoperability with various devices
- Various types of media, including WORM and read/writable
- Some form of archiving, data portability
- Standalone device or robotic library and autoloaders
- Support bootable formats

The challenge with optical media is that while costs have gone down, the relative capacities have not increased compared to other magnetic or semiconductor-based technologies. Tape, on the other hand, continues to be enhanced in capacity, performance, and functionality [e.g., compression, encryption, linear tape file system (LTFS) and partitions, bit error rate], which help make it cost-effective. Benefits and attributes of magnetic tape, in general, include a good bit error rate per capacity, high capacity for low cost, low power when in use, zero power used when off-line, and good streaming access for bulk reads and writes.

With LTFS enhancements, a small directory, index, or metadata repository is placed at the beginning of a tape cartridge. LTFS-enabled drives and associated software are then able to store location information in the LTFS header or directory on the tape cartridge. LTFS-enabled software is then able to tell the tape drive where to position itself on the tape to start reading or writing data.

Tip: Use cases for tape include bulk data movement, interchange, master backups, archives, BC/DR, and serialized access by which data is streamed (read or written) from fast storage. There is a myth that tape is slow, although some of the older generations can be. Often, however, the issue with tape is that it is only as fast as the HDD or SSD that is feeding it, or as fast as the network speed at which the data is moving.

On the other hand, while tape is fast from a streaming read or write throughput, it is not optimized for concurrent random access like HDD and SSD media. In addition to backup and archive, tape is also good for a long-term, cold storage medium, as well as for moving data off-site, or for supporting bulk data movement to or from a cloud or other data center location. Whereas HDD and SSD storage are often compared via an activity such as IOP or response time latency, along with bandwidth, tape's key metric from a performance standpoint is bandwidth or throughput.

Two main tape formats today are Enterprise tape from IBM and Oracle, as well as Linear Tape Open (LTO). Enterprise tape is a bit of a misnomer in that LTO is also used in enterprise environments, as well as small/medium business. However, Enterprise tape, also known as high-end tape, which supports both IBM mainframe and open systems, is a higher-end technology used at a larger scale in big data centers.

As of this writing, Enterprise tape products support Oracle StorageTek T10000D, which has 8.5 TB of native capacity and 252 MBps speed. IBM TS1150 supports 10-TB native capacity per cartridge with native performance of 360 MB/sec (i.e., 2880 Gbps or 5760 Gbps with 2:1 compression). With new generations, there are also processing improvements including faster compute, as well as enhanced algorithms, which is how compression has increased with tape.

As with any speed or specification, the information given here may be dated or changed by the time you read this. Various LTO generations and specifications are shown in Table 7.6.

> *Tip:* Note that there is also plenty of FUD with tape, both for and against, similar to SSD vs. HDD. Some apples-to-oranges comparisons include tape long media life vs. disk drive replaced at 3–5 years, as opposed to how long media on a HDD can be used. What needs to be discussed more often, however, is being able to read older generations of SSD, HDD, or tape media and devices, both physically as well as via software. You can of course preserve a physical device and associated cabling, and controllers for reading older data and devices, but what about the software that formatted the data—will that still exist? It is not uncommon to hear stories that a tape (or disk) can't be read, so data was lost. Or that the data can be pulled off of the tape or disk, perhaps even put onto SSD, except the software to read the data format (i.e., how it was defined) no longer exists.

Tape is often criticized as being slow, which it can be for trying to do random serial operations. However, for streaming data such as a bulk backup, archive, or a save operation from fast HDD or SSD, tape is a very good performer. And the same is true for bulk restores.

Table 7.6. Linear Tape Open (LTO) Speeds and Feeds

	LTO Generation							
	3	4	5	6	7	8	9	10
Native capacity	400 GB	800 GB	1.5 TB	2.5 TB	6 TB	12.8 TB	25 TB	48 TB
Compressed	0.8 TB	1.6 TB	3 TB	6.25 TB	15 TB	32 TB	62.5 TB	120 TB
Compression	2:1	2:1	2:1	2.5:1	2.5:1	2.5:1	2.5:1	2.5:1
Native speed (MBps)	80	120	140	160	300	472	708	1100
Compressed speed (MBps)	160	240	280	400	788	1180	1770	2750
Partitioning			Yes	Yes	Yes	Yes	Yes	Yes
Encryption		Yes	Yes	Yes	Yes	Yes	Yes	Yes
WORM	Yes	Yes	Yes	Yes	Yes	Yes	Yes	Yes

Tape is also sometimes considered unreliable, but it has a bit error rate (BER) per hard error of 10 ^ 19 vs. 10 ^ 16 or 10 ^ 17 for some HDD and SSD. Hence, tape is better together with other media, and what is best for you depends on what your application PACE needs are.

Tip: Keep not only a copy of your older software around; also do a physical-to-virtual (if not already virtual) conversion of your systems and software, then archive the virtual machine.

Another tip: While you may create an archive on tape, optical, HDD, or SSD and put it on the shelf for years or decades, keep in mind that these archives require regular maintenance per the manufacturer's recommendations. Also, do a data or media migration every decade or so. It might cost you more or change the economics of a long-term archive, but it will also change the economics in terms of actually being able to use the data.

7.8. What's in Your Storage Device Toolbox?

Some tools among others that I have in my storage device toolbox include Acronis, Aomei, Clonex, Clonezilla, Cosbench, Dstat, Disk2vhd, fdisk, Fio, Gpart, Grub, Hdparm, Iometer, Microsoft (Diskpart, Diskspd, Diskperf, Perfmon, wmic), Paragon, Rufus, Samsung Magician, Seagate (Discwizard, Seatools), Treesize, Vdbench, and various OS and hypervisor tools.

Also, in addition to software tools and utilities, other items in my storage device toolbox include various drive connector cables and other hardware devices. Oh, and I also have some virtual machines that are powered down containing an "active archive" of older backup and data protection systems; just in case I need to bring data back, I have the software ready.

7.9. Common Questions and Tips

How many times can a tape, HDD, or SSD be used before it wears out? It will depend on the specific make, model, and media type; consult your manufacturer's recommendations.

What is the best storage medium to avoid UBER and bit rot? To protect against unrecoverable bit error (UBER) and bit rot, while good storage media devices are important (check their bit or block error rate specifications—the higher the number, the better), proper data protection techniques are also essential. This includes making sure you have implemented the 4 3 2 1 rule: at least *four* copies of *three* or more versions spread over *two* different systems (or devices) with at least one of those off-site. Also perform regular file system integrity checks and repairs to detect problems proactively, as well as leverage other data integrity features including checksums. We'll talk more about data protection in later chapters.

What is the difference between a NVDIMM and NVM SSD? Both use semiconductor solid-state persistent memories, but different packaging. They could be NAND flash, PCM, 3D XPoint, or other SCM-based memories. However, NVDIMM are traditional DIMM form factors as opposed to a drive or PCIe AiC. Note that there are early-generation NVDIMM that are DRAM on one side and NAND flash on the other. They look like and behave like DRAM to the server, yet have persistence similar to an SSD.

Why use drive form factors vs. PCIe AiC? For some situations, it is a matter of what exists for adding storage, extra PCIe card or other slots, or drive slots. There can also be considerations of hot-swap and accessibility, interface, and performance, among others.

What's the best storage medium? That depends on what your needs are. Thus, the best type of storage medium is what does what you need it to do.

Why cover HDD and tape when the future is all about SSD? At least well into the next decade (i.e., the 2020s), HDD will still need to be around (pun intended) to support storage capacity needs alongside SSD and other technologies. Thus, they remain applicable, although where those HDD are located, such as at or in a cloud, will change. You may not have any HDD at your site, but if you are using a cloud or other service, the chances are that they will be using HDD in some shape, form, role, or function.

What's the difference between regular memory and storage class memory? Regular RAM is volatile, meaning if you remove power, you remove or lose the data. SCM have performance closer to RAM with the persistence of traditional NVM. SCM will initially fit from a PACE standpoint between legacy RAM and traditional NVM, perhaps over time taking over more of each of those roles, or at least changing them.

For example, as SCM become more cost-effective and assuming NAND flash costs come down further, and capacity goes up, these will take over more of what HDD are used for; then HDD will assume more of what tape has been used for. Likewise, as SCM become faster, they will take over some RAM functions, enabling more persistent fast memory to be deployed in different scenarios including as cache or buffers.

Is there a difference between SATA HDD and other drives? There can be, in that some lower-end drives are only available in SATA, or perhaps with a USB or Thunderbolt interface, while other higher-end or enterprise-class drives have SATA as well as SAS interfaces. Be careful judging a drive by its interface alone, because it's what's inside that determines the type of drive and functionality, as well as other PACE attributes.

How can an SSD wear out if there are no moving parts? While NVM SSD such as NAND flash have no mechanical moving parts, they do have semiconductor "floating" gates that will wear out over time. The amount of time varies based on the generation and type of device, but can be in the range of tens to hundreds of thousands of program/erase (P/E) cycles per cell. A good metric to look at is how many TBW or DWPD have been done to a device vs. what the manufacturer-specified metric is for a device.

What's wrong with using RPM to gauge storage performance? RPM can be a good indicator of HDD performance. However, old rules of thumb need to be updated as to how many IOP a drive of a given RPM can perform. There are many factors that go into HDD performance, RPM being one of them, along with drive form factor, read-ahead, firmware acceleration, read/write cache, as well as enhanced cache or turbo modes.

Do IOP exist in file and object storage solutions? Yes, although they may be referred to by different names, such as gets, puts, reads, writes, lists, deletes, or other forms of activity. Keep in mind that an IOP is simply a measurement of some I/O input, output, or other command operation per second, which can be block, file, or object, including transactions among others.

Why can't everything simply go to SSD? There are some technical reasons, along with people preferences among others. However, most reasons simply come down to economics, which is the cost for NVM SSD (which is always improving), as well as availability, or how much product can be produced by the manufacturers.

I have a heavily utilized server and application that is using flash SSD already. Could NVMe help reduce the CPU bottleneck? Generally speaking, NVMe uses less CPU per I/O, which means that, doing the same amount of work, your I/Os should go faster while using less CPU. It is possible that attaching your SSD via NVMe to a server could help to free up CPU to do more work or make your applications more productive.

Why use SAS or SATA when NVMe is now available? Good question! If you have the need for speed and sufficient budget, then load up on as much NVMe as you can. On the other hand, if you need to stretch your budget to get as much NVM as possible, then use NVMe for fast, low-latency connectivity as well as where you need to improve CPU usage. Then use SAS and SATA for lower-cost, lower-performing, higher-capacity, fast NVM as well as traditional storage in a tiered access and tiered storage environment.

What can you do to reduce or manage flash memory wearing out? The first thing is to be aware that flash memory does wear out with time. The second thing is to look for solutions that support a large number of TBW, or full drive writes per day, as well as have good optimizations to extend useful life. Besides TPW and DWPD, other metrics to look at include program/erase (P/E) cycles (not all vendors provide this), as well as the bit error rate (BER). In other words, look beyond the cost per capacity, keeping other metrics that matter in mind. Also, have good backup and data protection practices in place, and test the restore!

7.10. Learning Experience

Some useful Linux commands to display disk (HDD or SSD) information include *hdparm* to display information, *dd* (read or write to a device), *sudo lshw −C disk* (displays disk information), *gparted* and *fdisk* (device partitioning), *mkfs* (create a file system) and *gnome-disks* (device configuration, SMART metrics, and basic benchmark test tools). Windows commands include *wmic, diskspd, diskpart,* and *perfmon,* among others.

If you need to enable disk performance counters on Windows Task Manger, simply close Task Manager, then, from a command prompt as Administrator, enter *diskperf−Y,* then reopen Task Manager; you should see disk performance stats. Figure 7.11 shows examples of two different tools to monitor HDD and SSD performance.

The Windows *fio.exe* command below (in a single line) performs 70% reads and 30% writes, random with a 4K I/O size with 4K block alignment to a test file called TESTfile.tmp on device X: with a size of 100,000 MB. Some of the other parameters include 32 jobs with an I/O depth of 16, ramp-up time of 3600 sec, a run time of 14,400 sec, and results written to a file called FIO4K70RWRAN.txt in the local directory.

Figure 7.11   (Left) Samsung Magician. (Right) Ubuntu storage test tools.

```
fio --filename="X\:\TESTfile.tmp" --filesize=100000M --direct=1
--rw=randrw --refill_buffers --norandommap --randrepeat=0
--ioengine=windowsaio --ba=4k --rwmixread=70 --iodepth=16
--numjobs=32 --exitall --time_based --ramp_time=3600
--runtime=14400 --group_reporting --name=FIO4KRW --output
FIO4K70RWRAN.txt
```

Here is an example of using Microsoft Diskspd to test device file X:TESTfile.tmp for 14,400 sec, display latency, and disable hardware as well as software cache, forcing CPU affinity. The test does 8 KB random 100% reads with I/Os to the test file sized 100 GB with 16 threads and 32 overlapped I/Os.

```
diskspd -c100g -b8K -t16 -o32 -r -d14400 -h -w0 -L -a0,1,2,3
X:\TESTfile.tmp
```

The following is a bit more involved than the previous example and shows how to run a synthetic workload or benchmark. This uses vdbench (available from the Oracle website) as the workload generator and a script shown below. You can learn more about vdbench as well as other server, storage I/O performance, workload generators, and monitoring tools at www.storageio.com/performance.

With vdbench downloaded and installed, the following command (as a single line) calls a vdbench file called seqrxx.txt (shown below) with these parameters. Note that this workload assumes a file exists that is larger than 200G (i.e., the number of spaces specified) on the volume being tested.

Also, note that this assumes a Windows device; for Linux or Unix, other parameters would be used (see Vdbench documentation). The results are logged to a folder called NVMe_Results (or whatever you call it) at intervals of 30 sec, with each iteration or step running for 30 min. The steps are for 8K, 128K, 1M using 0, 50, and 100% random of 0, 50, and 100% reads.

```
vdbench -f seqrxx.txt dsize=200G tthreads=256 jvmn=64
worktbd=8k,128k,1m workseek=0,50,100 workread=0,50,100
jobname=NNVME etime=30m itime=30 drivename="\\.\N:\TEMPIO.tmp"
-o NVMe_Results
```

where seqrxx.txt is a file containing:

```
hd=localhost,jvms=!jvmn
#
sd=sd1,lun=!drivename,openflags=directio,size=!dsize
#
wd=mix,sd=sd1
#
rd=!jobname,wd=mix,elapsed=!etime,interval=!itime,iorate=max,fort
hreads=(!tthreads),forxfersize=(!worktbd),forseekpct=(!workseek),f
orrdpct=(!workread),openflags=direction
```

7.11. Chapter Summary

An industry trend is that more data is being generated and stored. Another trend is that some of that data is more active than others, requiring more I/O performance. Also, data is finding new value over time, which means it needs to be protected, preserved, and served for future use when needed, requiring even more, storage.

What this means is that while there is a focus on space capacity for some applications, there is also a need for improved performance. However, storage media such as NVM SSD and magnetic forms (HDD and tape) are not increasing in performance, capacity, and economic attributes in a balanced way.

SSD performance is much better on both reads and writes than HDD. Likewise, DRAM and SCM have better performance than NAND flash and some other SSD. NVM continue to improve on durability/endurance, as well as increasing capacity per areal density footprint together with cost reductions.

However, while HDD lag SSD in performance, they continue to ride the same curve of increased capacity, lower cost, and better availability, thus staying ahead yet following tape in some areas. Likewise, tape continues to improve in costs per capacity and availability, but even with improvements such as LTFS, it still lags SSD and HDD in various performance areas.

All this means leveraging these technologies in ways that complement each other in hybrid uses to derive maximum value in addressing application PACE requirements.

General action items include:

- Avoid judging or comparing a drive by only its cover, interface, cost, and capacity.
- Know your application PACE needs to align the applicable type of storage medium.
- Leverage different tiers and categories of storage for various PACE needs.
- There is more to determining HDD performance than the traditional RPM focus.
- Keep in perspective storage I/O performance vs. space capacity needs.
- The best storage is the one that meets your application needs.
- Memory is storage, and storage is persistent memory.

Bottom line: As long as there is a need for information, there will be requirement need for more storage, which is persistent memory.

Chapter 8

Data Infrastructure Services: Access and Performance

Data services are the tools for enabling data infrastructures.

What You Will Learn in This Chapter

- Data services enable data infrastructure functionality and management
- Common data services and feature functionality
- Where data services and functionality are implemented
- Everything is not the same, even across data service features
- Breadth of features along with depth of functionality
- What to look for and consider to compare features
- Access, performance, and management services

This is the first of four chapters where we look at various data services that enable data infrastructures. Keep in mind that data infrastructure includes legacy, Software-Defined Infrastructure (SDI), and Software-Defined Data Centers (SDDC), as well as Software-Defined Data Infrastructures (SDDI). Data services are the tools that enable data infrastructure functionalities, which in turn support various applications. Common data service functionalities include access, performance, availability, and capacity, as well as management that is implemented in various locations. The focus of this chapter is data infrastructure services including access, performance, and general management. Chapter 9 covers availability (data protection, RAS, and RAID); Chapter 10 covers data protection including backup, snapshot, CDP, and security; while Chapter 11 looks at capacity and data footprint reduction (DFR), also known as storage efficiency. Key themes, buzzwords, and trends addressed in this chapter include access, abstractions, block, file, object, and application program interfaces (APIs) as well as performance, productivity, and effectiveness, among others.

8.1. Getting Started: What's in Your Server Storage I/O Toolbox?

Data services are the tools for enabling, configuring, allocating, or provisioning access and consuming, accelerating, protecting, monitoring, tuning, troubleshooting, remediating, reporting, and, generally speaking, managing data infrastructure resources and services for supporting IT or business applications. The focus of this chapter is access, management, and performance for legacy, SDI, SDDC, and SDDI environments. In other words, in the data infrastructure toolbox are data services for managing resources, including hardware, software, and services as well as other tools. Data services are implemented in various locations ranging from a higher altitude (i.e., in the layers or stacks) shown in Figure 8.1 as well as at a lower level or altitude in the components.

Data services are the features (Table 8.1), functionality capabilities that enable data infrastructures to support upper-level applications and IT or cloud services. Since everything is not the same across different environments, sizes and types of organizations or applications, data services also vary. Some data services are focused on size and scale, others on smaller sites.

Data services are implemented in various locations including applications, databases, file systems and volume managers, hypervisors, and other software-defined storage management solutions. In addition to existing in servers, either as a dedicated appliance or as part of a general-purpose, converged, hyper-converged, virtual, cloud, or other software-defined implementation, data services also exist in storage systems and appliances, as well as in some networking devices.

Where is the best place for data services feature functionality to exist? That depends on what you need to do, the type of solution and the market or customer usage scenario being focused on. Data services can be found in from individual device components to servers, storage systems, and appliances, network devices, physical, virtual, container, and cloud servers, or in services integrated, or layered on top of, other functionalities. Locations where various data services are implemented, shown in Table 8.2, range from the lower level (lower altitude) to higher altitude further up the stack in applications on a physical, virtual, container, or cloud server.

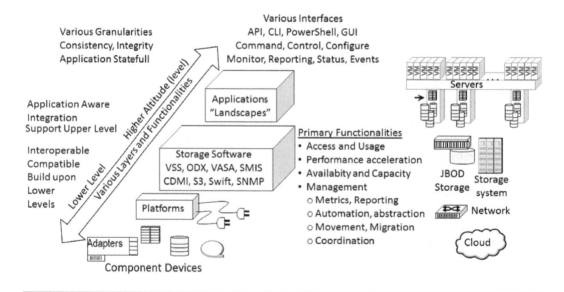

Figure 8.1 Data infrastructure data and application services.

Table 8.1. Common Data Services and Data Infrastructure Functionality

Access	Performance	Availability	Capacity	Economics	Management
Application	Acceleration	Backup	Allocation	Bundles	Analytics
Block	Activity	BC, BR, DR	Archiving	Capex	Automation
Cloud	Bandwidth	Consistency	Compress	Chargeback	Coordinate
Collaboration	Caching	Copy	De-dupe	Convergence	Metrics
Emulate	Effectiveness	Dispersal	Efficiency	Energy	Orchestrate
File and API	Latency	Durable	Optimize	Licenses	Planning
Namespace	Micro-tiering	Protection	Provisioning	Maintenance	Remediate
Object	Productivity	Security	Thick/thin	OpEx	Reporting
Unified	Queues	Sharing	Tiering	Show back	Tenancy
Virtual	Response	Synchronize	Usage	Subscription	Virtualization

In addition to being implemented in different locations, how the data services get implemented also varies. For example, there are lightweight implementations where the focus is on speed or robustness of the functionality. Then there are heavier implementations where the focus is less on speed, more on functionality extensiveness and capabilities. Both have their place in meeting different needs, with some doing both.

Also keep in mind that data services have interrelated PACE attributes; for example, availabity has elements of capacity, performance, access, management, and economics. Similarly, Performance needs availability, access, capacity, management, and economics.

Besides being located in different locations, data services and data infrastructure functionality also varies depending on the type of solutions positioned for various environments. For example, solutions or services positioned for supporting consumer or SOHO will have different data services than those for SMB, ROBO, or SME, which can vary from those used by large enterprises, managed service providers (MSP0, or cloud providers.

What also differs is the extensiveness, breadth, and depth of data services functionality across the different environments and solutions along with associated management interface tools.

Some solutions have many features but their functionality may be limited, while others may have the same or fewer features but are more robust and have extensive capabilities. For example, various solutions may show a checkbox that indicates they support snapshots. One solution may support limited snapshots with no application, operating, or file system integration. That solution may also require reserving space capacity, along with limits on how many snapshots are kept. Meanwhile, another solution may show a checkbox indicating it has snapshot support but also supports multiple concurrent snaps with little space overhead or reservations less (no space needs to be reserved in advance). Also, the snapshots can integrate with

Table 8.2. Where Data Services Are Implemented

Components	Devices, Drives, Cards, and Adapters
Network devices	Switches, directors, routers, bridges, gateways, firewalls, and adapters
Servers: physical, virtual, cloud, converged, hyper-converged	Applications, middleware, databases, messaging, and email, hypervisors, operating systems, file systems, volume managers, device drivers, plug-ins, utilities, containers, and software-defined
Storage systems appliances, software	Storage virtualization appliances, virtual storage appliances, storage systems, storage services (cloud, managed service provider, hosting), hybrid and all-flash arrays (AFAs), software-defined storage

hypervisors such as VMware, Hyper-V, KVM, and Xen, as well as different operating systems and applications.

> *Tip:* Look beyond the basic checkbox that indicates a feature in general or broad terms; also look into what are the capabilities and extensiveness of that functionality as well as whether it applies to your data infrastructure environment needs.

Data services considerations include:

- How robust and extensive are the data services features and functions? For example, snapshot capabilities (read-only or read/write, clones of snapshots)
- Application, database, operating system hypervisor or storage system integration
- Number of concurrent operations supported along with number of instances
- Management tools, capabilities, dependencies, or other requirements

8.1.1. Tradecraft and Context for the Chapter

Fundamental data services and feature functionality tradecraft topics include some context concerning different terms that have various meanings, as well as a reminder of some items discussed in previous chapters. Reminder: Data infrastructures span SDI, SDDC, and SDDI environments.

In Chapter 2 we discussed trends in application programs consisting of data structures and algorithms that are defined to enable different functionalities. Also in Chapter 2 we discussed trends about how data (files, objects, images, videos, blobs, and other entities) are getting larger (i.e., size of the data), as well as how the quantity of data items is increasing (i.e., the volume). As a result, I/O operations to store or write data to persistent memory (storage), get or read, update, list or status, and delete are also increasing in size. What this means is that software gets defined by software, and data infrastructure (also defined by software) technologies are enabled by software, hardware, and services feature functionalities to support IT applications. This applies to both the primary data as well as associated metadata (data about the data, and configuration). As a by-product of data getting larger along with changing I/O characteristics, how data gets protected, preserved, and served is also changing. For example, in the past data was broken up, chunked, shardded, or spread across different storage devices, systems, or file system extents in relatively smaller stripe sizes of a few to several kilobytes. That has evolved to sizes of several hundred kilobytes or megabytes, with some file systems, repositories, volume managers, or storage systems and software stacks today working with extents and allocations, not in kilobytes or megabytes, but gigabytes.

Hence, everything is not the same. Tradecraft and context for this chapter include everything that is not the same with different applications, environments, technology solutions (hardware, software, services) and configuration options, vendor, or service provider options. There are similarities between various types of data services and data infrastructure management tools that will vary by specific vendor implementation, as well as how they are configured and used.

Some tradecraft context items include the following.

- *Storage system.* There are physical storage systems and appliances as well as software-defined and virtual ones, some of which are tin-wrapped (packaged with commodity hardware).

Besides physical, software-defined, virtual, cloud, or container places where data is stored on different mediums (HDD, SSD, tape, optical), there are also repositories called storage systems. Examples of repositories include file systems, databases, and key value and object stores, among others. For example, some file systems, including HDFS, NFS, and ReFS, among others, are referred to as storage systems. Likewise, some databases (SQL and NoSQL) and key value repositories such as SQL Server, Oracle, and MySQL, Cassandra, HBase, ClearDB, Kudo, Mongo, and Parquet are referred to by some as storage systems. Other examples include Alluxio, Ceph, Swift, and Panzura, among many others, which to some people may not be a storage system while to others they are.

- *Marketing and vendor hero numbers.* These include benchmark, tests, simulations, performance, or functionality comparisons whose primary application objective is to support marketing and awareness workloads. In other words, they may or may not be relative to what you and your environment are doing or need to do. However, they can sound impressive. The key point is to keep in mind what applies to your needs and requirements. Also, context matters regarding how the speeds and feeds or comparisons were done, what was the workload or tool, how were things configured, applicable workload I/O size and profile to help make apples-to-apples vs. apples-to-oranges comparisons. For example, with a given workload, were any data services and background features normally used in a production environment active, or was the solution put into a "test mode"?

- *Peak vs. steady state.* A performance topic is whether a solution, service, server, storage, or network device and software being compared or profiled was running in a normal steady state, or in a start-up surge, peak, or some other mode. In a later chapter we will discuss metrics, measurements, benchmarking, and comparisons in more detail. Keep in mind the context of what is being compared.

- *Deterministic and predictive (QoS).* Tied to performance as well as availability are whether the service, functionality, or resources are performing or available in a predictable manner. In other words, in a given steady state, how deterministic or predictable is the service: Are there extreme fluctuations under different workloads conditions? Are there abilities to adjust buttons, dials, and knobs, API commands to configure quality of service (QoS) regarding bandwidth, activity, latency, availability or resiliency, among other attributes?

- *Stateless and stateful.* Stateful means that the current state and consistency of the data with respect to how applications or software are using it are preserved and integrity is maintained. This also means that if power is lost, data is preserved and non-volatile or persistent with integrity. Stateless means simply that data can be used, for example, as a cache, but its state is protected and preserved with integrity on some other persistent storage medium. In addition to data and storage being persistent and stateful, network and I/O adapters can provide stateless (i.e., pass through with some optional cache) functionality. On the other hand, some PCIe adapters or cards such as SSD devices that store data (beyond use as a cache) can be stateful. HBA and NIC, for example, are usually stateless, with the upper-level application, software, or drivers handling statefulness.

- *Telemetry data.* Similar to metadata, there is two general categories and context for telemetry. One is data from IoT, IoD, and other devices and sensors collected for analysis (i.e., application context). The other context for telemetry is data infrastructure data collected from servers, networks, storage systems, and client devices. Telemetry data includes health, status, activity, and events, among other items, that can be used for data infrastructure analytics, reporting, troubleshooting, diagnostics, event correlation, modeling, and forecasting. Other common names for data infrastructure telemetry data are log, activity, and event data.

Containers need some context, as we discussed back in Chapter 4 and will continue to do so as our journey takes us through this and other chapters. Containers include compute microservices such as Linux and Docker, or Microsoft Windows, among others. However, containers can also refer to database, file system, or repository and volume Managers. Containers and buckets should not be confused with Docker-like compute containers. Containers can also be virtual hard disks such as a VHD/VHDX or VMDK, among others. Hence, context matters.

A couple of other topic and terms where context matters includes *vaults and vaulting,* where a vault can be a physical place where data gets stored, or it can be a logical software-defined data structure such as a backup, archive, saveset, bucket, or container where data gets saved in a particular format. Vaulting can also mean physically placing data in a vault, or moving data logically into or out of a vault or storage device or cloud service functioning as a vault.

Sync is another term where context matters, as it can refer to synchronous or real-time updates of data being transmitted vs. asynchronous or time-delayed data (i.e., eventual consistency). On the other hand, sync can refer to file, data, volume, database, or other forms of synchronizing to make data copies. Another context is synchronous vs. asynchronous communications.

DPM can mean data protection management in general, or can mean product technology–specific, such as Microsoft Data Protection Manager. Similarly, *SRM* can mean storage (or system) resource management (or monitoring), including resource usage reporting. However, SRM can also mean VMware Site Recovery Manager as part of data protection. Hence, context matters. Also keep in mind our discussion in Chapter 5 about server I/O data and management traffic flows, including north–south, and east–west.

8.2. Virtual Disks, Endpoints, Namespaces, and Access

A fundamental data service is being able to access and store data that will be persistent and consistent for use when needed. This means that any data saved and intended to be persistent will save exactly as what was stored, without changes, lost or damaged bits, bytes, blocks, blobs, along with any associated metadata when it is accessed. The data and storage access can be local inside of a server, direct attached and dedicated to a server, or external and shared multi-host on a direct or networked storage device, appliance, system, or server. The data and storage can also be remote, at another data center, MSP, or cloud service provider. Note that access can also be within software-defined virtual, container, and cloud VMs, as well as instances.

Other basic data access services include accessing physical along with virtual and cloud storage, that is, block, file, object, or application-based service endpoints and namespaces. That access can be via a local dedicated private namespace, or via a shared global overlay namespace, either directly or via a proxy server. Storage access includes block, file, and object, the application shared along with unified and converged access, multi-protocol, in some cases even to the same data item.

Another attribute for data storage access is serial or one user at a time, concurrent and parallel or multiple users at the same time of the same data for reads, or reads and writes. For example, being able to access an item as block or file, NFS, CIFS, or HDFS for the file, or via different object access methods enabling you to park the data, then change how it is accessed and used.

8.2.1. Access, Personality, and Protocols

In Chapters 5 and 6 we discussed various server I/O access, interfaces, and protocols, and devices in Chapter 7. Building on the material already covered, in addition to access interface and protocols are the personalities, abstraction of underlying resources and characteristics of how servers access and use storage and data infrastructure resources.

Figure 8.2 shows an application at the top left that makes requests to data infrastructure storage and data services on the right. These requests or access are to store, read, manipulate, or manage data. Applications can make calls or access data and storage services via higher-level abstractions such as databases, key value stores, file systems, or other repositories, as well as application services.

Other access in Figure 8.2 can be to in-memory, memory-mapped I/O, data services, socket services, or other namespaces and endpoints as discussed in Chapters 4, 5, and 6. Moving down the stack on the right includes access via API, service calls, object, file, block, and raw I/O. The I/O request can be serial or parallel (not to be confused with low-level parallel data transmission) using various protocols and interfaces.

Different data and storage resource endpoints and services are accessed via various interfaces and protocols, while also having different personalities. The personality is how the service or function behaves or is optimized independent of the interface or protocol. For example, a disk or SSD backup and data protection storage device can have a GbE (or 10 GbE) port that supports a TCP connection. That connection could be iSCSI or NFS. However, the functionality and thus the personality are how the storage system or service behaves.

Part of the personality provided by the underlying data services and storage system or data infrastructure server includes emulation. Emulation includes appearing as a virtual tape drive or virtual tape library (VTL), or as a virtual volume, specific device type such as an IBM "i" system D910, among others.

While NFS, as an example, can be used for file sharing, a data protection appliances with de-dupe designed for backup or archive may be optimized for sequential streaming (writing or reading) of data, as opposed to a general-purpose file server. Likewise, a general-purpose file

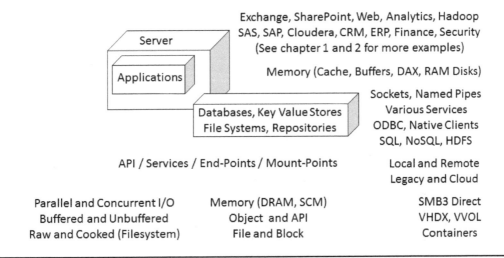

Figure 8.2   Access to resources with data services and PACE attributes.

server with an NFS interface and access can have de-duplication. However, it might be optimized for general file serving as opposed to backup or archiving.

Another variation is that a Lustre storage system providing file sharing may be optimized for large sequential streaming of reads or writes, including parallel access over multiple interfaces from a few servers. On the other hand, a different storage system providing file sharing can be optimized for small random reads and writes from many concurrent users.

Tip: Some file servers or software are optimized for large sequential and parallel access for high-performance computing, supercomputing, or big data analytics. Those systems may be better optimized for storing and retrieving very large files or objects measured in GB or TB as opposed to many smaller ones measured in KB or perhaps a few MB.

Table 8.3 shows various characteristics and examples of data services access.

Tip: It is important to look beyond the interface and protocol and understand what is the functionality and personality of the storage device or server, including what it is optimized to be used for. Table 8.4 lists common data infrastructure and data services access protocols and interfaces.

Table 8.3. Data Services Access Characteristics

	Characteristics	Examples
Access	How data infrastructure storage systems or services respond to I/O requests via API, application service, socket, object, file, block or raw.	Read, write, get, put, list, input, output, status, select, status, inquiry
Back-end	How and where the storage system communicates and accesses physical, virtual, or cloud storage resources using different interfaces and protocols.	SAS, NVMe, SATA, iSCSI, and NFS via IP via GbE, cloud storage via IP
East–west	How storage systems and software communicate with each other for management and other traffic implementing data services functionality. This may be over separate or shared front-end or back-end or management ports and interfaces.	Replication and mirroring, data dispersal, intersystem tiering, data migration and time services, load balance, cluster data movement
Front-end	How and where servers (or other storage systems) access and communicate with the storage system and data services. Interfaces, protocols, personality, and data services functionality. Can be block, file, object, or API accessed.	SAS, Fibre Channel (FCP or FICON), NVMe, Thunderbolt, iSCSI using GbE, NFS using IP via GbE, S3 object via IP via GbE, cloud access via IP
Interfaces	Physical and virtual I/O network mediums used by front-end, back-end, and management ports that support access using different protocols.	1/10/50/100 GbE, PCIe, InfiniBand, SAS, Fibre Channel, USB, Thunderbolt
Personality	How a protocol and interface function to support different activities. For example, NFS is a general-purpose file access protocol. The underlying functionality and personality of the service can be optimized for specific workloads such as general file sharing, transactional, streaming, backup, or archiving, among others.	General file sharing, video or content streaming, reads or writes, append or updates, database or data warehouse, transactional or large stream processing, parallel I/O for HPC, or small I/O updates
Protocol	Logical I/O protocols for server-to-storage, server-to-server, storage-to-storage, and cloud access. There are different layers of protocols from block to file to object and API.	TCP/IP, UDP, NFS, HDFS, CIFS/SMB, SCSI (FCP, SAS, iSCSI, SRP), NVMe, S3, CDMI, Swift, XML

Application personalities (abstractions) and functionality of storage include optimization for primary active performance and bulk, secondary, and copy data, along with protection copies (archive, backup, HA, BC, DR). Additional characteristics are performance optimized for reads or writes, sequential and append or random updates, concurrent, large parallel bandwidth I/O.

Table 8.4. Common Data Infrastructure Access Protocols and Interfaces

Access	Characteristics	Type
CIFS SMB	Microsoft Windows file sharing, also is known as SAMBA	File
Cinder	OpenStack block storage access and management APIs	Block
FCIP	Fibre Channel (FCP/FICON) over IP for distance	Block
FCoE	Fibre Channel native on Ethernet (not FCIP or iFCP) coexists with IP	Block
FICON	Fibre Channel ULP for IBM "Z" mainframe servers and storage	Block
HDFS	Hadoop Distributed File System used with big data and analytics	File
HTTP	URL access for management cloud, object storage services endpoints	Various
IP/UDP	TCP support for various ULP such as NFS, CIFS/SMB, iSCSI, HTTP	Various
iSCSI	SCSI protocol command set implemented on IP	Block
iSER	iSCSI with RDMA for low-latency, high-performance storage access	Block
Kinetic	Key value object storage drive protocol for accessing storage devices	Object
LTFS	Linear Tape File System abstracts underlying tape to speed random lookup of what's on a tape, then sequential stream to the data on tape	Block Tape
NBD	Network Block Device, Linux alternative to iSCSI for storage sharing	Block
NFS	Network File System for file, data, and storage sharing	File
NVMe	NVM Express low-latency protocol optimized for NVM and SSD	Block
NVMf	NVMe implemented over a fabric protocol, alternative to FCP/SCSI	Block
ODX	Microsoft Offloaded Data Transfer, supported by some systems	Various
pNFS	Parallel NFS for large sequential streaming of file server data	File
S3	AWS Simple Storage Service object access defacto standard	Object
SAS	Serial Attached SCSI implements SCSI protocol command set	Block
SATA	Serial ATA, also known as AHCI command set, can coexist with SAS	Block
SCSI	Command set protocol implemented on various interfaces	Block
SCSI FCP	SCSI protocol command set implemented on Fibre Channel, aka FCP	Block
SMB3 Direct	Peer to SMB instead of file access provides low-latency, deterministic, high-performance block-like storage access using RDMA	Various
Socket	Various local and networked services for accessing data services	Various
SRP	SCSI RDMA protocol for low-latency access over InfiniBand	Block
Swift	Object access storage part of OpenStack, similar to AWS S3	Object
T10 DIFF	SCSI Data Integrity Field, for data integrity consistency	Block
VAAI	VMware storage array integration and offload	Various
VASA	VMware for Storage Awareness and management	Various
VHDX	Virtual hard disk supported by Microsoft Windows and other systems	Block
VVOL	VMware Virtual Volume provides optimization and abstraction along with additional data service management enablement	VMware Object

Access can be optimized for the bandwidth of a single or multiple streams, as well as for activity such as IOPs, gets, puts, reads, writes, or pages served. Another performance consideration is for low latency, as well as CPU usage for handling I/O operations.

Besides performance, other access topics include availability, such as redundant paths for active/active load balancing as well as failover, or active/passive (standby). Security is another consideration, including endpoint protection, access control and authentication, and encryption of data, as well as for management access.

Some solutions require special adapters or NIC hardware along with unique firmware and device drivers for hardware devices. Others need special device drivers that are not in the box (i.e., part of the operating system or hypervisor build), along with other software such as volume managers, file systems, and related management tools.

Tip: Pay attention to what operating systems or hypervisors and their versions are supported for access storage and network resources. Also pay attention to what additional hardware, software, or platforms and tools are needed for managing the solution.

Common access considerations across block, file, and object and application access include:

- Open Systems, IBM zSeries mainframe, among other platforms
- Open or proprietary protocols, associated device driver and software plug-ins
- Jumbo frames, maximum transition unit (MTU), and other performance topics
- Operating systems ranging from *NIX to Windows, among others
- Dedicated and private access vs. shared access to device and data items
- Singular serial vs. concurrent and parallel access to a device or data item
- Active/active or active/passive via redundant access paths
- Security access controls, encryption of data at rest (and in-flight)
- Metadata that varies in size and scope from simple basic to advanced
- Emulation of devices, characteristics, protocols, and functionalities
- Platform support (OpenStack, VMware, Windows, Hyper-V, KVM, Xen)
- Host I/O limits and other performance and availability QoS
- Physical port types and logical protocols supported

Block Access (or DAS or SAN)

Beyond the access protocol and interfaces, for example, NBD, SCSI, Thunderbolt, SATA, or NVMe command sets via iSCSI, Fibre Channel, FC-NVMe, NVMeoF, InfiniBand, or SAS, as well as PCIe, there are other basic access considerations. For IBM zOS mainframe systems, protocols include FICON and Count Key Data (CKD) and associated volumes and emulated device types.

Block storage access for open systems considerations include how many LUNs, partitions, volumes, or virtual disks can exist in a given namespace. What is the maximum size of an LUN or virtual device that can be created and presented to a host server (i.e., front-end or northbound), and are they 512 bytes or 4KB in sector size?

Another consideration is how many back-end (south-bound) devices, including physical, cloud, or virtual, can be used to create a LUN, and what is their corresponding size. This also includes whether those devices are local JBOD, RAID arrays, third-party storage or remote, as well as whether they are 512 bytes or 4KB in sector size.

Tip: Another variation for storage block size is systems that use 520-byte or larger sectors. The larger sectors wrap the normal 512 bytes with additional data integrity information. Note that the server accessing the storage will see standard 512-byte or 4KB sector size, as the additional data is used by the underlying controller, storage system, or software-defined storage functionality. Also, note that when looking at data services access and storage systems or software, there is front-end size and back-end underlying storage size, so you need to pay attention to context.

Other attributes for block storage include dedicated or shared access with applicable pathing software to maintain data consistency across servers in a cluster. Is there ability to dynamically expand or shrink a LUN or volume, either on-line while in use or off-line with data remaining in place? Also, can that LUN or volume be accessed in raw device mapped (RDM) or pass-through mode, or via a local file system or volume manager?

In addition to being able to expand or shrink a volume's space capacity, does it have the ability to be configured for thick or thin provisioning as well as converting from one format type to another, including physical to virtual, or from one virtual format to another? This also includes the ability to aggregate and virtualize local physical and virtual storage, as well as external third-party storage.

Another option is having to create a new volume and then do a copy or export from the original to the destination. Conversion can be from a low-level MBR to GPT device format, to Virtual Hard Disk (VHD) to VHDX, basic to a dynamic disk (physical or virtual), as well as from one logical volume manager or file system to another.

Block access security considerations include encryption, access controls, mapping, and masking of LUNs and volumes to different hosts to restrict visibility of those storage resources and their access endpoints. In the past, a relatively safe assumption, at least in the open systems world, was that most storage was based on 512-byte sectors. Today there are 512-byte and 4KB sector size devices, as well as of LUNs, volumes, and virtual disks. Also tied to block storage access is how a LUN or volume is accessible via redundant paths on storage controllers and server servers.

Asymmetric Logical Unit Access (ALUA) is a feature found on many storage systems, appliances, or servers that have a redundant access path or controller. ALUA uses the command part of the SCSI-3 protocol to enable hosts to gain insight and manage path access to a storage resource such as a LUN or volume. A LUN or volume may be accessible via multiple storage paths but have an affinity (be "owned") by only one controller or storage server at a time. While the LUN or volume can be accessed via either path, one will be faster, having an optimized path to the data, while the other is non-optimized (the I/O needs to be shipped or passed off to the other storage controller peer).

Some software, such as VMware among others, has knowledge or insight into whether a storage resource is ALUA and the preferred direct path. VMware also has Most Recently Used (MRU) and Round Robin (RR) in addition to ALUA support as part of its Path Selection Policy (PSP) pathing support. Different software platforms such as hypervisors and operating systems support various pathing drivers, either built-in or via third parties.

Another consideration is for VMware environments is support for SCSI secondary LUN ID to support VMware VVOL, which requires applicable device driver and adapter or storage system firmware. VVOL also depends on VMware management interfaces including VASA (VMware API for Storage Awareness).

Block access considerations include the following.

- Some servers and systems use 520, 528, or other nonstandard block size.
- Front-end and back-end block sector size.
- Interface port interface and protocol.
- Number and size of LUNs for front-end and back-end.
- Number of single or dual attached hosts initiators.
- Device emulation, compatibility, and interoperability.
- Device drivers, plug-in, or other software tools.
- Parallel or concurrent storage access.
- Authentication including IPsec security.
- Device queues and buffer configuration options.
- Optional software for cluster sharing and data consistency.

File Access and NAS

There are many different types of file systems and various types of access. These include general-purpose file systems, file servers, and file shares along with NAS servers, as well as purpose-specific file systems such as backup and archive targets. Some file systems are optimized for parallel file I/O of large files and objects, while others are general-purpose.

Common to different file systems (and storage in general) is metadata, for which context needs to be understood. There is metadata for the actual data being stored (what's in the data), and metadata about where and how the data is being stored (where it is located) in a storage system, file system, database, or repository, as well as system configuration and status metadata.

Besides protocol and interface considerations such as CIFS/SMB/SAMBA, FTP/SFTP, torrent, NFS or pNFS v3 or v4, HDFS, CephFS, Posix, and Fuse, among others, there are also other file access considerations, for example, how the underlying file system (e.g., cxfs, lfs, ext, btrfs, xfs, zfs, ReFS, NTFS, PVFS, GPFS, GIT LFS, Lustre, and union, among others) behaves or is optimized for different workloads and usage scenarios.

Other file access considerations include the size of the file system namespace, the number of files in a file system or namespace, the size of files, metadata and access control capabilities, and quotas. Integration with other namespaces, concurrent and direct I/O as well as parallel I/O capabilities are supported by some databases. Concurrent and direct I/O may be referred to by different names, but regardless of what is called, this functionality enables properly configured and enabled databases, operating systems, and NAS devices to perform block-like I/O to an NFS file systems.

File system fundamental features and services include endpoint management and access control security and encryption, and use of various underlying file systems, tuning for large vs. small I/Os and files or objects, short or long file names, various character codes and language support.

Another feature is the ability to have data accessible via multiple protocols such as CIFS/SMB, NFS, and HDFS concurrently while preserving the semantics and attributes associated with each of these access methods. In addition to metrics related to file access space, other considerations include activity, such as creates, deletes, reads, and writes, along with bytes transferred. Other features can include encryption at rest and while in flight, along with active directory, Kerberos along with other endpoint and user or system authentication, and access rights management.

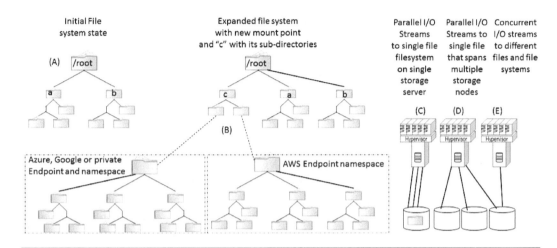

Figure 8.3   File access, file systems, endpoints, and overlays.

Different storage software and data services have various support for managing namespaces and endpoints. Some solutions and software support overlaying (aggregating) and abstracting underlying or different file systems, namespaces, and their endpoints. Those various namespaces can be local, remote, or cloud and object storage–based. Some storage systems have added the ability to abstract or overlay namespaces along with storage software that can be installed on servers to perform similar functions.

Figure 8.3 shows various file system namespaces, including simple (A), with overlays of local and remote including cloud (B). On the right are shown (C) servers that are doing parallel I/O streams across multiple interfaces to a single file on a file or storage server. Also shown are (D) a server doing parallel I/O to a single file across multiple interfaces on several storage servers (i.e., the file is spread across multiple servers for performance). A general-purpose concurrent I/O example is also shown (E), where multiple I/O streams exist to the same or different files in a file system, such as for home directories.

Tip: For some big data and HPC-type applications that require streaming very large files, parallel access can be a benefit. On the other hand, for general-purpose access of large and small files, traditional (nonparallel) access is commonly used.

Tip: Context here: how much data is moved vs. the number of concurrent operations operating in parallel. With parallel access, data is broken up, chunked, or sharded and written across multiple interfaces to one or more storage targets (devices, volumes, file systems, servers).

Other file access topics include 32- or 64-bit file system and addressing, pluggable and removable file systems, data services such as footprint reduction, protection, security, antivirus, and search, among others. Common functions include background file system integrity checks and repair, shared read and write by concurrent clients on different servers, and using various protocols (e.g., NFS and CIFS/SMB).

File access considerations include the following.

- Number of files, file systems, files per file system, file and file system size
- Sharing via different protocols such as NFS and CIFS/SMB access the same file

- Quotas, limits, permissions, and rights access attributes
- Data metadata along with file system and storage system metadata
- Plug-ins, drivers, additional client or server and management software
- File system, namespace, and endpoint overlays

Object, API, and Application Services Access

As mentioned in Chapter 3, there are many different types of object access and protocols, including S3, Swift and Centera, VMware VVOL, as well as other HTTP REST-based methods. Also keep in mind that context matters with object storage in that there are storage access, management, and underlying architecture as well as how they are implemented.

Why it is important to keep context in perspective is that some storage has an object-, HTTP/REST-, or API-based management with traditional access (block, file, or other). Likewise, some storage solutions, software, and services support various forms of object access (along with block or file) as well as management. There are solutions that are implemented using object-oriented architecture and techniques under the coverages, but with non-object access. Some of these are for accessing data or data application and infrastructure services, while others are for management.

Like block and file access, different object access methods are optimized (personalities) for various workloads or applications. Some are optimized for sequential streaming of reads and writes, large or small, with updates appended. Others are optimized for smaller random reads and writes such as with a database or metadata service. Thus, as with block and file, look beyond the type of object interface to understand its personality—what type of underlying functionality it provides and is suited for. For example, an object access storage that is optimized for archiving large images may work fine for storing virtual machines as well as videos, but it may not be suited for database or other dynamically changing data. What will also vary is how large an object can be created or how many can be in a given namespace. Some marketing material may tell you that a particular solution is unlimited in the number and size of objects supported—which might be true. Trust yet verify vendor marketing claims.

> *Tip:* Do your homework to avoid surprises by understanding what is the theoretical vs. architectural vs. what is implemented in the software functionality vs. what was tested and what is supported by the vendor. Currently (subject to change by the time you read this), many object storage systems support objects in the 5–10 TB size, with some larger, although these numbers, as with file systems and block LUNs, can be expected to increase over time.

In addition to size or a number of objects, also understand what are the performance characteristics besides whether optimized for sequential streaming or random activity. Some object services including cloud providers may have limits on how much data can be read or written per request. For example, some cloud and object storage APIs or access implementations need data to be broken up into multiple smaller chunks that then get reassembled when stored.

This also means that traditional tools for testing block or file system performance may not be applicable for testing different object solutions. Likewise, since there are different object solutions and access methods, there are different tools, as will be discussed in a later chapter.

Like file accessed storage, some object access storage supports bucket or container hierarchy so your data appears to be organized similarly to a file folder directory structure. Other considerations for object access storage include what protocol is used: Is it a bit or byte stream such as AWS S3 or OpenStack Swift, or associated with a database such as using ODBC or a native client? Alternatively, is the object access tied to a hypervisor such as VMware VVOL?

What additional software drivers, agents, plug-ins or management tools are needed on servers or systems to access and use the object accessed storage or service. How is endpoint security and encryption handled for the user or system access of data or resources? Also, how are public accessible object and database or other data services charged for? Are fees determined only by capacity used, or are there gets, puts, lists, deletes, data movement, and bandwidth charges?

Pay attention to whether a system or solution that mentions HTTP/REST does so for management interface purposes, or for actual data access. Some solutions do support HTTP/ REST management, but do not support object or HTTP accessible storage. This might not seem like a big deal, but if you are reading a vendor's data sheet and simply glance at the speeds and feeds, you might see HTTP/REST or object mentioned without paying attention to whether it is for management or actual data.

On the other hand, some vendors may get creative in how they list their speeds and feeds, which should prompt you to ask what they support, and the capabilities of what it is that they provide.

Object and API access considerations:

- API and object access protocol and personality
- Endpoint security, keys and access credentials
- Gateway or API data size limits, chunking, and other I/O limits
- Data format and wrapping, integration, key management
- Plug-ins, drivers, and other access software or tools

8.2.2. Migrating Server Storage

There are different types of server, storage, and data migration and conversion. This is also where some context comes into play, for example, there is converting or migrating a server from one physical device to another without changing the operating system and data such as during a hardware upgrade, and there is converting from one version of an operating system or application to another on the same or different hardware.

Another variation is physical-to-virtual (P2V) or physical-to-cloud (P2C) or moving from one type of virtualization hypervisor such as VMware to another such as Hyper-V (V2V). There are also other variations of data movement and migration, including replication, clone, copy or sync, data sharing and distribution, data protecting, data import or export, and tiering.

Migrations and conversions can occur locally (intrasystem) or remotely (intersystem). A local migration example is using a tool such as Microsoft Disk2VHD to create a virtual hard disk (VHD or VHDX) of a live running system. Once the VHD is created, it can be moved to another system by copying it like any other file, then mounting it for use or attaching it to a hypervisor such as Hyper-V and using it as a virtual machine's storage.

Table 8.5 shows common server and storage migration between different physical, virtual, and cloud systems. Many tools are available for performing P2P, P2V, V2V, and P2C, among other variations, some of which are free. Some examples include Acronis, Clonezilla, virt-v2v, Microsoft Virtual Machine Converter (MVMC), Disk2VHD, and VMware Converter. Many

Table 8.5. Conversion and Migration Terms and Modes

Term	Mode	Description
P2P	Physical to physical	Migrating from one operating system to another, converting from MBR to GPT partition format, or from one file system format type to another
P2V	Physical to virtual	Migrating from a physical machine to a virtual machine or a physical disk to a virtual disk such as VHD, VHDX, and VMDK. among others
V2V	Virtual to virtual	Migrating from one hypervisor virtual machine format such as VMware vSphere ESXi to Microsoft Hyper-V, or VMware and Microsoft to KVM/Xen
V2P	Virtual to physical	Migrating from a hypervisor virtual machine to a physical machine, for example, if you need to run a particular operating system image bare metal
V2C	Virtual to cloud	Migration from one hypervisor such as VMware VMDK or Hyper-V VHDX to clouds such as AWS, Azure, Google, Softlayer, or others
P2C	Physical to cloud	Migrating from a physical server image to a cloud instance such as a managed service provider, DPS, or VPS

backup and data protection tools can also perform conversions as part of protecting data with the ability to restore to different platforms and architectures.

Why would anybody ever need or want to go back from virtualization or even the cloud to physical? Sometimes there is the simple need to move a system image or instance to physical for troubleshooting, performance tuning, or other operational considerations. In other cases, you might need a copy of an image that is virtual, so instead of doing a fresh install, simply do a seed from the virtual, make applicable adjustments, and be on your way.

Migrating server considerations include:

- Wrappers and encapsulation
- Sysprep items, tools, and integration
- BIOS/UEFI conversions
- Partition changes and alignment
- Virtual driver conversion

8.2.3. Data Movement and Migration

Remember that server applications, settings, and data get stored on some storage device. How decoupled the server is from the storage device where hypervisors, operating systems, drivers, applications, settings, and data are stored will influence how easy it is to swap servers or make other changes. In the case of a virtualized server, once converted from physical or initially created, the virtual machine, or its virtual disk and associated configuration settings, can be moved around with relative ease.

On the other hand, in the case of a bare-metal server, if the system drive or devices can be either removed and moved to another physical server, or accessed via a shared storage device, there is also portability. The point is that servers and networks can be easily virtualized and components swapped in or out along with other configuration changes, because their data lives elsewhere on storage.

Storage too can be virtualized. However, its role is to store data, which still needs to be moved or accessed via networks. It is the flexibility of the storage and its ability to be accessed via different networks that enable servers to have the flexibility they do.

As mentioned, there are different types of data movement and migration, from the cache, tiering, and micro-tiering discussed a little later in this chapter to movement of folders, directories, file systems, databases, or applications. This movement and migration occur at different granularities, so keep context in mind when comparing movement, migration, and associated data services features: Watch out for apples-to-oranges comparisons.

Some additional data migration and movement topics include intrasystem (within a system or software-defined storage namespace), or intersystem (across systems and solutions) or federated across dissimilar solutions. Data movement can be a simple copy or clone, or can involve more integration including data conversion, format changes, data cleaning, or other functions.

Different solutions have various approaches for seeding and synchronizing data movement over distances, or for large quantities of data. For example, some solutions enable data to be moved or copied in steps, with an initial copy or replication, then subsequent copies, and final synchronization to bring data to a consistency point. Another approach involves making a copy of data and physically shipping (seeding) it to another location or cloud service. Then any changes or deltas from the source site are copied to the remote or cloud site and synchronized until consistent.

Data movement and migration can be done on-line while data is active for reads and writes, or on-line for reads but no updates, or off-line. The data movement can occur by accessing mount points or how storage is presented to a server, or via APIs. Some data movement simply migrates data and basic permissions along with attributes. Other migrations also move associated data protection policies, snapshots, or backup, among other information.

Another variation is that some data is simply moved intact from source to destination as opposed to performing data format conversion or other operations. Data conversion can be from one character encoding to another application or operating system format, among other variations. How data movement is done also varies, as some approaches work in parallel streams while others are serial.

Data movement and migration considerations include:

- Granularity of data migration (blocks, files, folders, volumes, databases, system)
- Move and convert, copy and convert, copy and leave copy behind
- Live data movement including virtual machine or physical conversion
- Import and export of virtual hard disks or other containers
- Automated, semi-automated, or manual movement and interaction
- Command line and PowerShell scripting along with GUI and wizard
- Cloud and remote site seeding via physical shipment of storage, then resync
- Data footprint reduction (DFR) including compression

8.2.4. Volume Managers and Virtual Disks

Besides hypervisors and operating systems, other sources for software-defined virtual disks include RAID implementations (card, controller, storage system, appliance, or software), logical volume managers (LVM) and software-defined storage managers (i.e., storage virtualization tools), storage systems and appliances (LUN, volumes, and partitions), and virtual storage appliances (VSA).

Other examples include Microsoft Windows Storage Spaces and Storage Spaces Direct (S2D), Disk Groups (DG), virtual disks (VD), virtual volumes (VV), virtual arrays (VA), VHD and VHDX, VMDK and VVOL, logical volumes (LV), LUN and partitions. Note that in addition to blocks, volume managers and abstraction layers can also support file and object storage for physical, virtual, cloud, and containers.

Logical disk managers (LDM) and logical volume managers (LVM) along with other software-defined storage managers manage underlying storage and, depending on where or how implemented, may be referred to as software-defined or virtual storage.

LVM exist as part of operating systems such as Windows Storage Spaces or Storage Spaces Direct (S2D) as well as various *nix systems. Also, LVM can be added separately, obtained via different venues. Note that in addition to leveraging host-based storage, containers such as Docker can also use tools such as Rexray from Dell EMCcode for local storage.

Functionality and data services of LVM range from RAID mirroring to stripe variations of parity and erasure code protection, thick and thin provisioning, compression and de-dupe, dynamic or static volume expansion or shrinking, snapshots and replication. LVM and LDM can be tightly or loosely integrated with different file systems.

Figure 8.4 shows a generic example of applications on servers (physical or virtual) accessing storage devices, volumes, file systems, or namespaces presented by some form of software-defined storage manager, LVM, or similar functionality. Also shown in Figure 8.4 are, on the left, various data services and functionalities that are implemented as part of the solution (center), along with other management capabilities (right). Across the bottom are various pools of underlying storage devices, systems, or services. The storage pools can be NVM and SCM SSD, HDD, as well as cloud or object storage services.

Smaller environments and systems are more streamlined than larger ones. In larger environments, there can be more layers of abstraction from the underlying physical. For example, a storage system can aggregate multiple back-end (i.e., southbound) HDD or SSD together into a RAID stripe, mirror, or parity/erasure code–protected volume, which in turn gets allocated as LUN to a host server (northbound).

The host server may have a RAID adapter that works in a pass-through mode, allowing the upper-level hypervisor, operating system, or volume manager to manage the LUN. On the

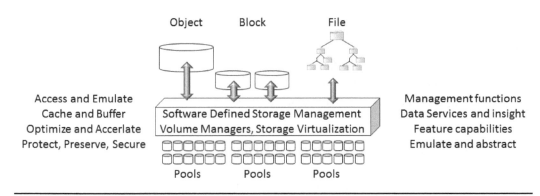

Figure 8.4 Software-defined storage management and volume managers.

other hand, the RAID adapter might be deployed to aggregate the various underlying physical or virtual LUN into a DG, VD, or simply as a pass-through JBOD volume. Further up in altitude or the I/O and data infrastructure software stack, a volume manager or other software-defined storage manager can aggregate and implement various data services on the devices that sit below it (physical, virtual, or cloud).

While virtual disks are more common, there are also virtual tape library (VTL) with virtual tape drives and virtual tape cartridges. In addition to physical and virtual servers and operating system implementations, clouds and containers also support software-defined virtual disks that can be block-based, such as an OpenStack Cinder or AWS Elastic Block Storage (EBS). Note that Microsoft Azure supports VHD, as do other hypervisors found in various cloud software-defined data center stacks.

Physical, software-defined, virtual, and cloud servers can access both virtual and physical disks. For example, Microsoft Windows (servers and workstations) can create and access Virtual Hard Disks (VHD) and newer-variant VHDX, while VMware vSphere ESXi uses VMDK along with virtual volumes (VVOL). A point of clarification is that VVOLs are volumes that contain objects or items such as VMDK.

In object storage parlance, a VVOL would be similar to a bucket or container and the VMDK like objects or blobs. However, to be clear, VVOL are *not* object storage entities in the same sense as an OpenStack Swift container, AWS bucket, Azure or Google container, and blob or object. In addition to VHD, VHDX, VMDK, and VVOL, there are other formats for different platforms and uses. Note that some hypervisors and operating systems besides Windows also support VHD.

Figure 8.5 shows examples of various types of virtual storage and disks. At the lower left (A) virtual disks (VD) are created from RAID adapters (or software) using disk groups (DG) or disk pools. At the lower center are (B) LUN, volumes, virtual disks, and VVOL being served from storage software on an appliance (physical or virtual) or a storage system via disk groups or disk pools.

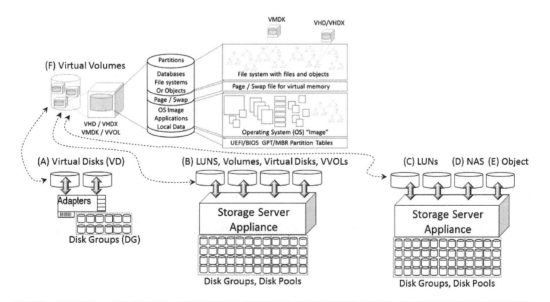

Figure 8.5   Virtual disks and virtual storage types.

The lower right in Figure 8.5 shows (C) block LUN, (D) NAS file systems, and (E) object namespace created from disk groups and disk pools. The top left of Figure 8.5 shows (F) a virtual volume (or file system or object bucket) that contains VHD, VHDX, and VMDK, which are files that in turn are virtual storage. For example, the VHDX can be mounted and used by physical as well as virtual servers, or as primary storage for virtual machines.

Microsoft has published an open specification for VHD (2TB) and VHDX that scales to 64 TB. Similar to physical disks, VHD and VHDX can be expanded, shrunk, compressed, and optimized, including in some situations on-line while in use, a functionality that varies by operating system version, among other considerations. By way of example, when this was written, older versions of VMware VMFS (VMware File System) supported 2-TB volumes, while VMFS 5 supported up to 62 TB, and by the time you read this, capacity may be even higher.

Depending on operating system and hypervisor version, some virtual disks can also be shared between servers for clustering or other uses. VMware also has system developers' kits with definitions for VVOL and VMDK, as do other vendors. In addition to the size of the volumes or containers, as well as how many items, files, folders, objects, or entities they can hold, other considerations include the size of space allocation, which can be in megabytes rather than the kilobytes found in traditional systems.

Other virtual disk and image formats supported by various operating systems, hypervisors, and cloud software stacks include qcow, img, and ISO, among others. Using various tools in your storage device toolbox, software-defined and virtual disks can be created, shrunk, extended, compressed, and optimized, as well as converted from one format to another. Other tools enable you to convert the contents of a physical storage device to a virtual device, initialize, and format and mount partitions, volumes, and file systems.

Following is an example of Windows PowerShell commands displaying VHD attached to a Hyper-V virtual machine HVIOWS2012R246 and then increasing the size of one of the VHD to 250 GB. The first command displays the VHD, their controller type, number, and location, which are used in the second command to resize the volume.

```
Get-VM "HVIOWS2012R246" | Get-VMHardDiskDrive

Get-VM "HVIOWS2012R246" | Get-VMHardDiskDrive -ControllerType
SCSI -ControllerNumber 0 -ControllerLocation 2 | Resize-VHD
-SizeBytes 250GB
```

Following is a variation of the above to get and display a specific VM VHD access path.

```
$x = Get-VM "HVIOWS2012R246" | Get-VMHardDiskDrive
-ControllerType SCSI -ControllerNumber 0 -ControllerLocation 1

write-host $x.Path
```

Volume manager, virtual, and software-defined storage device considerations include:

- Block, file, object, API, database, or service
- Data services features, functionality, and management
- Local or remote, endpoints and namespace
- Size and number of devices, front-end, back-end

8.2.5. Multi-tenancy and Virtualization

The fundamental attributes of virtualization are abstracting together with emulating underlying resources. Those attributes and capabilities, in turn, are used for aggregating (consolidating) as well as enabling agility, flexibility, and additional functionalities. While consolidation is commonly associated with virtualization and that is a popular use, not everything needs to be or can be aggregated. However, even without consolidation, most applications can be virtualized, even if that means dedicating a physical host server to a virtual machine or guest. Why do such a dumb thing, which seems to defeat the purpose of virtualization? While it may defeat the purpose of aggregation via consolidation, enabling a single VM to be the guest of a dedicated server enables agility, flexibility to move that VM to another while in use as well as for HA, BC, BR, and DR, among other operational needs. Besides enabling agility as well as consolidation to increase utilization where applicable, virtualization also masks and abstracts underlying physical resources to simplify management.

Another form of virtualization is emulation, with a storage device appearing as a different type of storage device for backward compatibility, or a disk appearing as a tape drive or tape cartridge with associated functionality (capacity, geometry, or number of blocks and sectors).

Figure 8.6 shows various types of tenancy, including (A) multiple applications on (B) the same physical server, (C) a virtual machine with a hypervisor (D), and a Docker, Linux, or Windows container (E).

While virtualization is often used for subdividing resources such as servers, storage, or networks for use, it can also be used for aggregating resources. In addition to subdividing a server, storage, or network device into multiple smaller resource units for use to support consolidation, multiple resources can be aggregated. For example, two or more servers can be aggregated and clustered to create a scale-out server, and the same applies to storage and networking resources.

Multi-tenancy is a function of virtualization to enable resources to be subdivided while providing some level of isolation to the different tenants. In Chapter 4 we looked at server virtualization and how different virtual machines, containers, or partitions could be isolated, yet physically share a common environment.

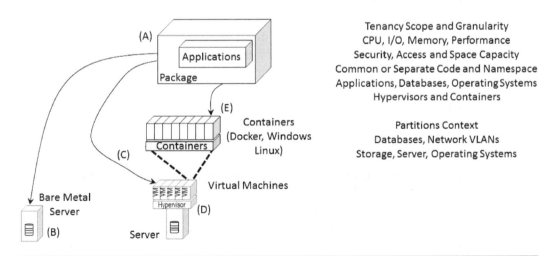

Figure 8.6   Examples of data infrastructure tenancy scenarios.

There are various approaches for isolating data in storage from other users or applications, including LUN and partitions, file systems, folders, logical volumes, and database instances, among others. Some of these approaches provide some level of security and authentication along with encryption for multi-tenancy.

However, another aspect of multi-tenancy is sharing performance resources as well as availability and being able to make changes without impacting others. There is also storage sharing vs. data sharing, where sharing of data requires some coordinated access control via software running on servers or storage servers.

Another variation of multi-tenancy with storage that ties to data services is the ability for a hypervisor running inside a storage system to host applications. These might be general-purpose applications, or databases or other data infrastructure applications, for example, focused on replication, data protection, or cloud or object storage gateway functionality, among others. Many vendors have made their storage system software available as a software or virtual storage appliance to run on virtual machines and cloud instances in addition to selling as tin-wrapped software (i.e., an appliance or storage system).

Variations of the above are converged infrastructure (CI), hyper-converged infrastructure (HCI), clustering-in-box (CiB), and solution bundles, which, along with other storage systems, are covered in Chapter 10. A few other considerations and features include VLAN, VPN, and VPC (virtual private cloud) network and security support.

Tenancy considerations include:

- Storage systems and software can support multi-tenant hypervisors and containers.
- Storage systems and software can also host multi-tenancy functionality.
- Tenancy can be for performance, data protection, or isolation.
- There are various plug-ins, drivers, and management software tools
- Storage sharing vs. data sharing and maintaining data consistency

8.3. Performance (Productivity and Effectiveness) Services

Part of performance comes from underlying hardware resources and capacity. Another part of performance comes from how software defines, uses, and otherwise unlocks the potential of the underlying hardware, as opposed to becoming a bottleneck. Fast servers, storage, and I/O hardware, as well as services, need fast software. Fast software needs fast hardware and resources, all of which are configured properly for effective performance.

Performance-enhancing data infrastructure software and services enable and unlock the potential of more servers as nodes, which mean more processor sockets, cores, threads, memory, and I/O capabilities. However, simply having more processors, ports, sockets, drive bays, or PCIe slots does not mean more effective performance. As more nodes, controllers, servers, and processors are tightly or loosely coupled together, overhead to communicate and maintain data consistency can also increase. That is where well-defined and well-implemented software along with performance-enhancing data services come into play.

In addition to servers and controller motherboards for processing, there are also off-load devices ranging from PCIe and GPU cards to ASIC and FPGA. Off-load devices enable general-purpose servers, controllers, or processing nodes to perform normal functions while handing off compute-intensive data services functions.

Compute-intensive data services functions can include data footprint reduction (compression, de-dupe, thin provisioning) as well as data protection tasks such as RAID, erasure code, and copy, among others. Further, some implementations that rely on off-loading capabilities can also run that software code on general-purpose processors, saving costs, albeit at lower performance in smaller environments.

Storage efficiency is often focused on and thus seen as referring to space capacity optimization. The corollary, then, to efficiency is storage effectiveness, referring to performance, productivity, and enabling work to be done. Performance relies on availability in that if a resource is not available, or is only intermittingly accessible, there will be an impact on productivity and effectiveness.

Conversely, if there is a lack of performance, that too can result in intermittent availability issues. Capacity has an impact on performance in that lack of space can cause applications to stall or encounter intermittent performance or availability issues. That includes space from memory in a server to persistent non-volatile memory and storage.

Performance is associated with, as well as a function of, the underlying hardware (server, storage, I/O adapters, and network paths), software and services basic capabilities, speeds and feeds. This can mean either getting the same amount of work or data moved in a shorter amount of time (reduced wait time delays), moving more data or doing additional work in the same time, improving overall resource use effectiveness (not just utilization), or a combination of the above.

Likewise, there are different focus areas of performance, including response-time latency, bandwidth or data transferred, I/O activity such as IOP, frames, packets, blocks, files, objects, and transactions per second, among others.

Performance-related data services and features include:

- Persistent and nonpersistent memory, RAM cache, and RAM disks
- Host I/O limits, I/O quotas, and other performance QoS attributes
- Parallel and multithreaded applications and I/O
- Reporting, real-time and historical trending and monitoring
- Policies and storage QoS templates, distance enablement
- Load balancing across resources

Quality-of-service controls and capabilities apply to servers, storage, and I/O networks, including from a performance perspective as well as availability. For example, QoS can mean performance such as IOPs or other forms of activity, bandwidth or latency, as well as the reliability of a service.

In addition to QoS capabilities that enable deterministic performance, policy managers and other tools can automate processing, implementing limits, quotas, throttling, load balancing, and fairness algorithms (policies) based on user- or system-defined controls, settings, and templates. Some solutions also support self-learning and adjusting, along with application integration awareness capabilities to enhance performance.

Servers with multiple cores and sockets support many threads that if the application or data infrastructure software can leverage, enables improved concurrent and parallel I/O streams.

Table 8.6. Performance Metrics and Matters

Metric	Characteristics	Description
Activity rate	Work of a particular type being done in a given amount of time, e.g., read or write, block or file IOps, TPS, FPS, PPS, input, output, update, object get, put, list, web pages or videos served. Different size activity (i.e., I/O): Activity rate = data moved/IO size (work to be done).	How much work is done, or productivity in a given amount of time
Data movement	How much data is moved in a given amount of time, including bandwidth, throughput, kilobytes or megabytes per second. Data moved = activity rate × IO size.	How fast or how much data is moved.
Errors	Re-transmissions, missing or dropped packets, and timeouts	Rework
Latency response wait time	How much time is spent waiting for work to be done or data to be saved, retrieved, or accessed, or for some task to complete. May be database transaction or report query time, file, video or object upload or download, backup or restore time.	Time lost due to waiting, or productivity lost as a result
Rebuild time	Time required to repair, recover, rebuild, restore, and reconstruct a failed or lost device, file, or system. During the rebuild time, there can be performance impacts.	Time exposed until full protection
Rebuild impact	Impact on other applications, workloads, or resources during a rebuild, restore, and recovery process.	Service delivery impact

These streams can be largely parallel streams for bandwidth, or many small independent concurrent I/Os or transactions. The features to look for include how many queues and device buffers are supported, along with protocol implementation such as NVMe to boost performance while reducing server CPU wait times and overhead.

General optimization features and data services include for flash SSD SCSI UNMAP and SATA TRIM to perform garbage cleanup and improved performance. TRIM and UNMAP may be disabled by default on some operating systems. In the case of Windows, you do a manual TRIM or UNMAP by selecting a device, going to Properties, Tools, and selecting Optimize. For non-SSD devices, Optimize performs a file system defragmentation.

Table 8.6 shows various data services performance metrics and related topic matters. Metrics can be looked at for different periods of time, such as on a yearly, quarterly, monthly, weekly, daily, hourly, minute, or per-second basis.

Defragmentation is another form of optimization in that as files get added, updated, and deleted, the allocated space becomes fragmented (refer to Chapter 3). Note that fragmentation is not a function of HDD or SSD but of the file system. However, slower HDD may appear to be a problem vs. faster SSD. A common myth is that HDD slow down as they fill up; it is actually the file system that slows down as it fragments.

Another form of performance optimization is write groupings such as full-stripe writes, group writes, and write gathering. These functions are often implemented in conjunction with RAID and SSD to help reduce the number of back-end (southbound) writes that occur on a storage device. Similarly, some solutions support read-ahead and pre-fetch to buffer data that may be used in the future to reduce actual disk I/O.

> *Tip:* When looking at server I/O and application performance, split I/Os can be an indicator of the need to optimize a storage system, device, file system, or database. Other signs include long response times and queues. Different device drivers can also make a difference in performance for physical as well as cloud and virtual environments, with some being more optimized while others are more

> generic for interoperability. For example, the VMware SAS driver is a good general-purpose disk driver for guest VM. However, the VM Para Virtualized disk driver can show an improvement in performance while using less CPU.

Application offload can be used, in which some of the workload or processing that is normally done by a server is handled instead by a storage system, appliance, or another device. Some examples include VMware VAAI and VVOL, where intelligence and awareness are leveraged and work such as copies and other functions are handled by the storage arrays. If the storage arrays do not support the offload, then the functionality is handled via the hypervisor on the server. Another example is copy offload in Microsoft ODX, which is supported by some storage solutions.

Oracle has converged database solutions in which database offload functionality is handled by back-end storage that has awareness into front-end database requests. For example, instead of a database asking a storage system for 10 TB of data from which it will select what matches criteria, the selection criteria can be pushed down to the database storage system. The result is that functionality and data select occur close to where the data resides, and less data has to move over the network. The caveat is that not all general-purpose solutions provide this feature.

To improve network traffic flow, jumbo frames can be enabled and configured to move more data effectively, along with using NIC teaming, load balancing, and other network techniques.

Physical, logical, and virtual resources include processors and their cores, DRAM memory, I/O connectivity and PCIe bus slots, and network and storage devices. Consider how many ports or connections are available of a given type and speed. Also, if these ports share a common connection, there is a potential point for a bottleneck, as discussed in Chapter 6 about adapters.

Other resources include hardware processor, ASIC, FPGA, GPU, or other forms of offloading and application (or feature function) acceleration. These range from graphics to network package inspection to RAID, parity, and erasure code calculations or cache.

In addition to physical connectivity, there are also logical considerations, including the number of device buffers and queues for performing operations. Logical connectivity also includes some LUN, file system, or objects that are accessible. How those resources get defined via management software tools into various configurations will impact performance. Performance-related data services and features also include load-balancing capabilities to address bottlenecks, as well as leveraging available resource capabilities.

> *Tip:* A technique that on the surface may not seem as intuitive for boosting performance is archiving. Many people associate archiving with regulatory and compliance or as a means of data footprint reduction (DFR), which it is. However, archiving in its many variations, along with automated and manual tiering, can clean up and reduce the amount of data in an active file system, database, key value, or other repository. When combined with data deletion or removal as well as cleanup, compression, and optimization, archiving can be a simple technique used in new ways to address old data issues that cause performance problems.

> *Tip:* Another performance optimization technique associated with cache and tiering is data pinning. Data pinning is a technique and feature enabled in different places to lock or force a copy of data to exist in a cache, buffer, or other high-performance memory and storage location tier. Depending on the specific functionality, data can be pinned at different granularities and locations. These include blocks or sectors on an HDD or SSD, LUN, or volume in a storage controller, adapter, or device driver, as well as files, folders, and objects.

Some solutions pin data for reads, while others can support read and write operations. The difference between pinned data and nonpinned data in a cache or buffer is that nonpinned data can be bumped out by other active data. Pinned data stays in the cache or buffer as a means of boosting performance, QoS, and enabling the most deterministic behavior.

> *Tip:* Another performance consideration is what background or data infrastructure tasks, including copy, data protection, and maintenance activity, is occurring while primary production data is being accessed. Even more important is what is the impact of those functions on primary data access. If there is no downside or performance impact to data services and features running while they provide a benefit, that is a good thing. On the other hand, some implementations of data services and features can cause performance bottlenecks to primary data access while in use.

Another performance consideration is support for quick or fast format of a device, database, or repository vs. slower, complete initialization. A quick format offers shorter time up front to make the storage accessible, but initialize performance may suffer for subsequent initializes before or during use. Different storage systems and software provide flexibility to choose various options including background, time deferred, quick, or full format, including zeroing out storage space.

General performance considerations include:

- Optimization and reorganization of file systems and databases
- Rebalance and load balance of capacity when storage is added or removed
- Amount of RAM and persistent NVM or SCM memory
- Number and type of nodes, devices, and I/O interfaces

8.3.1. Server Storage I/O Acceleration (Cache and Micro-tiering)

Reminder that the best I/O is no I/O, and second best is the one with least impact that leverages application workload and data locality of reference. This is where cache and related data performance acceleration services come into play.

Caching and micro-tiering can be done at various granularities, including database table column, table, database instance, application, file systems, file, volume, device, or block basis. In addition to different granularities, caches can be local or global, read, write, or a combination. For example, you can have a read-ahead cache that enables pre-fetching to help anticipate data to be accessed next, or simply collect some extra data near the data being read.

Caching and micro-tiering can occur in several places, including:

- Application middleware, databases, repositories, and file systems
- Logical volume, storage managers, operating systems, and hypervisors
- Device drivers, third-party software, cache cards, and adapters
- Memory RAM cache, network devices, storage systems. and appliances

General cache considerations include:

- Granularity, type, and mode of cache
- Location, size, and cache algorithm
- Configuration, policies, and management options
- Application workload being cache-friendly or not

Figure 8.7 shows locations where caching can be done.

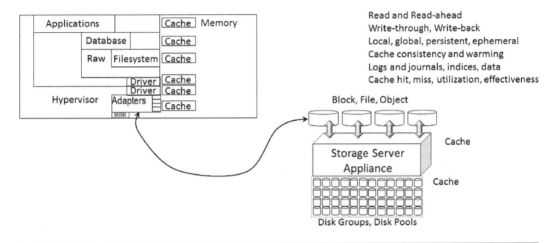

Figure 8.7   Where cache data services are implemented.

Cache-friendly data is data that has a high degree of locality and use, resulting in cache hits and effectiveness. The opposite of cache-friendly is data and—more important—application workloads that result in cache miss due to lack of cache size, inefficient or improper cache configuration or algorithm, or poor locality of reference.

Tip: A common mistake is to focus on cache utilization without considering cache hit or effectiveness (i.e., how many I/Os are resolved from cache) vs. misses (having to go outside the cache). A good cache will have a balance between keeping frequently accessed data while replacing stale data with new data and avoiding cache thrashing. Cache thrashing results in high utilization, low cache hit, and frequent cache flush, turn, or data being updated yet not useful.

Figure 8.8 shows various storage performance acceleration features, including (A) no performance optimization, (B) adding some cache (read or write), (C) migrating to fast NVM, SCM, or SSD tier with cache, and (D) implementing some form of tiering (macro or micro). Cache, micro-tiering, and tiering configuration options include performance QoS settings and policies based on reading, writing, read/write, IOP or bandwidth, timers, write gathering and read-ahead buffer size, and flush intervals, along with granularity (block, page, sector, LUN, volume, file, object, table, column). Other cache configurations include aggressive, normal, proactive, and self-learning modes. Cache consistency and coherency options include data integrity checks, logs and journals, rewarm (reload) of nonpersistent cache after shutdown and start-up.

Cache (and tiering) algorithms include least recently used (LRU), first in, first out (FIFO), last in, first out (LIFO), and round robin, among others, along with various storage performance

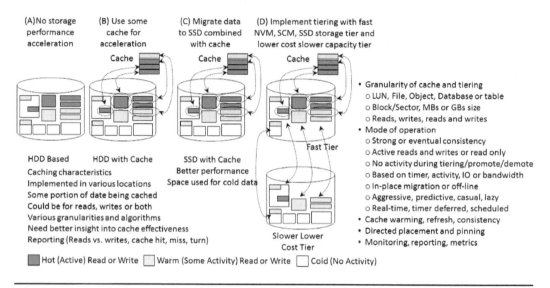

Figure 8.8   Cache, micro-tiering, and macro-tiering.

QoS policies. Granularity of cache and tiering promotes (move up into faster tier or cache) or demotes (move down or out of fast tier or cache) can be in kilobytes, megabytes or gigabytes based on various time intervals or other policies.

> *Tip:* When looking at cache and tiering solutions, look beyond space capacity optimization benefits and understand what is the performance and processing overhead when tiering, migrating, promoting, demoting, or warming to data being moved and other workloads.

Various vendor and third-party tools are available for analyzing applications, databases, file systems, operating systems, and hypervisors as well as storage systems to determine hot files, objects and volumes that are candidates for cache, tiering, or migration to other storage mediums.

Cache features and modes to help application workload performance include read-ahead, write-behind, predictive cache warming, directed placement, and pinning or locking data into the cache. The benefit of pinning data into a cache is that it is in the cache; the downside is that cache is not available for other uses.

Which is best to use, cache, tiering, or micro-tiering? That depends on your applications workloads, among other considerations. However, the solution does not have to be all or nothing, as the different technologies and tool implementations can be complementary.

Table 8.7 provides general characteristics of cache, micro-tiering, and tiering.

Figure 8.9 shows an example of micro-tiering using Enmotus FuzeDrive on a Windows server. On the left is a 1-TB FuzeDrive presented to Windows server for use. The FuzeDrive consists of a 400-GB NVMe PCIe SSD card for a fast tier and a 600-GB RAID 1 (2 × 15K HDD) as a slower, higher-capacity tier. Data gets promoted based on policies and activity across the fast and slow tiers in real time; data can also be pinned to the fast tier.

On the right, Figure 8.9 shows the performance (1,037 70% Read Random 4K IOP) of the RAID 1 (line graph) slow storage (before FuzeDrive) and its 600-GB capacity (solid bar). Moving left to right, next is a FuzeDrive providing 206,216 IOP with an aggregate capacity of

Table 8.7. Cache, Micro-Tiering, and Tiering

	Characteristics
Cache	Different granularities, types, modes, and locations. Performance tier is separate from capacity tier; data may be ephemeral, backed by persistent storage in a different tier or storage class. Cache may require rewarm and refresh for consistency.
Micro-tiering	Moves data at a finer granularity at different small intervals based on various policies. Data exists in a fast-performance or lower-cost capacity tier that can be optionally cached in RAM. Capacity and performance tiers are aggregated into a common namespace or address pool. Provides a balance of space capacity and performance. Implemented in various locations from servers to storage systems. SSD or fast devices do not have to be dedicated (they can also be used for other data), as is the case with many cache-based solutions. Examples include Enmotus FuzeDrive and Windows Server 2016 S2D cache, among others.
Tiering	Also known as information lifecycle management (ILM) and archiving, macro-tiering moves data on a coarser granularity basis at different frequencies and granularity. Data is moved discretely on a block, file, object, or volume basis between tiers on some interval or policy basis. There can be a time delay from time to first byte and subsequent data that has not been promoted and still resides on slower storage. The overhead of moving data can be a concern. There are many examples of server-based as well as a storage system and storage appliances.

1 TB (600 GB slow + 400 GB fast). Next is a 400-GB SAS SSD with no FuzeDrive providing 156,879 IOP, and last on the right, a 400-GB NVMe PCIe card with no FuzeDrive providing 222,332 IOP. The difference with the micro-tiering is a larger effective space capacity pool along with performance improvement providing a best of or a hybrid between traditional cache and tiering.

In addition to having a read-ahead to help future reads, some solutions also support a dedicated read cache that can also be configured as a write-through (i.e., read) cache, or write-back

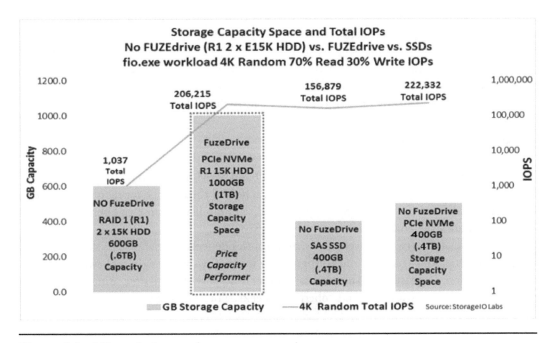

Figure 8.9 Micro-tiering performance example.

(actual write cache). The write-back cache can be faster in that the application does not have to wait for data to be acknowledged as being written (i.e., it is synchronous). Write-through cache does not improve write performance; however, data is written synchronously for data integrity while keeping a copy in the cache for subsequent reads.

Other cache considerations include whether the data is stateless or stateful, and how consistency is maintained. For global and shared cache, consistency and coordination (to avoid stale or inconsistent data), particularly for writes, is important to understand what a given solution does or supports.

Another cache consideration is cache warming, where, after a system start-up, how data gets repopulated back into the cache so the data is warm. Also, understand how long the cache warming or refresh operation takes. In the case of battery-protected cache, how long does data reside in the cache before being destaged? And what is the battery life (endurance)?

Cache and micro-tiering tools should provide tuning and monitoring tools along with QoS policy managers and automation capabilities. Reporting and monitoring should include cache hits and utilization, but also how effective the cache or micro-tiering is in improving performance. Simply watching cache hits and utilization may not give the entire picture; look also at how the overall application is being enhanced, as well as address hot and cold data spots.

Some considerations regarding devices used as cache include whether they are dedicated to that function, or they can be partitioned and shared. For example, does an entire SSD get dedicated to being used as a cache, or can a second or alternate partition be created where part of the SSD gets used for cache, while another portion is used as a general SSD?

Sharing a device can be more applicable for a smaller environment, while larger environments have the economy of scale so devices can be dedicated. Also, can the cache be turned on and off, resources released or configured, while the system is in use? Can data be migrated in place and while in use to a cached environment, or does a separate migration and conversion need to occur involving application downtime? Are any special device drivers or application plug-ins needed to use the cache or micro-tiering, or are they transparent to block, file, and object storage and application access?

Other considerations include what type of QoS, server compute, storage, and network I/O performance constraints and quotas exist or can be enabled—for example, limits on IOP or bandwidth, read or write, I/O size, as well as space reservations for snapshots and other protection items. Also, are there any limits on the size of the cache and how it is shared or dedicated?

Cache performance considerations include:

- Some solutions require dedicating NVM SSD and SCM devices to be cache devices.
- Some solutions allocate part of an SSD as a cache and the balance for regular data use.
- RAM and SCM with flash SSD help to increase durability lifetime and performance.
- NVM and SCM enable persistent caches to exist.
- Hardware and software drivers, tools, and utilities dependencies.
- Monitoring, management, and configuration tools providing insight and awareness.
- Application, database, file system, operating system, hypervisor, and storage interoperability.
- Cache refresh, consistency, and coherency checks, along with rewarming.
- Some applications, databases, and file systems can provide hints to cache tools.
- RAM disks appear as a disk or storage device yet operate at memory speeds.

8.4. Economics, Analytics, and Reporting (Insight and Awareness)

There are a couple of aspects to data infrastructure and data services economics including costs. One aspect is the cost to acquire, purchase, lease, rent, subscribe capital expenditure (CapEx) and operating expenditure (OpEx) items. This includes up-front, one-time as well as recurring costs and fees to use those resources as part of data infrastructure services delivery. Another aspect of economic costs is what you charge users of data infrastructure services (chargeback), which could be an actual invoice bill, or an informative statement of use (show back).

Another dimension of costs is component or unit costs, as well as aggregate services costs. For example, there is the cost of a server, storage, or network device, hardware, software, or service. Then there are the burden costs of additional hardware, software, configuration, management, staffing, and facilities, among other fees required to use that resource.

Another variation is how that resource and components are then used to support service delivery, or provide data infrastructure background support functions such as data protection, management and reporting, and service delivery, among others.

Key to costs and understanding usage, activity, and other aspects of how the data infrastructure and its services are running as well as being used are various metrics. Metrics include the various telemetry data from the data infrastructure, including usage, activity, events, and service-level objectives (SLO) being met or missed.

8.4.1. Metrics That Matter and Insight Awareness

There are many different types of metrics, from the measurement of actual usage and consumption of resources, events, alarms, and other telemetry data. Other metrics are estimated or compound (created by combining various metrics), all of which can span different time intervals, with varying focus.

Metrics get transformed into information via different program software tools for reporting, monitoring, analytics, troubleshooting, and planning, as well as chargeback and show back. Some metrics are measured actual usage based on some time interval from monthly to weekly, daily, hourly, by minute, even sub-second. These include resources used as well as activity counters (data transferred, API calls, I/O operations, events, and alarms) as well as estimated fees.

Analytics and metrics insight items include:

- Resource usage (CPU, I/O, storage space), APIs, network addresses, license fees
- Alarms for events, errors, failures, security access violations
- Measured standalone and compound as well as estimated
- Resources allocated, powered up, paid for but not used (i.e., orphan and ghost instances)
- Reporting, insight, and awareness tools, dashboards, policy rules and alerts
- Real-time display and recording collection, along with replay and historical

Understand your costs. What are your costs for given SLO delivery of a service backup/restore of a VM, database, table, mailbox, home directory, or other entity? Do you know what your costs are or should be? Have you looked at your cloud or managed service provider's invoices lately to see what you are using along with additional fees?

Do your costs include other items such as facilities fees, power, cooling charges, taxes, insurance, staffing, and security, among others? Do you have a bill of materials for delivering a resource or service that includes cost breakdowns of components, hardware, software, services, staffing, facilities, and other overhead items?

8.4.2. Metrics and Data Infrastructure Cost Considerations

Some metrics are focused on performance, others on availability, capacity efficiency, effectiveness, productivity, events, alarms, and supporting decision making or problem resolution. Various metrics can be combined to support related economic data infrastructure activities, including cost to support a given function to a specific service level.

Costs considerations include:

- Up-front hardware, software, staffing, training, migration, and conversions
- Recurring operating costs, insurance, staffing, rents, subscription, maintenance
- Power, cooling, floor space, insurance, and other facilities fees
- Staffing, training, consulting, and other professional services

Additional costs and fees considerations (your expenses and your users) include:

- Space usage, data moved, network access, number of database and table instances
- Cost of restoring, rebuilding, reconstructing a failed server, storage, or network device
- Resource usage overhead while data services are running in background
- Cost for storing data as well as accessing (get, put, list, directory)
- Level of resiliency and durability, along with performance limits
- IP and address, network gateway, DNS, router, and other fees
- How services or software fees are charged: usage, stored, retrieved, by user
- Are per-user fees based on a number of active or site-based?
- Fees for secondary, standby, BC/DR sites, including software licenses
- Space capacity based on raw or usable protected
- Cost or fees for alerts, rules, dashboard instances

Everything is not the same in the data center, across organizations, applications, and data infrastructures. There are various formats and types of invoices, chargeback and show backs. Figure 8.10 shows a generic sample chargeback or invoice for data infrastructure services. The example is simplistic, including application-centric services, monitor and alarms, compute, backup/data protection, security, network, and storage resources. Different services and tools may have even simpler, or more complex, reporting and invoices.

Data infrastructure resources (servers, storage, networks) consume power to perform work, which also means cooling to remove heat. With power costs and more commonly availability of both primary and secondary or standby power being a concern, there are different options.

UnlimitedIO (UIO) Invoice for Data Infrastructure Services   Invoice: #070116163055082
Client Account: StorageIO   Billing Period: June 2016

Invoice Summary		
UIO Service Charges (Metered usage and recurring fees)		$39.26
Payment amount received for current invoice		$39.26
Total for period		$39.06

Data Infrastructure Services Detail	Quantity	
Application Services		$0.00
Application Services Total		$0.00
Monitor Alarms	4	$0.00
Monitor Alarms Total		$0.00
Compute		$6.93
Compute Total		$6.93
Backup/Data Protection	5TB	$10.00
Backup/Data Protection Total		$10.00
Security		$0.00
Security Total		$0.00
Network		
Routing		$0.51
IP Addresses	2	$6.00
Bandwidth fees	28.132GB	$0.00
Network Total		$6.51
Storage		
Online	483.6GB	$15.81
Storage Total		$15.81

Figure 8.10   Data infrastructure sample invoice.

One option is to avoid energy consumption by not doing work and powering things off. However, that could be a barrier to productivity. On the other hand, if good productive work vs. overhead is being done, and you have efficient and effective resources, then you should also have a good service delivery per watt of energy.

Depending on objectives, there are various energy power and cooling options, including:

- Balancing productive with avoidance to a given level of PACE requirements.
- For performance and productivity, use fewer, faster, more energy-effective devices.

- For lower-performance applications, use power-savings mode and other techniques.
- Check your servers, network, and storage software settings to vary the power mode.
- Seek out and find ghost or orphan storage, VM, and other entities that are not being used.
- Leverage smart power management, cooling, speed step, and other techniques.
- Gain insight into service delivery, application PACE needs, and resource consumption.

General data services and functionality economic considerations include:

- What data services and management features are bundled as part of a solution?
- Are there extra software or hardware dependencies to use the solution?
- Are there limits on usage or consumption of a service or feature?
- What are the resource usage and monitoring reporting capabilities. Are there threshold alerts?

Additional cost considerations include:

- What are the costs (up-front and recurring) for the dependency items?
- What are the optional (not dependent) items available, their functionality and costs?
- How are costs broken down, up-front initial, maintenance, license, and support?
- If there are resource usage limits, what are the overage fees?
- How fee are charged—metered, estimated, by periodic audits, or an honor system?
- What happens to your data and applications if you forget to pay or renew your license?
- What you buy today may not be economical in the future for its intended role; can it be redeployed to a different class of service in a cost-effective manner, or simply retired?

8.4.3. Analytics, Discovery, and Search

Some context is needed regarding analytics, discovery, and search, as there are tools and functionalities in the data infrastructure and associated resources, as well as upper-layer applications. Some search functions are to determine what types of data exists on various server and storage systems to know what you have and how it is being used. Other search tools focus on electronic discovery (eDiscovery) in support of regulatory and other compliance, as well as data analytics.

There are upper-level application search and discovery tools that focus on looking through and analyzing data in databases, key value stores, file systems, and their contents. Then there are tools for searching and discovery what files, objects, or types of items exist, where, how much space they use, and when they are accessed, among other items. Similarly, there are Hadoop, SAS, Splunk, and other tools that support business and data analytics. Those tools can also be used for analysis and processing data infrastructure event, activity, and security logs, among other telemetry items. Some tools focus on metadata about actual data (i.e., the contents of what is stored), while others are focused on data infrastructure metadata (where and how data is stored as well as accessed).

When it comes to data services that are tied to search, is the context for the data infrastructure and associated metadata, or the application or business context? Tools range from those provided via operating systems such Linux or Windows to vendor-supplied and third-party tools. Some third-party tools include JAM TreeSize Pro, Splunk, Solarwinds, Virtual Instruments, Space Sniffer, Foldersize, StorageX, and RV, among many others.

Considerations include the following:

- Keep in mind the scope and context of analytics, search, and discovery.
- Some tools focus on the data infrastructure, while others focus on business data.

8.5. Management Services

Data infrastructure management services span server, storage, I/O, and networking, from hardware to software-defined cloud, virtual, container, and other areas. Management has different context and meaning from low-level device to broader, higher-level overall service delivery. Also, management can mean functionality, such as access, aggregation, performance, data protection, monitoring, and reporting, depending on the point of view and focus. Data services and software functionality can also be thought of as management services. Thus, there are different types of management tools and interfaces.

Management data services topics include:

- Interfaces and functionalities
- Automation and policies
- Analytics, discovery, search
- Third-party and cloud access
- Monitoring, reporting, insight

8.5.1. Management Interfaces

We began this chapter by discussing data infrastructure resource accessing including storage, and now we'll move to management access interfaces. There are different types of management interfaces, also known as command control, control plane, or fabrics, and—borrowing from the energy utility infrastructures industry—supervisory control and data acquisition (SCADA). There are different types of management software, tools, and interfaces.

Management interfaces include, among others:

- GUI, CLI, PowerShell, API, and management plug-ins
- Wizards and templates, web-based interfaces including HTML5 and REST
- Phone, call, text, or email home or to-cloud service
- Element, device, system, and framework manager dashboards
- Vendor supplied, third-party, and open-source provided

Management interfaces are used for:

- Command, control, communication, insight, awareness, analytics
- Applying patches, upgrades, new software, firmware, BIOS
- Configuration add, change, remove, save, backup
- Event, health status, performance monitoring and reporting

- Establishing policies, templates, thresholds
- Application-focused data search and metadata functions
- System and data infrastructure search and metadata functions

Some management interfaces are built into or embedded as part of a server, storage system, network device, card or adapter, operating system, hypervisor, file system, or other data infrastructure software or service. These range from framework and orchestration tools to dashboards, element managers, event and activity logs, and component-specific tools. For example, some tools work only with VMware, or Hyper-V, OpenStack, Windows, *nix, or something else, while other tools work with many different technologies.

There are tools that are platform-specific, such as VMware VASA, VAAI, VVOL, vSphere, and vCenter plug-ins, among others. Then there are OpenStack Cinder, and Microsoft ODX, among others, as well as SNMP management information blocks (MIB), CDMI, and SMIS. Some tools are for monitoring and event reporting, while others are for configuration or provisioning at a higher level (altitude), or at a lower level such as updating device software or firmware. For example, their sniffers, analyzers, hardware, and software probe for monitoring or collecting "traffic" information "traces" of protocol commands, data movement, or other activity. Some of these probes look only at header information to determine reads, writes, gets, puts, lists, source, and destination, while others go deeper to look at data content. These tools can focus on an application, database, file system, block, file, object, API, server, storage, or network.

Figure 8.11 shows various management interfaces for different functions spanning physical, software-defined, virtual, and cloud. What these have in common is that their focus is on interacting, command, control, configuring, diagnosing, provisioning, monitoring, metering,

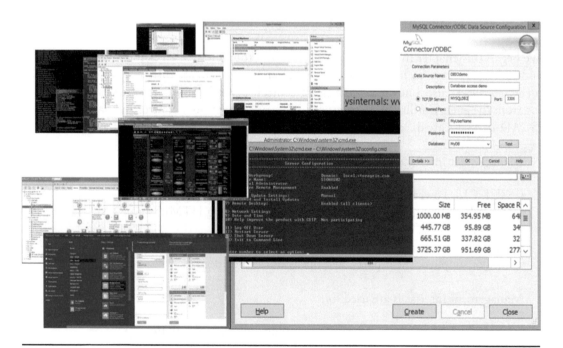

Figure 8.11 Some management interfaces and functions.

and reporting on resource usage, activity, and events. Some of these interfaces are one-way, meaning they simply provide health status or event notification, while others can also provide ongoing status, performance, and other insights.

Some management interfaces are in-band, meaning that the command control traffic is intermixed with data traffic. In other situations, the command and control are out-of-band, meaning that a separate management interface network, fabric, or control plane path exists. The benefit of an in-band approach is that both user data (north–south traffic; cf. Chapter 5) and management data (east–west traffic) use a common interface, reducing cost and complexity.

On the other hand, having separate data (north–south) and management (east–west) data paths, planes, or fabrics reduces congestion or performance impacts of mixed traffic. For smaller environments having mixed traffic might be ok, while for larger, more involved environments, that larger scale may require separate data and management networks (paths, planes, fabrics).

As mentioned, in-band means that the management function, or the data services features and functionalities, sit in the data path between the application host server and the target or destination resource such as a storage system or cloud service. The benefit of this approach is relative ease and speed of deployment along with interoperability flexibility. The caveat is that having something in the data path is like having something on the highway during rush hour: It slows down all the traffic, adding latency or possibly introducing performance bottlenecks. Another way of looking at this is a manager who likes to micro-manage or who gets in the way while you are trying to do your job.

Out-of-band, along with variations such as fast-path and control-path, move all unnecessary activities out of the data path. Unlike in-band, with an out-of-band or fast-path approach, management or the functionality gets involved only when needed, then gets out of the way. Think of this as like removing the bottleneck from the roadway, or having a manager who is there to help when needed, then knows to get out of the way. The result should be lower latency, better productivity, and improved performance. The caveat is that out-of-band and fast-path control path approaches take longer to develop and implement, and there can be interoperability dependencies on certain software, adapters, or operating systems and hypervisors.

Another management interface consideration is agent vs. agentless. Both have their pros and cons. For example, an agent that is lightweight (low overhead impact on performance and memory) should add value with extra functionality. The downside is that some agents may not behave properly, consume CPU, memory, or storage, causing network traffic if they are "chatty."

Some agents are more difficult to install and maintain across multiple systems if they have not been designed for large-scale deployment. Another consideration is how quickly the agent can support a new version of the operating system or system that it is installed as a part of.

Agentless has the benefit of not having to be installed and managed on each system, but there may be trade-offs. The choice between agent or agentless should come down to what they will do or not do for your environment: Will they work for you, or will you and your environment have to work to support them (i.e., adding complexity)?

Most management interfaces rely on some TCP-based communication to an API, socket, port, agent, or port-based service. Management interfaces can be used to access the health of the resource, such as availability, errors, and other events, as well as space capacity usage and performance activity. Different management tools along with data infrastructure resources (servers, storage, networks) provide data and information in various formats.

Figure 8.12   Dell EMC ECS dashboard.

Some management tools and interfaces support export as simple text files or displays, Comma Separated Value (CSV), XML, and JSON, among others. Common tools, interfaces, and protocols include product- or technology-specific APIs and programmatic bindings, command line

```
cephadmin@CEPHIOADMIN:~/my-cluster$
cephadmin@CEPHIOADMIN:~/my-cluster$ ceph health detail
HEALTH_OK
cephadmin@CEPHIOADMIN:~/my-cluster$ ceph status
    cluster 6a077c9a-545d-452f-9d46-e27d29808dbf
      health HEALTH_OK
      monmap e1: 1 mons at {CEPHIOADMIN=192.168.1.100:6789/0}, election epoch 2,
quorum 0 CEPHIOADMIN
      osdmap e36: 8 osds: 8 up, 8 in
       pgmap v56: 192 pgs, 3 pools, 0 bytes data, 0 objects
             305 MB used, 127 GB / 127 GB avail
                 192 active+clean
cephadmin@CEPHIOADMIN:~/my-cluster$ ceph osd tree
# id    weight  type name       up/down reweight
-1      0.1599  root default
-2      0.03998         host CEPHIOODS1
0       0.01999                 osd.0   up      1
1       0.01999                 osd.1   up      1
-3      0.03998         host CEPHIOODS2
2       0.01999                 osd.2   up      1
3       0.01999                 osd.3   up      1
-4      0.03998         host CEPHIOODS3
4       0.01999                 osd.4   up      1
5       0.01999                 osd.5   up      1
-5      0.03998         host CEPHIOODS4
6       0.01999                 osd.6   up      1
7       0.01999                 osd.7   up      1
cephadmin@CEPHIOADMIN:~/my-cluster$ []
```

Figure 8.13   Ceph health detail and status.

Figure 8.14 VMware vSAN dashboard.

interfaces (CLI) including shell and power shells for scripting, GUIs and wizards, SNIA CDMI and SMIS, OpenFlow, SNMP MIBs, HTML5, HTTP/ REST, ODBC for databases, Telnet, FTP, RDP, SSH, VNC, call, text, or email home, and WebDAV, among others.

A note about HTTP/Rest is that context is important. If a vendor mentions support as part of their data sheet, spec sheet, or other sales/marketing material, ask whether they are referring to management or data access. Likewise, ask whether GbE LAN ports are only for management (i.e., out-of-band) use or they are also for data (i.e., north–south access).

Additional tools and platforms for managing and processing data include R, Splunk, Spark, SAS, and Hadoop, along with the various operating system, hypervisor, and stack tools such as those for OpenStack, among others. Various management suites include those for SQL Server, Oracle and SAP among others.

The following are some example management interfaces for various data infrastructure components. Figure 8.12 shows a dashboard for Dell EMC ECS (Elastic Cloud Storage), which provides object and content storage services.

Figure 8.13 shows Ceph health detail and status of a four-node cluster.

Figure 8.14 shows another example, a VMware vSAN dashboard that displays various health and status alerts.

Figure 8.15 shows various OpenStack command lists, including endpoints, users, and roles.

Figure 8.16 shows an OpenStack dashboard displaying compute instances and images.

Figure 8.17 shows a VMware SDDC discovery (VMware fling) topology display.

Figure 8.18 is a VMware vSphere Web Client view of a vCenter dashboard and various resources in a software-defined data center and data infrastructure.

Figure 8.19 shows management interfaces on a Windows Server 2016 system, including PowerShell on the left and Server Manager dashboard on the right.

Management interface considerations include:

- There are different types and layers of management tools for various functions.
- Understand the context of management tools and interfaces for a particular need.
- What will be using the management interface, a person or software tool?
- Other software (or hardware) dependencies.

```
greg@OSIOCONTROL:~$ openstack endpoint list
+----------------------------------+-----------+--------------+--------------+---------+-----------+-------------------------------------------+
| ID                               | Region    | Service Name | Service Type | Enabled | Interface | URL                                       |
+----------------------------------+-----------+--------------+--------------+---------+-----------+-------------------------------------------+
| 78fe3598d59e49d598dee0fb994fdef6 | RegionOne | glance       | image        | True    | public    | http //osiocontrol 9292                   |
| bc38c6a079fd4595affb57c3ed3aaed6 | RegionOne | glance       | image        | True    | internal  | http //osiocontrol 9292                   |
| ec19aaa40ae54d448209c2ab6f099ba6 | RegionOne | glance       | image        | True    | admin     | http //osiocontrol 9292                   |
| 7084fb6f63304cdbab59b5ea74017943 | RegionOne | keystone     | identity     | True    | public    | http //osiocontrol 5000/v3                |
| 862af6f38992429bac6b939e0775ee13 | RegionOne | keystone     | identity     | True    | internal  | http //osiocontrol 5000/v3                |
| 3b4becf6e49c4830b641dbdcb45708a0 | RegionOne | keystone     | identity     | True    | admin     | http //osiocontrol 35357/v3               |
| e28f07331f6c4b1e9b60ced7a5e86bc5 | RegionOne | nova         | compute      | True    | public    | http //osiocontrol 8774/v2 1/%(tenant_id)s|
| 7700694a821a409aa67b83dca5bdba02 | RegionOne | nova         | compute      | True    | internal  | http //osiocontrol 8774/v2 1/%(tenant_id)s|
| 6202fbc19db94338a7d98541f548f443 | RegionOne | nova         | compute      | True    | admin     | http //osiocontrol 8774/v2 1/%(tenant_id)s|
| 0073cd20f45448118c1a7d3de98de18a | RegionOne | neutron      | network      | True    | public    | http //osiocontrol 9696                   |
| 01c2d61eb4a945fca5bca1bb0e13f033 | RegionOne | neutron      | network      | True    | internal  | http //osiocontorl 9696                   |
| 1be7932f5ee94c19894712ffb676e0a7 | RegionOne | neutron      | network      | True    | admin     | http //osiocontorl 9696                   |
+----------------------------------+-----------+--------------+--------------+---------+-----------+-------------------------------------------+
greg@OSIOCONTROL:~$ openstack user list
+----------------------------------+---------+
| ID                               | Name    |
+----------------------------------+---------+
| 7c9fcfb3deed4cba8e24095312bac4ba | admin   |
| 5002082d6d314fadbe93f42e54e1c23a | demo    |
| 390df2be0e104696951db9e66e7d8728 | glance  |
| f6a48aa55e654fdfb559a91adfb68fb0 | nova    |
| e7e68950f83c47d99035cfefdfbb046f | neutron |
+----------------------------------+---------+
greg@OSIOCONTROL:~$ openstack role list
+----------------------------------+-------+
| ID                               | Name  |
+----------------------------------+-------+
| 1fc147a8262345a1b33e23e8143df60e | admin |
| 4204132b461148c88adcb6999a7fbec8 | user  |
+----------------------------------+-------+
greg@OSIOCONTROL:~$
```

Figure 8.15   OpenStack command line status.

8.5.2. Management Functionalities

Many types of management services and functionalities are part of data infrastructures. These include the higher-level application and data infrastructure resource allocation and orchestration spanning local and remote clouds, as well as physical and virtual environments.

Figure 8.16   OpenStack dashboard compute status.

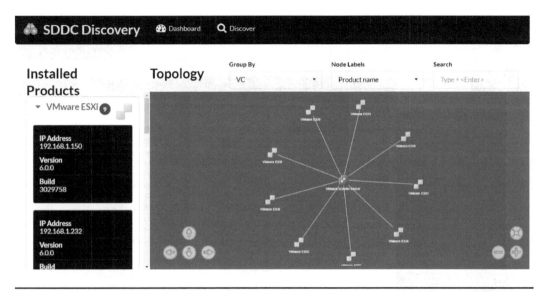

Figure 8.17   VMware SDDC topology view.

Management services functionality can be focused on:

- Application, database, middleware, and file systems
- Operating systems, hypervisors, stacks, and device drivers
- Servers, storage, network devices, systems, and components
- Services such as cloud and network bandwidth, among others
- Automation, template, workflows, and policies tasks
- Reporting, monitoring, alerts, usage and activity, and billing

Management services can be homogenous, focused on a specific platform or technology, for example, Linux or Windows, physical or cloud, a specific cloud service such as AWS, Azure, or Google, or a hypervisor such as VMware, Hyper-V, KVM, Xen, or OpenStack, among others.

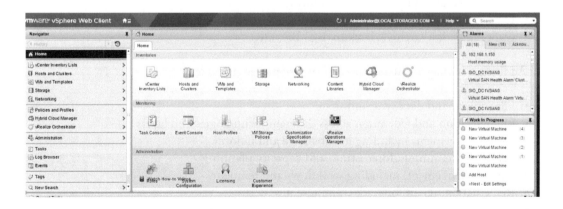

Figure 8.18   VMware vSphere Web Client display.

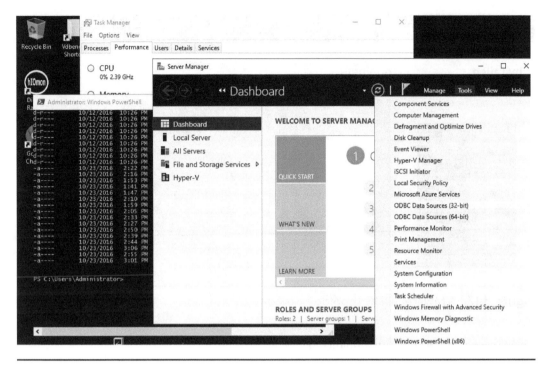

Figure 8.19   Microsoft Windows Server management interfaces.

On the other hand, management services, tools, and interfaces can also be heterogeneous, meaning they work across various technologies and platforms.

Common management services and functionalities include:

- Access control, identity management, encryption, and security
- Change and configuration management
- Time management and coordination
- Dashboards and interfaces for monitoring
- Data validation, verification, and consistency checks
- Data protection (archive, backup, BC, BR, DR)

Additional management services and functionalities include:

- File, object, and data sync and share
- Orchestration and federation across different technologies and layers
- Path management, including load balance and failover
- Performance, tuning, forecasting, and capacity planning
- Policy management, including workflows and templates
- Remediation, repair, upgrades, and updates
- Reporting, monitoring, analysis, alerts, and notification
- Self-service portals for provisioning of resources
- Service management (SLO, SLA, QoS)

Management functionality considerations include:

- Dashboards, key performance indicators, service management
- Self-service portals and service menus for resource and service allocation
- Hardware, software, and services dependencies as well as costs
- Plug-ins, drivers, APIs, utilities, and other software modules

8.5.3. Third-Party Storage and Cloud Support

Some data services and features exist within or as part of public cloud services as well as in cloud software solutions such as OpenStack and AzureStack, among others, for enabling hybrid and private clouds. Another data service and functionality that have evolved from standalone dedicated hardware and software-based gateways is cloud access. Cloud access ranges from accessing blob or object storage such as AWS S3, Google GCS, Azure Blobs to block, file, database and tables, and queues, among other services.

It is a common perception that all cloud storage access is via object such as S3, Swift, or some other HTTP/REST protocol. While many cloud services are in fact based on HTTP/Rest S3 or Swift, there are also those based on iSCSI block as well as NAS file, including NFS and Windows (CIFS/SMB) as well as API, among others. Some of these access functionalities are internal to a cloud, while others can be accessed from outside the cloud as well. Most object storage services from public clouds are externally accessible, while block- or file-based services may be more limited. Microsoft Azure, for example, supports file-based access both from inside Azure clouds as well as externally via Windows Server (file sharing).

How the different cloud services are used and accessed can also vary; for example, overlays and gateways can sit in front of object stores, providing an iSCSI block of NAS file sharing. I use S3FS as a plug-in on some of my Linux systems, which enables me to access and mount an AWS S3 bucket that appears as a local file system. There are many other tools including those built into data protection, backup, storage systems, and appliances, among others, that can also help with cloud as well as object storage access.

Other cloud access includes SQL and NoSQL databases or tables, queues, files, and virtual machine migration as well as virtual private clouds. Many storage systems or appliances, as well as software tools, include some form of cloud access or gateway-type functionality. The functionality ranges from enabling or supporting replication, snapshots, block, file, or object access along with endpoint and namespace management including security.

Data protection tools tend to support different cloud services as well as cloud and object storage as a target for archive, backup, snapshots, and replications. Some solutions can also run or be hosted within a cloud environment such as on an AWS EC2 instance, enabling data not only to be protected to a cloud, but also restored there. Some vendors are porting their storage systems and software-defined storage management stacks to run on cloud instances and function as peers to their physical or virtual implementations.

Tip: Look for support of popular interfaces and protocols such as AWS S3 and OpenStack Swift for object access, along with access via other standards including iSCSI, NFS, HDFS, and SQL, among others. Other features and functionalities to look for include endpoint security including management of keys and credentials for authentication.

Considerations include:

- Understand the various fees associated with cloud services, beyond cost per capacity.
- Cloud fees can include access (get, put, delete, list) as well as other activity costs.
- Know what the performance limits and quotas are for various services and their APIs.
- What tools, software, drivers, plug-ins, overlays, or devices are needed for access.
- How security credentials and access keys will be handled and managed.

8.6. What's in Your Toolbox?

This chapter is about tools in the context of data services and features that in turn get defined (using other software management tools) to enable your data infrastructure and the associated IT applications it supports. Refer to Chapter 9 for more toolbox tools.

8.7. Common Questions and Tips

What are marketing or hero numbers? These are metrics or benchmarks optimized to show a vendor's technology in the best light or to be a hero. Sometimes they may be relevant, many times they are relevant only for vendor marketing bragging rights.

What is a tin-wrapped data service? Tin-wrapped simply means that the software functionality is packed and delivered with a physical server as a turnkey appliance, "tin" referring to the physical hardware metal. A cloud-wrapped service would be one delivered via a cloud instance, just as a virtual or software-wrapped appliance or data service is delivered via a virtual machine such as a VSA, among other variations.

How do you read and compare vendors' data sheets? Keep context in mind, and look for extensiveness in what is being shown. For example, a vendor may simply show a checkbox for snapshots, as opposed to providing additional information such as the number of concurrent snapshots occurring, the total number of snaps that can exist, application and management integration, as well as whether the snaps are read only, or read and writable.

Where is the best place to locate or have data services provided from? That depends on your environment, what you are trying to accomplish, product preferences, and other considerations about how your data infrastructure supports different application PACE needs.

What are some examples of architectures, solutions, systems, and services that implement various data services? Great questions; answers are just a few pages away, in the following chapters.

How come you did not say more about backup/restore, archiving, security, using snapshots, and replication along with other data protection topics? Those topics are covered in Chapter 9.

8.8. Chapter Summary

This wraps up the first of four chapters on data services that enable data infrastructures functionality. Data services vary in functionality and capabilities, as well as where they are implemented, spanning block, file, object, and API.

General action items include:

- Everything is not the same; there are different data services for various functionalities.
- Different data services are implemented in various ways as well as locations.
- The breadth and extensiveness of data services implementations vary.
- There are different performance and access needs for data infrastructure services.
- Data infrastructure performance and access enable storage effectiveness (productivity).

Continue reading about data services, including availabity (data protection) in Chapters 9 and 10, and capacity including data footprint reduction (DFR), also known as storage efficiency, in Chapter 11.

Bottom line: Access and performance are key capabilities of data infrastructures.

Chapter 9

Data Infrastructure Services: Availability, RAS, and RAID

Data infrastructures store, protect, secure, and serve information.

What You Will Learn in This Chapter

- Reliability, availability, and serviceability (RAS), along with data protection
- Accessibility, availability, durability, security, and consistency
- Importance of 4 3 2 1 data protection rule for enabling availability
- Basic and high availability (HA), clustering, and other resiliency topics
- Copy data management (CDM), insight, awareness, and analytics
- Breadth of features along with depth of functionality
- What to look for and consider to compare features

This chapter is the second of four focused on data services that enable data infrastructure services capabilities. The focus of this chapter is availability (data protection), with access, management, and performance covered in Chapter 8. Chapter 10 continues with availability, focusing on recovery-point protection (e.g., backup, snapshots) and security. Chapter 11 covers capacity and data footprint reduction (DFR), also known as storage efficiency. Key themes, buzzwords, and trends addressed in this chapter include block, file, object, and application program interfaces (APIs), along with availability, durability, security, replication, RAID, erasure code, snapshots, and versions, among others.

9.1. Getting Started

A recurring theme is that everything is not the same in the data center, information factory, data infrastructure, various environments, and applications. However, there are similarities and commonalities both regarding threat risks or things that can and do go wrong from technology failure, human error including misconfiguration, to accidental and intentional acts of man, as well as acts of nature. Remember, if something can go wrong, it probably will at some point—either on its own or with the help (or neglect or mistake) of somebody.

In this chapter we look at both the data services technology and tools and the techniques as well as data protection tips for data infrastructure (and applications). Keep in mind that the fundamental function of data infrastructure is to store, protect, secure, and serve data for applications that transform it into information. This means having *availability,* the A in PACE. With that theme in mind, let's look at common data services and feature functionalities that support availability (protect, preserve, secure) and capacity (how data is processed and stored).

9.1.1. Tradecraft and Tools

Let's start the tradecraft and tools discussion for this chapter with *backing store.* Backing store has different context and meanings, but there are common themes. The common themes are some software with a back-end persistent, nonephemeral, non-volatile memory or storage where data is written (stored). Data is stored on different physical media or mediums such as NVM, SCM and other SSD, HDD, tape, and optical. The backing store can also be a system or appliance (physical or virtual) where data is stored, as well as a service such as cloud storage.

Another context for backing store can be the software that provides abstraction via different tools, applications, and associated data infrastructure components. This can range from a backing store driver or plug-in for Docker containers enabling persistent storage to persistent memory and storage for in-memory databases, virtual memory, storage back-ends for databases, volume managers, file systems, key value repositories, and other software.

Context matters to understand what is being discussed—hardware, software, cloud, virtual, or other backing store device, tool, application, or utility.

Beyond backing store, some other tradecraft conversations and context include remembering that delta refers to change, a point in time or the difference between different times, or copies of data. Coverage has a data protection context in what is protected or skipped (i.e., not covered).

Another term you may hear tied to availability, resiliency, data protection, business continuity (BC), business recovery (BR), and disaster recovery (DR), among other activities, is *belt and suspenders.* This simply means having both a belt and a pair of suspenders to be sure that if either fails, the other should keep your pants from falling. The idea is that for some applications, multiple types of data protection, availability, and resiliency are needed to complement each other, as well as in case something fails—for example, using both higher-level application or server software replication along with lower-level storage system or appliance mirroring and replication.

A number of 9's availability or downtime, for example, 99.99 (four nines), indicates the level of availability (downtime does not exceed) objective. For example, 99.99% availability means that in a 24-hour day there could be about 9 seconds of downtime, or about 52 minutes and 34 seconds per year. Note that numbers can vary depending on whether you are using 30 days for a month vs. 365/12 days, or 52 weeks vs. 365/7 for weeks, along with rounding and number decimal places as shown in Table 9.1.

Service-level objectives (SLO) are metrics and *key performance indicators* (KPI) that guide meeting performance, availability, capacity, and economic targets—for example, some number

Table 9.1. **Number of 9's Availability Shown as Downtime per Time Interval**

Uptime	24-hour Day	Week	Month	Year
99	0 h 14 m 24 s	1 h 40 m 48 s	7 h 18 m 17 s	3 d 15 h 36 m 15 s
99.9	0 h 01 m 27 s	0 h 10 m 05 s	0 h 43 m 26 s	0 d 08 h 45 m 36 s
99.99	0 h 00 m 09 s	0 h 01 m 01 s	0 h 04 m 12 s	0 d 00 h 52 m 34 s
99.999	0 h 00 m 01s	0 h 00 m 07 s	0 h 00 m 36 s	0 d 00 h 05 m 15 s

of 9's availability or durability, a specific number of transactions per second, or recovery and restart of applications.

A *service-level agreement* (SLA) specifies the various service-level objectives such as PACE requirements including RTO and RPO, among others. SLA can also specify availability objectives as well as penalties or remuneration should SLO be missed.

Recovery-time objective (RTO) is how much time is allowed before applications, data, or data infrastructure components need to be accessible, consistent, and usable. An RTO = 0 (zero) means no loss of access or service disruption, i.e., continuous availability. One example is an application end-to-end RTO of 4 hours, meaning that all components (application server, databases, file systems, settings, associated storage, networks) must be restored, rolled back, and restarted for use in 4 hours or less.

Another RTO example is component level for different data infrastructure layers as well as cumulative or end to end. In this scenario, the 4 hours includes time to recover, restart, and rebuild a server, application software, storage devices, databases, networks, and other items. In this scenario, there are not 4 hours available to restore the database, or 4 hours to restore the storage, as some time is needed for all pieces to be verified along with their dependencies.

Recovery-point objective (RPO) is the point in time to which data needs to be recoverable (i.e., when it was last protected). Another way of looking at RPO is how much data you can afford to lose, with RPO = 0 (zero) meaning no data loss, or, for example, RPO = 5 minutes being up to 5 minutes of lost data.

Data loss access (DLA) occurs when data still exists, is consistent, durable, and safe, but it cannot be accessed due to network, application, or other problem. Note that the inverse is data that can be accessed, but it is damaged.

A *data loss event* (DLE) is an incident that results in loss or damage to data. Note that some context is needed in a scenario in which data is stolen via a copy but the data still exists, vs. the actual data is taken and is now missing (no copies exist).

Data loss prevention (DLP) encompasses the activities, techniques, technologies, tools, best practices, and tradecraft skills used to protect data from DLE or DLA.

Point in time (PIT) refers to a recovery or consistency point where data can be restored from or to (i.e., RPO), such as from a copy, snapshot, backup, sync, or clone. Essentially, as its name implies, it is the state of the data at that particular point in time.

9.2. Copy Data Management

Just as there are different reasons for having copies of data, there are also various approaches, technologies, tools, and data services features. These various copy management data services features range from management configurations to policy definition, monitoring, and reporting, as well as making copies. Some data is considered an authoritative, master, gold, or primary version from which others can make copies or clones.

Other reasons for making copies, as shown in Figure 9.1, include business analytics, data mining, and protection copies, among other uses. Protection copies include different versions from various time or recovery-point intervals and durability to maintain availability. Other copy uses include import/export of items into databases, file systems, or object repositories, as well as cache for local and remote collaboration.

Data copies can be accomplished using backup/restore, archiving, versioning, point-in-time (pit) snapshot, continuous data protection (CDP), copy, clone, sync, and share, along with replication and mirroring. What this means is that there are different reasons and types of data copies, as well as various tools to support those functions, so comparing apples to apples is important for aligning the applicable tool.

The dilemma with data is that, on one hand, if the data has value (or unknown value, as discussed in Chapter 2), there should be a business or organizational benefit to having copies to support various functions. On the other hand, there is a cost and complexity associated with storing, protecting, preserving, and serving data for longer periods of time. Those costs are not just for the underlying storage capacity (hardware or cloud service), but also for associated servers, networks, software tools, and people management time. In other words, do you also need to protect against adapter, controller, software, file system, volume failure, or loss and damage to files and objects? The scale and type of environment will come into play, and everything is not the same, so why use the same approach or technique for everything unless that is all you have, know, or want to use?

Tip: Some copy data management (CDM) questions include: Where do those copies need to exist: locally, on a single server or shared storage device, remote or in the cloud? If local, is there a copy of data in memory or cache as part of an active working set, or on fast NVM SSD or another storage device? If remote, is there a copy kept in a content cache somewhere on the network?

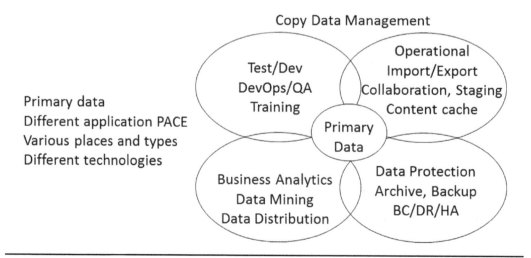

Figure 9.1 Data Infrastructure supports various data copy uses and applications.

CDM fundamental questions include:

- How many copies do you have (or do you know)?
- Where are they located (centralized, cloud, distributed, mobile)?
- How many do you need and would like to have?
- How will those copies be used (read, read/write, shared)?
- What type of insight, analysis, and reporting do you currently have?

Also, will multiple people or applications use those copies concurrently? How big are the copies, how often do they change or get updated, and what percent of the data changes, as well what type of files, objects, or data is being copied? Are copies and versions made to help expedite recovery or resumption of service in the event of a device, system, or data center failure?

Additional CDM questions include:

- Are copies used for in-turn making other copies (full or partial clones)?
- How will you synchronize downstream and upstream copies?
- Are the copies of the same or different versions (points in time)?
- What are the copies of (files, database, documents, attachments, virtual machines)?
- Do you have an existing file or document management or collaboration system?

Tradecraft leveraging skill sets and experiences is to find a way to have more effective copies and versions of data while removing cost overhead without adding complexity or other compromises. In Chapter 11 we'll look at some of the data services features for implementing data footprint reduction (DFR) to minimize overhead while having more data copies.

Tip: Keep in mind that having multiples copies and versions is not a bad thing if they provide or enable some benefit to the organization, as they might simply be part of the cost of doing business. What this also means is having tools that enable you to know what data exists where, how many copies and versions there are, along with other insight to have awareness for copy management.

Considerations:

- Integration with solutions such as Documentum or SharePoint, among other applications
- Reporting, insight, and analytics into what you have, where it is located, and versions
- Policy-based and user-defined management policies as well as configuration
- Multivendor, heterogeneous storage and data as well as application support

9.3. Availability (Resiliency and Data Protection) Services

The A in PACE is *availability*, which encompasses many different things, from accessibility, durability, resiliency, reliability, and serviceability (RAS) to security and data protection along with consistency. Availability includes basic, high availability (HA), BC, BR, DR, archiving, backup, logical and physical security, fault tolerance, isolation and containment spanning systems, applications, data, metadata, settings, and configurations.

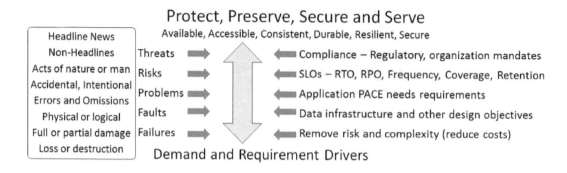

Figure 9.2 Data infrastructure, applications, and data availability points of interest.

From a data infrastructure perspective, availability of data services spans from local to remote, physical to logical and software-defined, virtual, container, and cloud, as well as mobile devices. Figure 9.2 shows various data infrastructure availability, accessibility, protection, and security points of interest. On the left side of Figure 9.2 are various data protection and security threat risks and scenarios that can impact availability, or result in a DLE, DLA, or disaster. The right side of Figure 9.2 shows various techniques, tools, technologies, and best practices to protect data infrastructures, applications, and data from threat risks.

What this all means is keeping in mind that there are different threat risks or things to protect your data infrastructure, applications and data from, but you also have options. You have the option to treat everything the same or to align the appropriate protection and resiliency to different applications based on their PACE requirements and SLO needs.

There is an industry sales and marketing saying that one should never let a good (or bad) disaster go to waste. This means that when something bad happens and appears in the news, it is no coincidence that you will hear about a need for local or cloud backup, BC, BR, DR, security and antivirus protection, endpoint protection, and resiliency, among other related themes. That is because availability has traditionally been seen as a cost overhead similar to insurance: a necessary expense of doing business rather than a business enabling asset. As a result, some vendors who sell availability and data protection hardware, software, or services tend to do so based on fear of something happening, or what the cost and expense will be if you are left exposed. There is merit to this approach, but so too is selling not just on fear, but also on availability being a business asset and enabler.

Tip: Instead of selling on fear of what downtime will cost you, how about turning the equation and tables around: What is the business-enabling value of being available, accessible, data intact and consistency ready for use while others are not? With that in mind, let's take a look at some data services that support various aspects of availability from residency to access, durability, protection, preservation, and security, as well as applications, settings, and related items.

Figure 9.3 builds on Figure 9.2 by showing various data infrastructure and application points of interest needing data protection and availability. Also shown in Figure 9.3 are various general techniques that can be used at various data infrastructure layers in combination with others.

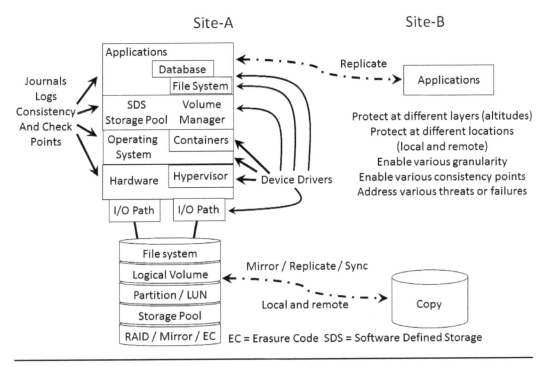

Figure 9.3   Points of availability, accessibility, and durability concerns.

9.3.1. Revisiting 4 3 2 1—The Golden Rule of Data Protection

Everything is not the same; different threat risks require different data protection technologies, tools, and techniques. These tools, technologies, and techniques address threats, risks, and availability concerns at different layers in the data infrastructure stack. Some may be seen as competitive. However, they can also be complementary, for example, using upper-level application replication for small, narrow, tight, or synchronous RPOs while using lower-level storage systems, appliances, and software for broader RPOs leveraging asynchronous communications.

Data protection focus is tied to different threat risks or things that can happen, or impact data infrastructure as well as applications and data. For example, things that can go wrong or happen include physical or logical, from damage to deletion and distraction of data or systems. This includes intentional or accidental device and software failures, network and service provider outages, and configuration errors.

> *Tip:* Data loss can be complete (all gone) or partial (some data gone or damaged). Scope, scale, and focus of data protection ranges from geographic region locations to sites and data centers, systems, clusters, stamps, racks, cabinets, shelves, and individual servers or other components (hardware and software). This includes software-defined storage, file systems and volumes, databases, key value repositories, hypervisors, operating systems, and associated configuration settings.

What are your data protection priorities?

- Speed of protection and impact on production work
- Performance impact during recover, rebuild, repair
- Speed of reconstruction, restore, restart, resume service
- Low overhead (performance, space capacity) and cost
- Retention, compliance, regulatory, corporate mandates, disposition
- Least amount of data loss and application impact
- Portability, mobility, flexibility, and relocate ability of data
- Number and types of failures or faults to tolerate (FTT)

Granularity for data protection has different focus areas:

- Application, transaction, database, instance, table, row, or column
- File system, folder, file, bucket or container, blob or object, mailbox
- Data and metadata, application software and settings
- Operating system, virtual machine, hypervisor
- Device, drive, volume, system, appliance, rack, site

Tip: Keep in mind that availability is a combination of accessibility (you can get to the data and applications) and durability (there are good consistent data copies and versions). The *4 3 2 1 golden availability protection rule,* also known by a variation of grandfather, father, son (i.e., generations), encompasses availability being a combination of access, durability, and versions.

What this means is that availability sometimes has a context of simply access: Can you get to the data? Is it there? This is enabled with high availability and fault tolerance, along with mirroring, replication, and other techniques. This might be measured or reported in the number of 9's, or how many hours, minutes, or seconds of downtime per year. Figure 9.4 shows how the 4 3 2 1 rule encompasses and enables SLO, availability and data protection QoS, as well as RTO, RPO, and accessibility.

There are different types of availability data services to enable reliability, accessibility, serviceability, and resiliency (Table 9.2). This means enabling the 4 3 2 1 principle of data protection: having at least four copies (or more) of data, with at least three (or more) versions on two (or more) systems or devices, and at least one of those is off-site (on- or off-line).

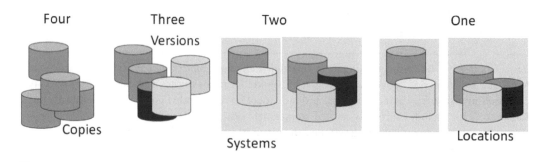

Figure 9.4 The 4 3 2 1 data protection best practice rule.

Table 9.2. Fundamental Availability and Resiliency Topics

Reliability, availability, serviceability (RAS)	Power-on self-test (POST), self-healing, automatic failover, active/passive or active/active redundant hardware (servers, controllers, power, cooling, I/O ports and paths, management). Fault isolation and containment, proactive data integrity checks and repairs. Error correction code (ECC), parity, and other protection and integrity for the block, file, object data checksums.
Enabling recovery-point objective (RPO)	Versioning, point-in-time (pit) copies, snapshots, time interval protection, continuous data protection, backup, consistency point, transactional, application, crash, checkpoint, archive, consistency points, log or journal switch
Enabling recovery-time objective (RTO)	High availability, fault tolerance, replication, mirroring, and RAID, including advanced parity such as Erasure Code (EC) and forward error correction, among other approaches.
Data and storage security	Physical, logical, access control, encryption, endpoint protection

One aspect of enabling availability along with facilitating copy management is having multiple copies of data in different locations. However, if those different copies in different locations are all corrupt, damaged, infected, or contain missing or deleted data, then there is a problem—thus the need for multiple versions and timer interval protection.

9.3.2. Availability, FTT, and FTM Fundamentals

A component of availability besides access is the durability and consistency of the data. This means that there are versions and copies of data in different places (the 4 3 2 1 rule) that are stateful and have integrity. Something to be aware of is that some cloud and other solution providers might mention 13, 15, or more numbers of 9's of availability. Some people might tell you that is not possible or practical; given the proper context—access only or durable only—that could be true.

> *Tip:* In the context of combined access plus durability, 13, 15, or more 9's are possible and practical, depending on your specific needs. Hence keep context in mind when looking at availability and associated metrics. The fundamental point with data services and features is to keep in mind what are you protecting against, and whether all threat risks are applicable across your different applications.

Another availability topic is the number of faults (or failures) to tolerate (FTT) for an application, site or data center, system, appliance, server, or other data infrastructure component. An FTT = 0 means there is no fault tolerance or resiliency, whereas FTT = 1 (one copy or replica exists, i.e., $N + 1$) means one fault or failure can be tolerated. Note that FTT can apply to the entire stack from the application to the site, to the system, subsystem, server, appliance, or specific component.

> *Tip:* There can be different FTT for various layers, such as FTT = 1 at the database layer, relying on database replication, and FTT = 2 at the storage system layer. Another option is to use FTT = 0 at the database layer while the underlying storage system might have a FTT = 2 or 3. Another variation is converged infrastructure (CI) and hyper-converged infrastructure (HCI) solutions, where FTT applies to application servers as well as underlying storage and networks.

Tip: Knowing the data protection priorities, requirements, granularity, and focus helps to determine what type of fault or failure-tolerant method (FTM) and data protection technique to use. For some scenarios, there will be an application, database, host or upper-level protection focus using replication and other techniques. Other scenarios will place more emphasis on the lower-level storage system, server clustering, and other FTM and data protection techniques. Still others will use hybrid approaches across different applications and service needs.

Tip: There are different FTM and data protection availability techniques and technologies that can be used in various ways. For example, is your primary focus on protecting against an HDD or SSD failure, bit error rate (BER), recoverable BER (RBER), or unrecoverable BER (UBER), among other considerations. There can be a focus on RAID or erasure code for protection and availability, including for normal operations or when something fails.

Bit rot can occur on magnetic or NVM-based storage mediums as well as in virtual and logical file systems, databases, and among solutions. This phenomenon results in bits degrading or being damaged over time, compromising data integrity. The solution is to have file systems, databases, and other software-defined storage perform background data integrity checks and repair as needed. Another approach is to perform E2E data integrity checks including write verification.

For example, there can be the primary focus on a particular RAID approach such as RAID 10, or, a combination of techniques using different RAID levels along with replication or snapshots. Alternatively, if your focus is on protecting against device failures, also look into using better drives or devices and components along with software tools.

Tip: There should be a focus on having protection copies, replicas, and other approaches for when other things fail, from file systems to object stores, front-ends, gateways, networks, and so forth. Not to mention supporting different granularity. In other words, are you building a data protection approach focused around something failing; then, as part of the solution, look at what is going to fail and what alternatives may exist to using that technology. For example, what are your needs and requirements for all applications and data, or different categories and tiers such as faults to tolerate, to determine the applicable fault tolerance mode (FTM)?

Faults to tolerate (FTT) can be specified for a component device, system, rack, cabinet, cluster, or site (i.e., the fault domain) with varying values. For example, some components or layers in the data infrastructure stack may have higher FTT vs. others, such as a number of HDD or SSD that can fail, or host adapters or servers, among other resources. This also applies to what types of failures and their locations to be resilient toward and protect against. The FTM indicates what type of protection techniques and technologies to use to support FTT as well as RPO and RTO.

Considerations include:

- Where to implement availability and data protection
- What FTM to use when and where to meet FTT needs
- How much availability (or tolerance to downtime) is needed
- Keep availability regarding accessibility and durability in mind
- Keep in mind that if something can go wrong, it probably will, so be prepared

9.3.3. Common Availability Characteristics and Functionalities

When it comes to availability, including access, durable data protection (backup, BC, BR, DR), and resiliency, no one approach (other than common sense and tradecraft experience) fits all environments, applications, and scenarios. There are different applications with various PACE attributes and QoS requirements, and there are big and small environments. Likewise, there are different threat risks that have various applicability and impact (i.e., what matters to your specific needs).

In addition to various threat risks (things that can go wrong) impacting data infrastructure, the application, as well as data availability (and security), there are also different approaches for protection. Some approaches leverage more resiliency in the underlying server and storage layers, while others rely on applications and upper levels of protection.

> *Tip:* Another approach is hybrid, using the best of an upper-level, higher-altitude application, along with a lower-level storage system, as well as network techniques and technologies for data infrastructure reliability, availability, and serviceability (RAS).

What this means is that since everything is not the same, yet there are similarities, different techniques, technologies, tools, and best practices should be leveraged using your tradecraft experience. In other words, use different approaches, techniques, tools, technologies, and solutions for various applications, environments, and scenarios.

Still another approach is use many lower-cost, less reliable components and subsystems instead of relying on upper-level applications, database, and/or software- defined storage for resiliency. Another is a hybrid of the above on an application or system-by-system basis.

What are you protecting for/from?

- Location impact based (geographic region, metropolitan, local site)
- Physical facility damage or loss of access (physical or logical network)
- Logical, full or partial impact (loss, corrupted, deleted, inconsistency)
- Data loss event (DLE) or lost access (DLA) to safe, secure data
- Data intact and consistent, data infrastructure or application software damaged
- Technology failure or fault (hardware, network, electrical power, HVAC)

Additional considerations that determine what type of availability and resiliency mode or method to use include planned (scheduled) downtime, unplanned and unscheduled downtime, recurring or transient failures and issues (problems), application service PACE requirements, and SLO, RTO, and RPO needs.

Availability and protection quality of service (QoS) is implemented via:

- Reduced and standard redundancy along with high-availability
- Redundant components (hardware, software, power, cooling)
- Clusters and failover (active or standby, local and remote)
- Archives, clones and copies
- Backups, snapshots, CDP, and backdating
- Replication, mirroring, RAID, and erasure codes

Availability, resiliency, and data protection management includes:

- Testing and verification of processes, procedures, and copies made
- Grooming and cleanup, removing expired protection copies
- Change management, planning, implementation, and fallback
- Contingency planning and preparing, testing, remediating, communicating
- Priority of queues, tasks, jobs, and scheduling of protection
- Support for various data sizes, data volumes, data values, and applications
- Insight and awareness into data access patterns and usage

Availability, resiliency, and protection solutions, in general, should support:

- Long and complex file names along with deep directory folder paths
- Application integration, open files, plug-ins, and pre-/postprocessing
- Consistency checkpoints, quiescence of applications or file systems
- Filters to select what to include, exclude, and schedule for protection
- Options on where to protect data along with different format options
- Support for multiple concurrent data protection tasks running at the same time

Figure 9.5 shows how data is made durable and protected by spreading across different resources to eliminate single points of failure (SPOF). The top left shows (A) an application leveraging a data infrastructure configured and defined for resiliency. Data is spread across

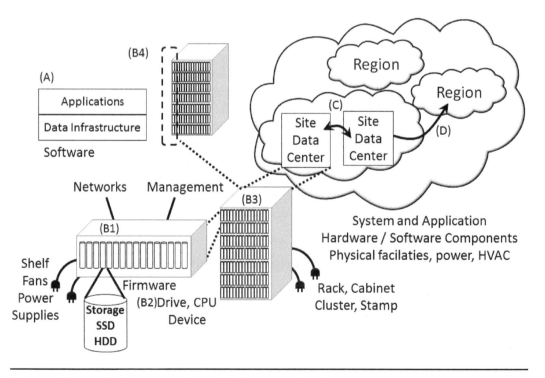

Figure 9.5   Fault domains and availability considerations.

storage shelves (B1) with redundant power supplies, cooling fans, dual-path I/O and network interfaces, and management ports. The storage devices themselves (B2) have various resiliency and fault isolation and repair capabilities, combined with data infrastructure and software-defined storage tools. Multiple shelves exist in a cabinet or rack (B3) as a fault domain, with data being placed on different shelves (B4) for resiliency.

Tip: For additional resiliency, data can be spread across multiple racks or cabinets, across data centers as a higher-level fault-domain and across sites (Figure 9.5). Data can also be spread across systems, clusters, or stamps spanning data centers, sites, and regions for geographic dispersal. How much resiliency and number of faults or faults to tolerate (FTT) will depend on different application PACE and SLO requirements.

Data infrastructure availability can be implemented in various places, including:

- Applications, databases, middleware, file systems, and repositories
- Volume managers and other software-defined storage places
- Operating systems, hypervisors, and third-party add-on software
- Storage systems, appliances, networks, cloud, and service providers

Figure 9.6 shows an example of the *4 3 2 1 data protection rule* implementation with at least four copies, at least three versions, on at least two different systems or storage devices with at least one of those being off-site. Also shown in Figure 9.6 is a storage pool implemented using software-defined or traditional storage using some underlying resiliency such as mirror and parity-based RAID including RAID 5, RAID 6, or various Reed-Solomon erasure code (EC) approaches. In addition to showing storage availability and resiliency, also shown are protection copies in different locations as well as point in time (PIT) recovery and consistency-point copies.

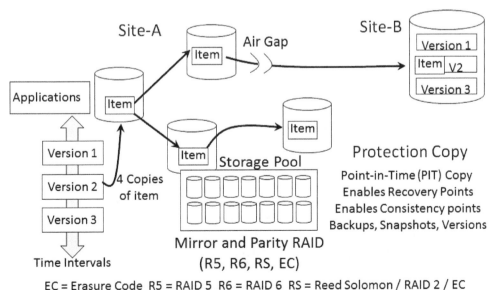

Figure 9.6   Combining various availability techniques.

Also shown in Figure 9.6 is an air gap between Site-A and Site-B. The air gap provides a level of protection to protect against something happening to Site-B, such as a virus or other threat. For a true air gap, Site-B should be isolated from on-line Site-A.

Considerations include:

- Restore to alternate locations and formats (P2V, V2V, V2P, V2C)
- Protect against accidental overwrite and selection filters
- Test and dry-run modes to verify intended operations
- Proactive, preventive, pre-emptive vs. reactive or after-the-fact protection
- Granularity of protection and application integration
- Probability of system and component failures or subsequent faults
- How to protect against full vs. partial data loss or damage
- Workspace capacity for logs, backup staging, import/export, catalogs, conversions

9.3.4. Reliability, Availability, Analytics, and Data Protection Management

Data protection management (DPM), as its name implies, pertains to managing data protection and related data infrastructure activities. For some, DPM can include actual protection tools and technologies such as backup (e.g., Microsoft DPM), while for others the focus is on insight (informing), reporting, monitoring, notification, and configuration.

Additional DPM tasks and activities include discovery, diagnostics, coverage, compliance, SLO, trending, fault and failure analysis, and troubleshooting.

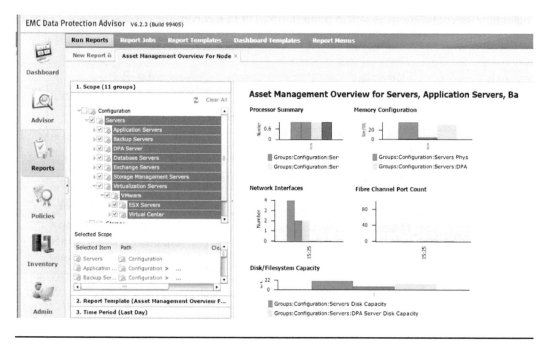

Figure 9.7   Data protection management and analysis software example.

DPM includes and can be found as built-in tools as part of applications, databases, file systems, volume managers, operating systems, hypervisors, software-defined and traditional storage systems or appliances, along with cloud or service providers. Additional software tools are available from third parties, open source, as well as in-house custom-developed solutions that can be applied at different points in the data infrastructure stack.

Some other questions that DPM addresses are whether backups or other data protection tasks are completing faster than normal because there was more data that should have been protected. Did the protection task run fast because not all data was protected, indicating an error or problem somewhere?

Figure 9.7 shows an example of data protection management and analytics tools (Dell EMC Data Protection Advisor). Some tools focus just on backup, snapshot, replication, host-based, virtual, storage system, or cloud-focused DPM. DPA is an example of a tool that looks across physical and virtual servers, and at different applications such as SQL Server, Exchange, and Oracle in terms of data protection insight, configuration, and assessment.

General reliability, availability, and data protection management topics include:

- Interoperability with various vendors hardware, software, and cloud services
- Activity, event, and data protection–related telemetry collection and reporting
- Analytics and correlation engines to sift through raw data to see causes
- Agent or agentless, plug-ins, APIs, lightweight commands (cmdlet), scripts, browser
- What should be protected, what was protected, when, and success or failure
- How many protection copies you have, what type, versions, where located
- Success and accuracy of protection as well as recovery operations
- Notification via dashboards (local or cloud-based), email, text, and other alerts

Various tools from different vendors, as well as those built into operating systems, hypervisors, storage systems, appliances, and cloud services, are available. There are also several add-on tools from different vendors, including Dell EMC, Microsoft, NetApp OnCommand Insight (OCI), Quest, Turbonomic, Virtual Instruments, and VMware, among many others.

> *Tip:* Just as there are different tools from various sources including open source, functionality also varies. Some look at performance or scheduling of data protection; some focus on a server, storage, or network and can also provide data protection insight. Some tools focus on data management and provide insight into what files, data, or objects are located where, along with optional performance or access information.

Metrics that matter include:

- Applicable application PACE indicators of usage, activity, and availability
- Annual failure rate (AFR), which is 1/(MTBF/8760) * 100 (expressed in %)
- SLOs such as RTO, RPO, success, failure rates, ratios, and percentages
- Mean time between failures (MTBF) and mean time to repair (MTTR)
- Planned, unplanned, scheduled, and unscheduled downtime, for availability
- Probability of a device, component, hardware, software, or service failing
- Probability of a second failure in a short period and its impact

Note that probability of loss (pL) of something rated for 1000 hours is 1/1000 or 0.001. If it takes 3 days to replace, repair, rebuild, or reconstruct a failed hardware, software, service, or other data infrastructure component, the pL (second failure) becomes 0.003.

Considerations include:

- Event and activity logging and monitoring
- Notification, insight, awareness, alerts, and reporting
- Analytics, event correlation, coverage, and compliance analysis
- Interoperability with various hardware, software, and applications
- Local, remote, and cloud managed service capabilities

9.3.5. Enabling Availability, Resiliency, Accessibility, and RTO

> *Tip:* Key to resiliency is being able to isolate and contain faults (hardware, software, services, technology, and configuration) from spreading and resulting in larger-scale issues, problems, or disasters. As part of resiliency, how is cached or buffered data rewarmed (reloaded) from persistent storage after a system shutdown or crash? Can you resume a stopped, paused, or interrupted operation (snapshot, copy, replica, or clone) from where it last left off, or do you have to start over again from the beginning?

Another variation is the ability to resync data and storage based on some rules or policies to add changes (or deletions) without having to recopy everything. This could be to resynchronize a mirror or replica after a system or device is removed for planned or unscheduled maintenance, hardware, or software failure. As part of resiliency and failover to another system, do applications have to roll back (go back to a given RPO) to a consistency point, or can they simply resume operations where they left off?

Still another consideration is what type of fallover is required—for example, a system or application restart, or can it resume from the last transaction (i.e., is there application and transaction consistency)? What is required, for example, simple, quick repairs or more extensive rebuild and reconstruction to file systems, other software-defined entities along with failed RAID, mirrors, replicas, and erase coded automated using spares, or is manual intervention necessary?

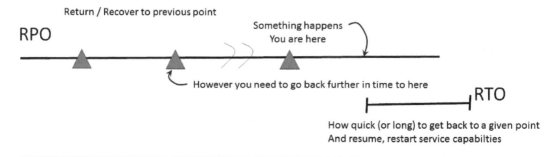

Figure 9.8   Resiliency including availability, RTO, and RPO.

In addition to automated or manual attention, what is the repair or rebuild overhead in terms of space or systems capacity, network, server, and storage I/O performance, and impact on applications? Figure 9.8 shows points in time where data is changed as well as protected (i.e., consistency, checkpoint or recovery point), also known as delta (i.e., something changes).

Also shown in Figure 9.8, besides the RPO or points in time from where data can be recovered, or applications and systems rolled back to resume processing, are times to recover (or restore). The time to recover or restart represents the RTO of how much time is needed to resume, restart, restore, rebuild, or recover, along with several other things that start with R (e.g., repair, reload, reconstruct, rollback, or roll-forward).

> *Tip:* Keep in mind that RTO needs to be seen in context: Is it end to end (E2E) for all data infrastructure components, or is it for a single item. For example, a single component might have an RTO for just storage or just database or servers, where E2E includes total time for applications, database rollback, data restore, storage, or other configuration and repairs.

Availability and resiliency topics, technologies, and techniques include:

- Redundant components and paths (hardware, software, services)
- Power, cooling, I/O, and management network access
- Configuration for RAS, basic and HA, to meet FTT and PACE needs
- Active/active or active/passive (standby) failover capabilities
- Manual, semiautomatic, and automatic failover (and failback)
- Mirroring, replication, sync, and clone
- RAID (mirror, parity, Reed-Solomon erasure code)
- Rebuild, reconstruction, restore, restart, repair, recovery
- Combine and complement with point-in-time protection techniques

Figure 9.9 shows resiliency- and availability-enabling technologies along with techniques used at different layers as well as locations to meet data infrastructure and application PACE

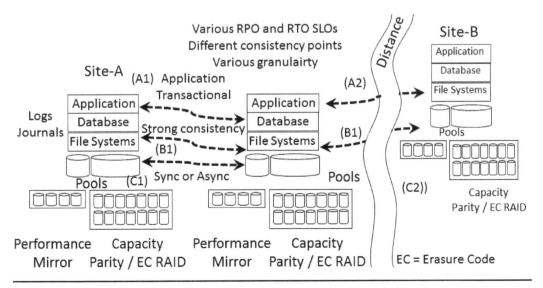

Figure 9.9 Combing multiple availability techniques.

needs. This includes local (Site-A) resiliency across systems and applications using different techniques that in turn protect to a secondary site (Site-B).

Some of the systems in Figure 9.8 implement application-level protection using local as well as remote replication. Other applications rely on lower-level replication via database, file systems, operating system, hypervisor, and storage system or appliances. Data is also protected at different levels for various granularities using snapshots, backups, versions, and other point-in-time techniques. For example, application- or database-level point-in-time protection enables application and fine-grained transaction consistency.

On the other hand, lower-altitude data protection in the data infrastructure stack such as a file system, operating system, hypervisor, or storage represents context and coarser granularity for those given levels. Figure 9.9 is an example of how different techniques should be combined to address different needs aligned to your environment.

Considerations include:

- What threat risk scenarios your data infrastructure needs to protect against
- Application PACE characteristics requirements for normal and abnormal times
- Balancing speed of restore or restart with protection overhead
- Heterogeneous mixed vendor, hardware, software, services, and technology
- Geographical location needs to meet SLO and other requirements
- Whether you will treat all applications and systems the same or use various techniques

Reliability, Availability, and Serviceability (RAS)

A resilient data infrastructure (software-defined, SDDC, and legacy) that protects, preserves, and serves information involves various layers of technology. These technologies enable various layers (altitudes) of functionalities, from devices up to and through the applications themselves. Some applications need a faster rebuild; some need sustained performance (bandwidth, latency, IOPs, or transactions) with the slower rebuild; some need lower cost at the expense of performance; others are ok with more space if other objectives are meet. The result is that since everything is different yet there are similarities, there is also the need to tune how data infrasture protects, preserves, secures, and serves applications and data.

General reliability, availability, serviceability, and data protection functionality includes:

- Ability to manually, or automatically via policies, start, stop, pause, and resume protection
- Adjust priorities of protection tasks, including speed, for faster or slower protection
- Fast-reacting to changes, disruptions or failures, or slower cautious approaches
- Workload and application load balancing (performance, availability, and capacity)

RAS can be optimized for:

- Reduced redundancy for lower overall costs vs. resiliency
- Basic or standard availability (leverage component plus)
- High availability (use better components, multiple systems, multiple sites)

- Fault-tolerant with no single points of failure (SPOF)
- Faster restart, restore, rebuild, or repair with higher overhead costs
- Lower overhead costs (space and performance) with lower resiliency
- Lower impact to applications during rebuild vs. faster repair
- Maintenance and planned outages or for continues operations

RAS techniques, technologies, tasks, and topics include:

- Automatic fault detection, isolate, contain, and repair where possible
- Protected journal and log files for transactional (or system) integrity
- Data infrastructure and component health status check
- Data integrity and checksums at various locations
- Local, remote, physical, virtual, cloud, and hybrid across all of them

Additional RAS functionality and topics include:

- Power-on self-test (POST) for hardware, software, networks, and services
- Redundant $N + 1$ power supplies, cooling, I/O connections, and controllers
- Mirrored cache with battery backup (BBU) power protection
- Background data integrity and consistency checks and repairs (self-healing)
- Hot and on-site spare components such as HDD, SSD, and server cluster nodes
- Nondisruptive code load (NDCL) and nondisruptive code activation (NDCA)

A technique and technology for enabling resiliency, as well as scaling performance, availability, and capacity, is clustering. Clusters can be local, remote, or wide-area to support different data infrastructure objectives, combined with replication and other techniques.

Another characteristic of clustering and resiliency techniques is the ability to detect and react quickly to failures to isolate and contain faults, as well as invoking automatic repair if needed. Different clustering technologies enable various approaches, from proprietary hardware and software tightly coupled to loosely coupled general-purpose hardware or software.

> *Tip:* Care should be taken to configure clusters for resiliency; that is, in addition to the cluster enabling data infrastructure resiliency, make sure the cluster is not an SPOF. For example, make sure there are adequate and redundant components, I/O and network paths for east–west traffic including heartbeat and health status, repair, and other functions, along with general north–south access. Other considerations include the ability to tune for faster failover and repair, balanced with how to fallback when things return to normal.

Another clustering topic is partitioning, for which some context is needed. To enable resiliency and data integrity, the cluster needs to be protected from split-brain (the cluster divides into two or more independent clusters) or partitioned mode. In a cluster context, *partitioned* refers to how the cluster operates as a single entity, as opposed to how a disk is subdivided and formatted.

Most clusters require some number of member nodes with a vote to establish a quorum (i.e., the cluster can exist). Some clusters require an odd number such as three or more nodes to function, although this attribute varies by implementation. Additional proxy, witness, or surrogate quorum voting nodes (or devices) can be used to create a cluster quorum and prevent a partition or split-brain scenario.

Clustering characteristics include:

- Application, database, file system, operating, hypervisor, and storage-focused
- Share everything, share some things, share nothing
- Tightly or loosely coupled with common or individual system metadata
- Local in a quad (or larger) server chassis (e.g., multiple unique nodes in a box)
- Local in a cabinet or rack as part of a CI, HCI, CiB, or stamp
- Local in a data center, campus, metro, or stretch cluster
- Wide-area in different regions and availability zones
- Active/active for fast failover or restart, active/passive (standby) mode

Additional clustering considerations include:

- How does performance scale as nodes are added, or what overhead exists?
- How is cluster resource locking in shared environments handled?
- How many (or few) nodes are needed for quorum to exist?
- Network and I/O interface (and management) requirements
- Cluster partition or split-brain (i.e., cluster splits into two)?
- Fast-reacting failover and resiliency vs. overhead of failing back
- Locality of where applications are located vs. storage access and clustering
- Does the clustering add complexity or overhead?

RAS considerations:

- Application higher-level, fine-grained, transactional consistency protection
- Lower-level, coarse crash consistency protection
- Leverage good-enough components with higher-level resiliency
- Leverage more reliable components with less higher-level resiliency
- Use a combination of techniques and technologies as needed
- Leverage lower costs to have more copies and resiliency
- How you will protect applications, data, and data infrastructure metadata settings

Mirroring and Replication

Mirroring (also known as RAID 1; more on RAID later) and replication creates a copy (a mirror or replica) across two or more storage targets (devices, systems, file systems, cloud storage service, applications such as a database). The reason for using mirrors is to provide a faster (for normal running and during recovery) failure-tolerant mode for enabling availability, resiliency, and data protection, particularly for active data. Figure 9.10 shows general replication scenarios. Illustrated are two basic mirror scenarios: At the top, a device, volume, file system, or object bucket is replicated to two other targets (i.e., three-way or three replicas); at the bottom is a primary storage device using a hybrid replica and dispersal technique where multiple data chunks, shards, fragments, or extents are spread across devices in different locations.

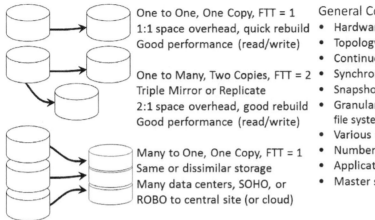

One to One, One Copy, FTT = 1
1:1 space overhead, quick rebuild
Good performance (read/write)

One to Many, Two Copies, FTT = 2
Triple Mirror or Replicate
2:1 space overhead, good rebuild
Good performance (read/write)

Many to One, One Copy, FTT = 1
Same or dissimilar storage
Many data centers, SOHO, or
ROBO to central site (or cloud)

General Considerations
• Hardware, software, hybrid based
• Topology (1 to 1, many to 1)
• Continuous or Scheduled
• Synchronous, Asynchronous
• Snapshot, PIT, CDP, RPO protection
• Granularity (system, volume, device,
  file system, file, block, object)
• Various chunk/shard sizes
• Number of mirrors/replica
• Application and management Tools
• Master slave, both sides active

Figure 9.10   General mirror and replication configuration considerations.

Tip: Note that mirroring, while commonly associated with replicating storage and data, is also used in servers, networks and storage systems, and appliances, as well as in clouds for cache and other memory consistency. Some solutions require an external battery-backed unit (BBU) power supply to protect mirrored cache and memory buffers for data integrity and persistence.

For normal running modes, depending on the implementation, it is possible for reads to be resolved from either local mirror member (solution-dependent). In the case of a failure, to reconstruct or rebuild a failed mirror or replica, data is simply read from remaining member(s). Unlike erasure codes and other parity-based protections, no extra reads or calculations are required, just a simple copy.

Normal and degraded as well as rebuild performance aside, the other benefit is that by having "$n + 1$" mirrors or replicas, where n is a number of mirrors, you can tolerate n failures. For example, in a $1 + 1$ (two-way) mirror, if one of the replicas fails, you still have data access; with a $2 + 1$ (three-way or triple) mirror, two of the replicas can fail and you still have data access; and so forth.

There is a caveat, which is that more space capacity and associated resources are needed. For example, if you are mirroring or replicating an 8-TB device, you need at least another 8-TB device or storage space (assuming no data footprint reduction is used). Similarly, if you are doing a three-way (triple) mirror or replica, you need 3×8 TB or at least 24 TB of space.

Figure 9.11 shows System-A on the left with Volume (or file system)-A that is mirrored to System-B. In addition to mirroring, there are also snapshots which provide point-in-time (PIT) copies to support recovery-point objectives.

Tip: Keep in mind that mirroring and replication by themselves are not a replacement for backups, versions, snapshots, or another time-interval (time-gap) protection. The reason is that replication and mirroring maintain a copy of the source at one or more destination targets. What this means is that anything that changes on the primary source also gets applied to the target destination (mirror or replica). However, it also means that anything changed, deleted, corrupted, or damaged on the source is also impacted on the mirror replica (assuming the mirror or replicas were or are mounted and accessible on-line).

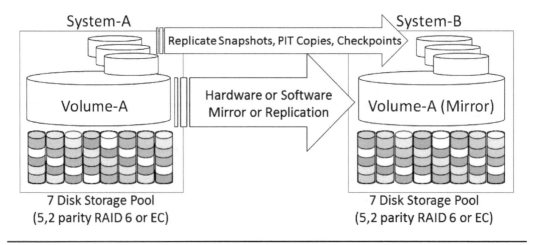

Figure 9.11 Mirroring and replication, with snapshots.

Another variation of the pros and cons is that with mirror and replication, you can transfer 100% of the bits, bytes, blocks, and blobs from the primary site to the replica, unless data still in memory buffers is flushed, quiescent, or made consistent to the source storage, not all of the data for integrity may be replicated (i.e., only what was on disk at the time of mirroring). The solution is to combine mirroring and replication with other parity-based RAID, including erasure codes, with some time-interval (RPO) protection (snapshots, versions, or backups; more on these later).

> *Tip:* Mirroring and replication can be done locally inside a system (server, storage system, or appliance), within a cabinet, rack, or data center, or remotely, including at cloud services. Mirroring can also be implemented inside a server in software or using RAID and HBA cards to off-load the processing.

Operating systems, hypervisors, volume managers and other software-defined storage can also perform the mirroring and replication functions. Besides local and remote, mirroring and replication can also be done between physical and physical, physical and virtual, virtual and virtual, virtual and cloud, and cloud and cloud, among other permutations.

Common features and functions of mirroring and replication include:

- Between two or more devices, systems, appliances, or services
- Scheduled and ad-hoc or on-demand
- Full or partial (resync or catch-up) copies
- Pause and resume
- Background, time-deferred, or immediate
- Block, file, object, or application focused

Another option supported by some solutions and services is seeding or staging data for bulk movement between different locations. In some scenarios, data is mirrored, replicated, or synced to another storage system or a server that gets shipped to different location. Once at the destination, changes are applied by resynchronized the systems via a network.

Implementations in various locations (hardware, software, cloud) include:

- Applications and databases
- File systems and volume managers
- Software-defined storage managers
- Third-party storage software utilities and drivers
- Operating systems and hypervisors
- Hardware adapter and off-load devices
- Storage systems and appliances
- Cloud and managed services

In addition to where they are implemented, mirroring and replication can also be done to the same (homogenous) or dissimilar (heterogeneous) devices, systems, or services, depends on the particular solution. Replication can be unidirectional (one-way) or bi-directional (two-way). Some solutions require a master and slave or client relationship. For example, Device-A at Site-A is primary mirroring to Device-A at Site-B. Meanwhile, a second mirror can be Device-B at Site-B as primary replicating to Device-B at Site-A. Other solutions allow updates from either site to be replicated to the partner. Still others support read and write from one system (the primary), with reads on the secondary site unless there is a failover.

There are various granularities, depending on what is being mirrored or replicated, including entire applications, databases, systems, file systems, storage systems, or appliances; or selected databases or object (logs, journals, tables) resources, files, folders, LUNs, or volumes on an object, file, or block basis.

In addition to where and how implemented, along with granularity, there are also many different types of mirroring and replication topology configuration options to address different needs. Mirror and replication topologies include:

- One to one (peer to peer), also known as point to point
- One to many (fan-out): Replicate to multiple sites, systems, or locations
- Many to one (fan-in): Replicate from many to a single location or system
- Cascade (one to one to one): Multi-hop across systems or locations

---

Tip: Besides various locations where and how mirroring and replication can be implemented, along with topologies, other options include synchronous or real time (strong consistency), as well as asynchronous or time-delayed (eventual consistency). There are also hybrid modes such as semi-sync, where synchronous data transfers are done between a pair of systems or devices; then a secondary copy is done asynchronously to a third location—for example, synchronous replication between a primary and secondary systems, devices, or applications that are in the same building, campus, or metropolitan area, and then asynchronous replication between the secondary and a third, more distant location.

As we discussed in Chapters 5 and 6, synchronous provides strong consistency, although high-latency local or remote interfaces can impact primary application performance. On the other hand, asynchronous enables local and remote high-latency communications to be

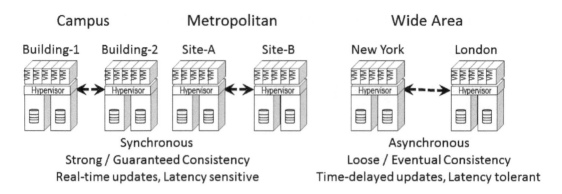

Figure 9.12   Examples of replication and mirroring across locations.

spanned, facilitating protection over a distance without impacting primary application performance, albeit with lower consistency.

Functionality includes activity status reporting and monitoring along with estimate to completion, amount of data moved, average bandwidth being used, and effective data transfer rate. Other features include the ability to pause and resume, stop or cancel tasks, as well as to schedule on-demand or other tasks on a predetermined schedule or event basis.

In addition to scheduling, other functionalities to look for include pre- and postprocessing scripts, exits, and call-outs. These provide the ability to involve a script, an API call, or execute a command before mirror and replication operations occur, as well as when they are done. For example, before replication quiesce file systems and force a database checkpoint to occur for consistency. Also look for the reverse, which is the ability for the mirror and replication operations to be called from an application or via a script, via either executing a command or an API call.

Figure 9.12 shows various mirroring and replication examples, including local, remote, and synchronous as well as asynchronous, implemented in various ways to meet different needs. On the left of Figure 9.12 are local examples, including inside a data center or campus, as well as metropolitan using synchronous replication. On the right of Figure 9.12 is an example of asynchronous replication for spanning wide areas across different geographies, or over shorter high-latency distances.

There are various modes for synchronizing, and clones can also apply to replication as well as mirroring. These include, among others:

- *Synchronize.* New and updates to data are copied in both directions; renames and deletes are performed in both locations. The benefit is that both sides are synchronized to each other and updates from either are copied both ways. The caveat is that if a mistake, deletion, or problem occurs on one side, it is replicated or synchronized to the other. Use in conjunction with versioning or another RPO capability for complete data protection.
- *Echo.* New and updates to data are copied from one site to the other; renames and deletes are also copied from one site to the other, for example, left to right. The benefit is that accidental changes on one side (e.g., the right) are not propagated back to the other (the

left). The caveat is that any changes on the destination side are not synchronized with the primary.

- *Contribute.* New and updates to data, along with renames, are copied from one site to the other as with echo. However, deletes are not done on the destination. The benefit is that destination data becomes cumulative, with no deletions; the caveat is that extra space is needed.

Mirroring and replication considerations include:

- Built-in or external (Cron, Task Scheduler, Rabbit) schedule manager
- Can replica be actively read, or read and write
- Use along with time-interval protection (CDP, snapshot, versions)
- How many total and simultaneous operations (copies and impact)
- Distance, synchronous and asynchronous data consistency
- Where and how implemented, granularity and application integration

Additional considerations include:

- Integration with RPO and consistency point tools such as snapshots
- Application and data protection management tool integration
- Monitoring, reporting, and analytics
- Seeding and pre-stage, pause, resume, restart of replication
- Schedule and policy-based, along with ad-hoc on-demand
- Bandwidth and distance optimization

Sync and Clone

Functionality wise, sync and clone are similar to mirroring and replication, sharing many common characteristics, attributes, featurea, and considerations. What can differ is how and when they get used together for different purposes as well as at various altitudes (layers) in the data infrastructure stack. Another difference is the tools and where they are implemented, which enables various granularities and application integrations spanning block, file, and object.

Some tools and services include AWS, Google, Azure, Dropbox, Robocopy, Xcopy, Synctoy, Rsync, Microsoft Windows, VMware, and many different enterprise file sync share (EFSS) solutions, among others.

In addition to file and object sync, other variations include cloning a virtual machine or database. For VM, the functionality and features vary by the hypervisor, however generally you can clone to make a new copy with unique network MAC addresses among other items. Likewise, you can clone a VM to create a template from which many new VM can be based.

Another variation is linked clones, where a master gold VM is created which is used as the basis for child or sibling VM. Linked clones enable VM to share common data such as applications to reduce the space size capacity needs, similar to how Docker and other containers share programs and libraries.

The following is an rsync example where I synchronize files, images, photos, and videos from a NAS-mounted namespace to a cloud bucket-mounted namespace. In the example, an "include" file is used to indicate what items to select or skip as part of the sync operation. I am using S3FS to mount one of my AWS S3 buckets to a Linux system, which in turn enables it to be accessed as a general file accessible namespace. Note that as part of system start-up; my AWS S3 credentials are used by S3FS to authenticate and gain access to my buckets.

Tip: Something else to note with cloud services, besides data space capacity costs, is other fees. For example, some services allow free data transfers into their clouds, but there is a fee for accessing. Likewise, some include API calls (e.g., Put, Copy, Post, List, Get, Delete) at no charge, while others have different fees for various service requests. For example, there may be no charge for a Delete of data, then one fee for Get (i.e., reading data), another for Put, Copy, Post, or List.

```
rsync  -av  --exclude-from='/home/user/Exclude _ S3 _ List.txt'  "/
myNAS/Myphotos/My Pictures/2016" "/myClouds/s3mnt/Myphotos"
```

For example, the above rsync command creates synchronous files from one of my NAS shares with those stored as objects in one of my AWS S3 buckets. The S3 bucket is mounted to a Linux system using S3fs (among many other tools). In the example, besides data being moved, there are also corresponding API calls, including List to determine what to update.

Tip: The obvious caveats are to pay attention to your Get, Put, Copy, and Post or similar operations (e.g., saving, retrieving data, and metadata). Also, pay attention to List, as on a traditional physical or virtual storage system or server there are no fees or costs associated with doing a directory (dir or ls) operation. However, some cloud services charge for those functions. This is important to know if you are doing functions such as rsync to update files and objects, or backups stored as individual files or objects. Those functions can have an impact on your cloud storage costs.

Considerations include:

- Can the tool do a smart fast-scan (it knows what's been copied) or a full scan?
- Unbuffered I/O copy operations can speed up large bulk data transfers.
- Use in combination with other data protection and availability techniques.
- Pay attention to synchronous mode (see mirror and replication discussion above).
- Leverage test "dry run" modes to verify what will be copied and to where.
- rsync and other tools can cause excessive Gets and List operations (be cloud-aware).
- Utilize checksums for data integrity, along with selection (and exclusion) filters
- Be aware of hard and soft links, relative and absolute path name addressing

RAID, Parity, and Erasure Codes

RAID (Redundant Arrays of Independent Disks) is an approach to addressing data and storage availability and performance. RAID as a technique and technology is about 30 years old and has many different types of implementations in hardware and software. Everything is not the same, and different RAID types and implementations can be used in various ways to meet various performance, availability, capacity, and economic (PACE) needs.

RAID remains relevant today, with many improvements, variations and implementation options to meet different data infrastructure and application needs. RAID including mirroring and parity provides availability protection but by itself is not a replacement for backup. Instead, RAID along with replication should be combined with some point-in-time protection technique such as snapshots, backups, versions, and CDP, among others.

> *Tip:* Note that some people associate mirroring with being local as well as hardware-based, as in a storage system, HBA, or RAID adapter card. Likewise, some people associate replication with being remote as well as software-based. Keep in mind the context of what is being done and the objective (to have one or more replica copies or mirrors of data, which can be local or remote). Likewise, the implementation can vary, including a mix of hardware and software.

What has changed and evolved with RAID include:

- Various implementations in hardware as well as software
- Different approaches from mirroring to parity including erasure codes
- Different application workloads with various PACE needs
- Increased size of data items resulting in larger allocation and I/O size
- Wider, larger (more devices) mirror and parity groups as well as larger devices
- Geographically dispersed and hybrid approaches

The original premise of RAID was to address the issue of reliability of lower-cost, general-purpose (what we call commodity today) HDD that were appearing to replace legacy proprietary solutions. It is ironic that here we are 30 years later defining data infrastructure using lower-cost, higher-capacity HDD as well as SSD that need protection using different techniques.

Many variations as well as differences in implementation have occurred since the original (classic) RAID definitions appeared. Some of these variations include combining different RAID levels to create hybrid such as 1 (mirror) and 0 (stripe) resulting in 10, as well the inverse, that is, 01. Mirroring and replication are better for some applications such as those with frequent writes or updates vs. others with less frequently updates, primarily read or cold data that lend themselves to parity-based solutions including erasure codes.

There are also many parity-based approaches and their derivatives, including RAID 2 (Reed Solomon [RS]), and derivatives known as erasure codes (EC) as well as local reconstruction codes (LRC) and shingled erasure codes (SHEC), among others. Additional parity-based approaches include RAID 3, 4, 5, 6, and their various derivative implementations.

Certainly, HDD and SSD have become more reliable in terms of annual failure rate (AFR) and bit error rate (BER), as well as faster. Further, the capacity of the devices has jumped from hundreds of megabytes with early RAID drives to a few gigabytes in the 1990s and early 2000s, to 12 GB (and larger) for HDD, with 64 TB and larger SSD.

What this means is that with more drives, even if they are more reliable, the chance of one failing increases statistically with the population size. My point is that HDD and SSD, like any technology component, can and will fail. Some are more reliable than others; sometimes you get what you pay for. For resiliency to meet your particular environment needs, invest in good technology (drives and SSD) along with associated hardware as well as software.

RAID is commonly implemented and found in:

- Hardware using general-purpose processors or ASIC
- Software, built in or via optional third-party tools
- HBA and RAID cards, workstation and server motherboards
- Adapter and RAID cards, store systems and appliances
- Hypervisors, operating systems, volume managers, file systems
- Software-defined storage solutions and virtual appliances
- Block, file, object, and cloud solutions

In some of the above, among others, it might be called replication (i.e., mirroring) or erasure code (i.e., parity) to differentiate from legacy hardware–based storage systems RAID.

> *Tip:* This is a good place to mention some context for RAID and RAID array, which can mean different things. Some people associate RAID with a hardware storage array, or with a RAID card. Other people consider an array to be a storage array that is a RAID-enabled storage system. Also, a trend is to refer to legacy storage systems as RAID arrays or hardware-based RAID, to differentiate from newer implementations. Context also comes into play in that a RAID group (i.e., a collection of HDD or SSD that is part of a RAID set) can also be referred to as an array, a RAID array, or a virtual array.

So keep context in mind, and don't be afraid to ask what someone is referring to: a particular vendor storage system, a RAID implementation or packaging, a storage array, or a virtual array. Also keep the context of the virtual array in perspective vs. storage virtualization and virtual storage. RAID as a term is used to refer to different modes such as mirroring or parity, and parity can be legacy RAID 4, 5, or 6 along with erasure codes.

In addition, some people refer to erasure codes in the context of not being a RAID system, which can be an inference to not being a legacy storage system running some other form of RAID.

RAID functionalities vary by implementation and include:

- Dedicated or global hot spare for automatic rebuild of failed devices
- Horizontal, across a shelf, along with virtual multiple shelf drive groups
- Read-ahead, read and write cache, write gathering
- Chunk, shard, or stripe size, from few a few kilobytes to megabytes or larger
- Import and export (save) of configurations
- Pause, resume rebuild, data integrity scan, initialize functions
- Flash SSD TRIM and UNMAP, and other wear-leveling optimization
- Battery backup units (BBU) to protect mirrored and nonmirrored cache

Additional common RAID concerns, including mirror and parity features, include:

- Support for two or more drives in a mirror, RAID, stripe, or parity set
- Mirroring or replicating to dissimilar or different target devices

- Ability to combine various-sized HDD into a RAID group
- Variable chunk size, or how much data is written per drive at a time
- Number of drives or drive shelves supported in a RAID set
- Wide stripe size across drives, in the same or different shelves
- Background initialize, rebuild, as well as data integrity and parity checks
- Number of concurrent drive rebuilds
- Hardware acceleration using custom ASIC or off-load processors
- Support for Data Integrity Format (DIF) and device-independent RAID
- On-line and while-in-use RAID volume expansion

Figure 9.13 shows various RAID approaches.

Different RAID levels (Table 9.3) will affect storage energy effectiveness, similar to various HDD performance capacity characteristics; however, a balance of performance, availability, capacity, and energy needs to occur to meet application service needs. For example, RAID 1 mirroring or RAID 10 mirroring and striping use more HDD and, thus, power, but will yield better performance than RAID 6 and erasure code parity protection.

Similar to legacy parity-based RAID, some erasure code implementations use narrow drive groups while others use larger ones to increase protection and reduce capacity overhead. For example, some larger enterprise-class storage systems (RAID arrays) use narrow 3 + 1 or 4 + 1 RAID 5 or 4 + 2 or 6 + 2 RAID 6, which have higher protection storage capacity overhead and fault=impact footprint.

On the other hand, many smaller mid-range and scale-out storage systems, appliances, and solutions support wide stripes such as 7 + 1, 15 + 1, or larger RAID 5, or 14 + 2 or larger RAID 6. These solutions trade the lower storage capacity protection overhead for risk of a multiple

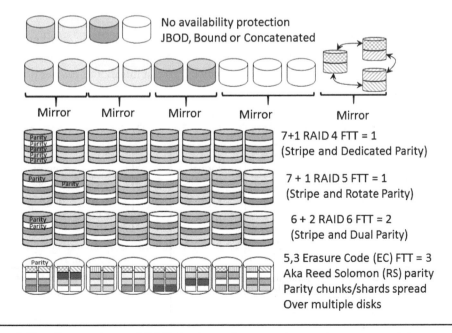

Figure 9.13 Comparison of RAID (stripe, mirror, parity, and erasure code).

Table 9.3. RAID Characteristics

	Normal performance	Availability	Performance overhead	Rebuild overhead	Availability overhead
RAID 0 (stripe)	Very good read & write	None	None	Full volume restore	None
RAID 1 (mirror or replicate)	Good reads; writes = device speed	Very good; two or more copies	Multiple copies can benefit reads	Resynchronize with existing volume	2:1 for dual, 3:1 for three-way copies
RAID 4 (stripe with dedicated parity, i.e., 4 + 1 = 5 drives total)	Poor writes without cache	Good for smaller drive groups and devices	High on write without cache (i.e., parity)	Moderate to high, based on number and type of drives	Varies; 1 Parity/N, where N = number of devices
RAID 5 (stripe with rotating parity, 4 + 1 = 5 drives)	Poor writes without cache	Good for smaller drive groups and devices	High on write without cache (i.e., parity)	Moderate to high, based on number and type of drives	Varies 1 Parity/N, where N = number of devices
RAID 6 (stripe with dual parity, 4 + 2 = 6 drives)	Poor writes without cache	Better for larger drive groups and devices	High on write without cache (i.e., parity)	Moderate to high, based on number and type of drives	Varies; 2 Parity/N, where N = number of devices
RAID 10 (mirror and stripe)	Good	Good	Minimum	Resynchronize with existing volume	Twice mirror capacity stripe drives
Reed-Solomon (RS) parity, also known as erasure code (EC), local reconstruction code (LRC), and SHEC	Ok for reads, slow writes; good for static and cold data with front-end cache	Good	High on writes (CPU for parity calculation, extra I/O operations)	Moderate to high, based on number and type of drives, how implemented, extra I/Os for reconstruction	Varies, low overhead when using large number of devices; CPU, I/O, and network overhead.

drive failures or impacts. Similarly, some EC implementations use relatively small groups such as 6, 2 (8 drives) or 4, 2 (6 drives), while others use 14, 4 (18 drives), 16, 4 (20 drives), or larger.

Table 9.4 shows options for a number of data devices (k) vs. a number of protect devices (m).

Table 9.4. Comparing Various Data Device vs. Protect Device Configurations

k (data devices)	m (protect devices)	Availability; Resiliency	Space capacity overhead	Normal performance	FTT	Comments; Examples
Narrow	Wide	Very good; Low impact of rebuild	Very high	Good (R/W)	Very good	Trade space for RAS; Larger m vs. k; 1, 1; 1, 2; 2, 2; 4, 5
Narrow	Narrow	Good	Good	Good (R/W)	Good	Use with smaller drive groups; 2, 1; 3, 1; 6, 2
Wide	Narrow	Ok to good; With larger m value	Low as m gets larger	Good (read); Writes can be slow	Ok to good	Smaller m can impact rebuild; 3, 1; 7, 1; 14, 2; 13, 3
Wide	Wide	Very good; Balanced	High	Good	Very good	Trade space for RAS; 2, 2; 4, 4; 8, 4; 18, 6

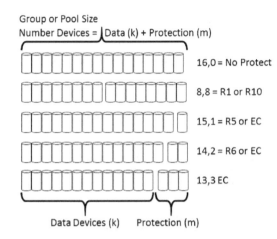

Group or Pool Size
Number Devices = Data (k) + Protection (m)

16,0 = No Protect

8,8 = R1 or R10

15,1 = R5 or EC

14,2 = R6 or EC

13,3 EC

Data Devices (k)   Protection (m)

- Everything is not the same
- Meet your PACE requirements
- Constraints (e.g. number devices)
- Amount and size of data (allocation)
- Application and data I/O size
- (Chunk, Shard, Slab size)
- FTT and other SLOs (RPO, RTO, RAS)
- Normal performance and space use
- Rebuild, reconstruct time and overhead

Figure 9.14   Comparing various data drive to protection devices.

> *Tip:* Note that wide k with no m, such as 4, 0, would not have protection. If you are focused on reducing costs and storage space capacity overhead, then a wider (i.e., more devices) with fewer protect devices might make sense. On the other hand, if performance, availability, and minimal to no impact during rebuild or reconstruction are important, then a narrower drive set, or a smaller ratio of data to protect drives, might make sense.

Figure 9.14 shows various options and considerations for balancing between too many or too few data (k) and protect (m) devices. The balance is about enabling particular FTT along with PACE attributes and SLO. This means, for some environments or applications, using different failure-tolerant modes (FTM) in various combinations as well as configurations.

Figure 9.14 top shows no protection overhead (with no protection); the bottom shows 13 data drives and three protection drives in an EC configuration that could tolerate three devices failing before loss of data or access occurs. In between are various options that can also be scaled up or down across a different number of devices (HDD, SSD, or systems).

Some solutions allow the user or administrator to configure the I/O chunk, slabs, shard, or strip size, for example, from 8 KB to 256 KB to 1 MB (or larger), aligning with application workload and I/O profiles. Other options include the ability to set or disable read-ahead, write-through vs. write-back cache (with battery-protected cache), among other options.

The width or number of devices in a RAID parity or erasure group is based on a combination of factor, including how much data is to be stored and what your FTT objective is, along with spreading out protection overhead. Another consideration is whether you have large or small files and objects. For example, if you have many small files and a wide stripe, parity, or erasure code set with a large chunk or shard size, you may not have an optimal configuration from a performance perspective.

> *Tip:* The tradecraft skill and trick is finding the balance, knowing your applications, the data, and how the data is allocated as well as used, then leveraging that insight and your experience to configure to meet your application PACE needs.

Table 9.5. RAID Characteristics and What to Use When

	When and What to Use for
RAID 0 (stripe)	Data that can tolerate the loss of access to data until restored. Many hardware and software implementations. Before flash and SSD popularity, multiple HDD were striped together for aggregate performance. Stripe by itself is a single point of failure, as loss of a drive breaks the aggregate stripe. Context is that stripe spreads data and performance equally across drives, whereas a bound or concatenated volumes appends space independent of other drives, with no aggregate performance benefit.
RAID 1 Mirror	Good for I/O-intensive writes, as well as reads such as OLTP, transactional, database, metadata, logs, and journals. Can be used on a local- or wide-area basis, depending on implementation tool. Some implementations enable reads to be accelerated by using a different mirror or replica members. Fast rebuilds, as extra parity I/Os are not needed for reconstruction, however at the cost of space capacity.
RAID 4	Using read and write cache, well suited to file-serving environments.
RAID 5	Good for reads including web or files; write performance affected if no write cache. Good fit for when using a smaller number of drivers in a RAID group (stripe), such as 3 + 1, 4 + 1, 5 + 1. Balance lower cost, larger capacity drives with time to rebuild and the chance of a second failure. Some solutions implement write optimization including grouped writes to reduce back-end read, modify writes.
RAID 6	Good general purpose when combined with write cache. Second parity provides additional protection, enabling wider or large disk groups (stripes). Various implementations; some can accelerate rebuild time using underlying disk pools vs. legacy RAID 6. Well suited for larger-capacity drives including HDD and SSD.
RAID 10	I/O-intensive applications requiring performance and availability. Good option if you have four or more drives and do not need to set up two separate RAID 1 sets.
Reed-Solomon erasure codes	RAID 2 (Reed-Solomon) saw limited commercial deployment. However, its many derivatives are widely deployed today in the form of erasure codes. Erasure codes protect against erasure (delete, damage, loss) of a data item similar to other parity protections. Different EC implementations optimized for lower storage capacity cost, faster rebuild with less local and network I/O, among others. Well suited for bulk and large content, as well as object data as opposed to active transactional. Good fit for backup, archive, active archive, bulk data, big data, and content repositories with wide drive groups. Keep in mind that erasure codes, like parity, require CPU and extra I/O to create as well as reconstruct data. Note that also like other parity RAID, more parity or recovery codes spread across devices enables more resiliency against faults or failures.

Table 9.5 shows when and where to use different types of RAID.

In addition to those mentioned in Table 9.5, others include RAID 3, RAID-DP (e.g., NetApp variant of RAID 6), RAID 7, Disaster Tolerant Disk Subsystem (DTDS) and DTDS+ among different RS erasure code variations, as well as vendor-specific implementations, features, functionalities as well as packaging.

Tip: Some apples-to-oranges marketing occurs when vendors position erasure codes such as a 16, 4 having a lower capacity overhead vs. a 3 + 1 RAID 5, hence context matters. Similar to other parity RAID protection, how many drives to include in a group along with parity members is a balance of keeping your storage overhead cost low vs. risk of device failure and speed of rebuild and reconstruction.

Tip: Note that some RAID implementations (hardware, software, or storage systems) differentiate between stripe RAID 0 and JBOD mode. For example, some vendors' implementations present a single drive as a RAID 0 JBOD, while others treat a single-drive JBOD as a pass-through device. The difference can be whether the RAID implementation puts its wrappers around the device and data, or simply treats the device as a raw device.

Other variations include 5 (stripe with rotating parity) and 0 (stripe), resulting in 50; and 6 (stripe with dual parity), where there are variations of dedicated or rotating parity, along with 60 (RAID 6 + RAID 0), similar in concept to combining RAID 1 and RAID 0. Still other variations include vendor-specific variations such as dual parity (DP), which has been used in some NetApp implementations.

In addition to the above, there have also been improvements in how the various RAID approaches are implemented, as well as in functionalities. For example, some implementations of RAID do not group writes, leveraging mirrored and battery-backed cache to reduce the number of actual back-end disk I/Os. Solutions that have good write optimization can group writes together in protected cache until a full stripe write can occur (or a data integrity timer expires), resulting in fewer back-end writes.

Some vendors have made significant improvements in their rebuild times, along with write-gathering optimizations with each generation of software or firmware able to reconstruct a failed drive faster. For example, write gathering and grouped writes can reduce the overhead of legacy read, modify, and write along with associated extra I/O operations.

In addition to reducing write amplification from extra I/Os and subsequent wear on flash SSD, many solutions also implement wear-leveling to increase lifetime durability (i.e., TBW) of non-volatile memory such as flash.

Tip: If disk drive rebuild times are a concern, ask your vendor or solution provider what they are doing as well as what they have done over the past several years to boost their performance. Look for signs of continued improvement in rebuild and reconstruction performance and a decrease in error rates or false drive rebuilds.

Many solutions, including hypervisors, operating systems, CI, and HCI, among others, are using a combination of RAID 1 mirroring or replication for active data, and parity-based solutions for other, less frequently updated data. Figure 9.15 shows an example where local data is written to a mirror or replica (hardware- or software-based) of SSD that include a journaled log for data integrity. Local data and storage is protected across sites and systems using hardware- or software-based mirroring or replication. The remote storage can be a hybrid mix of hardware- and software-based RAID, parity, and mirroring.

As data cools, it gets de-staged or tiered to lower-cost, slower storage protected with parity-based solutions including erasure codes. Data is also mirrored or replicated across locations using synchronous and asynchronous modes. The mirror, replication, parity, and erasure code is done in different data infrastructure locations, including software.

Figure 9.16 shows an example of creating a software RAID 6 using six devices with the lsblk command. The Linux mdadm utility is then used to create the RAID 6 volume, followed by displaying the /proc/mdstat configuration.

Additional steps to make the RAID 6 volume usable could be, for example,

```
sudo mkfs.ext4 -F /dev/md0
sudo mkdir -p /media/md0
sudo mount /dev/md0 /media/md0
echo '/dev/md0 /mnt/md0 ext4 defaults,nofail,discard 0 0' | sudo
tee -a /etc/fstab
```

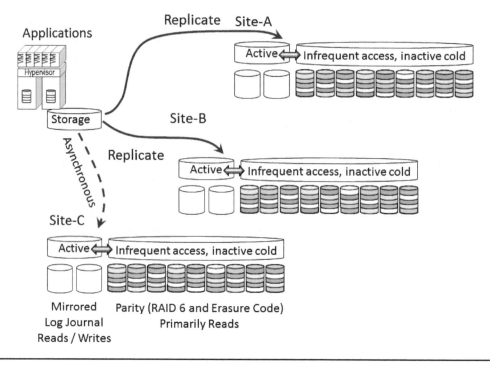

Figure 9.15 Using multiple (e.g., belt and suspenders) RAID techniques, local and remote.

Figure 9.17 shows an example of an Avago (formerly known as LSI) RAID Storage Manager dashboard that can be used for physical (hardware devices and RAID card) and logical configuration of storage.

Figure 9.18 shows yet another example of creating a RAID mirror set using Microsoft Storage Spaces on Windows Server 2016.

```
greg@DOCKIOO5:~$
greg@DOCKIOO5:~$ lsblk -o NAME,SIZE,FSTYPE,TYPE,MOUNTPOINT
NAME     SIZE FSTYPE TYPE MOUNTPOINT
sda      20G         disk
├─sda1   19G         part /
├─sda2   1K          part
└─sda5   1022M       part [SWAP]
sdb      2G          disk
sdc      2G          disk
sdd      2G          disk
sde      2G          disk
sdf      2G          disk
sdg      2G          disk
sr0      1024M       rom
greg@DOCKIOO5:~$ sudo mdadm --create --verbose /dev/md0 --level=6 --raid-devices=6 /dev/sdb /dev/sdc /dev/sdd /dev/sde /dev/sdf /dev/sdg
mdadm: layout defaults to left-symmetric
mdadm: layout defaults to left-symmetric
mdadm: chunk size defaults to 512K
mdadm: size set to 2095616K
mdadm: Defaulting to version 1.2 metadata
mdadm: array /dev/md0 started.
greg@DOCKIOO5:~$ cat /proc/mdstat
Personalities : [raid6] [raid5] [raid4]
md0 : active raid6 sdg[5] sdf[4] sde[3] sdd[2] sdc[1] sdb[0]
      8382464 blocks super 1.2 level 6, 512k chunk, algorithm 2 [6/6] [UUUUUU]

unused devices: <none>
greg@DOCKIOO5:~$
```

Figure 9.16   Sample software RAID creation in Linux.

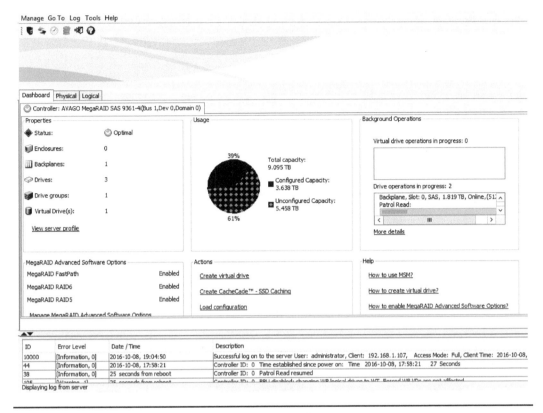

Figure 9.17   RAID adapter card management software.

Consider:

- Number of drives (width) in a group, along with protection copies or parity
- Balance rebuild performance impact and time vs. storage space overhead savings
- Ability to mix and match various devices in different drive groups in a system
- Management interface, tools, wizards, GUIs, CLIs, APIs, and plug-ins
- Different approaches for various applications and environments
- Context of a physical RAID array, system, appliance, or solution vs. logical

9.4. What's in Your Toolbox?

Besides having an adequate number (and type) of persistent NVM, SSD, and HDD storage devices, various software tools, documentation, notes, tip sheets, as well as copies of keys, certificates and credentials are part of the toolbox. You may also need to keep any older tape drives, CD/DVD readers, or devices required to be able to read backed-up and archived data. This also means keeping a copy of the software that you used to format and store the data, including applicable application software.

Part of your data infrastructure toolbox should be an emergency kit containing digital copies as well as hard copy of important keys, credentials, or passwords that are kept in a safe, secure location for when you need them. Other toolbox items include Acronis clone and

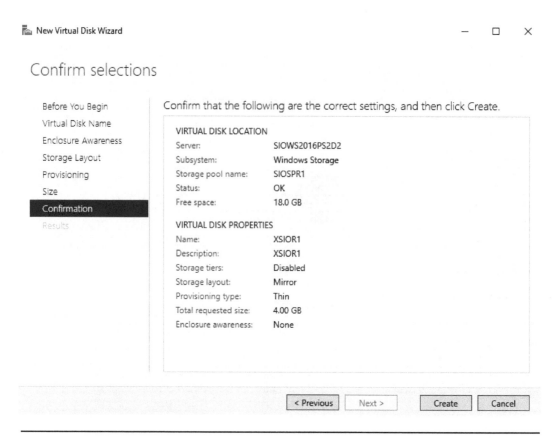

Figure 9.18   Microsoft Windows Server 2016 Storage Spaces RAID 1.

image, AD migration, AWS CLI, benchmark factory, Camtasia (screen capture and record), CDMI, Clonezilla, cloudberry, cosbench, curl, Cyberduck, dstat, dd, fdisk, fio, foglight, gdisk, gpart, gpg2, graphviz, gsutil, hammer, htop, ifconfig, iometer, iorate, iostat, iotop, iperf, Jam Treesize, and mdadmin.

Additional tools include Microsoft (Azure tools, Disk2vhd, Diskpart, Diskperf, Diskspd, Excell, Hyper-V manager, ipconfig, Microsoft Virtual Machine Converter [MVMC], sigcheck, Nttcp, Powershell, SQL Server Management Studio, Storage Spaces, Storage Spaces Direct (S2D), System Center Operations Manager [SCOM], Windows Management Instrumentation Command [WMIC], Windows Manager Interface [WMI]), dstat, lshw, Mydump, netstat, ntop, ping, R, RDP, rman, Robocopy, Rsync, Rufus, RV Tools, S3 browser, S3FS, S3fs, s3motion, SAR, Seatools, shareplex, SMIS, spacesniffer, Spotlight on Windows / *nix, SSH, ssh-keygen, Sysbench, Sysmon, topio, Updraft, Vdbench, Visualesxtop, VMware (VAAI, VASA, vSphere/ESXi, and many more), among others.

9.5. Common Questions and Tips

I heard that RAID is dead and bad? There are some RAID implementations in hardware or software that may be dead or dying, just as there are new emerging variations, particularly around erasure codes. Keep in mind that erasure codes such as SHEC and LRC, among others, are derivatives of Reed-Solomon (RAID 2). Likewise, there are some not-so-good implementations of RAID, as well as configuration decisions to use the wrong tool, technology, or approach for a given solution or scenario.

Why would you want to do replication in a storage system vs. in an application, hypervisor, or software-defined storage? It depends on what your application PACE requirements are, along with SLO as well as technology capabilities. For example, your hypervisor, CI, HCI, software-defined storage, or storage system may only support asynchronous replication, while your database might support synchronous.

Why not use erasure codes everywhere, particularly with SSD? The simple answer is for performance: Erasure codes are good for inactive, or mainly read-based, data. A good strategy for many deployments is to use a mix of mirroring such as on SSD pools for updated intensive including logs, journals, and databases, and then erasure code and parity protection for less frequent, higher-capacity storage pools.

Is backup dead, as some people say? In Chapter 10 we'll have more to say about recovery-point protection including backup, snapshot, and CDP. Some backup products, tools, technologies, and techniques may be dead, but others simply need to be used in new ways.

Can data be used while it is being copied from one system to another? That depends on the solution, system, software, or service being used. Some, such as laptop or workstation drive clones moving data from a HDD to a SSD require or recommend that you not update or save data while the copy is ongoing (i.e., work in read-only mode). Other tools will lock out or prevent any activity during a copy or clone.

Other solutions enable you to continue working, including making updates which are tracked and then applied. In addition, some solutions also enable a copy from one storage system to another to occur, and at some point in time begin using data on the destination storage system. The result is an effectively shorter recovery time in that some data can begin processing while other data is still in-flight or being moved.

What are some legal considerations regarding data protection, resiliency and availability? There are governmental as well as industry regulations, which vary. For example, there are various laws that determine what data can be moved or stored outside of a country for privacy, among other concerns.

9.6. Learning Experience

The following are some questions to use as a learning exercise based on what was covered in this as well as preceding chapters.

- Do you know what legal requirements determine where data can exist?
- What is copy data management (CDM)?

- What does the number of 9's indicate, and how would you use it?
- Why do you want to implement availability and protection at different layers?
- Do you know how much data as well as storage you have?
- How many versions as well as copies do you have in different locations (4 3 2 1 rule)?
- Can you identify when your data was last updated or modified as well as the last read?
- Explain the role and where to use RTO and RPO as well as FTT and FTM.
- When would you use mirror and replication vs. parity and erasure code?
- What's the difference between hardware- and software-based RAID?
- Explain the difference between the gap in coverage and an air gap?
- Why is RAID (mirror, replication, parity, erasure code) by itself not a replacement for backup? How or what would you do to leverage some form of RAID as part of resiliency, data protection, and an approach to address backup needs?

9.7. Chapter Summary

This wraps up the second of four chapters on data services that enable data infrastructure functionality. Data services vary in functionalities and capabilities, as well as where they are implemented—block, file, object, or API. Likewise, there are many aspects to availability for data infrastructures, including faults to tolerate (FTT), SLO, and application PACE to consider. RAID, mirroring, and replication are not a replacement for backup and need to be combined with recovery-point objective time-interval copy techniques.

General action items include:

- Everything is not the same; there are different applications, SLO, etc.
- Understand the 4 3 2 1 data protection rule and how to implement it.
- Different tools, techniques, and technology, support the 4 3 2 1 rule and FTT.
- Balance rebuild performance impact and time vs. storage space overhead savings.
- Use different approaches for various applications and environments.
- Context of a physical RAID array, system, appliance, or solution vs. logical.
- What is best for somebody else may not be best for you and your applications.

Continue reading about availability in Chapter 10 (recovery point, backup, snapshot, and CDP) along with security. Additional data services include access, performance, and general management (Chapter 8), capacity and data footprint reduction (also known as storage efficiency) in Chapter 11.

Bottom line: Availability and data protection enable data as well as information to be accessible, consistent, safe, and secure to be served.

Chapter 10

Data Infrastructure Services: Availability, Recovery-Point Objective, and Security

---

If you cannot go back in time, how can you restore in the future?

What You Will Learn in This Chapter

- Backup, continuous data protection, replication, and snapshot point-in-time protection
- Applications, data, and data infrastructure security (physical and logical)
- Protecting applications and data at various levels, intervals, and granularities
- Breadth of features and depth of functionality
- What to look for and consider in comparing features

This chapter is the third of four focused on data services that enable data infrastructure services capabilities. The focus of this chapter is availability from the perspective of recovery-point objective (RPO)—that is, backup, snapshot, and continuous data protection—along with security. Access, management, and performance are covered in Chapter 8. Chapter 9 looks at availability from a RAS, RAID, and replication perspective. Chapter 11 covers capacity and data footprint reduction (DFR), also known as storage efficiency. Key themes, buzzwords, and trends addressed in this chapter include block, file, object, and application program interfaces (APIs) along with availability, durability, security, RPO, change block tracking (CBT), snapshot, continuous data protection (CDP), and security, among others.

10.1. Getting Started

Recall that everything is not the same across various sizes of environments with different application workload profiles. Granted, there are some similarities that common tools, technologies, and techniques can be leveraged to provide availability of applications, data, and data infrastructure resources. Remember, if something can go wrong, it probably will at some point—either on its own or with the help (or neglect or mistake) of somebody.

Let's continue expanding our availability and data protection discussion to recovery points. Recovery points are also known as point in time (PIT), checkpoint, consistency point, protection interval, snapshots, versions, or backups to address various service-level objectives (SLOs) and PACE requirements. Recovery-point objectives are some points in time that are fixed or scheduled intervals; or they can be dynamic, with different granularities (how much data is protected). In other words, if you do not have recovery- or consistency-point protection, how can you go back in time to recover, restore, roll back, or resume operations? Likewise, if you cannot go back, how can you go forward?

10.1.1. Tradecraft and Tools

Table 10.1 shows examples of various data movements from source to destination. These movements include migration, replication, clones, mirroring, and backup, copies, among others. The source device can be a block LUN, volume, partition, physical or virtual drive, HDD or SSD, as well as file system, object, or blob container or bucket. An example of the modes in Table 10.1 include D2D backup from local to local (or remote) disk (HDD or SSD) storage, or D2D2D copy from local to local storage, then to remote.

Table 10.1. Data Movement Modes from Source to Destination

Mode	Description
D2D	Data gets copied (moved, migrated, replicated, cloned, backed up) from source storage (HDD or SSD) to another device or disk (HDD or SSD)-based device.
D2C	Data gets copied from a source device to a cloud device.
D2T	Data gets copied from a source device to a tape device (drive or library).
D2D2D	Data gets copied from a source device to another device, and then to another device.
D2D2T	Data gets copied from a source device to another device, then to tape.
D2D2C	Data gets copied from a source device to another device, then to cloud.

Tip: In the past, "disk" usually referred to HDD. Today, however, it can also mean SSD. Think of D2D as not being just HDD to HDD, as it can also be SSD to SSD—or S2S if you prefer.

10.2. Enabling RPO (Archive, Backup, CDP, Snapshots, Versions)

RAID, including parity and erasure code along with mirroring and replication, provide availability and accessibility. These by themselves, however, are not a replacement for backup or sufficient for complete data protection. The solution is to combine resiliency technology with point-in-time tools that enable availability and facilitate going back to a previous consistency time.

Recovery-point protection is implemented within applications using checkpoint and consistency points as well as log and journal switches or flush. Other places where recovery-point protection occurs include in middleware, database, key-value stores and repositories, file systems, volume managers, and software-defined storage, in addition to hypervisors, operating systems, utilities, storage systems, appliances, and service providers.

> *Tip:* There are different approaches, technologies, techniques, and tools, including archive, backup, continuous data protection, point-in-time copies, or clones such as snapshots, along with versioning.

In the context of the 4 3 2 1 rule, enabling RPO is associated with durability, meaning number of copies and versions. Simply having more copies is not sufficient because if they are all corrupted, damaged, infected, or contain deleted data, or data with latent nefarious bugs or rootkits, then they could all be bad. The solution is to have multiple versions and copies of the versions in different locations to provided data protection to a given point in time.

Timeline and delta or recovery points are when data can be recovered from to move forward. They are consistent points in the context of what is/was protected. Figure 10.1 shows on the left vertical axis different granularities, along with protection and consistency points that occur over time (horizontal axis). For example, data "Hello" is written to storage (A) and then (B), an update is made "Oh Hello," followed by (C) full backup, clone, and master snapshot or a gold copy is made.

Next, data is changed (D) to "Oh, Hello," followed by, at time-1 (E), an incremental backup, copy, snapshot. At (F) a full copy, the master snapshot, is made, which now includes (H) "Hello" and "Oh, Hello." Note the previous full contained "Hello" and "Oh Hello," while the new full (H) contains "Hello" and "Oh, Hello." Next (G) data is changed to "Oh, Hello there," then changed (I) to "Oh, Hello there I'm here." Next (J) another incremental snapshot copy is made, date is changed (K) to "Oh, Hello there I'm over here," followed by another incremental (L), and other incremental (M) made a short time later.

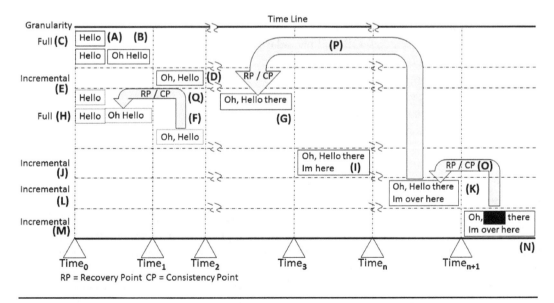

Figure 10.1   Recovery and consistency points.

At (N) there is a problem with the file, object, or stored item requiring a restore, rollback, or recovery from a previous point in time. Since the incremental (M) was too close to the recovery point (RP) or consistency point (CP), and perhaps damaged or its consistency questionable, it is decided to go to (O), the previous snapshot, copy, or backup. Alternatively, if needed, one can go back to (P) or (Q).

> *Tip:* Simply having multiple copies and different versions is not enough for resiliency; some of those copies and versions need to be dispersed or placed in different systems or locations away from the source. How many copies, versions, systems, and locations are needed for your applications will depend on the applicable threat risks along with associated business impact.

The solution is to combine techniques for enabling copies with versions and point-in-time protection intervals. PIT intervals enable recovering or access to data back in time, which is a RPO. That RPO can be an application, transactional, system, or other consistency point, or some other time interval. Some context here is that there are gaps in protection coverage, meaning something was not protected.

Another gap in protection coverage can be a time interval enabling RPO, or simply a physical and logical break and the distance between the active or protection copy, and alternate versions and copies. For example, a gap in coverage means something was not protected. A PIT gap occurs when a snapshot, copy, or version is made on some time interval—for example, hourly.

A protection air or distance gap is having one of those versions and copies on another system, in a different location and not directly accessible. In other words, if you delete, or data gets damaged locally, the protection copies are safe. Furthermore, if the local protection copies are also damaged, an air or distance gap means that the remote or alternate copies, which may be on-line or off-line, are also safe. Hence context is important when discussing data protection gaps.

Data services and features for enabling RPO and time-interval copies include:

- Application, AD/DC/DNS and system restore points or backups
- Point-in-time snapshots, CDP, and back-dating
- Scheduled, ad-hoc, on demand and dynamic
- Checkpoints, consistency points, scheduled
- Archives, backups, clones, and copies
- Versioning, journals, transaction and redo logs
- Backup, protection, and consistency groups

Point-in-time copies can be implemented in different locations from applications, databases, key-value and object repositories, file systems, volume managers, operating systems, and hypervisors to storage systems, appliances, and cloud services.

In addition to where the functionality is located, how those capacities are delivered also varies. For example, some are bundled as part of an application, database, operating system, hypervisor, or storage system, while others are add-ons, either for a fee or free. Some approaches implement these features as an add-on software module, or in a storage system or appliance. Other implementations are delivered via a software appliance model that is provided as a virtual appliance, cloud appliance, or as tin-wrapped software (software delivered with hardware).

Another variable is the granularity, consistency point, and application integration of PIT copies. Some approaches are tightly integrated with an application for transactional data consistency, others with databases or file systems. This means that the granularity can vary from fine-grain application transaction consistent, time interval, entire database or file system, volume, or system and device basis for crash consistency.

> *Tip:* Simply having a snapshot, for example, of a particular time, may not be sufficient even if 100% of what is written to disk is protected if 100% of the data has not been written to disk. What I mean by this is that if 5% of the data, including configuration settings, metadata, and other actual data, is still in memory buffers and not flushed to disk (made consistent), and that data is not protected, then there is not a consistent copy. You might have a consistent copy of what was copied, but the contents of that copy may not be consistent to a particular point in time.

What's your objective: to resume, roll back or roll forward, restore, reload, reconstruct or rebuild, restart, or some other form of recovery? Different applications (or portions of them) along with various environments will have different PACE criteria that determine RTO and RPO as well as how to configure technologies to meet SLOs.

Figure 10.2 shows on the left various data infrastructure layers moving from low altitude (lower in the stack) host servers or PM and up to higher levels with applications. At each layer or altitude, there are different hardware and software components to protect, with various policy attributes. These attributes, besides PACE, FTT, RTO, RPO, and SLOs, include granularity (full or incremental), consistency points, coverage, frequency (when protected), and retention.

Also shown in the top left of Figure 10.2 are protections for various data infrastructure management tools and resources, including active directory (AD), domain controllers (DC),

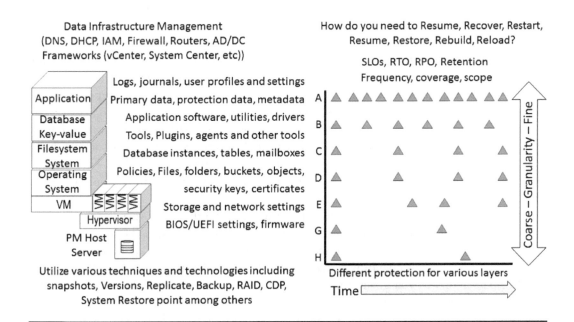

Figure 10.2 Protecting data infrastructure and enabling resiliency at various layers.

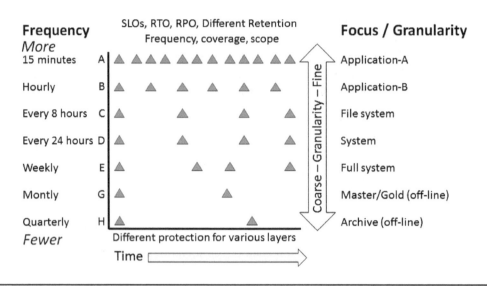

Figure 10.3   Protecting different focus areas with various granularities.

group policy objects (GPO) and organizational units (OU), network DNS, routing and firewall, among others. Also included are protecting management systems such as VMware vCenter and related servers, Microsoft System Center, OpenStack, as well as data protection tools along with their associated configurations, metadata, and catalogs.

The center of Figure 10.2 lists various items that get protected along with associated technologies, techniques, and tools. On the right-hand side of Figure 10.2 is an example of how different layers get protected at various times, granularity, and what is protected. For example, the PM or host server BIOS and UEFI as well as other related settings seldom change, so they do not have to be protected as often. Also shown on the right of Figure 10.2 are what can be a series of full and incremental backups, as well as differential or synthetic ones.

Figure 10.3 is a variation of Figure 10.2 showing on the left different frequencies and intervals, with a granularity of focus or scope of coverage on the right. The middle shows how different layers or applications and data focus have various protection intervals, type of protection (full, incremental, snap, differentials), along with retention, as well as some copies to keep.

Protection in Figures 10.2 and 10.3 for the PM could be as simple as documentation of what settings to configure, versions, and other related information. A hypervisor may have changes, such as patches, upgrades, or new drivers, more frequently than a PM. How you go about protecting may involve reinstalling from your standard or custom distribution software, then applying patches, drivers, and settings.

You might also have a master copy of a hypervisor on a USB thumb drive or another storage device that can be cloned, customized with the server name, IP address, log location, and other information. Some backup and data protection tools also provide protection of hypervisors in addition to the VM, guest operating systems, applications, and data.

The point is that as you go up the stack, higher in altitude (layers), the granularity and frequency of protection increases. What this means is that you may have more frequent smaller protection copies and consistency points higher up at the application layer, while lower down, less frequent, yet larger full image, volume, or VM protection, combining different tools, technology, and techniques.

Tip: Look for data services and features that integrate with, for example, SQL Server, Oracle, SAP, SharePoint, or Exchange, as well as operating systems such as Windows (e.g., VSS) and Linux along with hypervisors (e.g., Hyper-V, VMware ESXi/vSphere, KVM, Xen, and OpenStack). For example, in addition to integration, are there capabilities to interact with the application or database, file system, operating system, or hypervisor to quiesce, flush data that is in memory, or at least take a copy of what's in memory, as well as what's on disk?

Management considerations for enabling recovery-point protection include:

- Manual and policy-enabled automated tools
- Insight, awareness, analytics, reporting, and notification
- Schedules, frequency, coverage, clean-up, grooming, and optimization
- Monitoring, diagnostics, troubleshooting, and remediation
- Interfaces, plug-ins, wizards, GUIs, CLI, APIs, and management tools

Figure 10.4 shows how data can be made available for access as well as durability using various techniques combined on a local, remote, physical, and virtual and cloud basis. This includes supporting recovery points in time as well as enabling RTOs.

Considerations include:

- Local and remote as well as geographic location requirements
- Change block tracking, fast and intelligent scans to determine what data changed
- Impact of protection on normal running systems and applications
- Impact on applications during degraded or rebuild scenarios
- Heterogeneous mixed vendor, technology, tool, and application environments
- Hardware, software and services dependency and interoperability
- How many concurrent copies, clone, mirror, snap, or other protect jobs

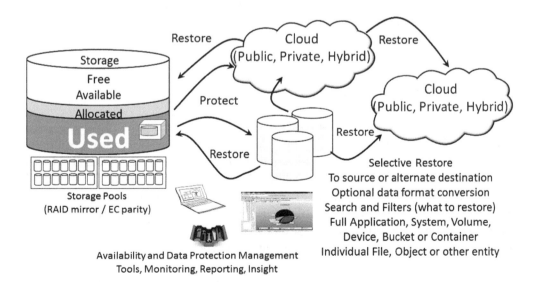

Figure 10.4   Protecting and restoring data to/from various locations.

10.3. Backup and Archiving

Backup and archiving have similarities along with differences. Both archive and backup provide protection to a given point in time, yet have different objectives and intended uses. In some cases, they rely on and use different software tools with common target destination storage. On the other hand, some solutions software can be defined to implement separate backup and archive policies.

Archives can be cold and off-line, representing data (and applications) at a point time or for future needs (e.g., value), including compliance regulations. In addition to off-line and cold, archives can also be on-line and active; that is, the data does not change. However, it becomes a write once/read many (and often) or a reference repository.

Backup and data protection policies topics include:

- SLOs—RTO, RPO, FTT, for protection and recovery
- What gets protected: include or exclude, granularity and coverage
- When it gets protected: schedule and frequency, application consistency
- Where it gets protected: destination where data is protected to
- How it gets protected. Full, incremental, snapshot, image or file based
- Retention: how long to keep items, how many copies to keep
- Maintenance: data integrity checks and repair of data and media
- Disposition: how protection copies and media are securely removed or recycled
- Security: encryption and access authorization to copies and metadata

Backup and PIT RPO protection can be:

- File, object, file system, volume, device, database, or application focused
- Fine-grain granularity such as database table, email mailbox or attachment
- Different consistency points (time intervals) to meet various SLOs
- Full, partial, incremental application, server, or storage system based
- Protect to local, remote, cloud, or other destination in various formats

10.3.1. Local and Remote Copies

Traditional backup and PIT protection copies to support 4 3 2 1 and other application data as well as data infrastructure needs to be focused on making a local copy to another medium such as D2T. Over time, backups and PIT protection copies evolved to D2D and D2D2T using both HDD as well as SSD. The newest evolution has been to D2C (disk to cloud), D2D2C (disk to disk to cloud) as well as D2D2C2T (disk to disk to cloud to tape), among other variations. Note that local copies made to removable media such as tape or even portable HDD and SSD can be kept on-site or off-site in a logical or physical vault.

A vault, in general, is a place where data and copies such as backups, clones, and copies can be kept in an organized safe place. A logical or virtual vault can be a robotic tape library located in a computer room or other safe location in a data center. Another option is to have a separate safe physical place on-site or off-site, where copies are physically transported by vehicle to establish an air gap between the protection copy and the primary data.

Note that an off-site (encrypted) copy can be shipped to a facility or service such as Iron Mountain, among others, or to a managed service provider including cloud services such as AWS, among others.

Copies of data in the vault can be off-line and near-line such as with a tape library (TL), or on-line such as with a virtual tape library (VTL) or disk library (DL), as well as other "bulk storage."

Note that a VTL presents endpoints including NAS (NFS or CIFS/SMB) as well as block (FC, iSCSI, and SAS, among others) that emulate tape drives, tape cartridges (virtual tape), and robotics. Even though a VTL or DL may have as an endpoint NAS or iSCSI, its personality (refer to Chapter 8) is optimized for streaming (reads or writes) backups, archives, active archives, clones, and other CDM as well as protection copies.

Also note that a DL, VTL, or other "bulk storage" repository can be located on-site or off-site, including at an MSP or cloud service provider where data is encrypted while being sent over a network, as well as once it lands at the destination.

In addition to local to off-site, including local to the cloud, there are also cloud-to-cloud, as well as cloud-to-local scenarios. The idea is that if you have data moved to, or born (i.e., created) in a public cloud such as AWS, you may need or want the flexibility (or requirement) to have a copy located at another provider such as Microsoft Azure or GCS, among others.

10.3.2. Walking the Protection Talk

As an example of combing the above, for my environment, I have scheduled protection copies (backups) of data kept on RAID protected storage that gets copied to local HDD storage. There are also scheduled protection copies that go to one of my AWS S3 buckets. Also, a couple of times a year I do a full copy of important items to a removable HDD-based device that also goes off-site to a safe, secure location. Besides these protections, the local on-site backup copies along with the data protection metadata, settings, configuration, catalog, and backups are cloned to a separate device that is not mirrored.

Some have asked me why not simply mirror devices together instead of using a separate clone? The answer is that a mirror or RAID protected device would be good and make sense for increasing availability and resiliency of the primary protection target. However, the RAID mirror, parity, or erasure code would not provide protection for an accidental deletion, corruption, virus, or other logical damage to the data on the device, including a file system problem. By making a separate clone to another device, not only is there another copy on a different device, there is also protection against logical and physical errors. Of course, if your environment is smaller, scale the approach down; if it is larger, then scale up accordingly, as everything is not the same. However, there are similarities.

Not all my applications and data are on premise, as I also have some running in the cloud, such as email servers, the web, blog, and other items. At the time of this writing, those applications and data reside on a dedicated private server (DPS) at my service provider located three or four states away. In addition to regular PIT protection copies done of my DPS and data at the service provider, I also have scheduled as well as ad-hoc backups done to AWS (i.e., cloud to cloud).

For example, my WordPress site gets protected using a tool called UpDraft Plus Backups, which copies all metadata, settings, configurations, and plug-ins as well as MySQL database items needed for recovery. The UpDraft Plus Backup results are then sent to one of my AWS S3 buckets (plus I have a copy locally). On a larger scale, you could use a technology tool such as Dell EMC Spanning to protect your Office 365, Salesforce, and Google among other "cloud-borne" data to another service or back to your own on-site location.

10.3.3. Point-in-Time Protection for Different Points of Interest

Figure 10.5 shows backup and data protection focus, granularity, and coverage. For example, at the top left is less frequent protection of the operating system, hypervisor, and BIOS/UEFI settings. At the middle left is volume- or device-level protection (full, incremental, differential), along with various views on the right ranging from protecting everything, to different granularities such as file system, database, database logs and journals, and OS and application software, along with settings.

In Figure 10.5, note that the different focus points and granularities also take into consideration application and data consistency (as well as checkpoints), along with different frequencies and coverage (e.g. full, partial, incremental, incremental forever, differential) as well as retention.

> *Tip:* Some context is needed about object backup and backing up objects, which can mean different things. As mentioned elsewhere, objects refer to many different things, including cloud and object storage buckets, containers, blobs, and objects accessed via S3 or Swift, among other APIs. There are also database objects and entities, which are different from cloud or object storage objects.

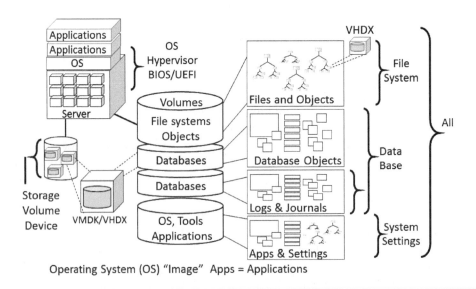

Figure 10.5   Backup and data protection focus, granularity, and coverage.

> *Tip:* Another context to keep in mind is that an object backup can refer to protecting different systems, servers, storage devices, volumes, and entities that collectively comprise an application such as accounting, payroll, or engineering, vs. focusing on the individual components. An object backup may, in fact, be a collection of individual backups, PIT copies, and snapshots that combined represent what's needed to restore an application or system.

On the other hand, the content of a cloud or object storage repository (buckets, containers, blobs, objects, and metadata) can be backed up, as well as serve as a destination target for protection.

Backups can be cold and off-line like archives, as well as on-line and accessible. However, the difference between the two, besides intended use and scope, is granularity. Archives are intended to be coarser and less frequently accessed, while backups can be more frequently and granularly accessed. Can you use a backup for an archive and vice versa? A qualified yes, as an archive could be a master gold copy such as an annual protection copy, in addition to functioning in its role as a compliance and retention copy. Likewise, a full backup set to long-term retention can provide and enable some archive functions.

10.3.4. Point-in-Time Protection and Backup/Restore Considerations

The following are some tips and considerations, as well as features and functionalities, to keep in mind about PIT protection copies including backup (as well as restore). While these focus on backup and restore, many also apply to other PIT recovery-point protection, including snapshots and CDP among others.

Some general archiving and backup considerations include:

- Application context and awareness, including software, settings, metadata, and data
- Application integration for consistency-point integrity
- Application and data infrastructure FTT needs and SLOs
- Scope and coverage: full vs. incremental or delta changes; retention and frequency
- Balance speed of protection copies vs. overhead and speed of restoration
- Optimize to minimize time needed for protection vs. speed of recovery

Backup and archiving activities and techniques include:

- Management: tracking (copies, catalogs, metadata, devices, and media), monitoring tasks and jobs, schedules, reporting, policy definition and configuration, compliance monitoring.
- Data movement: how data gets moved, copied, and replicated from source to destination using server or server-less (e.g., NDMP) solutions, among other approaches. Application integration with snapshots, consistency points, and checkpoints, along with destination access.
- Policies: schedules, frequency, coverage, retention, governance, where kept, who has access to the copied data as well as metadata, disposition, recycling, secure erase, and audits.

- Destination targets: where, how protected data is stored, file format, open or proprietary save set or backup container, on SSD, HDD, tape, optical, or cloud, virtual or physical tape library, block, file, or object format.
- Security: physical and logical, encryption, access control, authentication, audit access.
- Data footprint reduction (DFR): compress, consolidate, de-dupe, optimize, reorganize.

How backup and RPO data protection is done:

- Dump, export and unload, import and reload
- Backup or data protection tool
- Snapshot integration, full or partial scan, CBT
- Use standard built-in tools and utilities
- Vendor, third-party, or open-source tools
- DIY: build your own scripts and tools

Backup functionality and features include:

- Hot, open file, snapshot, application integration
- Read verification and data integrity checks
- Path to tape and tape as well as optical, HDD, or cloud migration
- Encryption and compression, various formats
- Application integration with databases, email, file systems, hypervisors
- Clones of backups and protection copies
- Management tools and interfaces from GUI, Wizard, CLI, and PowerShell

Various data footprint reduction (also known as storage efficiency and optimization) techniques are used along with backup (and data protection in general). DFR techniques include compression and de-dupe, among others (see Chapter 11). Table 10.2 shows various common types of backups and copies to support RPO protection. Which approach to use depends on

Table 10.2. Various Types of RPO (Backup, PIT Copy, and Snapshot) Protection

Protection	Characteristics
Differential	Aggregate cumulative of all changes since last full backup, replica, or protection copy. Enables faster recovery time compared to applying many incrementals for a complete restore operation, while leveraging changes since last full.
Full	Coarse granularity from RPO time standpoint, complete data for coverage. Useful for periodic full protection such as weekly, monthly, quarterly protection or to speed up full restoration. The benefit is complete protection, fast full restore, and the caveat is that it takes up more time and space.
Incremental	Fine-granularity RPO time standpoint, changes since last full or incremental protection. The benefit is faster and smaller, less space than full, the caveat is many incrementals along with a full may be needed for complete recovery.
Incremental forever	After an initial copy, backup, or snapshot, incrementals are leveraged to create a virtual full along with incremental protection, similar to synthetic.
Synthetic progressive	Using previous incrementals (and initial full or synthetic), the equivalent of a full is created without having to perform an actual full backup.

your application PACE requirements and SLO needs. Needless to say, there are many variations of the techniques shown in Table 10.7, along with different implementations. Which is best for your needs will depend on your data infrastructure along with application PACE as well as SLO requirements

In addition to DFR, another consideration is encryption of data at various layers or levels. This includes data that is already encrypted, encrypting the target backup or protection device, volume, file, save set, vault, or repository.

Backup and protection performance considerations include:

- Available server memory, I/O, and CPU resources
- Network and I/O interface bandwidth
- HDD, SSD, magnetic tape, or optical
- On-line, near-line, or off-line, including removable
- Block, file, object, API, or other access and format
- Purpose-built backup appliance (PBBA), backup server, and backup proxy
- Physical, virtual, or cloud server, storage system, or appliance

Tip: In general, there is more overhead processing many small files vs. the same amount of data spread over fewer but larger files. The overhead is from metadata handling, lookups, and opening and closing files or objects before streaming data. With larger files and objects, less time is spent on overhead operations vs. doing data streaming, which can consume less CPU overhead.

Additional backup and protection performance topics include:

- Proprietary or open format (tar, gzip) or native file system
- S3, Swift, and other cloud or objects, buckets, and containers
- Targets, including block, file, object, and API
- Virtual tape library, disk library, or optical library
- Physical and logical labels, RFID and tracking
- Throttling and performance priorities to enable protection quality of service
- Processing offloads and drivers for compression, de-dupe, and other functions
- Cloud, object, API, chunk and shard size issues, file split

Backup technology and implementation architecture considerations include:

- Fast scan, smart scan, and change block tracking (CBT)
- Agent and agentless, server and server-less
- Network Data Management Protocol (NDMP) backups
- Local and remote client (desktop, mobile) as well as server backup
- Serial, multiplex, and parallel processing of backup or copy streams
- Instant mount of a backup to enable boot of virtual machine
- Import and export of databases such as Dumpit and Mydumper
- Snapshot and PIT copy integration
- Fixed and removable or portable media and device tracking
- Vendor-neutral archives (VNA), DICOM, IL7, among other support

Change block tracking (CBT; not to be confused with Computer-Based Training) can be implemented in operating systems, hypervisors, as well as other data infrastructure locations. CBT is not unique to backup, but it is a common use case as a means to speed up protecting data, as well as reducing CPU, I/O, and memory demands. Fast and smart scan as well as CBT can be found in various solutions including Microsoft Hyper-V, Oracle, and VMware vSphere, among many others.

Figure 10.6 shows an example (top) of how, without CBT, once an initial copy (or snapshot or protection copy) is made, a scan of dirty, updated data is needed to determine what has changed. Depending on how much data or storage exists, the scan can take a long time while consuming CPU, memory, and I/O to perform. Once the identify changes are known, then those bits, bytes, blocks, and blobs can be copied.

The bottom half of Figure 10.6 shows how, with CBT, the scan is skipped, instead looking to the log or journal of tracked changes to know what data to copy. Some file sync and share tools can also maintain metadata on what files or objects have been copied to speed up subsequent operations, as well as reduce the number of cloud or object storage API calls such as LIST.

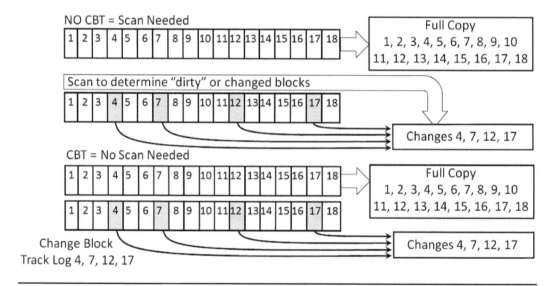

Figure 10.6 Change block tracking.

Backup and data protection management data services features include:

- Data management, including what you have protected and where it exists
- Application integration such as Oracle RMAN, among others
- Tagging of data, files, backups, other metadata items
- Optimize for back-end cloud access, dir, list, etc.
- Scheduling, automation, and policy management
- Authentication and passwords, access controls
- Restore media and logical restore test and verify
- Call-outs, exits, pre/postprocessing, Shell, API, scripts, dashboard monitor

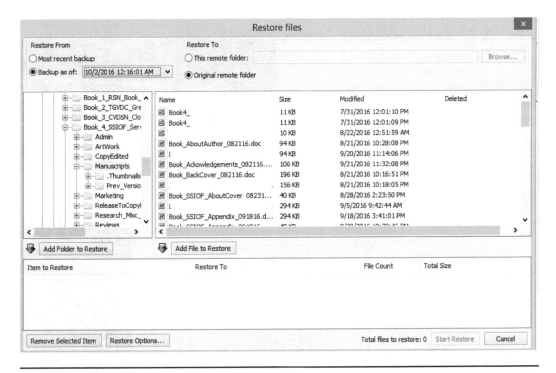

Figure 10.7   Sample file restore from the cloud.

Figure 10.7 shows an example of a restore from a cloud protection copy (backup) being able to select a specific folder, file, and date range for restoration. In this example, I am specifying files in a folder for this book that was backed up on October 4. I can specify which files to restore as well as where to restore them to (besides their default location). Besides selecting files, I can also select entire folders as well as use other selection filtering criteria.

Another example is shown in Figure 10.8 using Microsoft SQL Server Studio to select restore (or backup) options for a given database. In addition to using GUIs such as Microsoft Studio for SQL Server (or other databases), the same functions (backup, restore, checkpoint, snapshots) can be done via command-line or PowerShell scripts.

Additional management and related functionalities include:

- Quotas, policies, templates, scripts, and configuration settings
- One-click and quick restore, including launching VM directly from backup
- Long-term retention holds, secure disposition, and recycling
- Backup vault and repository maintenance and rotation, free space, and verification
- Grooming, cleanup, optimization, and restructuring catalogs
- Recover, repair, rebuild, protect catalogs, and backup sets
- Protect, save, export as well as import settings, schedules, and scripts
- Software and device driver maintenance and updates
- License admission, audit and access logs, reporting, notifications, alerts, alarms

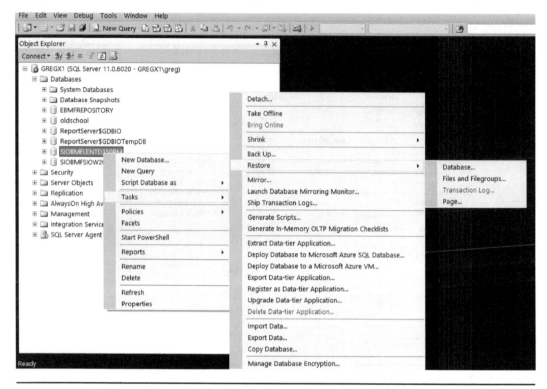

Figure 10.8   Microsoft SQL Server Studio restore (and backup) example.

Considerations include:

- Performance processing overhead of many small files vs. few large ones
- Throttle and QoS to meet PACE requirements and policies
- Archive VM as well as data, along with copies of software and metadata
- Protect your data protection catalogs, settings, metadata, repositories
- Reporting and monitoring, analytics, clean-up and grooming
- What software, plug-ins, modules, or optional tools are included
- Are optional software tools free, part of maintenance, or for a fee?
- What additional hardware, software and service dependencies exist

10.4. Snapshots, CDP, Versioning, Consistency, and Checkpoints

Part of backup or data protection modernization that supports data footprint reduction includes space-saving snapshots. Space-saving or space-efficient PIT copies can be used for more than data protection. As part of CDM, PIT copies also support making copies of production data that can be used for quality assurance or testing, development, decision support, and other uses.

> *Tip:* A fundamental premise of recovery-point protection is that at some point in time a copy of the data is made so that recovery can be accomplished to that point. What varies is how those copies are made, for example. like with traditional backups, or with snapshots among other techniques. Another variation besides the time interval or frequency is granularity, along with whether data is quiesced (i.e., quiet) or changing.

PIT copies include:

- Backups (full, incremental, differential)
- Snapshots, consistency points, and checkpoints
- Versioning, journaling, and CDP
- Change management of applications and data infrastructure software

PIT copies can be done via:

- Applications, databases, key-value repositories, email systems
- File systems, volume managers, operating systems, hypervisors
- Storage systems, appliances, cloud services

> *Tip:* The importance of a space-saving snapshot is to reduce the overhead of extra space needed every time a copy or snapshot of a snapshot is made. First-generation snapshots, which have been deployed in various systems for many years, if not decades, have continued to improve in terms of performance and space efficiency. The next wave was to enable copies of copies using change tracking and redirection on write tehniques to reduce the amount of storage required while enabling fast copies to be made. The importance of space-saving, as well as traditional, snapshots is that, as part of a data protection modernization, changing the way information is copied can reduce the overhead of storage needed, enabling a smaller data footprint. The object is to be able to store and retain more data in a smaller, more economical footprint.

Another aspect of PIT copies including snapshots, checkpoints, and consistency points (and for that matter, backups) along with versions is time interval (i.e., how often they are done). The time interval can be a defined fixed schedule or ad-hoc. The interval might be monthly, weekly, daily, end of shift, hourly, or every few minutes (or more often). A longer interval between PIT copies is referred to as coarse recovery points, narrower or shorter intervals being fine-grained.

> *Tip:* Why not just have continuous replication or mirroring as discussed in Chapter 9? Replication and mirroring provide a continuous update of what has changed. Recall that the good news here is that if something good gets changed, the copy is instantly (or with only a slight time delay with asynchronous mode) made. The bad news is that if something bad happens, then that also occurs to the mirror or replica unless there is some form of PIT recovery point.

A variation of snapshots is known as CDP, which combines the best of replication and mirroring along with PIT RPO copies and the ability to run at a fine grain with low overhead (e.g., under a minute). For many environments, simply running frequent snapshots—every few to several minutes—is sufficient and what some refer to as near-CDP.

> *Tip:* With some CDP and other PIT-based solutions, besides fixed static intervals, there is also the ability to back-date to dynamically create a recovery point after the fact. What is best for your environment and applications will vary depending on your SLOs including RPOs and PACE requirements.

Besides frequency and time interval, another aspect is granularity of the protection, similar to backup disused in the previous section. Granularity includes from an application transactional perspective, database instance or table, file system, folder or file, virtual machine, and mailbox, among other resources, as well as software change management. Granularity also connects to what you are protecting against, as well as the level of data consistency needed. As with backup, application interaction, is another consideration, for example, how snapshots, CDP, and consistency points are integrated with databases such as SQL Server and Oracle, among others.

Databases also leverage their own protection scheme using checkpoints or consistency points to a checkpoint, journal, or log file. These checkpoints, consistency points, or log switches are timed to the desired RPO. The checkpoint, consistency point, and log or journals can be used as part of database recovery to roll back transactions or the database to a known good point.

DBAs will find a balance between too many, too long a consistency point that meets RPO requirements as well as space along with performance implications. Note that as part of a data protection scheme; a database can be replicated as well as backed up to another location. So too can its log files, which, depending on the database and configuration, can be applied to a standby copy as part of a resiliency solution.

Snapshots, checkpoints, and consistency points can be organized into groups that refer to the resources for an application or some other function. For example, a consistency group of different snapshots, versions, or PIT copies spanning various times and multiple volumes or file systems can be created for management purposes.

Figure 10.9 shows two examples of hypervisor VM PIT protection with a Hyper-V checkpoint (snapshot) on the left, and VMware vSphere snapshots on the right. Note that the

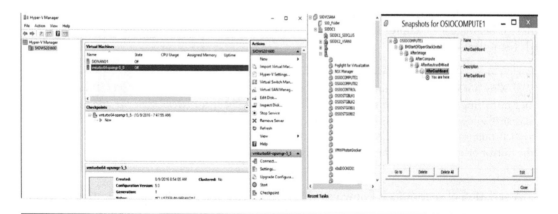

Figure 10.9   Hyper-V checkpoint and VMware vSphere snapshots.

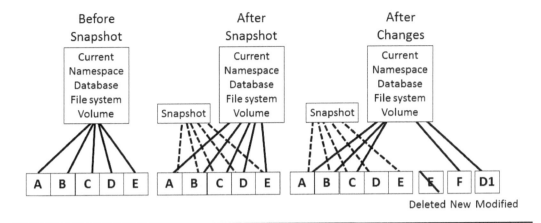

Figure 10.10   Simple view of how snapshots work.

granularity of the snapshot is the contents of the VM by default. The snapshot (or checkpoint) can be done "hot" on the VM or by queiessing the file systems to a known state for consistency. What this means is that, by default, the snapshot of the VM without any other configuration will be similar to a snapshot of a running system, providing a copy as of that point in time. Should additional granularity such as within a database or specific file need recovery, additional recovery-point protection will be needed (refer to Figure 10.5).

Figure 10.10 shows on the left the current namespace (file system, bucket, container, volume, or storage device) before the first snapshot containing "ABCDE." The center of Figure 10.10 shows the snapshot (dashed lines) pointing to the current namespace and data "ABCDE."

On the right of Figure 10.10, changes to the active namespace are shown (E deleted, F added, and D changed to D1). The snapshot still points to "ABCDE," which can be restored if needed. A subsequent (child or decendent) snapshot would contain "ABCFD1." The snapshot is a "picture" or "view" as of a point in time of how the data looked at a given consistency, incremental or recovery point.

Functionality varies with different solutions and where they are implemented. However, there are some common considerations, including application, server, storage, and cloud services integration via plug-ins or other interfaces; granularity by system, application, database, table, file, folder, or device; performance impact when copies are made, along with how many concurrent operations, as well as total number of copies per item protected, and in total. Other commonalities include scheduling, retention, the scope of what to protect, and security in terms of who can access the data.

PIT copy concerns include:

- Snapshot, CDP, and consistency-point functionality
- DFR, including compression and de-dupe
- Whether and how log or journal files for databases can be copied to other systems
- Integration with mirroring, replication, and other data protection tools
- Whether there are space reservation requirements for the copies, and what the overhead is
- Whether the snapshots, copies, or versions can be accessed for read/write, or marked as WORM

- What are the dependencies and prerequisite software or hardware and management tools
- Import and export of snapshots, repair, rebalance, and integrity checks

Checkpoint, consistency-point, and snapshot management issues include:

- How protection copies, snapshots, and versions are managed as well as tracked
- User vs. administrator scheduled snapshots and restore checkpoints
- Scheduling, clean-up and merging, reverting to the previous point in time
- Log file and journal shipping, copying, or replication
- Apply (or replay) on-line or off-line log files to secondary database
- How many different versions exist and what is their overhead

General considerations for PIT copies include:

- Data integrity checks to prevent stale and inconsistency data
- Granularity of the PIT copies, for example, database, table, mailbox, file system, or file
- Different schedules and policies for various granularities (volumes, file systems, folders)
- Ability to have PIT copies made locally or remote as well as export to cloud
- Security including access control, read, write, list, and delete, as well as encryption

Other considerations include:

- Space-saving and efficient snapshots and CDP can be part of a DFR strategy
- How much space needs to be reserved for snapshot and consistency-point data and metadata
- CBT and fast scans along with metadata tracking of what has changed
- Number of concurrent operations that can occur at the same time with different priorities
- What is the overhead to implementing snapshots along with snap management
- Are the versions full copies, or delta copies dependent on previous versions
- How are permissions and other security attributes preserved

10.5. Data Infrastructure Security (Logical and Physical)

The why of data infrastructure security is to protect, preserve, and secure resources, applications, and data so that information services can be accessed and served. This means protecting and repelling various threats against unauthorized access to applications and data infrastructure components along with associated data, as well as metadata.

Examples of why and what data infrastructure security addresses include:

- Enabling privacy of information, applications, and configuration
- Protecting against nosey neighbors, eavesdropping, and unauthorized access
- Guarding against theft, damage, or destruction of data and applications
- Defending against ransomware, malware, viruses, and rootkits
- Deterring hackers, intruders, ransomware, malware, viruses and rootkits
- Facilitating regulatory, industry, self- and compliance requirements

Data infrastructure security includes both physical as well as logical and software-defined as well as being associated with data protection (backups, BC/DR/BR).

Data infrastructure and related application security occurs in or at:

- Physical, software-defined virtual, cloud, container, and converged resources
- Servers, storage, I/O, and networking devices along with associated protocols
- Hypervisors, operating systems, file systems, volume managers, and device drivers
- Data infrastructure applications including DNS, DHCP, AD/DC, and other components
- Local, remote, mobile across various devices including tablets and phones
- Utilities, databases, repositories, middleware, and applications
- Facilities, including on-premise as well as third-party or managed cloud services

Security tasks and activities are ongoing, performing checks, virus, and other data integrity safety scans, leveraging digital forensics to determine what happened, why, and how—for example, looking for digital (or physical) fingerprints of what happened, perhaps by whom, or, in some cases, looking at or for the absence of fingerprints vs. known and what should exist to determine that something happened.

Security is implemented when:

- Before something happens (proactive and preventive)
- While something is happening
- After something happens

Security is implemented via:

- Access control, identity management
- Encryption and checksums
- Physical and logical

10.5.1. Data Infrastructure Security Implementation

There are several different security focus areas for data infrastructures. These include physical as well as logical, spanning legacy, software-defined, virtual, cloud, and container data infrastructures. Similarly, security topics span from applications to the database, file systems and repository, mobile and fixed devices for accessing application and data infrastructure resources. Additional focus areas include servers, storage, I/O networks, hardware, software, services, and facilities, among others.

Logical data infrastructure security topics include:

- Identity access management (IAM) controls, roles and rights management
- Username, password, PIN, roles and access controls
- Controlling access for creating, read, update, and delete (CRUD) of data and metadata

- Securing primary, secondary, and protection copies of data and applications
- Audit, event, and activity logs for analysis, reporting, monitoring, and detection
- Encryption of data-at-rest and in-flight (when moved), key management
- Grant, remove, suspend, configure rights, roles and policies
- Intrusion detection alert, notification, remediation
- Policies, logical organizational units (OU), single sign on (SSO)

10.5.2. Physical Security and Facilities

As its name implies, physical security spans from facilities to access control of hardware and other equipment along with their network and management interfaces. Physical data infrastructure security topics include:

- Biometric authentication, including eye iris and fingerprint
- Locks (physical and logical) on doors, cabinets, and devices
- Asset-tracking tags including RFID, barcodes, and QR codes
- Tracking of hardware assets as well as portable media (disk, tape, optical)
- Logging and activity logs, surveillance cameras

Physical security items include:

- Physical card and ID if not biometric access card for secure facilities
- Media and assets secure and safe disposition
- Secure digital shredding of deleted data with appropriate audit controls
- Locked doors to equipment rooms and secure cabinets and network ports
- Asset tracking, including portable devices and personal or visiting devices
- Low-key facilities, absent large signs advertising that a data center is here
- Protected (hardened) facility against fire, flood, tornado, and other events
- Use of security cameras or guards

Physical security can be accomplished by addressing the above items, for example, ensuring that all switch ports and their associated cabling and infrastructure including patch panels and cable runs are physically secured with locking doors and cabinets. More complex examples include enabling intrusion detection as well as enabling probes and other tools to monitor critical links such as wide-area interswitch links (ISL).

For example, a monitoring device could track and send out alerts for certain conditions on critical or sensitive ISLs for link loss, signal loss, and other low-level events that might appear as errors. Information can then be correlated with other information to see whether someone was performing work on those interfaces, or whether they had been tampered with in some way.

10.5.3. Logical and Software-Defined Security

Fundamental security and permissions include account, project, region, availability zone (AZ) or data center location, group, and user, among others. Additional permissions and roles can be by company, division, application, system, or resource, among others. Basic permissions to access files, devices, applications, services, and other objects should be based on ownership and access. In addition to standard access, roles can be assigned for elevated permissions via the Linux *sudo* command and administrator user modes for performing certain functions needing those rights.

Basic UNIX and Linux (*nix) file system permissions specify owner, group, or world access along with read, write, execute a file, device, or resource. Read (r) has a value of 4, Write (w) has value of 2, eXecute (x) has a value of (1). Each category (owner, group, and world) has a value based on the access permission, for example, owner (rwx=7), group (rx=5), world (rx=5) has value 755 and appears as -rwxr-xr-xr.

As an example, the LINUX commands

chmod 755 -R /home/demo/tmp
chown greg:storageio -R /home/demo/tmp
ls /home/demo/tmp -al

will set files in all subdirectories under /home/demo/tmp to be rwx by owner, rx by group, and rx by world. Chown sets the ownership of the files to greg, who is part of group storageio. The command *ls /home/demo/tmp.al* will show directories with files attributes including ownership, permissions, size, modification, and other metadata.

Figure 10.11 shows various data infrastructure security–related items from cloud to virtual, hardware and software, as well as network services. Also shown are mobile and edge devices

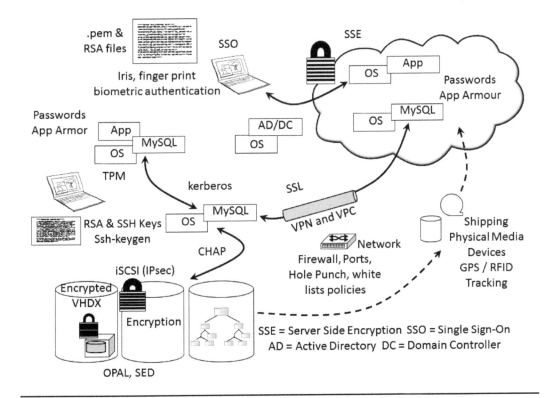

Figure 10.11   Various physical and logical security and access controls.

as well as network connectivity between on-premise and remote cloud services. Cloud services include public, private, as well as hybrid and virtual private clouds (VPC). Access logs are also used to track who has accessed what and when, as well as success along with failed attempts.

Other security-related items shown in Figure 10.11 include Lightweight Direct Access Protocol (LDAP), Remote Authentication Dial-In User Service (RADIUS), active directory (AD), and Kerberos network authentication. Also shown are virtual private networks (VPN) along with Secure Socket Layer (SSL) network security, along with security and authentication keys, credentials for SSH remote access including Single Sign On (SSO).

Encryption is shown in Figure 10.11 for data in-flight as well as while at rest on storage, along with server-side encryption. Server and storage I/O access includes port prohibits, port blocks, volume mapping and masking, along with other endpoint protection. Additional items include SSH and RSA key files, Linux App Armor, Challenge-handshake authentication protocol (CHAP), Internet Protocol Security (IPsec), and Trusted Platform Module (TPM).

10.5.4. Checksum, Hash, and SHA Cipher Encryption Codes

There are many different aspects to data infrastructure security. However, common services and functionalities include encryption of data at rest as well as while in-flight. Encryption can be done via applications, databases, file systems, operating systems and hypervisors, as well as network, storage system, and cloud services.

In addition to encryption, passwords, and access security certificates, hash, secure hash algorithms (SHA), and other codes are also used as checksums to enable data integrity checks of files, objects, folders, and buckets. Hash and SHA codes are also used to create index keys as well as in support of DFR de-dupe, among other uses. Keys, certificates, and other security certificates are created and managed in different locations with various tools for traditional, software-defined virtual, cloud, and container. Some tools include ssh-keygen, putty gen, and sigcheck, among many others built into various data infrastructure hardware, software, and services or as add-ons.

Figure 10.12 shows, on the left, input data or messages which get processed through a hash function (algorithm), with the resulting code (digest) shown on the right.

Figure 10.13 shows an example of a 5.4-GB ISO download (VMware Integrated OpenStack Appliance, VIOA) on the right, along with corresponding data integrity validation checksums. On the left is shown output from sigcheck tool on a Microsoft Windows system to compare the MD5 and SHA checksums to verify that the file is consistent (no lost or dropped bits) as well as safe and secure (no virus or malware added).

Hash and cryptographic (cipher) functions (algorithms) transform input data (message) into a code value (digest). For example, SHA can be used to create hash keys used in compression and de-dupe for computing, storing, and tracking "fingerprints" or "thumbnails" along with digital signatures, checksums, and as an index to reduced data fragments (shards or chunks).

Tip: De-dupe and other DFR technologies, discussed in Chapter 11, also leverage SHA to create a hash. Similarly, some RAID implementations use mathematical algorithms to generate parity and erasure codes.

Input data   **Resulting code**

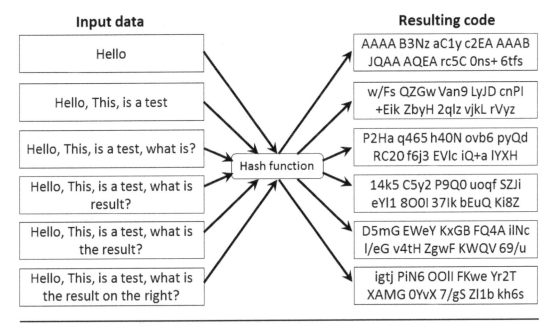

Figure 10.12   How data or messages get transformed into a code digest.

SHA can also be used to validate the integrity and consistency of an item such as an ISO file or object download. The integrity check can be used to verify that it has not been tampered with from a security standpoint, as well as to ensure that bits were not dropped or lost (i.e., the data is lossless) during transmission. Encryption algorithms (aka cipher codes) encrypt and decrypt data using a key. Both can be used together to ensure data integrity, as well as security and privacy, or they can be used separately.

Several secure hash algorithms have been published by the National Institute of Standards and Technology (NIST) as part of the U.S. Federal Information Processing Standard (FIPS). These include SHA-1, 160 bit (20 bytes, 8-bit words), also known as FIPS PUB 180-1; SHA-2, including SHA-256 (32-bit words); and SHA-512 (64-bit words), also known as FIPS PUB 180-2, along with variations. Another option is SHA-3 with similar length as SHA-2.

A question that comes up sometimes is what happens if there is a hash or SHA code collision resulting in a conflict (i.e., a duplicate). Algorithms and their implementations should be able to detect and address collisions or minimize the chance of them happening, as well

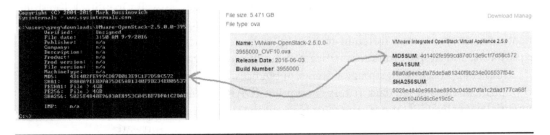

Figure 10.13   Checksums and SHA data integrity.

as resolving them should they occur. Vendors may indicate what they guarantee and or have tested; some will simply say architecturally, or in conceptual theory, what should or should not happen. If you are concerned and not sure, ask your vendor or provider what they do and how they do it.

Tip: Note that there is a difference between FIPS certified and compliant: Ask the vendor whether they are in fact certified or simply compliant. Some vendors may be compliant and working or planning on becoming certified, as that is a longer process.

Other encryption and hash functions include MD5, as well as Advanced Encryption Standard (AES), also known as FIPS-197, which uses variable key lengths of 128, 192, and 256 (i.e., AES 256). Another variation of encryption keys and security certificates or password files is Open Authorization (OAuth and OAuth 2.0). OAuth and OAuth 2.0 are open standards for authenticated support of HTTP access to various sites and services that leverage authorization servers and tokens instead of sharing passwords.

The key to encryption (pun intended) is key management. Keep in mind that these keys are not just for encryption, as there are also certificates and other access control items to be secured and managed. Similar to physical keys that might be attached to a key chain or ring, or placed in a safe, secure location, logical or cyber keys also need to be managed, organized, and secured. There are standards such as Key Management Interoperability Protocol (KMIP) along with its variations and implementations. Examples of enterprise key management (EKM) include SafeNet KeySecure, among others.

Encryption can be done on data at rest (i.e., once it is on disk, local or remote), while in-flight (i.e., moving over a network), on portable and mobile devices, as well as a combination of all the above. For data at rest, there are self-encrypting devices (SED) including HDD and SSD that secure the data on a drive internally via its firmware per OPAL standard. The advantages of SED drives is that if they are removed or unbound from their host server, laptop, or storage system, they are effectively shredded or encrypted. When the drives reattach or rebind to their known host, the data becomes usable again.

What this means is that, in addition to having secured encrypted devices, it is also possible to reduce the amount of time to secure erase for many general environments. Granted, depending on your environment, you may need to run additional secure erase patterns on the devices. The downside to drive-level encryption is that it needs the server, laptop, storage system, or appliance to support their key management.

Keys, certificates, and access control can be managed using data infrastructure resources such as Microsoft Active Directory and its companion resources, as well OpenStack Keystone among others. Encryption key management can be dedicated to a specific application or implementation, or span across different technologies. Features include the ability to import and export keys, and support for key certificates from different locations including self-generated.

Another approach is for storage systems, appliances, and software-defined storage software to implement the storage data at rest encryption (DARE) function. The downside to this is potential increased CPU processing to implement the compute-intensive encryption functions. The upside is tighter integration with the storage and management software.

Besides DARE, server-side encryption (SSE), as its name implies, is encryption for example on a service that is functioning as an application or data infrastructure node to ingest and store data. Data-in-flight encryption can be done in various places, including servers as well as network routers and other devices. Encryption can also be done in file systems, as well as in databases, email, and other locations as part of a layered security and protection defense strategy. Some examples of server software encryption include TruCrypt and Microsoft Bitlocker, among others.

10.5.5. Identity Access Management and Control

Beyond basic user passwords, group membership, resource ownership, and access control lists (ACLs), there are other aspects of identity access management (IAM) and control. These include assigning identifiers to resources (devices, applications, files, folders, services, metadata) such as a globally unique identifier (GUID) and other security identifiers (SID). The various granularities include access from the entire system, application, database, and file system to specific functions, files, or objects.

Figure 10.14 shows, on the left, various access controls for an SQL Server database, and on the right, Microsoft Active Directory (AD) user settings including roles, memberships, and other access (read, write, list, delete) on a local or remote basis.

Access and control policies include group policy objects (GPO), not to be confused with cloud and object storage objects. Organizational Units (OU) define entities such computers, applications, or other data infrastructure resources that users or groups can be members of, as well as have different access roles and rights. For example, a user can be a member of an OU with the role to grant to other users various rights, as well as monitor, revoke, or suspend access.

Another functionality is the ability to grant subordinated or delegated management or rights authority to others. Subordinated management rights could be for a subset of full administrator capabilities, or to enable management of a particular resource such as a VM. Another variation of rights and access control is being able to grant full access to the system administrator, yet restrict access to particular applications and or their data.

Organizational Units are part of Microsoft Active Directory, but there are other variations for different environments such as OpenStack Keystone or those found in VMware, among others. In the case of AD, there are various forests (a group of domains) and trees (domains) that collectively describe, organize, and provide trusted access control for organizations. Microsoft AD can run on-premise on physical or virtual machines, as well as in the cloud, such as Azure

Figure 10.14   SQL Server Studio database (left) and Microsoft Active Directory (right).

AD. While being Microsoft-centric, many Linux and other environments, as well as clouds, also provide hooks (interfaces) into AD.

10.5.6. Server, Storage, and I/O Networking

Normally, when you hear or think "data infrastructure," your thoughts tend to focus on the core physical, virtual, or cloud data center. This means a focus on servers, storage, I/O network hardware, software, and applications, among other related items already mentioned. However, it also means protecting out to the edge for fixed and mobile devices, as well as across sites and managed or cloud service providers.

A common trend is the perception that all threats come from outside the walls of an organization (external). However, studies show that many threats occur from within the walls (physical or virtual) of an organization, although some may be accidental. What this means is that having layers or rings of security is part of a strong defense. It means having robust physical as well as logical data services defined and configured to secure data infrastructures along with the applications they host. However, it also means avoiding creating barriers to productivity that might cause users to circumvent the security. Besides physical access control to physical servers and their management interfaces, security also involves access to virtual servers, hypervisors, and their associated management interfaces—for example, controlling access to VMware vCenter, Microsoft System Center or Hyper-V Manger, as well as OpenStack Horizon, among other dashboard management platforms.

> *Tip:* Another aspect of securing servers is having not only multi-tenancy with VM and containers, but also having shielded VM. For example, Microsoft Hyper-V can shield, isolate (quarantine) a misbehaving VM from impacting others, beyond normal multi-tenancy. Besides VM and cloud instance tenancy, there are also containers, among other options for isolating from noisy (loud) and nosey (snooping) neighbors in adjacent tenancies.

Keep in mind that VM and cloud instances, as well as containers, are stored on data storage, which means those items need to be protected. For example, what happens if you give me your laptop and I take the drive out, attach it to a cable and access it from one of my systems. Will I be able to read the data, or will it be secure via encryption or other security controls? How about if you have 1,000 or 10,000 or more SSD or HDD? Sooner or later, one of those is going to fail. How is this handled? What if one of those drives that is not encrypted were to fall into the wrong hands? Could somebody access data?

In addition to secure shredding (physical), there is digital or logical destruction, which can be done via self-encrypting drives, as well as software that "bleaches" the data. Recall from Chapter 3 that when you delete a file, the data may not be removed. Instead, the data may remain in place until some later garbage collection or erase function occurs, which also enables data to be undeleted. Likewise, using journals, data can also be un-deleted.

> *Tip:* By "bleaching" a file, folder, or device, the data is removed by writing repeated patterns to make sure no traces of the data remain. In the case of object storage that is immutable, a delete operation, like a file system, may not remove the actual data right away. Instead, a "tombstone" is placed marking the "dead" data to be removed during later operations. Note that different cloud and object storage implementations vary in how they handle deletes, similar to file systems and databases.

For accessing remote resources, network firewalls, virtual private networks, virtual private clouds (VPC), and associated appliances come into play. These resources can be used for managing incoming access, as well as facilitating access to remote cloud resources. For example, on your local physical, virtual, or remote cloud instances, specify what IP addresses can access which ports and services, among other resources. Optionally this can include specifying users' access roles and controls, security certificates for authentication, and endpoint protection.

10.5.7. General Data Infrastructure Security-Related Topics

There are many general security-related topics, tools, technologies, techniques, and data services for data infrastructure environments in addition to those mentioned. These range from definition, communication, and implementation of policies, terms, of service, or use, to digital forensics.

> *Tip:* Context is important: There are digital fingerprints such as those used by biometric finger scanners to gain access to resources, and digital fingerprints in the context of an SHA or hash key, as well as nefarious digital fingerprints left behind (via access and audit logs) after doing things. Additional general topics include white and black lists for granting or restricting access to resources, as well as policies based on time, location, or other criteria for resource access.

> *Tip:* Other activities and security data services include scans of media as well as data (applications and actual data along with metadata and settings) for integrity, as well as to proactively detect issues before they become problems. Scan media and data ranges from downloadable content from the web, or what is exchanged via USB, SD, and microSD cards, among other devices.

In addition to scans for virus, malware, ransomware, rootkits, or other nefarious items, integrity checks can be used or detect data (physical or logical) corruption or other inconsistencies. When infected or questionable items are detected, such as a file, folder, object, application, virtual machine, or other entity, it can be quarantined until properly handled.

> *Tip:* Something else to keep in mind is whether the software, files, folders, or objects that you receive are signed with a valid digital signature. This includes data infrastructure software from hypervisors to operating systems, device drivers, tools, firmware, plug-ins, and other utilities. For example, make sure that VIB, ZIP, ISO, OVA/OVF, VMDK, VHDX, EXE, BIN, and IMG, among other items, are validated with trusted signatures for safe installation.

Reporting and alerts along with alarm notification are also features and data services. These range from simple access and activity event logs to more extensive analytics and machine learning features. For example, some software is capable of learning what is normal and abnormal and detecting that a particular device or user accesses resources from a set of known locations. However, when that user suddenly appears to access from another unknown location, an alert or precautionary notification can be sent similar to what you might receive from Facebook or some other social media sites, or your credit card company.

An additional capability is the ability of detection software to send an alert when you access a resource from, say, Minneapolis at 10 AM central time, then an hour later from Singapore (with no time zone adjustments). Somebody might be accessing the resource via different proxy servers, but for the typical person, for their normal known device to be used in two different locations half a world apart within only an hour should at least trigger an alarm!

Some additional security as well as audit-related certifications and accreditations include SSAE16 SOC1, ISO 27001 and 27002, CISSP and CISA, and AT101 SOC2 and SOC3. Additional security-related organizations, groups, topics, and accreditations include Authorization To Operate (ATO) for public cloud providers to support government agencies, Federal Information Security Management Act (FISMA), Federal Risk and Authorization Management Program (FedRAMP), Internet Assigned Number Authority (IANA), National Registration Authority (NRA), as well as Provisional Authority to Operate (P-ATO)

Considerations include:

- Dedicated Private Services (DPS) at managed and cloud service providers
- Local and private cloud, hybrid, and public cloud-enabled security
- Physical, virtual, container, cloud, and mobile-enabled
- Support for IoT, IoD, BYOD, and legacy devices as needed
- Should be an enabler instead of a barrier that results in security issues

10.6. What's In Your Toolbox?

Besides having an adequate number (and type) of persistent NVM, SSD, and HDD storage devices, various software tools, documentation, notes, tip sheets, as well as copies of keys, certificates, and credentials are part of the toolbox. You may also need to keep any older tape drives, CD/DVD readers, or devices required to be able to read backed up and archived data. This also means keeping a copy of the software that you used to format and store the data, including applicable application software.

Part of your data infrastructure tool box should be an emergency (digital recue recovery) kit containing digital copies as well as hard copy of important keys, credentials, or passwords that are kept in a safe, secure location for when you need them. Other toolbox items include Acronis clone and image, AD migration, AWS CLI, benchmark factory, Camtasia (screen capture and record), CDMI, clonezilla, cloudberry, cosbench, curl, cyberduck, dd, fdisk, fio, foglight, gdisk, gpart, gpg2, graphviz, gsutil, hammer, htop, ifconfig, iometer, iorate, iostat, iotop, iperf, Jam Treesize, and mdadmin.

Additional tools include Microsoft (Azure tools, Disk2vhd, Diskpart, Diskperf, Diskspd, Excell, Hyper-V manager, ipconfig, Microsoft Virtual Machine Converter [MVMC], sigcheck, Nttcp, PowerShell, SQL Server Management Studio, Storage Spaces, Storage Spaces Direct (S2D), System Center Operations Manager [SCOM], Windows Management Instrumentation Command [WMIC], Windows Manager Interface [WMI]), dstat, lshw, Mydump, netstat, ntop, ping, R, RDP, rman, Robocopy, Rsync, Rufus, RV Tools, S3 browser, S3FS, S3fs, s3motion, SAR, Seatools, shareplex, SMIS, spacesniffer, Spotlight on Windows/*nix, SSH, ssh-keygen, Sysbench, Sysmon, topio, Updraft, Vdbench, Visualesxtop, VMware (VAAI, VASA, vSphere/ESXi, and many more), among others.

10.7. Common Questions and Tips

What is the difference between FIPS 140-2 compliant and certified? Compliant means that the solution meets the specification (complies) with the standard for protecting and managing crypto and cipher (encryption) modules. Certification means that the technology, solution, software, or service has undergone formal certification by an accredited organization. Pay attention to whether your requirements are for certification or simply compliance, as well as what vendors offer when they say FIPS 140-2 (among other standards).

What is the best technology and location to do encryption? The best technology and approach is the one that works for you and your environment, which enables encryption without introducing complexity or barriers to productivity. Your environment may require different solutions for various applications or focus areas from a single or multiple vendors. Avoid being scared of encryption for fear of losing keys, performance impacts, or increased complexity. Instead, look at different solutions that complement and enable your environment.

What are the legal requirements for where data can be protected to? There are many different industry, state, province, as well as federal government among other regulations that impact data protection. This includes where data can be protected to or stored, for example, whether it can be sent outside of a state or other jurisdictional area, or across country borders. In addition, there are industry-specific regulations such as for financials, among others, that determine how many copies as well as how far away they need to be located for safety. Besides the number of copies as well as versions in various locations, there are also requirements for being able to prove data integrity for a given level of a disaster to meet audit compliance requirements.

How long does data (primary and protection copies) need to be retained? That depends on your applications, industry, and other regulations. In addition, recall Chapter 2, where we discussed the role of data value (it has none, some, or unknown), which can also have a bearing on how long to retain data. In addition to understanding how long data must be retained, also learn how quickly it must be restored, accessible, and useful by applications to meet various regulatory requirements.

How do you know what type of data protection technique and technology to use and when? That is where tradecraft skills and experience come into play, including what you have learned (or refreshed) in this book. Keeping in mind that everything is not the same, there are similarities for different scenarios. Table 10.3 provides a general overview of different types of outages and incidents that can impact data infrastructures, applications, data, and organizations.

The impact of an outage such as those shown in Table 10.3 will depend on how your organization will function during or after an incident, as well as how your data infrastructure is configured to protect, preserve, secure, and serve applications along with their data. Will your organization try to conduct operations as best as possible, or shut down until applications and data, all or partially, can be restored? Similarly, the chance of something happening will vary by your location or proximity to danger.

In addition to what is shown in Table 10.3, Table 10.4 shows how to define your data infrastructure to protect, preserve, and secure applications and data against various threats. Note that this is not an exhaustive or definitive list, rather a guide, and your organization may have different mandates, laws, and regulations to follow.

Table 10.3. Disaster, Event, and Outage or Disruption Scenarios

Outage Result and Impact Reach	Impact to Data Infrastructure and Applications	Impact if Occurs	Chance of Occurring	How Organization Will Function During Outage
Local or regional impact: loss of site. Fire, flood, hurricane, tornado, earthquake, explosion, accidental, intentional act of man or nature.	Complete or severe damage to facilities and surroundings, data infrastructure components, applications, settings, and data.	High if no off-site (or cloud) copy, or no second site, or no BC/BR/DR plan.	Varies by location, region, as well as type of business.	Will the organization shut down for some time, or continue to function in a limited mode using manual processes?
Loss or damaged data or applications from virus, software bug, configuration error, accidental deletion.	Facilities and data infrastructure are intact, severe damage, loss of data has site useless.	High if PIT copies are not present and are not consistent.	Varies by industry and other factors.	Leverage strong security and data protection combined with change management testing.
Loss of power or communications. Power outage, power line or network cable severed.	Site and data infrastructure along with data intact, loss of primary power or network.	High if no standby power or network access.	Varies by location and system separation.	Leverage standby power, cooling, alternative networks, along with secondary sites and resilient DNS.
Failed component, server, storage, I/O network device.	Varies based on how much or what type of redundancy exists or is used.	High if no redundant devices configured.	Higher with lower-cost items.	Minimize disruption by implementing resiliency to isolate and contain faults.
File, object, or other data lost, deleted, or damaged	Varies based on your 4 3 2 1 rule implementation and ability to isolate and contain faults.	Varies, can be small and isolated or on a larger scale.	Relatively high and common occurrence.	Minimize impact with PIT copies for fast restore and recovery. If unattended, could lead to larger disaster.
Cloud or MSP failure. Loss of access due to network, DNS outage, or DDoS attacks. Loss of data due to bug in software or configuration error. Lack of DNS and cloud resiliency.	Varies depending on type of incident. Many cloud failures have been loss of access to data as opposed to actual loss of data. There have been data loss events as well.	Varies on how cloud is used and where your data as well as applications exist. Keep in mind 4 3 2 1 rule.	Varies. Some cloud MSP are more resilient than others.	If all of your data and applications are in the affected cloud, impact can be very high. How you use and configure the cloud and MSP to be resilient can mean being available.

How many failures do different types of protection and resiliency provide? Different configurations will tolerate various numbers of faults. Keep your focus on the solution, or a particular component or layer such as server, storage, network, hardware, or software. In Table 10.5, different numbers of servers are shown along with various storage configurations that provide varying FTTs. Focus beyond just server or storage or network, hardware or software, as availabity is the sum of the parts, plus accessibility, durability, and security.

For additional resiliency, configurations similar or in addition to those in Table 10.5 can be combined with replication across servers, racks, data centers, geographies, or other fault domains.

Keep the 4 3 2 1 rule in mind: Four (or more) copies of data (including applications and settings), with at least three (or more) different versions (PIT copies, backups, snapshots) stored on (or in) at least two (or more) different systems (servers, storage systems, devices) with at least

Table 10.4. How to Protect Against Various Threats

Outage	What to Do
Local or regional impact loss of site	DR, BC, and BR plan in place and up to date that implements 4 3 2 1 protection (or better). Secondary (or tertiary sites) that can be mix of active or standby to meet different RTO requirements. Data is protected to alternative locations per specific RPOs using replication, mirroring, and other forms of RAID as well as PIT copies with backups, snapshots, and archives. Balance protection to number and types of faults to tolerate along with needed speed of recovery and applicable regulations. Leverage DNS and network load balancer resiliency for local and cloud.
Loss or damaged data or applications	Have a DR, BC, and BR plan that can be used in pieces as needed to address various threats or incidents. Leverage various PIT copies as part of 4 3 2 1 protection for fast recovery. Have strong security and defenses along with change configuration management to prevent or minimize chance of something happening, while enabling rapid recovery.
Loss of power or communications	Standby power including generators as well as UPS batteries to carry load while switching from power sources. Look for and eliminate single points of failure (SPOF) and ensure proper battery, switch, and other facilities maintenance. Standby or alternative site to failover to, along with resilient DNS for network.
Failed component	Applications and data placed on resilient server, storage, and I/O paths. Leverage mirror, replicate, and other forms of RAID to meet specific application PACE and SLO. Align applicable fault tolerance mode (FTM) and technology to counter component faults for server, storage, network, hardware, and software. Isolate and contain fault with redundant components when needed to meet RTO, RPO, and other SLO as well as regulations.
File, object, or data damaged	Varies based on your 4 3 2 1 rule implementation and ability to leverage different consistency-point granularities.
Cloud or MSP failure	Implement 4 3 2 1 data protection along with other resiliency and availabity best practices in and across clouds. This means protecting critical data located in a cloud to another cloud or back to your primary site. Also leverage DNS and resilient network access.

Table 10.5. Numbers of Server and Storage Configurations and Faults to Tolerate

Server Nodes	Storage Protection	Storage Space Efficiency	Server FTT	Storage FTT	Characteristics
1	None	100%	0	0	No protection
1	Mirror N + N	50%	0	1	No server faults, one disk fault
1	Parity N + 1	Varies	0	1	No server faults, one disk fault Width, size, or number of disks varies
1	Parity N + 2	Varies	0	2	No server faults, two disk faults Width, size, or number of disks varies
2	Mirror N + N	50%	1	1	One server OR one disk, not both
3	Mirror N + N + N	33%	1+	2	Two faults, one (or more, depending on implementation) server, and two disk
4	Parity N + 1	50+%	2	1	Two (or more) server and disk faults Space varies depending on number of drives
4	Parity N + 2	Varies	2+	2	Multiple server and storage faults, varies by implementation and space overhead

one (or more) of those versions and copies located off-site (cloud, MSP, bunker, vault, or other location). As we have discussed, 4 3 2 1 combines PIT copies such as backups, snapshots, and archives to meet specific recovery-point intervals, along with clustering, mirroring, replication, and other forms of resiliency to meet recovery-time objectives.

10.8. Learning Experience

Some learning experiences include:

- What are your organization's terms of service for accessing resources?
- Have you recently read the terms of service for various web services and resources?
- Are your protection copies, associated applications, settings, and metadata protected?
- Does your organization meet or exceed the 4 3 2 1 backup and data protection rule?
- How can you improve your data protection and security strategy and implementation?
- Does your organization treat all applications and all data the same with common SLO?
- Do your RTO consider individual components or focus on end-to-end and the entire stack?
- How many versions and copies of data exist in different locations, both on- and off-site?
- Can you determine what the cost of protecting an application, VM, database, or file is?
- How long will it take you to restore a file, folder, database, table, mailbox, VM, or system?
- How long will it take to restore locally, using off-site media or via a cloud provider?
- Can you restore your data and applications locally or to a cloud service?
- Using Tables 10.3 and 10.4 as a guide, update them with what's applicable from a regulatory, legal, or organization policy for your environment.
- What regulatory, industry, or organizational regulation and compliance requirements does your organization fall under?
- For your organization, are there any geographical considerations for where your data can be placed (or cannot be placed or moved to)?
- When was the last time a backup or protection copy was restored to see if it is good and consistent, as well as that applicable security permissions and rights are restored?

10.9. Chapter Summary

This wraps up the third of four chapters on data services that enable data infrastructure functionality. Data services vary in functionality and capabilities along with where they are implemented, spanning block, file, object, and API.

Figure 10.15 shows how various availability data services, components, technologies (hardware, software, services), tools, and techniques, combined with tradecraft (skills, experience), provide resiliency against various threat risks. This includes mirror and replication along with parity, as well as erasure codes implemented in software packaged as software or hardware.

Also shown in Figure 10.15 are backups, local, and remote mirror and replication, snapshots, CDP, PIT copies, and encryption, among other items, as well as how different techniques are needed to support various RTO, RPO, as well as other SLO and application PACE

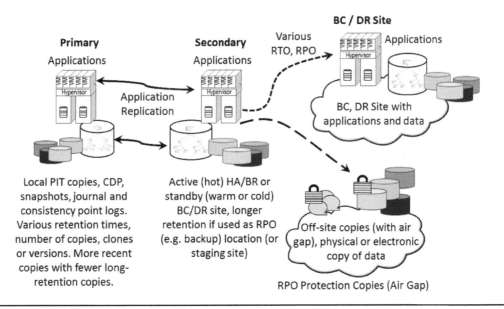

Figure 10.15   Mirroring, replication, and data protection techniques.

requirements. Keep in mind that everything is not the same, so why protect, preserve, secure, and serve applications and data the same as your data infrastructure?

General action items include:

- Everything is not the same, thus there are different availability techniques for different scenarios.
- Backup, CDP, replication, and snapshots enable PIT and RPO protection.
- RPO and RTO along with other RAS capabilities combine to enable availability.
- Availability is the sum of many pieces, including security, access, and durability.
- Protect applications and data at various levels, intervals, and granularities.
- Any technology, hardware, software, or service can fail; be prepared.
- For some, having a résumé up to date is a disaster recovery plan!
- The best DR plan may be the one that you never have to use, at least entirety.
- *If you cannot go back in time, how can you restore in the future?*

Bottom line: Availability and data protection enable data as well as information to be accessible, consistency, safe, and secure to be served.

Chapter 11

Data Infrastructure Services: Capacity and Data Reduction

There is no such thing as an information or data footprint recession.

What You Will Learn in This Chapter

- Data footprint reduction (DFR) and storage efficiency
- Thin provisioning and other capacity optimizations
- Compression, consolidation, and thin provisioning
- De-duplication (de-dupe) and single-instance storage (SIS)
- Role of different techniques, technologies, and tools
- What to look for and consider in comparing features

This chapter is the fourth of four focused on data services that enable data infrastructure services capabilities. The focus of this chapter is on capacity optimization and data footprint reduction (DFR), also known as data infrastructure and storage efficiency. Other data infrastructure services are covered in Chapter 8, where we discuss access, performance, and general management, and in Chapters 9 and 11, where we discuss availability (data protection). Key themes, buzzwords, and trends addressed in this chapter include block, file, object, and utilization, along with capacity optimization, DFR, compression, de-dupe, and thin provisioning, among others.

11.1. Getting Started

A recurring theme is that everything is not the same in the data center, information factory, data infrastructure, various environments, and applications. Note that in this chapter we look at both data services technology and tools, along with techniques as well as capacity tips for data infrastructures (and applications). Keep in mind that the fundamental function of data infrastructures is to store, protect, secure, and serve data for applications that transform it into information. This means having Capacity, the C of PACE. With that theme in mind, let's look at common data services and feature functionalities that support capacity (how data is processed and stored).

11.1.1. Tradecraft and Tools

Let's start the tradecraft and tools discussion for this chapter with the backing store. Backing storage has different context and meanings, but common themes. The common themes are some software with a back-end, persistent, non-ephemeral, non-volatile memory or storage where data is written (stored). Data is stored on different physical media or mediums such as NVM, SCM and other SSD, HDD, tape, and optical. The backing store can also be a system or appliance (physical or virtual) where data is stored, as well as service such as cloud.

Another context for the backing store can be the software that provides abstraction via different tools, applications, and associated data infrastructure components. This can range from a backing store driver or plug-in for Docker containers that enable persistent storage to persistent memory and storage for in-memory databases, virtual memory, storage back ends for databases, volume managers, file systems, key-value repositories, and other software.

Context matters to understand are based on what is being discussed: hardware, software, cloud, virtual, or other backing store device, tool, application, or utility.

Beyond backing store, some other tradecraft conversations and context include remembering that delta refers to change, a point in time or difference between different times, or copies of data. Coverage has a data protection context in terms of what is protected or skipped (i.e., not covered or not protected).

11.2. Capacity (Space Optimization and Efficiency) Services

There are several different capacity topic areas for data infrastructure, including storage space, I/O networking and bandwidth services, server CPU cycles, and available memory. Some other capacity topics include physical (or virtual) available connectivity ports, PCIe and expansion slots, drive bays, rack space, as well as power and cooling. Another capacity focus area is software licenses and services subscriptions.

> *Tip:* Keep in mind that different data has various value and application PACE characteristics. Those characteristics and attributes help define what type of resource capacity to use to meet service objectives. There is a strong relationship to availability and data protection as well as capacity. Data occupies space (capacity) which needs some level of protection depending on its value. Similarly, to protect data and make it available, some amount of overhead capacity is needed.

General capacity considerations include:

- Planning, analysis, forecasting, monitoring, and reporting
- Monitoring physical capacity resources (servers, CPU, memory, I/O, storage, networks)

- Number, type, size of LUN, volumes, file systems, files, buckets, and objects
- Policies, templates, analytics, and applications that enable automation
- Space for logs, import/export, load/unload of data, protection, and staging
- Leveraging insight, awareness, monitoring, reporting, and analytics tools
- Policies and templates to enable automation of capacity workflow tasks

Additional capacity considerations include:

- Aligning applicable techniques, tools, and technologies to application PACE needs
- Good overhead enables value to protect preserve and serve data cost effectively
- Balance efficiency (space and capacity optimization) with effectiveness (productivity)
- Expand efficiency optimization focus beyond hardware to software licenses
- Keep cloud and managed service capacity costs in perspective

11.2.1. General Capacity Items

Keeping in mind that our focus is on data services tools, technologies, and techniques, Capacity is the C in PACE. Capacity topics have been discussed in prior chapters with reference to server, I/O, networking, and storage components, as well as in upcoming about systems and solutions. From a software-defined data infrastructure perspective that spans hardware, software, services, cloud, virtual, and containers, there are several dimensions to capacity.

Figure 11.1 shows various data infrastructure capacity focus areas from hardware to software, the server (CPU and memory), storage I/O networking, local, virtual, and cloud. In addition to hardware and network services, other capacity considerations include software licenses and services subscription.

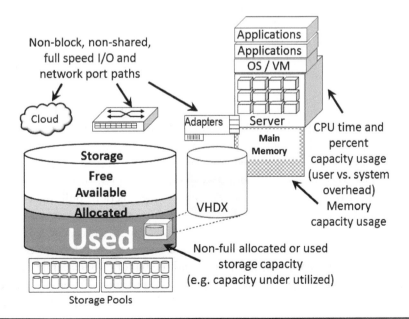

Figure 11.1   Data infrastructure capacity focus points.

In Figure 11.1, top left, network resources can be full-speed, non-block, non-oversubscribed solutions for providing connectivity from servers to the cloud and on-site systems including storage. Those same network resources (switches, routers, adapters) can also be over-subscribed where multiple ports share back-end connectivity, similar to a hub vs. a switch to reduce costs while boosting connectivity.

Also shown in Figure 11.1 is server memory, which can be increased with virtualization memory leveraging disk storage for page and swap space. There may also be tenancy in the server, such as with containers and hypervisors, and virtual machines to increase capacity usage (efficiency) and effectiveness (productivity) of CPU and memory. Storage space capacity is shown as free, available for allocation, as well as allocated and used.

Data infrastructure capacity resources and topics include:

- Physical, logical, virtual, and cloud resources (primary, secondary, and protection)
- CPU, memory and I/O interconnect, cabinet, rack, slot, or enclosure
- Power and cooling requirements, software license and subscriptions
- Raw, formatted, allocated, used, sparse, dense, contiguous or fragmented
- Assets and resource inventory, configuration management database (CMDB)
- Server, memory, and storage counting (see Chapter 3 about bits, bytes, blocks, and blobs)

Capacity tools, techniques, and technologies topics include:

- Allocation, provisioning, manual and automated, policy-enabled
- Insight, search, discovery, analytics reporting, and monitoring
- Dynamic resizing (expansion or shrinking)
- Resource tiering for performance, availability, and capacity
- Space reclamation and provisioning (allocated, thick, thin, oversubscribed)
- Protection, availability, durability, fault-tolerance modes
- Application, database, file system, operating system, hypervisor
- Server, storage system, appliance, network, or cloud service
- Space optimization, reorganization, load leveling, and rebalance

In addition to the above considerations, another capacity topic is how to move large amounts of bulk data in a timely manner from one location to another. For some scenarios, fast high-bandwidth networks may be available. For others, however, even with fast networks, the amount of data to be moved in a given amount of time can exceed bandwidth capabilities.

Thus, to support organization, application, data, data center, or technology moves and migration, or as part of implementing availability data protection, bulk data move options need to be considered. These may include making a copy of data to removable or portable media such as magnetic tape, optical, HDD, or SSD. In some scenarios, entire storage systems and servers may be moved. Some cloud vendors, among others, also offer services that include receiving individual or numerous storage devices and media for bulk data moves.

Amazon (as well as others) has a service where you can put your data on storage devices and send to them via physical transport such as FedEx or UPS, among others. Once the devices are received, data gets ingested into the AWS (or other) cloud.

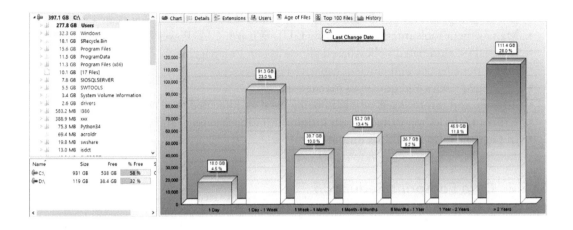

Figure 11.2   Example of gaining capacity and activity insight.

Another Amazon offering is a service called Snowball, where AWS sends an "appliance" with a given amount of storage capacity (more than a single or a few drives) to which you copy your data. The Snowball (or other) appliance is then shipped back to AWS for its contents to be ingested into their storage cloud. Considerations include encryption, tracking, and other logistics items. Note other providers also offer various similar services.

> *Tip:* Having insight and awareness into resources along with their usage along with activity enables analysis, among other management tasks. Figure 11.2 is an example using TreeSize Professional (JAM Software), showing the age of storage space capacity used. Caution should be used in looking at only when items were last updated or created vs. when they were the last read.

Referring to Chapter 2, recall that different data will have various value and changing access patterns over its life-cycle management (LCM) lifetime. In Figure 11.2 there is a large amount of data in the example that is over 2 years old; using a different view, chart, or report might show data that is frequently accessed (reads or writes). Another variation would be data that is old, yet frequently read only.

General considerations about capacity include:

- Number and size of pools, devices per pool or namespace
- Maximize capacity usage, minimize overhead, complexity, and cost
- Find and reclaim (clean up) lost storage space (orphan and forgotten)
- Find and archive or remove zombie virtual machines (powered on but no longer used)
- Free up allocated but unused physical and logical ports
- Space for primary, ephemeral temporary scratch and workspace
- Page and swap space for operating systems as well as hypervisor VM

Additional capacity considerations include:

- Over- or undersubscribed and allocated

- Allocation, provisioning, analysis, reporting, and insight management tools
- Management tools interface including GUI, Wizard, CLI, PowerShell, API, XML
- Manual vs. automated policy-based allocation and management
- Add more effective and efficient capacity resources
- Leverage policies and quotas to manage spaces
- Educate and inform users of what's being used vs. allocated (and cost)
- Utilize analytics reporting to gain insight into capacity usage and activity
- Implement various DFR techniques

11.2.2. Data Footprint Reduction

Data footprint reduction (DFR) is about enabling storing more data in a denser, less complex, cost-efficient, and effective footprint. This includes storing more data managed per person when the additional data being retained adds value or has yet-to-be-determined value to an organization. By reducing your data footprint, more data can be moved in the same or less time to meet service requirements. Additionally, to leverage cloud environments, data needs to be able to be moved effectively and promptly.

Tip: Another focus of DFR is to enable IT resources to be used more effectively, by deriving more value per gigabyte, terabyte, etc., of data stored. This also means alleviating or removing constraints and barriers to growth, such as storage and associated data infrastructure activities. IT resources include people and their skill sets, processes, hardware (servers, storage, networks), software and management tools, licenses, facilities, power and cooling, backup or data protection windows, and services from providers including network bandwidth as well as available budgets.

DFR benefits include:

- Deferring hardware and software upgrades
- Maximizing usage of existing resources
- Enabling consolidation for energy-effective technologies
- Shortening time required for data protection management
- Reducing power and cooling requirements
- Expediting data recovery and application restart for DR scenarios
- Reducing exposure during RAID rebuilds as a result of faster copy times
- Enabling more data to be moved to or from cloud and remote sites

11.2.2.1. What Is Your Data Infrastructure Data Footprint Impact?

Your data footprint impact is the total data storage needed to support your various business application and information needs. Your data footprint may be larger than how much actual data you have, as in the example shown in Figure 11.3. This example has 80 TB of storage space allocated and being used for databases, email, home directories, shared documents, engineering documents, financial, and other data in different formats (structured and unstructured) as well as varying access patterns.

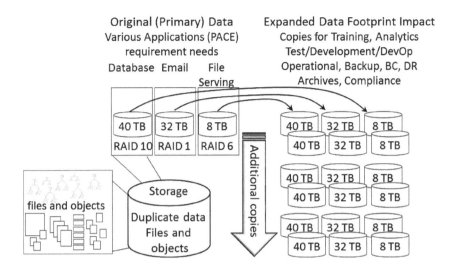

Figure 11.3 Expanding data footprint impact (80 TB becomes 320 TB).

> *Tip:* The larger the data footprint impact, the more data storage capacity and performance bandwidth is needed. How the data is being managed, protected, and housed (powered, cooled, and situated in a rack or cabinet on a floor somewhere) also increases the demand for capacity and associated software licenses.

For example, in Figure 11.3, storage capacity is needed for the actual data as well as for data protection using RAID (mirror or parity and erasure codes), replication, snapshots, and alternate copies, including backups. There may also be overhead in terms of storage capacity for applying virtualization or abstraction to gain additional feature functionality from some storage systems, as well as reserving space for snapshots or other background tasks.

On the other hand, even though physical storage capacity is allocated to applications or file systems for use, the actual capacity may not be being fully used. For example, a database may show as using 90% of its allocated storage capacity, yet internally there are sparse data (blanks or empty rows or white space) for growth or other purposes. The result is that storage capacity is being held in reserve for some applications, which might be available for other uses.

As an additional example, assume that you have 8 TB of Oracle database instances and associated data, 4 TB of Microsoft SQL data supporting Microsoft SharePoint, 8 TB of Microsoft Exchange email data, and 16 TB of general-purpose shared NFS and CIFS Windows-based file sharing, resulting in 36 TB (8 + 4 + 8 + 16) of data. However, your actual data footprint might be much larger.

The 36 TB simply represents the known data, or how storage is allocated to different applications and functions. If the databases are sparsely populated at 50%, for example, only 4 TB of Oracle data exists, though it is occupying 8 TB of storage capacity. The previous example could be a scenario for using thin provisioning, discussed later in this chapter.

Assuming for now that in the above example the capacity sizes mentioned are fairly accurate in terms of the actual data size based on how much data is being backed up during a full backup, your data footprint will include the 36 TB of data as well as the on-line (primary), near-line (secondary), and off-line (tertiary) data storage configured to your specific data protection and availability service requirements.

For example, if you are using RAID 1 mirroring for data availability and accessibility, in addition to replicating or mirroring your data asynchronously to a second site where the data is protected on a RAID 6 volume with write cache, as well as a weekly full backup, your data footprint will then be at least (36 TB × (2 × RAID 1 capacity)) + [36 TB + RAID 6 Parity (i.e., 2 drives)] + (36 TB for a full backup) = 144 TB + parity capacity overhead = data footprint impact. And this simple example assumes only one full backup. What if there are additional full and incremental, or differentail as well as other PIT copies including snapshots? Your data footprint might be even higher than 144 TB (plus parity or erasure code overhead).

Suppose you add daily incremental or periodic snapshots performed throughout the day to the data footprint impact example. Then add extra storage to support application software, temporary workspace, and operating system files including page and swap. Then add room for growth and whatever free space buffer is used for your environment. You can see how the data footprint impact expands beyond the actual data footprint.

Table 11.1 shows an example assuming an Oracle Database, SharePoint with SQL Server database, Exchange email (and attachments), along with general file-sharing applications and their workload. The databases might be Cachè, Cassandra, ClearDB, HBase, Kudu, MongoDB, MySQL, Riak, TokuDB, or Spark, among many others. The primary space allocated is shown in the table and assumes that 50% of the space is allocated and used.

Also shown in Table 11.1 are the protection for Site-A using mirroring or RAID 1, along with Site-B (data is replicated between sites). In addition to multisite mirror or replication, recovery PIT protection is done via three generations of full backups that leverage 2:1 data footprint reduction via compression. The net result is for the 36 TB of space allocated; the resulting data footprint impact is 171 TB for this example.

Next, let's change Site-B replica destination from a mirror or RAID 1 (i.e., $N + N$ or 2 × 36 TB) to an $N + 2$ parity or erasure code RAID scheme (i.e., 2 protection drives). Let's also assume that the target storage at Site-B has real-time compression, de-dupe, and thin

Table 11.1. Sample Traditional Availability, Data Protection, and Capacity Usage

	Primary Space	Site-A Space	Site-B Space	PIT Copy Protection and Space Needed	DFR Savings (Capacity Efficiency)
Oracle (.ORA and redo along with others)	8 TB	Mirror 2 × 8 TB	Mirror 2 × 8 TB	3 × full backup 50% space used 12 TB of PIT (full backups)	2:1 backup compression Reduce from 12 to 6 TB
SQL Server SharePoint and related items	4 TB	Mirror 2 × 4 TB	Mirror 2 × 4 TB	3 × full backup 50% space used 6 TB of PIT (full backups)	2:1 backup compression Reduce from 6 to 3 TB
Exchange email	8 TB	Mirror 2 × 8 TB	Mirror 2 × 8 TB	3 × full backups 50% space used 12 TB of PIT (full backups)	2:1 backup compression Reduce from 12 to 6 TB
File sharing	16 TB	Mirror 2 × 16 TB	Mirror 2 × 16 TB	3 × full backups 50% space used 24 TB of PIT (full backups)	2:1 backup compression Reduce from 24 to 12 TB
Subtotal	36 TB	72 TB	72 TB	54 TB	27 TB savings
Total	171 TB				

Table 11.2. Availabity, Data Protection, and Data Footprint Reduction Impact

	Before DFR Optimization	After DFR Optimization
Total space used (both sites)	171 TB	127 TB
Site FTT	1	1
Site-A Drive FTT	1	1
Site-B Drive FTT	1	2
PIT full copies	3	1
PIT incremental copies	0	5
PIT snapshot copies	0	15
Total PIT recovery points	3	21

provisioning enabled that combined results in a 50% (or better) space efficiency (data footprint overhead reduction).

Thus, instead of 72 TB needed (N + N or 36 + 36 and no DFR) at Site-B, only 18 TB plus 2 parity copies for protection are needed (36/2 = 18 TB via 50% DFR). To keep the math simple and allow some extra space, lets use 4-TB drives (HDD or SSD). Using an N + 2 parity or erasure code with 7 × 4 TB drives (5, 2 with 5 × 4 TB = 20 TB for data, 2 × 4 TB for parity). Let's also change the PIT protection scheme from 3 full, to 1 full plus 5 incremental (or you can use differentials among other approaches), and let's also add 15 snapshots.

Let's also assume that 10% of data changes per day, and we are looking at a 3-week window to keep things simple. For our example, then, assume 1 full (36 TB) plus 5 incremental (5 × 3.6 TB) plus 15 snapshots (15 × 3.6 TB) that we are assuming are not space-efficient, with more overhead than you might see in the real-world.

Another assumption is that the incremental and snapshot PIT copies are done at different times and retained for 3 weeks. Also assume that backups (full, incremental, or differential) and snapshots are reduced by at least 4:1 with real-time compression and de-dupe. What this means is that your actual results will vary and may be better than what has been shown here, depending on the type of data, configuration settings, and the specific solution used.

The net impact, shown in Table 11.2, is that the total space needed for the resulting (reduced) data footprint impact that supports enhanced availabity is 127 TB (or better). Note that the enhanced availabity includes supporting Site-B drive FTT of 2 instead of 1, boosting PIT copy intervals from 3 to 21, granted two fewer full PIT backup protection copies.

Tip: Note that in addition to the DFR approaches mentioned above, additional optimization and capacity efficiency can be achieved. For example, leverage Oracle and SQL Server database footprint reduction via compression, normalization, and other techniques. Other options include adjusting what gets protected when, coverage, and retention. Server-side file system, volume manager, and software-defined single-instance storage (SIS) can also be combined to further reduce your data footprint overhead impact.

Common IT costs associated with supporting an increased data footprint include:

- Data storage hardware and management software tools acquisition
- Associated networking or I/O connectivity hardware, software, and services

- Recurring maintenance and software renewal fees
- Facilities fees for floor space, power, and cooling, as well as IT staffing
- Physical and logical security for data and IT resources
- Data protection for HA, BC, or DR, including backup, replication, and archiving

Reducing your data footprint can help reduce costs or defer upgrades to expand the server, storage, and network capacity along with associated software license and maintenance fees. Maximizing what you already have using DFR techniques can extend the effectiveness and capabilities of your existing IT resources, including power, cooling, storage capacity, network bandwidth, replication, backup, archiving, and software license resources.

> *Tip:* From a network perspective, by reducing your data footprint or its impact, you can also positively impact your SAN, LAN, MAN, and WAN bandwidth for data replication and remote backup or data access as well as move more data using existing available bandwidth.

Figure 11.4 shows various points of interest and locations where DFR can be implemented to meet different application and data infrastructure PACE and SLO requirements. Some of the examples in Figure 11.4 include SIS and de-duplication (de-dupe) along with compression such as Zip implemented in applications, server operating systems, file systems, hypervisors, storage systems, appliances, and other data infrastructure locations.

Also shown in Figure 11.4 are linked clones, where a template, master, or "gold copy" of a virtual machine is created, and then subsequent clone partial copies are made. Instead of having a full clone copy, linked clones leverage common underlying software (bits and bytes) across the different VM, with unique data stored per VM. For example, instead of having 100

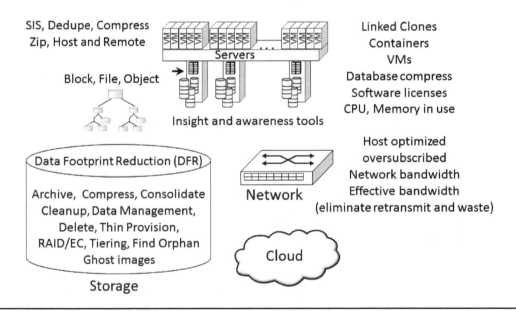

Figure 11.4   General DFR focus points of interest for different needs.

full copies of a VM for VDI deployment, there can be a master with 100 linked clones. Each of the linked clones is smaller, containing only changes or unique data, resulting in less storage space capacity being used.

11.2.2.2. Not All Data or Applications Are the Same

Data footprint reduction can be achieved in many ways with different implementations in different places to meet diverse application and IT requirements. For example, DFR can be done when and where data is created and stored, or it can be done after the fact, during routine data infrastructure tasks including backup/restore, archiving, or during periods of application inactivity.

Tip: Not all applications and data are the same in terms of access or usage patterns as well as life-cycle patterns. For example, some data is actively being read or updated, while other data is inactive (no reads or writes). On the other hand, some data has no updates, however is frequently read. Also, some data and applications have time-sensitive performance requirements, while others have lower demands for performance. Table 11.3 shows various applications and data characteristics as they pertain to DFR.

Tip: The importance of active vs. inactive data in the context of DFR is to identify the applicable technique to use to gain some data reduction benefit without incurring a performance penalty. Some data lends itself to compression, while other data is well suited for de-dupe. Likewise, archiving can be applied to structured databases, email, and Microsoft SharePoint and file systems, among others.

Table 11.3. Different Applications Have Various Data Footprint Reduction Needs

Storage Tier	Tier 0 and Tier 1 Primary On-Line	Tier 2 Secondary On-Line	Tier 3 Tertiary Near-/Off-Line
Characteristics focused	Performance focused Some capacity needed	Some performance Emphasis on capacity	Less performance Much more capacity
	Active changing data Databases, active file systems, logs, video editing, or other time-sensitive applications	Less active or changing data, home directories, general file shares, reference data, on-line backup and BC	Static data, infrequent access, active on-line archives, off-line backups, and archive or master copies for DR purposes
Metric	Cost per activity Activity per watt Time is money	Activity per capacity Protected per watt Mix of time and space	Cost per GB GB per watt Save money
DFR approach	Archive inactive data Space-saving snapshots, RAID (mirror, parity, erasure code) optimized, thin provisioning and I/O consolidation, real-time compression, de-dupe if possible and practical	Archive inactive data Space-saving snapshots, RAID (mirror, parity, erasure code) optimized, modernized data protection, thin provisioning, space and I/O consolidation, compression and de-dupe	Data management, the target for archived data, tiered storage including disk, tape, and cloud. De-dupe and compression. Storage capacity consolidation

11.2.2.3. Data Infrastructure DFR Techniques

As mentioned previously, there are many different DFR approaches and technologies to address various storage capacity optimization needs. Likewise, there are different metrics to gauge the efficiency and effectiveness of the various approaches, some of which are time (performance)-centric whereas others are space (capacity)-focused. Having insight and awareness into your environment enables informed decisions and assessments of what DFR technology to use when, where, and how to configure it.

In general, common DFR technologies and techniques include:

- Application and database normalization optimization
- Archiving (beyond compliance focus, structured, unstructured, big and little data)
- Compression and compaction including real-time, streaming, and post-processing
- Consolidation of data infrastructure resources, workloads, applications, and data
- Data de-dupe and SIS

Additional DFR techniques and technologies include:

- Data management, including cleanup and deletion of unnecessary data
- Finding, fixing, and removing overhead complexity, or moving elsewhere
- Network and bandwidth optimization (local and wide-area or cloud)
- RAID levels (mirror, parity, and erasure codes), discussed in Chapter 9
- Real-time, at ingest, time-deferred, or batch and post-processing
- Space and resource reclamation, finding lost or orphan capacity
- Space-saving snapshots and reclamation, discussed in Chapter 9
- Thin and thick provisioning, dynamic allocation, and storage tiering

Which DFR technique is the best? That depends on what you are trying to accomplish in terms of business and IT objectives. For example, are you looking for maximum storage capacity at the lowest cost, with performance not being a concern? Alternatively, do you need a mix of performance and capacity optimization? Are you looking to apply DFR to primary on-line active data or applications, or for secondary, near-line, inactive, or off-line data? Some forms of storage optimization reduce the amount of data or maximize available storage capacity. Other forms of storage optimization are focused on boosting performance or increasing productivity.

An issue to consider is how much delay or resource consumption you can afford to use or lose to achieve a given level of DFR. For example, as you move from coarse (traditional compression) to granular, such as data de-dupe or SIS, more intelligence, processing power, or off-line post-processing techniques are needed to look at larger patterns of data so as to eliminate duplication.

> *Tip:* If you are concerned enough to be evaluating other forms of DFR technologies for future use, including archiving with data discovery (indexing, eDiscovery) or data de-dupe techniques, leverage appliance-based compression technology for immediate relief to maximize the effectiveness and capacity of existing storage resources for on-line, backup, and archiving while complementing other DFR capabilities.

Tip: Consider your data footprint and its impact on your environment using analysis tools and/or assessment services. Develop a holistic approach to managing your growing data footprint: Look beyond storage hardware costs, and factor in software license and maintenance costs, power, cooling, and IT staff management time.

Leverage data compression as part of an overall DFR strategy to optimize and leverage your investment in your existing storage across all types of applications. In short, deploy a comprehensive DFR strategy combining various techniques and technologies to address point solution needs as well as your overall environment, including on-line, near-line for backup, and off-line for archive data. Table 11.4 shows how different DFR techniques and technologies can be used to address various needs and application PACE requirements.

What this means is that the focus of data footprint reduction is expanding beyond that of just de-duplication for backup or other early deployment scenarios. For some applications, reduction ratios are an important focus, so the need is for tools or techniques that achieve those results.

Protecting, preserving, and serving data and their associated applications is about time and space—time for processing and space to store the information. However, there is another dimension to time and space for data infrastructure, and that is the overhead to protect, preserve, and serve information and data.

Tip: Server, storage, and I/O networking efficiency is often assumed or referred to as meaning utilization, space or capacity reduction, and avoidance. On the other hand, effectiveness then refers to performance and productivity. Efficiency tends to focus on themes and metrics such as space used (or saved) along with reduction ratios. Effectiveness tends to focus on themes of performance, rates, response time, and enablement.

Data footprint reduction is a collection of tools, technologies, and techniques to reduce the impact of storing more data. DFR applies not just to storage space; it is also applicable to network traffic using compression and compaction techniques for data being sent locally and long distances. Applications also implement various forms of DFR, ranging from basic compression

Table 11.4. DFR Approaches and Techniques

	Archiving	Compression	De-duplication
When to use	Database, email, and unstructured data	Email, file sharing, backup or archiving	Backup or archiving or recurring and similar data
Characteristics	Software to identify and remove unused data from active storage devices	Reduced amount of data to be moved (transmitted) or stored on disk or tape	Eliminate duplicate files or file content observed over a period to reduce data footprint
Examples	Database, email, unstructured file solutions	Host software, disk or tape (network routers), appliances	Backup and archiving target devices and VTL, specialized appliances
Caveats	Time and knowledge to know what and when to archive and delete, data and application aware	Software-based solutions require host CPU cycles, impacting application performance	Works well in background mode for backup data to avoid performance impact during data ingestion

in databases and email to data formats such as photo image, audio, and images using JPEG, MPEG, AVI, or MOV, among many others.

For other applications, the focus is on performance with some data reduction benefit, so tools are optimized for performance first and reduction second. In response, vendors will expand their current capabilities and techniques to meet changing needs and criteria. Vendors with multiple DFR tools will also do better than those with only a single function or focused tool. Table 11.5 shows various DFR techniques, technologies, and tool attributes and characteristics.

Metrics that matter for DFR along with server, network, and storage tend to focus on utilization, efficiency, and reduction (or savings) ratios. In addition to ratios and percentage of savings, other metrics that matter include rates, latency, delays, and effectiveness of how productivity is impacted by DFR functions. Ratios indicate how much relative space, capacity, or resource saved while rates indicate the amount of work done or impact.

In addition to the DFR technologies, tools and techniques mentioned above, others include rethinking how and when data protection (backup, BC, BR, DR) are done. Another DFR technique that may not seem obvious at first is to use fewer, faster, and more reliable components.

Table 11.5. DFR Techniques and Technologies

Technique	Performance	Space or Capacity	Comment
Relocate: move problem elsewhere	If off-site, then network issues need to be considered	Keep QoS, SLO, and SLA in perspective vs. cost of service	Off-loading to someone else buys time while or until you can optimize
Archive	Maintain or improve general QoS for normal activities in addition to faster data protection or DCIM tasks	Recover space for growth, enhancements, or new applications; move archive data to another tier, including cloud or MSP	Applies to regulatory and compliance data or applications, including databases, email, and file shares; archive, then delete stale or dead data
Backup modernization	Reduce backup/restore or data protection time	Free up space for more backups/faster restore	Reduce overhead of data protection
Bandwidth optimization	Application or protocol bandwidth vs. latency	More data moved in the same or less time	May boost bandwidth vs. latency
Compression	Minimal to no impact on performance, depending on where and how implemented	Some capacity benefits over broad spectrum of applications with various algorithms	Application, database, operating, file system, network, storage system, or device based
Consolidation	Consolidate IOPs to fewer, faster NVM SSD	Large-capacity SAS and SATA SSD and HDD	Avoid bottlenecks as a result of aggregation
Data de-dupe	Some impact possible on performance in exchange for data reduction benefit	Good data reduction benefit over some types of data or applications	Verify ingest or de-dupe rate in addition to restore or reinflate rate
RAID (parity/erasure code)	May be better option for some data mirroring	Parity-based including erasure code has less overhead vs. mirroring	Look at number of HDD per RAID group
Space-saving snapshots	Make copies faster to enhance service delivery	Reduce overhead or space needed for copies	Data distribution, development/testing, analytics
Thin provisioning	Improve SSD utilization	Improve capacity usage	Avoid overbooking

The idea is to reduce the overhead of multiple lower-cost, slower components that might also be less reliable by replacing with fewer, faster, more resilient ones. For example, instead of 20 or 100 lower (or higher)-cost HDDs to meet performance needs that have low space utilization, leverage fewer, faster, more reliable NVM including SCM and SSD among others. On the other hand, to boost performance and effective performance as well as efficient utilization of slower devices, combine with a fast cache or tiering approach.

Having timely insight and awareness into how capacity resources are used is essential to making informed decisions, as well as implementing automated policy-driven management. Insight, awareness, and analytics tools need to show not only who or what is using capacity, but also activity including when last updated as well as when last read.

> *Tip:* Exercise caution in focusing only on when data was last changed, as that metric may not be reflective of use, such as in a write once/read many (WORM) environments.

Other DFR techniques, technologies, and tools include thin provisioning, archiving, consolidation and cleanup, and storage optimization such as using different RAID and erasure code approaches. Data and storage management aspects of DFR also include finding and dealing with ghost or orphan VM, objects, files, folders, or other resources allocated but not being used.

General DFR considerations include:

- Leverage analytics, insight, awareness, and reporting.
- Look at when data was last read in addition to when it was modified.
- Balance your focus on DFR ratios (how much saved) vs. rates (productivity).
- Efficiency is utilization and space reduction (ratios) or amount reduced by.
- Effectiveness is performance productivity (rates) or amount of data moved.
- Everything is not the same, so balance efficiency and effectiveness to PACE needs.
- Use the applicable technique (or combination) for different needs.

11.2.2.4. Archives and Archiving

Today many people believe or perceive archiving as for regulatory compliance, which is a common use case. Up until the early 2000s post-dot.com bubble burst and Enron era, archiving had a traditional focus of protecting, preserving, and data footprint reduction (i.e., storage optimization) across all types of data and applications. However, in the post-Enron era, the archiving focus has shifted toward and is now mostly associated with regulatory compliance needs and mandates.

Archiving and data footprint reduction has gone by different names, including ILM (It's Lots of Marketing, or Information Life-Cycle Management), system managed storage (SMS), automated storage management (ASM), hierarchical storage management (HSM), along with storage tiering, among many variations (and implementations) spanning automated, semiautomated, and manual processing. Refer to Chapter 2, where we discussed data value as well as life cycles and access patterns and other data infrastructure along with application PACE characteristics.

Archives and archiving can be:

- Project-focused, long or short term, entire system or application
- On-line or near-line (active, cold, or deep cold) as well as off-line
- Located on-site or off-site, including cloud or managed service provider
- Application-specific (email, database, CRM, ERP, document) or general file
- Regularity compliance-centric or general purpose
- Remove namespace entry or leave a pointer to new namespace location
- A system with applications and metadata (no data)
- Just data, with or without context metadata

If there is a magic silver bullet or software-defined aspirin to provide relief for many data infrastructure ailments (headaches or pains), it is archiving. The reason archiving in life after (or before) compliance is an aspirin for data infrastructures is that many challenges are tied to protecting, preserving, storing, and serving more data for longer periods of time.

Most tactical (quick) solutions are focused downstream at the target or destination where data ends up, using DFR techniques such as compression, thin provisioning, and de-duplication, among others. Likewise, those and other techniques are implemented upstream at or near the source, where data is initially ingested, created, processed, and stored.

Where archiving comes into play is to migrate cold, inactive data (no reads or writes) to another lower-cost, easier to protect, preserve, and secure storage tier (service, location, type of storage). Likewise, for static (no updates) and reference data that is read-only (WORM), with updates applied elsewhere, active archives can be created.

For some this can mean simply moving and masking the problem (i.e., data growth), while for others it can mean removing data from active namespaces (and subsequently associated data protection along with management concerns). The namespace entry could be an actual, hard or soft and symbolic link in a directory or folder (local or remote cloud accessible). The namespace can be file system or object based with different access.

The goal of archiving is to maximize the effective use of on-line or expensive resources by keeping those for active services delivery while preserving information that needs to be retained on lower-cost media. Backup, though similar, focuses on protecting data with a shorter retention period using tools for rapid restoration of a single file, folder, or file system. The archive has longer retention time and a focus on preserving the state of a collection of data as of a point in time for future or possible future access.

Tip: Archiving can have one of the greatest impacts on reducing data footprint for storage in general, but particularly for on-line and primary storage. For example, if it can be identified in a timely manner what data can be removed after a project is completed, what data can be purged from a primary database, or which older data can be migrated out of active email databases, a net improvement in application performance as well as available storage capacity can be realized.

Applying archiving as a form of DFR for cloud or virtualized environments enables more data to be retained in a denser, more cost-effective footprint and reduces the amount of data and associated resources. Archiving can be used to clean up and either discard data that is no longer needed or move it to another medium or resource where the associated management costs are lower. De-dupe and/or compression simply reduce the data's impact; while it may apply to some data, it is also similar to treating the symptom instead of the disease.

Archive applies to:

- Telemetry data from applications and data infrastructure tools
- Security and access logs as well as Splunk among others
- Machine learning, IoT and IoD long-term data of value
- Transactional structured and unstructured data
- Trade and event data for finance-related applications
- Medical EMR and healthcare environment (HIPPA, Hitech, PACS)
- Financial, including PCI, Sarbox, and CFR, among others

Additional data, applications, and archive topics include:

- Genomic sequence and research
- Energy and mineral exploration
- Email, instant messaging (IM), and voice mail messaging
- Energy and mineral exploration, as well as simulation modeling
- Databases and key-value repositories, storage back-end services and appliances
- Bulk storage including cloud, object, tape, optical, HDD, and SSD
- Collaboration and document management systems, including Microsoft SharePoint
- Engineering drawings, diagrams, and other records retention
- Home directories and file shares or project repositories
- Digital asset managers (DAM) and digital evidence managers (DEM) for security, dashboards and body cameras, video surveillance, and audio content

Archiving can be applied to structured databases data, semistructured email data and attachments, and unstructured file data. Archiving is evolving and important for preserving information, with clouds as a synergistic opportunity for parking or preserving data. For example, data archived from an on-line database, email, SharePoint, or file system can be migrated to lower-cost, high-capacity disk or tape, or to a cloud MSP.

> *Tip:* The key to deploying an archiving solution is having insight into what data exists along with applicable rules and policies to determine what can be archived, for how long, in how many copies, and how data may ultimately be retired or deleted. Archiving requires a combination of hardware, software, and people to implement business rules.

A challenge with archiving is to identify what data should be archived and what data can be securely destroyed. Also complicating matters is that knowledge of the data value is needed, which may well involve legal issues about who is responsible for making decisions on what data to keep or discard. If a business can invest in the time and software tools, as well as identify which data to archive, the return on investment can be very positive toward reducing the data footprint without limiting the amount of information available for use.

Tip: Do you also need to archive applications necessary for using the data? If you are archiving a database, are you simply removing rows of data after they have been copied to another table or destination, or do you also have to preserve the context of the data, including business rules with XML or JSON wrappers?

From a regulation or compliance perspective, what are the requirements of where data can be placed, and how many copies are needed? For example, if your objective is to move archive data to a cloud or MSP, first verify that there are no regulations stipulating in what geographic area your particular applications data can reside. Some local or national governments regulate what types of data can leave or cross borders into other states or countries.

Indexing or data classification can occur at the application layer via native or optional plug-in capability, via archiving software, or, in some solutions, in the target device, or even all of the above. For legal management systems support, redaction (blanking or marking out of certain data) may be needed, along with litigation hold of data to prevent accidental deletion or digital shredding. Other archiving features include encryption, access audit trails, and reporting.

An effective archiving strategy or deployment includes:

- Policies as to what to archive where, and for how long, including how to dispose of data
- Management buy-in and support for implementing policies and procedures
- Organizational involvement representing different interests from across the business
- Server, storage, or system resource management and discovery tools
- Application modules or plug-ins to interface with data movers and policy managers
- Archiving tools to interface with applications and target devices or cloud services
- Compliance and security (physical as well as logical) of data and processes
- Storage target devices or cloud and MSP services

General tips and comments:

- Factor in the total cost of ownership (TCO) and return on investment (ROI).
- Include time and cost for secure digital disposition of media (tapes and disks).
- Archiving is a useful technique for managing compliance and noncompliance data.
- Long-term data retention applies to all types of data that have business value.
- Implement media tracking along with data protection management.
- Adhere to vendor-recommended media management and handling techniques.
- Align the applicable technology—for example, storage tier—to the task at hand.

Some archive usage scenarios require electronic discovery (eDiscovery) for legal or other search purposes. Keep in mind the context for "search": Is it content-specific, such as searching through files, documents, emails, attachments, databases, and other items, or does it refer to data infrastructure search? Data infrastructure search is focused on metadata, including size, when created, last read or updated, ownership, retention, disposition, performance, availability, security, and other attributes.

Additional features and data services that may be applicable include WORM and litigation hold on a file, document, attachment, folder, directory, mailbox, database, volume, bucket, or container basis. Other features include application-specific data format, including images, video, audio, medical PACs and images, as well as scanned text.

Additional considerations include:

- Archive formats, including IL7 and DICOM, among others
- Different layers and aspects to archiving, besides storage mediums
- Different layers of archiving software, from application to storage managers
- Does solution leave a stub, pointer, index, or reference in existing namespace?
- Do you need to refer to different namespace to find archived items?

11.2.2.5. Compaction, Compression, and Consolidation

A caveat for consolidating and re-tiering is the potential to cause bottlenecks or aggravation as a result of aggregation. Reduce this possibility by consolidating active data onto fewer slower, high-capacity disk drives instead of smaller, faster devices. Remember that the objective is to reduce costs and maximize resources to support growth without introducing barriers to business productivity and efficiency.

Compression is a proven technology that provides immediate and transparent relief to move or store more data effectively, not only for backup and archiving, but also for primary storage. Data compression is widely used in IT and consumer electronics environments. It is implemented in hardware and software to reduce the size of data to create a corresponding reduction in network bandwidth or storage capacity.

If you have used a traditional or TCP/IP-based telephone or cell phone, watched a DVD or HDTV, listened to an MP3, transferred data over the Internet or used email, you have likely relied on some form of compression technology that is transparent to you. Some forms of compression are time-delayed, such as using PKZIP to zip files, while others are real-time or on the fly, such as when using a network, cell phone, or listening to an MP3.

Compression technology is very complementary to archive, backup, and other functions, including supporting on-line primary storage and data applications. Compression is commonly implemented in several locations, including databases, email, operating systems, tape drives, network routers, and compression appliances, to help reduce your data footprint.

11.2.2.5.1. Compression Implementation

Approaches to data compression vary in time delay or impact on application performance as well as in the amount of compression and loss of data. Two approaches that focus on data loss are lossless (no data loss) and lossy (some data loss for higher compression ratio). Additionally, some implementations make performance the main consideration, including real-time for no performance impact to applications, and time-delayed, where there is a performance impact.

Data compression or compaction can be timed to occur:

- In real time, on the fly, for sequential or random data, where applications are not delayed
- Time-delayed, where access is paused while data is compressed or uncompressed
- Post-processing, time-deferred, or batch-based compression

Data compression or compaction occurs in the following locations:

- Add-on compression or compaction software on servers or in storage systems
- Applications, including databases and email as well as file systems
- Data protection tools such as backup/restore, archive, and replication
- Networking components including routers and bandwidth optimization
- Cloud point of presence, endpoints, gateways, and appliances for moving or storing data
- Storage systems, including primary on-line storage, tape drives, disk libraries

With active data, including databases, unstructured files, and other documents, caution needs to be exercised not to cause performance bottlenecks and to maintain data integrity when introducing data footprint reduction techniques. In contrast to traditional ZIP or off-line, time-delayed compression approaches that require complete decompression of data before modification, on-line compression allows for reading from or writing to any location in a compressed file without full file decompression and the resulting application or time delay.

Real-time appliance or target-based compression capabilities are well suited for supporting on-line applications including databases, on-line transaction processing (OLTP), email, home directories, websites, and video streaming without consuming host server CPU or memory resources, or degrading storage system performance.

With lossless compression, compressed data is preserved and uncompressed exactly as it was originally saved, with no loss of data. Generally, lossless data compression is needed for digital data requiring an exact match or 100% data integrity of stored data. Some audio and video data can tolerate distortion in order to reduce the data footprint of the stored information, but digital data, particularly, documents, files, and databases, have zero tolerance for lost or missing data.

Real-time compression techniques using time-proven algorithms, such as Lempel-Ziv (LZ) as opposed to MD5 or other compute, "heavy-thinking" hashing techniques, provide a scalable balance of uncompromised performance and effective data footprint reduction. This means that changed data is compressed on the fly with no performance penalty while maintaining data integrity and equally for read operations.

Note that with the increase of CPU server processing performance along with multiple cores, server-based compression running in applications such as a database, email, file systems, or operating systems can be a viable option for some environments.

LZ is of variable length for a wide range of uses and thus is a popular lossless compression algorithm. LZ for compression involves a dictionary or map of how a file is compressed, which is used for restoring a file to its original form. The size of the dictionary can vary depending on the specific LZ-based algorithm implementation. The larger the file or data stream, combined with the amount of recurring data, including white spaces or blanks, results in a larger effective compression ratio and subsequent reduced data footprint benefit.

Real-time data compression allows the benefits associated with reducing footprints for backup data to be realized across a broader range of applications and storage scenarios. As an example, real-time compression of active and changing data for file serving as well as other high-performance applications allows more data to be read from or written to a storage system in a given amount of time. The net result is that storage systems combined with real-time compression can maximize the amount of data stored and processed (read or write) without performance penalties. Another variation is adaptive compression–based algorithms that can learn and adjust based on data patterns and performance trajectories.

Benefits of real-time compression for on-line active and high-performance DFR include:

- Single solution for different applications
- Improved effective storage performance
- Increased capacity for fast disk drives
- Enhanced data protection capabilities
- Extended useful life of existing resources

In some DFR implementations, a performance boost can occur as a result of the compression, because less data is being transferred or processed by the storage system and offsetting any latency in the compression solution. The storage system can react faster during both operations and take up less CPU utilization without causing the host application server to incur any performance penalties associated with host software-based compression.

This approach is well suited to environments and applications that require processing large amounts of unstructured data, improving their energy efficiency without sacrificing performance access to data. Some applicable usage examples include seismic and energy exploration, medical PACS images, simulation, entertainment and video processing of MP3 or MP4, MOV as well as JPEG and WAV files, collection and processing of telemetry or surveillance data, data mining, and targeted marketing.

Post-processing and deferred compression are often misunderstood as real-time data compression techniques. These processes dynamically decompress data when read, with modified data being recompressed at a later time (i.e., deferred). The benefit to this method is that static data that seldom changes can be reduced, freeing storage space, while allowing applications to read more data in a given timeframe.

The downside to this approach is that changing data, including file servers, email, office documents, design, and development, as well as database files, is written without the benefit of compression. The impact is that more space on the disk is required to write the data, with no performance improvement benefit during write operations plus the overhead of a subsequent read and rewrite operation when data is eventually recompressed.

Considerations:

- Fixed- or variable-sized de-dupe, adaptive compression algorithms
- Application-aware such as database and email compression
- Local and remote network bandwidth compression and optimization
- Formats and tools: CRAM, HDF5, BAM, bgzip, gzip, rar, split, 7zip, among others
- File splits of large items for parallel or faster network uploads
- Encryption and security controls, lossy vs. lossless compression
- Real-time vs. batch or other modes of operations
- Interfaces including CLI, script, batch, wizard, GUI, and API
- Self-extracting or extraction tool required

11.2.2.6. De-duplication and Single-Instance Storage

Figure 11.5 shows a simple example of how data footprint reduction occurs. Moving from left to right in the example, you can see how data changes and evolves, and how the data footprint impact can be reduced. On the left, no reduction is applied as "Hello" is saved (and protected) as "Hello", and then "Hello There" is saved and copied (protection) as "Hello There". In the middle, some reduction occurs, with "Hello" as well as "Hello There" being saved (and protected), but using points rather than a full copy. "Hello There" points to existing "Hello" as well as "There".

On the right in Figure 11.5 is more granular reduction, where unique individual letters are saved (and protected) rather than longer words. You can see on the bottom right how much space has been saved. Note that different DFR techniques implement various granularities from file, object, or database records, to blocks or other sizes (i.e., not just single characters).

Data de-duplication (de-dupe), also known as single-instance storage (SIS) is a technique or technology for normalizing (commonality factoring, differencing) that eliminates duplicate or recurring data. It is a more intelligent form of data compression.

De-dupe facilitates:

- Multiple versions and copies of with less space capacity overhead
- Faster local data protection backups and restores
- Boosting storage and network capacity and utilization
- Supporting remote office/branch office (ROBO or satellite office data protection

De-duplication normalizes the data being processed by eliminating recurring duplicate data that has already been seen and stored. Implementations vary, with some working on a file basis while others work in a fixed block, chunk, or byte boundary; still others can adjust to variable-size byte streams. For example, in a backup usage scenario, data that is being backed

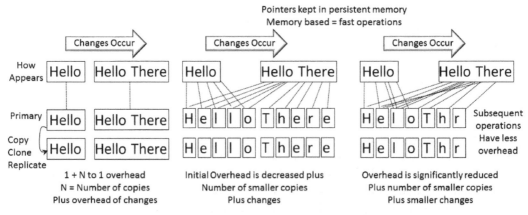

Figure 11.5   Simple data reduction example.

up is analyzed to see whether it has already been seen and stored. If the data has been previously stored, then an entry is made indicating where the data is stored and the new copy is discarded (de-duped).

If new data is seen, it is stored and a reference pointer is made along with an entry in the de-dupe database (otherwise known as a dictionary, index, repository, or knowledge base). How the incoming data stream is analyzed and what size or amount of data is compared varies by implementations, but most use some form of the computed hash value of data being analyzed to enable rapid lookup of known data.

This is where the intelligence comes in concerning de-dupe vs. traditional algorithms such as LZ, because the more data that can be seen over time and in different contexts, the more reduction can occur, as is the case with global de-dupe (discussed a bit later in this chapter).

While data compression essentially performs a coarse elimination of recurring data patterns, data de-duplication works on a more granular level and requires more processing power and intelligence. Data de-dupe builds on traditional coarse compression to reduce the data footprint, by adding intelligence, leveraging processing power, and awareness of what data has been seen in the past.

Essentially, de-dupe, regardless of where it is implemented, trades time (thinking and looking at data access patterns or history) for space (capacity) reduction. Some de-dupe–enabled solutions also combine basic data compression with de-dupe to further reduce data footprint requirements.

Current industry and market focus on de-dupe is targeted toward protection copies, including backups, snapshots, replicas, archives, and other copy data management (CDM)–related items. This is not to say that there are no other opportunities; some vendors are finding success with VM or VDI, where there are additional duplicates. Focus is also on ratios, where the need is to expand rates to enable transition to larger, more performance-sensitive environments that are still dominated by tape.

11.2.2.6.1. De-dupe and SIS Fundamentals

A common technique to check for duplicate data is to use a hash key lookup based on a checksum or chunks of data being seen. Hash keys, computed based on some amount of data being viewed, are compared to stored keys in a database, dictionary, knowledge base, or index of previously stored or known data. When there is a match of a hash of incoming data to known existing data, there is duplicate data. If there is a miss, then there is new data to be stored and a new hash to be added to the index or knowledge base. SHA-1 (Secure Hash Algorithm-1) has been used as an algorithm for creating a comparison or lookup hash in many de-dupe solutions. Other hashing algorithms include SHA-2, with SHA-3 in development, along with MD5.

The importance of algorithms being enhanced with more bits is to produce a unique hash key to span larger amounts of data without collisions, to maintain data integrity while boosting performance. For example, SHA-1 produces a 160-bit (20-byte) hash, which is adequate for many deployments; however, with larger amounts of storage and expanding data footprints, larger hash keys are needed to avoid collisions.

De-dupe has more intelligence than traditional compression because of extent of the dictionary, index, or knowledge base of what has been seen combined with the algorithms for computing hash values, along with available compute processing capability. The challenge with de-dupe, and why it trades time for space capacity savings, is that time is needed to compute the hash key and look it up to determine whether it is unique.

> *Tip:* The ingestion rate for de-dupe (how fast data can be processed) depends on the specific algorithms, the size of the available dictionary, and the pool of reduced data, along with available processing performance. As such, de-dupe is typically not well suited for low-latency, time-sensitive applications, including databases or other actively changing storage use scenarios.

Some data de-dupe solutions boast spectacular ratios of data reduction given specific scenarios, such as backup of repetitive and similar files, while providing little value over a broader range of applications. This is in contrast to traditional data compression approaches, which provide lower yet more predictable and consistent data reduction ratios over more types of data and applications, including on-line and primary storage scenarios. For example, in environments where there are few or no common or repetitive data files, data de-duplication will have little to no impact, while data compression generally will yield some amount of DFR across almost all types of data.

Figure 11.6 shows an example of how de-dupe DFR works as data is changed. On the left is an example of a file that is initially processed (ingested) by a generic de-dupe engine (actual reduction and compaction will vary with the specific vendor product). Depending on the specific implementation, the initial savings may be minimal, but after a copy of the first file is made, then some changes made to it and saved, there begin to be reduction benefits.

As additional copies and changes are made, moving from left to right as shown in Figure 11.7, more duplicate data is seen, additional reduction benefits occur, and the result is a higher reduction ratio.

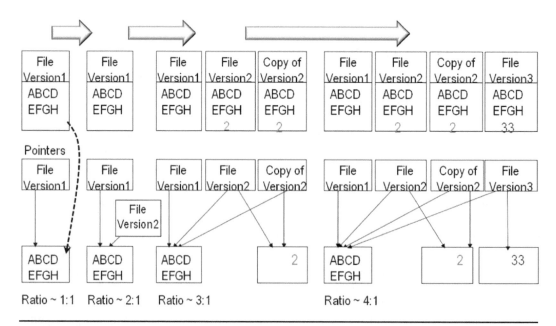

Figure 11.6   Example of de-dupe benefit on files with similar data.

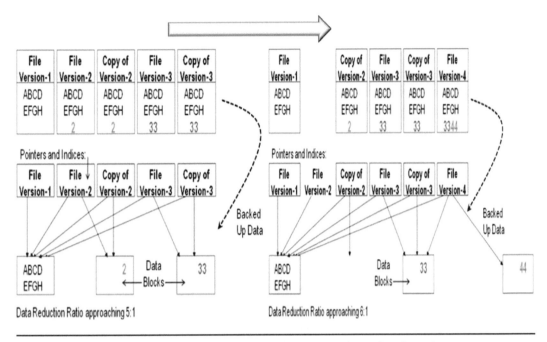

Figure 11.7 Continuation of de-dupe example removing redundant data.

Figure 11.8 builds on Figure 11.7 in that the data reduction or de-dupe ratio continues to increase over time as additional copies or duplicate data are seen by the de-dupe engine. As more copies of the data are seen, such as with daily backups, the potential for recurring data increases, and thus an opportunity for a higher de-dupe reduction ratio appears. This capability has

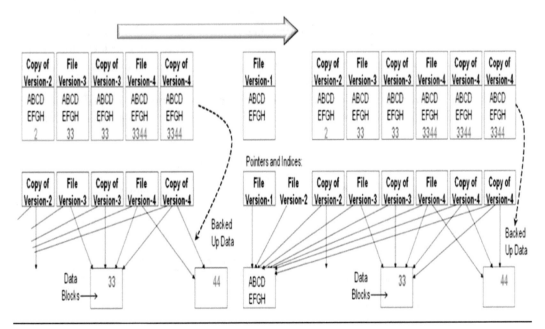

Figure 11.8 Continuation from Figure 11.7 of de-dupe data reduction.

contributed to why de-dupe has been initially targeted for backup/restore, given the potential for a high degree of duplicate data occurring over time and the resulting space-saving benefit.

Figure 11.8 extends the example by showing how the DFR benefit continues to improve as more copies of the same or similar data are seen. Depending on the specific de-dupe implementation, the reduction benefit may be greater or less, depending on factors including the specific algorithms, local or global view of data, type of data, file or block based, and others.

11.2.2.6.2. How and Where De-dupe Is Implemented

Where is the best place for doing de-dupe, and what is the best method? Unless you have a preference or requirement for a particular approach or product, the answer is "It depends." De-dupe as a functionality can be found implemented in:

- Servers, storage (primary and secondary), and I/O networking technology
- Hypervisors, operating systems, volume managers, file systems
- Software-defined storage, storage systems and appliances, cloud services
- Built-in and integrated without functions, optional layered add-on functionality
- Backup or other data protection software
- Agents, appliances, and gateways (physical and virtual)

In addition to when de-dupe is done (immediate or deferred), other variations include where it is implemented, such as in hardware storage systems or appliances as a target destination (Table 11.6), or in software as part of a source backup or data protection tool.

Different de-dupe–enabled solutions or software vary in feature functionality. Some provide granularity regarding how reduction is done, such as at file or object level, as well as in entire storage pools, or individual volumes. Data may be reduced on a file or block (chunk or shard)–sized (fixed or variable) basis. The dictionary where index and metadata about how data is reduced (i.e., where fingerprints or reduction hashes are stored) can be local and unique to a solution, or global and shared. Some solutions enable multiple dictionaries (tenancy isolation) for security purposes, for example, different dictionaries for various storage pools to prevent the risk of possible data bleed from one user or group to another. Global dictionaries enable more information to be stored and shared, helping to further reduce duplicate data.

Table 11.6. Where to Perform De-dupe

	Source	Target
Drawback	Disruptive to existing BC/DR and backup/restore tools. New software or upgrades may be required. CPU and memory are consumed on server where de-dupe is done.	Extra functionality added to VTL or storage system or via an appliance or gateway. Does not address reducing data footprint for data that must be moved over networks on a local or remote basis.
Benefit	Performs reduction closer to data source, less data moved, making more efficient use of networking resources for local as well as for sending from ROBO or sending to cloud-based target destinations.	Plug-and-play with existing BC/DR as well as backup/restore software and skill sets. When combined with global capabilities, a larger knowledge base can be leveraged for additional DFR benefits.

General characteristics of de-dupe include:

- It may be always on, selectable, policy-based, time-deferred, or post-processing.
- Solutions tend to be software-based (or positioned that way).
- Most solutions are "tin-wrapped" software (i.e., appliance/gateway).
- Data reduction occurs at or near data target/destination.
- Flexibility with existing backup/data protection software.
- Enables robust scaling options for larger environments.
- Multiple "ingest" streams from sources enable high reduction ratios.

Considerations about source- and target-based de-dupe (see Figure 11.9) include:

- The source can be standard backup/recovery software without de-dupe.
- The source can be enhanced backup/recovery software with de-dupe.
- The target can be a backup server that in turn de-dupes to the storage target.
- The target can be a backup device, storage system, or cloud service with de-dupe.

Note that in addition to source and target, there is also hybrid de-dupe and compression. For example, a client or agent such as Dell EMC DataDomain Boost (DDboost) can be installed on a server (source) to improve performance effectiveness and capacity efficiency working with the target storage appliance.

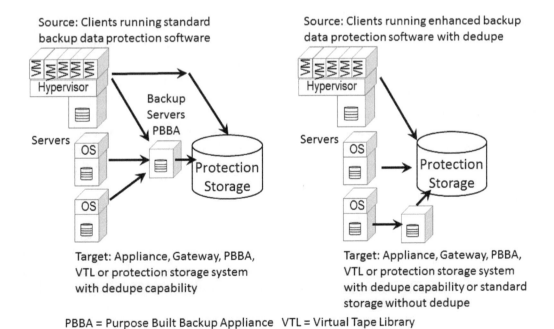

Figure 11.9   Source- and target-based de-dupe.

Common characteristics of source de-dupe include:

- Data reduction occurs via software, at or near the data source.
- Network (LAN, MAN, WAN) activity is reduced.
- Some backup software de-dupes at the client, some at the backup server.
- De-dupe is independent of the target device.
- Source de-dupe may not achieve the same reduction ratios as target de-dupe.

Common characteristics of target- or destination-based de-dupe include:

- Supports immediate, deferred, or policy mode to adapt to different needs
- Can be a backup server or a storage server, appliance, or gateway
- Optimized for backup/restore, archiving, or general storage needs
- Block, file, and object access along with tape emulation (VTL)
- Optional support for path to tape, including supporting Open Storage Technology (OST)

Some solutions have been optimized for single streams of data, others for dual or multiple streams; some for parallel, others for multithreading to support high concurrency; some are good at ingesting data and others at re-inflating it, while still others do immediate or post-processing or a mix. Some are optimized for scaling, with multiple nodes sharing a global memory, while others have multiple nodes yet are independent.

Some solutions have a single-target, others support complex topologies. De-dupe can be implemented in many different ways and locations, even in non-storage devices such as networking applications or protocol optimizers. Table 11.7 shows various de-dupe modes of operation.

As with immediate mode, implementations vary by different vendors' products, so "your mileage may vary" based on the type of data and usage. Another thing to keep in mind with post-processing mode is the restoration rate or performance for a single file as well as for larger amounts of data, including an entire volume or file system.

The amount of time delay before reduction occurs also varies by implementation. Some solutions wait until all data has been placed on the target destination, others start the post-processing almost immediately, after a short time interval, or after a certain amount of data is captured.

Local de-dupe is so named because the dictionary covers only what has been seen by a given de-dupe engine instance. A de-dupe engine instance might be either source- or target-based, functioning in either immediate or deferred mode. Since the de-dupe engine (the algorithm or software functionality) sees only what it processes, the view is localized to that instance. For example, on the left in Figure 11.10, three sources are backing up to a de-dupe–enabled device, a target storage system. This means that data can be reduced based only on the three streams or backup sources that are seen. As another example, if an environment has six servers being backed up, the de-dupe engine (on the left in Figure 11.10) does not know about the servers and their data on the right, and vice versa. If de-dupe is local, then for a de-dupe engine that has multiple nodes, each with a separate instance to boost scaling of performance, availability, and capacity, each instance or engine has its own knowledge base and is limited to only what it has seen.

Since de-dupe reduction ratios or benefits are a product of what data has been seen previously, a localized dictionary or knowledge base can limit the full DFR capabilities for a given environment, and the DFR is limited to the localized environment. This is not all that

Table 11.7. De-dupe Modes to Meet Different Service-Level Objectives

	Characteristics	Watch for	Benefit Result
Source de-dupe	Data is reduced as close as possible to source where it is created, via operating system or other software functionality.	Implementations that slow applications down, granularity such as file level, or lack of a global dictionary.	Data reduced upstream minimizes I/O and network traffic flowing downstream. Local dictionary provides security.
Target de-dupe	Data is reduced where it is stored, in target or cloud storage service.	Immediate vs. deferred processing, performance impact, global dictionary.	Sees more data from multiple sources, which can boost DFR impact.
Local de-dupe	Relatively easy to implement, local to node or system	How dictionary is made resilient with copies or repair/rebuild functions	Data is kept local, no need for coordinating with other nodes.
Global de-dupe	Dictionary or index data is shared among nodes to boost DFR impact.	Security concerns about data possibly being accessible from other users or groups.	Increased DFR impact by seeing more data from sources.
Immediate in-line, in-band	Data is reduced in real-time at point of ingestion as part of basic I/O handling, or in-line on-the-fly in the data path by a separate process.	Implementations that slow down application performance (queues or larger response times, decrease in data rates) as data is reduced. Watch for reduced read performance as data is re-inflated if solution is optimized for writes.	Data is reduced as or before it is stored to realize immediate space capacity benefit. Can benefit other functions such as snapshots.
Deferred	Data is reduced at some later point in time via time-deferred or postprocessing. Deferred time can be seconds, minutes, hours, or days.	Implementations needing extra space capacity for buffering data until reduction, performance impact when reduction finally occurs.	Eventual DFR and space-saving efficiency without impacting running time–sensitive applications. Reclaim space when convenient.

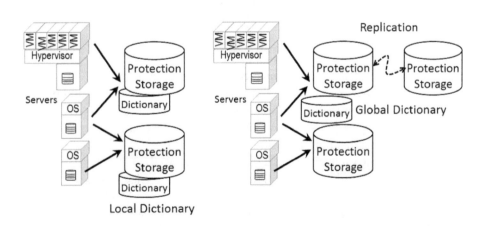

Figure 11.10   Local de-dupe (left) and global de-dupe (right).

different from basic compression algorithms, whose DFR benefit or ratio is limited to what they see. Figure 11.10 shows a local de-dupe dictionary (i.e., a dictionary local to that system) on the left and a global de-dupe dictionary (a dictionary shared across systems) on the right.

By having a shared or global database, knowledge base, or index where the results of what has been seen by different de-dupe engine instances can be compared, additional redundancies can be eliminated. For example, if different servers as shown in Figure 11.10 are configured and load-balanced across a pair of de-dupe engines (applications, nodes, backup nodes, or other targets), and they are backing up similar files or data that has duplicate occurrences across the nodes, de-duplication can be enhanced.

In addition to where (source or target), how (hardware or software, local or global), and when (immediate or deferred), other de-dupe considerations include how the solution works with other DFR techniques, including compression, granularity, or flexibility in data comparison size, support for replication, as well as retiring of data, including path to tape.

Another consideration is the different topologies that various solutions can support. One variation is many sources sending data to a single target destination (many to one). Another has one target destination that replicates to one or many other targets in the same or different locations, including those from managed service or cloud providers.

When looking at de-duplication solutions, determine whether the solution is designed to scale regarding performance, capacity, and availability, along with how the restoration of data will be impacted by scaling for growth. Other items to consider include how data is re-duplicated (i.e., re-inflated, re-hydrated, expanded), such as in real time using in-line or some form of time-delayed post-processing, and the ability to select the mode of operation.

For example, a de-dupe solution may be able to process data at a specific ingest rate in-line until a certain threshold is hit, and then processing reverts to post-processing so as not to cause a performance degradation to the application writing data to the de-dupe solution. The downside of post-processing is that more storage is needed as a buffer. It can, however, also enable solutions to scale without becoming a bottleneck during data ingestion.

Tip: Another consideration is fixed- vs. variable-length de-dupe solutions that can adjust to different types of data and access. For example, backups and data protection tend to work with larger amounts and allocations of data that may be better suited for variable length de-dupe. Higher performance access of smaller amounts of data may lend itself better to fixed-length approaches.

De-dupe considerations include:

- How and where the dictionary is protected from loss or damage
- What data safeguards exists for actual data blocks as well as metadata and indices
- If tools are provided for recovery or rebuilding damaged dictionary or reduced data
- Data integrity checks, repairs, and safeguards against silent data corruption
- Source- or target-based, application and software plug-ins
- Immediate real-time vs. post-processing and time-deferred
- Performance impact or delays to applications storing (or retrieving) data
- Granularity of data being de-duped (volume, file system, file, block)
- Ability to selectively enable de-dupe on a volume or file system basis
- Interoperability with snapshots, replication, and encryption
- Management configuration, monitoring, insight and awareness tools

11.2.2.7. Thin and Thick Provisioning

Thin provisioning is a technique that provides a benefit of maximizing usable space capacity. The benefit of thin provisioning is to simplify storage space allocation up front while not having to commit all of your actual storage capacity. This is handy for applications that want a given amount of space up front yet have low actual usage (i.e., sparse rather than dense, such as databases, file systems, folders, or other storage spaces). Storage thin provisioning is implemented in storage systems, appliances, software-defined storage, cloud, virtual, hypervisors, operating systems, and volume managers.

> *Tip:* Key to leveraging thin provisioning is similar to what airlines do with seats on a flight (they oversell): They utilize insight, awareness, and good analytics to know or make informed decisions. With knowledge of access and usage patterns, you can eliminate or reduce the chance of costly overbooking (the oversold or out-of-storage situation) or underbooking (no-shows or unused capacity). And in addition to knowing how applications use storage space, having insight into performance activity and access is also important.

Figure 11.11 shows (A) a 20-TB volume allocated from a storage pool (which could be a storage system, appliance, software-defined or virtual, HDD or SSD). The 20-TB volume (B) is allocated to (C) as 1 TB (thick provisioned), (D) and (F) as 5 TB thin, (E) as 15 TB thin, and (G) 5 TB thin, resulting in in 31 TB provisioned.

The top right of Figure 11.11 shows (G) an additional layer of thin provisioning being implemented via software such as a volume manager, file system, operating system, hypervisor, or software-defined storage. The 5 TB (G) are further carved up and allocated as (G1) 2 TB thick with dense storage needs; (G2) an unallocated thin 2-TB volume, and (G3) a thin 2-TB

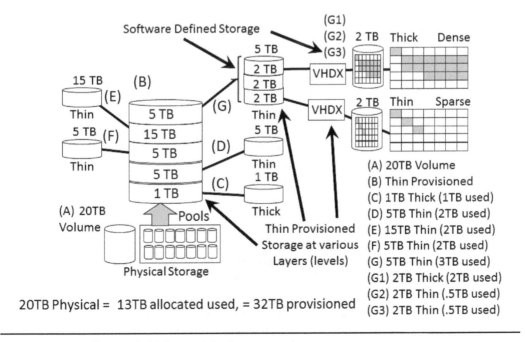

Figure 11.11 Thin and thick provisioning example.

volume. Note that (G3) has sparse allocated and used space, which places less actual storage demand on the underlying storage resources than if it were more dense.

In Figure 11.11, note that 20 TB of physical storage (the volume allocated from the underlying storage pool) gets provisioned as 31 TB, but only 13 TB are allocated and being used. If some of the other volumes increased their usage by 1 TB each, you would be close to being in an oversubscribed or overbooked, oversold scenario.

> *Tip:* Keep in mind that thin provisioning boosts storage space capacity effective usage but does not by itself boost performance. However, if you are using fast NVM and SSD technologies to boost performance, thin provisioning can in some situations improve storage space effective usage without introducing additional bottlenecks. On the other hand, similar to server consolidation, aggregation can cause aggravation if you simply combine space capacity without also increasing performance resources.

Another consideration with thin vs. thick provisioning is how space will be allocated—for example, thick lazy (space gets allocated and formatted in the background over time) or eager and performing the allocation up front. Depending on the implementation, there can be a performance impact, as some solutions will allocate (and format underlying storage) up front, making subsequent access faster.

Other storage thin provisioning implementations can make the storage available immediately, but full performance may not be realized until background processing is complete. Ask the vendor or provider how their allocation schemes work, along with any associated performance impact. Additional items to look into and consider include hypervisor integration such as VMware VAAI as well as Microsoft Windows along with Hyper-V.

Besides storage thin provisioning, similar benefits and techniques are used with memory, such as dynamic allocation along with other virtual memory management approaches. With memory, a server may be allocated 16 GB of DRAM yet be only using a fraction of that in a shared virtual server environment. Virtual memory with persistent backing storage can be used to boost effective memory space capacity for applications.

Another variation is networking ports, where multiple physical ports may share the aggregate of bandwidth (i.e., host-optimized or oversubscribed) to the core switch or ASIC, vs. each port having its full speed. For example, for simplified connectivity, a switch or device may have 4 × 10 GbE ports that share a common 10 GbE connection to a core switch or internal via ASIC or FPGA to the switch core. If all four ports are busy, they will not be able to operate at full speed because the link is limited to 10 GbE. On the other hand, if each port has its 10 GbE link to the core (switch, FPGA, or ASIC), then each of the four can run at 10 GbE (faster for full duplex) or 40 GbE of line-rate performance. However, a device or switch might be deliberately oversubscribed to reduce costs.

Considerations with thin and thick provisioning include:

- Convert dynamically (no export or import) from thin to thick, thick to thin
- Thick eager or lazy allocation with no performance impact
- Awareness and analysis tools for effective configuration
- Integration with hypervisors, operating systems, and other software
- Integration with storage system, appliances, and management tools
- Memory and network provisioning and performance allocation

11.2.2.8. Other DFR Techniques

In addition to those techniques and technologies already mentioned, another way to reduce your data footprint impact, at least from a storage space capacity standpoint, is to revisit RAID configuration. Different RAID levels (see Chapter 9) will have an impact on storage, I/O networking, and CPU capacity for normal running, as well as during rebuild and reconstruction.

Some RAID levels, such as mirroring, replicate data to two or more drives, systems, servers, or locations using hardware or software implementations. Other forms of RAID include traditional single-parity RAID 4 and RAID 5, along with dual-parity RAID 6. Erasure codes are advanced parity derived from Reed-Solomon codes (RAID 2) that are well suited for storing infrequently updated data at a low cost overhead. Note however that erasure codes incur write performance penalties, with different solution implementations varying in how they perform.

General notes and comments regarding RAID include the following.

- Larger RAID sets can enable more performance and lower availability overhead.
- Some solutions force RAID sets to a particular shelf or drive enclosure rack.
- Match RAID level performance and availability to the type of data, active or inactive.
- Boost performance with faster drivers; boost capacity with large-capacity drives.
- Drive rebuild times will be impacted by drive size for large-capacity SAS and SATA.
- Balance exposure risk during a drive rebuild with appropriate RAID level.

Besides evaluating different RAID options, along with various width or size of drive groups to optimize space capacity, other considerations include:

- Meet with end users, customers, and records retention personnel regarding data value.
- Generate awareness of the value of data and the importance of protecting it.
- Also generate awareness of the costs of storing data that has no value.
- Besides cost and value, along with protection, keep access of that data in mind.
- Keep your organization and business needs in mind when it comes to capacity.

Access to data by applications and their servers is important in that you can have a large pool, pond, or harbor of data, but if there is a narrow stream or path to it, that will become a bottleneck to productivity. Likewise, that narrow access path can be a barrier for accessing those data pools and ponds for maintenance, security, and virus checks as well as data protection.

Tip: Use a page out of the cloud storage and managed service providers' playbook to generate awareness of costs. If needed, issue invoices and charge-back reports as well as "show-back" reports to educate users about what they are using along with the associated costs. You might also find that cloud services can provide comparable service levels at a better cost value. On the other hand, you might also find that your costs are a better value when compared on apples-to-apples service level and other attributes.

With your customers, end users, and records retention personnel, review what data to keep. As part of those meetings, determine what data has value, no value, or unknown value. For data with no value, establish or follow policies for deletion of that information. Recall Chapters 1 and 2 data value discussion.

Data with known value can be looked at and determined whether it can be moved to a different tier, or its protection and other policies adjusted as needed. It might even be discovered that some data thought to have value no longer has value and can be disposed of. Unknown value data should be looked at periodically to determine whether it can be promoted to known value, demoted to no value and removed, or retained with unknown value until needed.

11.3. What's in Your Toolbox?

In addition to the tools mentioned in Chapters 8 and 9, additional tools include Jam Treesize, Quest, Turbonomic, Solarwinds, Spacesniffer, CRAM, BAM, SAR, Zip, 7zip, rar, and split, among others.

11.4. Common Questions and Tips

What is a de-dupe ratio vs. rate, and how does it compare to compression? Ratio is the metric to compare how much data is reduced by from its original size using compression and de-dupe. For example, a ratio might be 2:1 or to half of its original size, such as with some magnetic tape drives (some can also do 2.5:1) in a streaming mode. A 4:1 reduction ratio indicates that data is reduced to a fourth of its original size, and so forth. Rate refers to the speed and amount of data that can be ingested and reduced (or re-inflated) using de-dupe or compression. For example, the rate might be in megabytes or gigabits per second of data processed.

Note that some vendors get creative, using terms such as "effective" as opposed to actual data rates. What is important is how much actual data gets moved and reduced in a given amount of time if performance is an objective. As for the difference between compression and de-dupe ratios, de-dupe can have higher ratios when it is able to look at more and larger amounts of data, but performance should also be considered. Keep in mind that not all data de-dupes at the same amount or ratio.

What is effective vs. actual capacity? Effective capacity is what you are able to use in terms of storage capacity, memory, or other resources. For example, you may only have 16 GB of RAM, but with virtual memory your applications may have the impact benefit of 18 GB or more, depending on what technology is used. Similarly, using various data footprint reduction techniques and space optimizations, your application may have the effective benefit of 100 GB of storage space when only 50 GB of physical space exists. This latter example assumes that data footprint reduction such as compression, de-dupe, thin provisioning, and other techniques are able to cut the footprint size of the data in half, although of course there may be performance considerations.

Are snapshots and de-dupe the same or similar? On the surface, they may appear and address things in similar ways using pointers and other techniques to reduce data footprint impact, along with various approaches and objectives. Having said that, there is a trend of combining low-overhead de-dupe reduction at ingest (when data being stored) and in turn using metadata points to speed up a subsequent copy, clone, replicate, snap, or other operation.

How important is low cloud cost vs. higher all-inclusive cost, or cost to restore? Like many things IT and non-IT–related, look at the total cost for a given scenario. You might get a good deal on lower up-front costs, but you need to understand what additional fees apply when you need to access or restore data. For example, are there capacity charges along with network bandwidth fees or API calls (gets, lists), among others? Knowing your costs can help you decide whether bundling feeds or going à la carte is better value. Also keep in mind that if the cost per cloud capacity is low, you can use that for savings and budget reduction, or for enabling more durable copies, resiliency, and general availability.

Why not use de-dupe everywhere? Depending on your solution and its implementation, you may very well be able to have an always-on de-dupe without any performance impact, perhaps even for transactional solutions. Keep in mind, however, that not all data de-dupes equally, particular with different implementations. On the other hand, many traditional de-dupe implementations and solutions are not focused on performance effectiveness including for transactional activity, but rather on space efficiency and reduction ratios. Look at and test solutions with data services such as de-dupe enabled to see what the impact is on your running applications.

Why is data footprint reduction important for clouds or virtual environments? There are two main dimensions to why DFR is important: One is being able to move more data in a timely and affordable manner to public or cloud resources; the other is that clouds can serve as a target or medium for moving archived or other data to complement and extend the use of on-site resources. So, on one hand, it enables data to fit into a cloud, and on the other, it is about using clouds to off-load or enable existing environments to be cleaned up.

Storage is getting cheaper; why not buy more? It is true that the cost per gigabyte or terabyte continues to decline and that energy efficiencies are also improving. However, there are other costs involved in managing, protecting, and securing data. In addition to those costs, there is the issue of how many gigabytes or terabytes can be effectively managed per person. A data footprint reduction strategy enables more gigabytes, terabytes, or petabytes to be effectively managed per person.

What's the connection between DFR and data protection? By reducing your primary data footprint impact, you can also lower your data protection overhead. Likewise, by reducing your data protection overhead while enhancing resiliency, durability, and security, you can save more data and stretch your budget to support growth.

Are there any regulatory or legal ramifications pertaining to DFR and de-dupe? Generally speaking, there are different regulations pertaining to preserving the uniqueness of data and ability to restore it to its native form. For example, consider three data records, one each for Jon Doe of Stillwater, John Dowe Stillwater, and Jonathan Doe of Stillwater. These may be unique people or they might just be different misspellings. For some strict regulatory levels those three records, even if they refer to the same person, may need to be kept separate, and yet might be able to be linked.

Are there business or industry standards and compliance requirements that impact a data footprint reduction project? Yes, there can be different requirements for compliance for various industries. Learn about your specific organization and industry to find out what requirements are applicable.

What is the difference between efficiency and effectiveness? Efficiency is often used to infer capacity optimization or improvement on utilization, including reducing space needs along with lower data footprint impact on servers, storage, networks, and related items. Effectiveness refers to productivity, performance, and the benefit boost of being able to do more work or the same work in less time, or data moved, among other variations.

Doesn't the 4 3 2 1 data protection rule cause an increase in your data footprint impact along with extra costs? If done using traditional techniques, tools, technologies, and best practices, the answer would be yes. However, if using DFR techniques, technologies, new tradecraft skills, and experiences, best practices that have been covered here as well as in companion Chapters 8, 9, and 10, you can increase your availability and resiliency with 4 3 2 1 while reducing your data footprint overhead impact and costs.

Granted, everything is not the same with different applications and data, which will determine how much data can be reduced. Likewise, different application PACE requirements along with SLO will also have an impact. However, by rethinking when, where, why, and how data protection is implemented, leveraging new (and old) technologies in new ways, implementing a broad DFR strategy (not just compress or not just de-dupe or not just thin provision, all the above), you can enhance your recoverability via 4 3 2 1 while keeping costs and capacity overhead in check.

11.5. Learning Experience

Do you know what data infrastructure capacity resources you have, as well as how they are being used (or allocated and not used)? Do you know what your data footprint impact is, why it is what it currently is, as well as opportunities to improve it? What are your capacity forecast plans for the server (CPU, memory, I/O, software licenses), I/O networking, and storage (capacity and performance), along with associated data protection? Using the tools, technologies, tips, and techniques along with tradecraft from this as well as Chapters 8 and 9, can you address or start to find answers to the above among other questions?

11.6. Chapter Summary

This wraps up the fourth of four chapters on data services that enable data infrastructure functionality. Organizations of all shapes and sizes are encountering some amount of growing data footprint impact that needs to be addressed, either now or in the near future. Given that different applications and types of data along with associated storage mediums or tiers have various performance, availability, capacity, energy, and economic (PACE) characteristics, multiple data footprint reduction tools or techniques are needed.

For some applications, reduction ratios are an important focus, so the need is for tools or modes of operations that achieve the desired results. For other applications, the focus is on performance with some data reduction benefit, so tools are optimized for performance first and reduction secondarily. In response, vendors will expand their current capabilities and techniques to meet changing needs and criteria. Vendors with multiple DFR tools will also do better than those with only a single function or focused tool.

What this means is that the focus of data footprint reduction is expanding beyond just de-duplication for backup or other early scenarios, so have multiple DFR tools and techniques in your toolbox and tradecraft skills to know when, where, why, and how to use them.

General action items include:

- Align the applicable form of DFR to the application PACE characteristics.
- Develop a DFR strategy for on-line and off-line data.
- Different applications and data will need various tools and techniques.
- Different data services are implemented in various ways as well as locations.
- Everything is not the same; there are different data services for different functionalities.
- Have multiple tools in your DFR toolbox.
- Keep rates (performance and productivity effectiveness) in perspective (including time).
- Keep ratios (utilizations and efficiency) in perspective (i.e., including space).
- Look at data footprint reduction in the scope of your entire environment.
- Small percentage reductions on a large scale have big benefits.
- The breadth and extensiveness of data services implementations vary.

Bottom line: Capacity and data footprint reduction enable more data with varying value to be stored in a cost-efficient manner to support growth. Read about other data services including access, performance, and general management (Chapter 8), and availability (data protection) as discussed in Chapters 9 and 10.

Chapter 12

Storage Systems and Solutions (Products and Cloud)

What's the best storage system or solution? It depends. . . .

What You Will Learn in This Chapter

- What is a storage system, solution, appliance, or service
- There are many different types of storage systems, and solutions
- Storage for legacy, software-defined, cloud, virtual, and containers
- Various architectures, and packaging for different application needs
- How server, storage, and I/O components get combined with data services
- What to use as well as what to look at for various application scenarios
- Software-defined and wrapped vs. tin-wrapped, and appliance

In this chapter, we put together what we have covered in previous chapters as well as lay the framework for what's coming next. We look at storage systems and solutions spanning physical on-site, virtual, and software-defined as well as cloud services. Common storage systems include legacy high-end to low-end or entry-level arrays, appliances, and gateways. There are also software-defined, virtual, tin-wrapped, container, and cloud-based solutions. These include primary, secondary, performance, bulk, and data protection storage systems. Key themes, buzzwords, and trends addressed in this chapter include architecture, software-defined, tin-wrapped, bulk, performance, all-flash arrays (AFA), hybrid, SSD, virtualization, and cloud, among others.

12.1. Getting Started

Data storage is often taken for granted while not being well understood, particularly when there is no more room to save files, photos. or other data. Cost is also a concern, whether it is how much you have to pay to park your files, videos, or other data, or to have it backed up or to purchase more storage space. Some context for this chapter is that storage system can refer to traditional storage systems, appliances, as well as cloud services, and also software-defined storage solutions, along with functionality.

> *Tip:* Keep in mind that some people refer to databases, repositories, and file systems as storage systems, so context is important. In other words, storage system also means storage functionality.

Figure 12.1 combines themes that have been discussed so far in previous chapters. These include how algorithms (logical, rules, processes) plus data structures (bits, bytes, blocks, and blobs) get defined to support applications and data infrastructures. Also shown are how hardware needs software, and software needs hardware—at least somewhere in the solution or service stack.

Remember hardware, software, and services get defined using software tools to implement data infrastructure configured for different services and applications leveraging underlying resources. Underlying resources combined with data services and functionality enable the data infrastructure (local or cloud, physical, virtual, or container) to protect, preserve, and serve information.

Figure 12.1 shows various types of storage system solutions, implementations, and packaging approaches, along with data services functionalities. At the heart of the storage systems shown in Figure 12.1 is storage software (lower center) that defines the functionality, data services, PACE attributes, and other features of a solution. These include standalone storage systems, networked, and shared, block DAS, and SAN, file NAS along with object, and API accessed. Also shown in Figure 12.1 are storage functionalities implemented via software as part of a server (physical, virtual, container, or cloud) along with aggregated including HCI, and CiB (cluster-in-a-box) as well as disaggregated including CI.

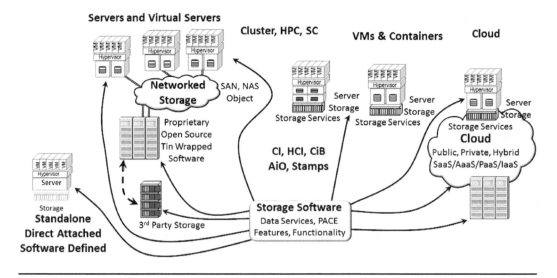

Figure 12.1   Where and how storage system software gets deployed.

The top of Figure 12.1 shows different types of storage systems and solutions (or services) packaged, presented, accessed, and consumed in various ways. The center of Figure 12.1 shows how various storage, and data services, features, functionalities, and management tools (aka the software), configure and manage the underlying resources (hardware, software, services) that get defined for use.

Common to all of the examples shown in Figure 12.1 is some combination of data services and storage functionality along with management defined in software (bottom center) along with storage resources. The resulting storage system architecture, data services, and functionality implemented in software (or optional hardware) gets packaged in various ways.

Data services (Chapters 8, 9, 10, and 11) are the feature functionalities that enhance underlying hardware, software, or other services used for implementing data infrastructure services. Various storage systems have different types of data services along with varying feature functionality extensibility (how extensive those features are). Different solutions vary by what is included, or as extra cost items.

> *Tip:* Ask vendors or service providers what is included (services, SLAs, capacities, quotas, or limits) as part of a solution's up-front costs. Also understand what the recurring costs will be, along with any additional fees for services to be used. In addition, ask what options and additional hardware or software are needed (or available) to use or enhance a service, along with their associated costs.

Different storage system implementations have various architectures as well as packaging, including where data services or other functionalities are located. For example, some storage systems focus on all data services in the solution; others rely on a hybrid combination of server software and hardware. Other solutions have all of their data services and storage feature functionalities implemented via software in a hypervisor, operating system, or layered software on a server, or as part of a cloud service.

> *Tip:* Which is the best type of storage system, architecture, implementation, and packaging approach for you? It depends on your application PACE needs along with other environment considerations, including what existing hardware and software resources you have, as well as preferences.

In addition to the feature functionality, what is also defined as part of the storage software architecture is its personality. The storage system's personality is how it is abstracted and optimized, and how it acts. For example, protection storage is optimized to support the needs of data protection applications or functionality. High-performance transactional storage has the personality to implement I/O optimization, quality of service (QoS), and low-latency access, among other capabilities. The key point is that a storage system (solution, service, or software) personality is more than an interface, protocol, or access method and underlying media (HDD, SSD, or cloud resource).

The storage system software can be tightly bundled as part of a CI, HCI, or CiB solution, or tin-wrapped as a storage system appliance or legacy array. Also, the storage software can be deployed as virtual storage appliances or arrays, as well as via firmware in storage systems, or as cloud services, among others approaches.

Figure 12.2 helps to clarify some context by tying together items we have covered so far, including bits, bytes, blocks, and blobs that get defined and wrapped (Remember the Russian

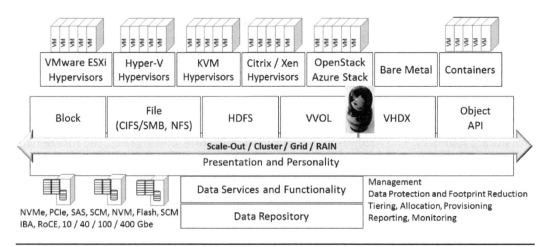

Figure 12.2   Storage system functionalities and components.

doll?) in layers as part of software stacks. In Figure 12.2 are shown, from the bottom up, hardware (servers, storage devices, I/O networks) and software (hypervisors, operating systems, storage, and other data infrastructure tools) that get defined as data infrastructure to support various applications. In the middle is storage functionality, and features including various data services for access, performance, availability, capacity, and management to meet different application PACE needs.

Figure 12.3 shows examples of different types of storage system implementation, architectures, packaging, access, and deployment. Everything is not the same; while there are general-purpose storage solutions, there are different solutions for various environments. In Figure 12.3 at top left are shown combined CI, HCI, and CiB solutions along with disaggregated clusters or standalone servers, and storage. The aggregated (HCI) solutions scale server compute

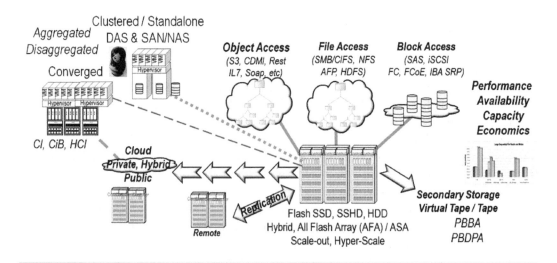

Figure 12.3   Various types of storage systems, implementations, and access.

and storage together. Disaggregated servers and storage scale independent of each other. Also shown are dedicated as well as shared direct access storage, along with SAN, NAS, and object shared networked storage, local and remote.

Primary and general-purpose storage are shown in Figure 12.3 along with secondary, bulk, and protection storage, local and cloud-based. Also shown are tin-wrapped (software bundled with hardware) along with traditional storage systems. Also shown are appliances, as well as software-defined with a hybrid mix of HDD and SSD, as well as all-SSD or all-Flash. Other storage systems shown include large-frame as well as modular solutions, disk, virtual tape, and tape libraries, as well as purpose-built backup appliances (PBBA) and purpose-built data protection appliances (PBDPA), among others.

12.1.1. Tradecraft, and Context for the Chapter

Some tradecraft and context for the chapter have already been mentioned, including how the term "storage system" can mean different things from legacy to emerging, hardware-centric to tin-wrapped, software as an appliance, software-defined, cloud, container, and virtual.

Tip: Avoid judging a book (or storage system) by its cover (packaging); rather, it's what's inside that counts. This applies to storage systems as well, meaning be careful when judging storage systems by their appearance, physical size, packaging, interfaces, protocols, capacities, or other attributes. What this means is that it is the sum of the pieces inside that matters—that is, the hardware and software resources, combined with PACE-enabling capabilities, data services, feature functionality, and implementation.

Tradecraft context for this chapter includes the following.

Aggregated – Refers to the consolidation of servers and storage resources to improve utilizations or introduce performance bottlenecks. Aggregated also refers to whether compute resources such as servers are tightly coupled with storage such as with HCI or disaggregated (separate). Another variation is an aggregate that is a unit of storage allocation used by some systems such as NetApp ONTAP, which is managed and presented to application servers for use.

Array – Can be a physical storage system or appliance with or without data services and other functionalities. A virtual array or storage device group from a storage controller, RAID card, or software-defined storage, and a volume manager.

Controller – Can be an adapter or RAID card, storage system or appliance controller, storage processor, server or node. Controller may also refer to an object storage device (OSD) or storage node, gateway, or service with some endpoint for access. Regardless of where or how a storage system is implanted and packaged, the controller is what implements the functionalities, features, data servers, and personalities of its architecture. Note that the controller can be a single card, device, server, or node, or multiple nodes in a controller. The controller functionality can also be split up across different nodes such as access, gateway, management, monitor, metadata, and actual storage functions.

Footprint – There are physical footprints such as floor space, rack, or cabinet space requirements. Then there are logical footprints such as an operating system or server instance, also known as an image. Data footprint is the overhead combination of actual data, metadata, as well as protection copies (backups, versions, snapshots, replicas). Data footprint also includes copies of the same file, photo, video, or data that exists in different folders, file shares, or systems.

Metadata – There are different types, classes, and layers of metadata. There is higher-level (higher-altitude) application-focused metadata describing data being used or worked with. There is configuration and setting metadata as well as rules, policies, and associated business logic. Then there is database, file system, or repository metadata, along with operating system, hypervisor, or cluster metadata. In a storage system, there can be higher-level metadata about the data being stored and applications using it. There is also system-level metadata for configuration settings and system state. Additional system metadata includes lower-level points and tracking information for snapshots, point-in-time copies, CDP, back-dating, replication, and other functions.

Nodes – Members of a cluster (or redundant system) that perform different functions, including as a storage controller. These are servers configured to perform some function such as compute, storage, management, gateway, monitor, and managing metadata, among other roles.

Stamps – Refers to a collection of server, storage, and network hardware, as well as software, configured as a unit of computing, storage, or converged functionality. For example, Microsoft uses the term *stamp* to refer to a cabinet or rack containing some number of servers and storage to support various Azure cloud functionalities. A stamp can be thought of as a CiB or node(s) in a cluster part of a fault domain, among other variations.

Storage system – Can be a turnkey storage solution with integrated hardware and software supporting DAS, SAN, NAS, and/or object, and API access to SSD or HDD. Other options include software that turns a server into a storage system, CI, HCI, as well as virtual and cloud arrays and appliances.

Tiers and tiering – There are different tiers and classes of storage solutions, as well as types and categories of storage device media for various application PACE needs. Another context for tiering besides types, classes, and categories of storage systems, solutions or devices, and media is a migration (promote, demote, or other movement) of data across tiers (manually, semi-automated, or automated).

Vaults – Can mean a storage system or protection storage used for backup, archiving, or storing protected data. Another variation of vaults can be persistent storage on HDD or SSD used to store system-level state, configuration, and system metadata.

12.2. Why So Many Types of Storage Systems?

Everything is not the same in data centers across different environments and applications. The result is that different types of storage systems and solutions are needed to meet various needs. Figure 12.4 shows different storage systems to meet various application PACE as well as environment needs. These include from small (lower center) to large scale (top center), do it yourself (DiY) (center left) or turnkey (center right). Other storage systems include modular, large-scale legacy frame-based, software-defined deployed on the physical server, virtual, or cloud system. Also shown in Figure 12.4 are CI, HCI, and CiB among other approaches from a single device shelf to multiple racks and cabinets.

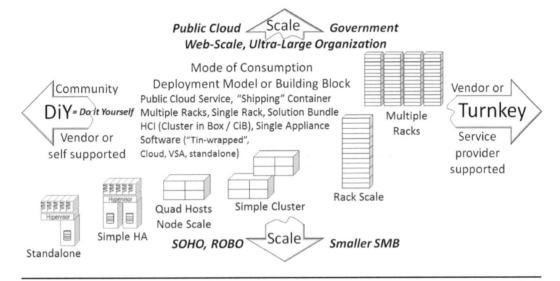

Figure 12.4   Different approaches, types, packaging, and storage scaling options.

Storage system considerations include:

- Type and size of environments
- Different applications and workloads
- Internal or external, dedicated or shared storage
- Aggregated or disaggregated compute servers
- Local on-site, remote off-site, and cloud
- Various PACE attributes capabilities
- General-purpose or specific function focus
- Legacy hardware, software-defined, and virtual

General storage system trends include:

- Server-side and software-defined storage
- Mix of all-Flash, all-SSD, all-HDD, and hybrid storage
- Traditional, CI, CiB, and HCI storage solutions
- On-site legacy physical arrays and appliances changing roles
- Software-defined, virtual, cloud, and container storage
- Scale-out bulk storage including file and object access

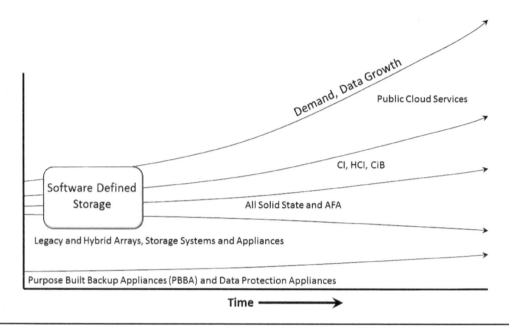

Figure 12.5   Server storage I/O solution trends.

Keep in mind that everything is not the same in different environments and various applications. Thus there is not one size, approach, packaging, or functionality storage solution for every scenario, although there are of course flexible multifunctional solutions that can meet many different scenarios.

Figure 12.5 shows a trend and a relative forecast of how different types of storage systems are being deployed to meet various needs today and in the future. Looking at Figure 12.5, growth areas include cloud and software-defined storage options for general-purpose, high-performance (bandwidth or transactional), as well as scale-out bulk storage. Note that Figure 12.5 shows only general industry trends.

What varies in Figure 12.5 is how the software that defines the storage functionality gets packaged. For example, some of the software will be deployed as tin-wrapped appliances or storage systems, others as CI, HCI, and CiB solutions. In other scenarios the storage software gets deployed on or in a cloud instance or virtual storage appliance.

The importance of Figure 12.5 is to set the stage for discussing current as well as future storage solution, system, and services options. Some of you may not ever have the need to work with legacy storage systems, while others will continue to do so.

On the other hand, some will be more focused on the new and evolving, while others will have the need to support hybrid environments. Hybrid environments include not only hybrid storage systems that support HDD and SSD, but that also support a mix of DAS, SAN, NAS, object or block, file, object, API appliances, virtual, cloud, and software-defined approaches.

Tip: In Figure 12.5, the actual adoption (deployment) by customers or service providers may be higher or lower for a given category. There are plenty of different market forecasts that you can find via Google to suit your particular needs. The important thing is to realize that things are evolving, for some more quickly than for others. This means different approaches for different organizations.

12.2.1. Different IT Environments

Everything is not the same: There are different applications, environments, size, scope, complexity, and focus. These range from the cloud and managed service providers to large enterprise, small/medium enterprise (SME), mid-market small/medium business (SMB) along with remote office/branch office (ROBO) down to small office/home office (SOHO) or very small environments.

Tip: In addition to different sizes as well as types of environments, there are various use cases and how these are used, some dedicated storage to applications while others are more general-purpose. Some use targeted, special-purpose storage, others use general-purpose solutions. For some environments, the focus for storage systems is performance, such as bandwidth, activity including transactions or IOPs, as well as latency response time. Meanwhile, other environments and applications have availability, and resiliency as a storage focus, while still others are in need of space capacity, or economic cost savings, or all of the above, among other considerations.

Other considerations for storage across different IT environments include the following.

- Some environments use one or more storage systems and all of their features' functions.
- Other environments use different storage systems for specific tasks and functions.
- Storage systems include general purpose to specialized or function-specific solutions.
- Not all application workloads are or will be virtualized (at least near term)
- There are different tiers, classes, and categories of storage, and systems.
- Applications have various PACE workload requirements and needs.

12.2.2. Application Workloads and Usage Scenarios

So, which is the best or right storage system? It depends on your environment, application, and associated PACE, as well as data infrastructure functionality needs. Looking back to Chapters 1 and 2, there are different application workloads with various PACE attributes based on how they are used. For some environments, those application workloads and their PACE attributes, along with other data servers, management, as well as functionality requirements can be addressed with a general-purpose storage system. For other environments or applications, however, specialized storage systems such as all-Flash arrays (AFA) or all-SSD arrays (ASA) are needed for performance.

Another workload with specialized storage systems needs is data protection. In addition to AFA, ASA, as well as hybrid (a mix of HDD and SSD) storage systems, there are also specialized solutions for database, video, and content serving, among other workloads.

As mentioned previously, in addition to physical interfaces such as Ethernet, FC, PCIe, Thunderbolt, InfiniBand, and SAS, among others, as well as logical protocols (SAS, NVMe, IP [iSCSI, NFS, CIFS/SMB, HDFS, object]), storage systems also have different personalities. Those personalities are how the underlying storage resources combined with data services along with other feature functionalities abstract and present storage to applications.

Personalities can be optimized for general-purpose block, file, or object, as well as for high-performance sequential streaming and bandwidth-centric. Large, sequential streaming, optimized storage personalities can be a good fit for some high-performance computing, high-productivity computing, or super-computing, and for big data analytics applications.

Other personalities can be optimized for small random (and sequential) reads and writes supporting transactional, database, key-value repository, or file-serving workloads. Some other personalities may be less focused on performance, with emphasis on bulk durable and available storage that is cost-effective. Another consideration is that some personalities, and how storage is presented or accessed, can be focused on concurrent activity, while others are optimized for non-concurrent activity.

> *Tip:* There is more to storage than simply cost per gigabyte or terabyte, as well as its packaging. Likewise, there is more to storage than just performance in IOPs or bandwidth to meet different application PACE needs and other requirements. And there is more to storage than just the physical interface or logical protocols along with PACE requirements.

Some applications work with general-purpose storage, but they may benefit from specialized storage systems. Various storage systems will have different data services, feature functionalities, and management interfaces. For example, if your environment, applications, and data infrastructure leverage hypervisors, then support for VMware vSphere, Microsoft Hyper-V, KVM, Citrix, Xen, OpenStack, and Azure stack, among others, should be part of your solution.

12.2.3. Storage System, Solution, and Service Options

Storage systems, solutions, software, and services include those for primary, secondary, tertiary, bulk, and protection use. Some solutions are optimized for performance, others for space capacity or availability, while still others are focused on low cost.

Storage system considerations beyond PACE requirements include:

- Aggregated or disaggregated attachment to compute servers
- Dedicated or shared storage, located locally or remote
- Data sharing (e.g., NAS file) vs. storage sharing (e.g., SAN or DAS block) or object
- Support for physical, legacy, virtual, cloud, or container environments
- What emulation, device or access, and functionality presentation are needed
- Data services and functionalities needed for applications
- Compatibility with existing data infrastructure tools, hardware, and software
- Third-party hardware or software compatibility support
- Structured, semi-structured, or unstructured; big or little data
- Storage usage: on-line primary, near-line, or off-line

Table 12.1 shows different storage system classifications.

Table 12.1. Storage System Categories

Storage	General Storage Characteristics
Bulk	Scale-out and durable elastic storage for storing large quantities of data at relatively low cost. Off-line bulk storage includes tape and optical, while on-line includes disk-based solutions that support block, file, and object access. Use cases include media and video content, DEMs and DAMs, object, and archives, among others.
General purpose	Storage that has the flexibility to deliver a mix of performance, availability, and capacity with various data services. Sometimes called "Swiss army knife" storage because they often support a mix of block SAN and DAS as well as NAS files and objects, along with other functionalities.
Primary	As its name implies, storage optimized for primary and general-purpose use, with a mix of performance, availability, capacity, and economics. For larger environments, primary storage may be a large storage system that has secondary storage as a companion for less critical applications. Note that what a larger environment considers secondary storage may be primary storage for a smaller environment. Think of primary storage as the place where you primarily place, and access, your data for commonly used applications.
Secondary	Storage that is not used as primary, for example, to support secondary applications that require less performance or different service levels at a lower cost. Can include bulk and scale-out storage with block, file, or object access. Application use includes email messaging, home and work folder directories and file shares, scratch and temporary staging areas for import and export of data, reference, and other items. May be used as protection for primary storage systems.
Tertiary protection storage	On-line, near-line, and off-line (tape, optical, or disk) storage for protection copies, including backups, clones, copies, snapshots, CDP, back-dating, replication, and archives. Can also be general-purpose bulk (high-capacity), with very low cost, low performance, and lower resiliency than primary, secondary, or general storage. Various data services and functionalities to reduce data footprint and cost, emulation for virtual tape libraries, and other features. Includes purpose-built backup appliances as well as general disk and tape libraries, and object or other bulk storage.

12.2.4. Storage Systems and Solutions—Yesterday, Today, and Tomorrow

Storage systems continue to evolve, in terms of both technology to address application PACE needs and feature functionalities. In addition, storage systems are also evolving via different packaging as well as being used in new or different ways.

Some changes and trends include traditional enterprise storage being used in new ways, such as by service providers for specialized as well as general-purpose consolation at scale. Traditional legacy enterprise storage is also being used for virtual, cloud, and container environments. Likewise, mid-range and software-centric storage solutions are taking on workloads traditionally handled by legacy systems.

Figure 12.6 shows how storage systems have evolved over time, from left to right. First on the left were servers with dedicated direct attached storage for mainframe and open system servers. Second from the left are external dedicated and basic shared storage. In the center are shown networked shared storage, including SAN and NAS as well as hypervisors for software-defined environments. On the right is a combination of SAN and NAS, as well as software-defined, cloud, and CI, HCI, and CIB solutions using HDD, SSD, and hybrid storage. Also shown are tin-wrapped software appliances, along with software-defined solutions with block, file, or object storage and data services.

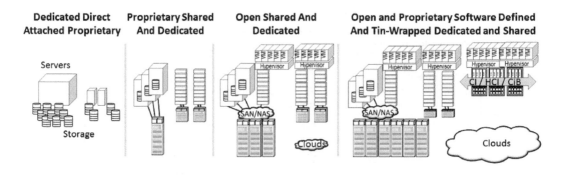

Figure 12.6   The evolution of storage systems and solutions.

12.3. Common Storage System Characteristics

Tip: Look beyond the number of slots, ports, processors, and physical interfaces to determine performance. Some system may appear to have more performance based on how many or how fast devices, ports, and interfaces are vs. their actual capabilities. The effective performance will be the sum of the parts minus any hardware and software bottlenecks. These bottlenecks can include how performance is impacted by different functionalities and data services being used.

Common physical storage system characteristics include:

- Some number of slots for storage devices (HDD, SSD, PCIe, or other cards)
- I/O access ports such as PCIe, SAS, SATA, Thunderbolt, NVMe, FC, GbE, or IBA
- Management access path and service ports for maintenance
- Processors or controllers that run storage software
- DRAM, SCM, and other NVM, including Flash SSD
- Power and cooling fans that are fixed or hot-swappable

Common logical storage system characteristics include:

- Storage software implementing features and data services
- I/O access protocols leveraging physical or virtual interfaces
- Management tools for maintenance, configuration, and provisioning
- Various data services, functionalities, personality, and PACE attributes
- Inter- and intra-system communications and management
- Plug-ins and APIs for interfacing with data infrastructure applications

Tip: Keep in mind that storage systems functionality, regardless of architecture and packaging, relies on software to implement different capabilities. That software runs on standard server processors or using custom FPGA and ASICs to offload servers. What is import to remember is that, like all applications, storage software is an application that requires CPU and memory resources to support different functions and data services. Likewise, CPU and memory are needed to set up and process I/O operations. Other CPU, memory, and I/O resources are consumed to support background data integrity checks and repairs, among other tasks.

12.3.1. Storage System PACE Considerations

A common theme throughout this book has been awareness and discussions of application PACE needs along with resource capabilities. Similar to servers and I/O networking, applications have storage PACE needs, including management considerations. Figure 12.7 shows how hardware, software, and data services feature functions along with management tools combine to address various PACE requirements.

Table 12.2 shows how different application PACE needs are enabled by storage systems, and their various feature functionalities, along with business or organizational outcome (benefit). Also shown are various metrics that matter for different storage system solutions.

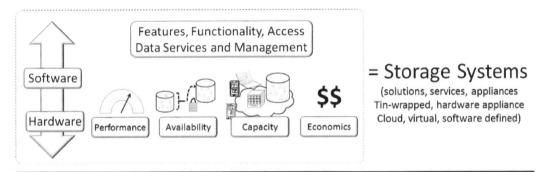

Figure 12.7   Common storage solution characteristics.

12.3.1.1. Performance and Productivity

For some applications and storage systems, performance is a primary priority. Also, keep in mind that it is difficult to have availability or use capacity without performance.

There are different aspects to storage performance, from access to applications, servers, and users, with various characteristics. These include reading, writing, and updating as well as getting the status of data or of metadata. Workloads can be random or sequential, large or small, based on IOP activity or focused on bandwidth data throughput movement.

Other considerations include latency response time, resource queues, and CPU usage or associated bottleneck delays. While storage performance and capacity utilization are often a focus, so too should be awareness of productivity and effectiveness. That is, storage efficiency, and utilization is important, but keep performance and productivity in perspective. For example, at some point, driving utilization too high might result in poor performance that reduces productivity—negating the benefits of higher utilization.

Tip: For storage performance, keep normal sustained, burst, and peak as well as fluctuations during start-up or background processing in perspectives. Leveraging compound (multiple combined) metrics helps to gain insight across server, storage I/O activity, and effectiveness. For example, instead of just looking at IOPs or bandwidth or space capacity used per cost, look at compound metrics such as available (or required) space capacity per IOP or transaction. Another example is IOP or transaction per watt of energy, CPU per I/O, among other variations discussed in Chapter 5 and elsewhere.

Table 12.2. Storage System PACE Attributes and Business Benefits (Outcomes)

	Attribute	Functionality	Outcome	Metrics That Matter
P	Performance Productivity Predictable	Cache, tiering, NVM and SCM, RAM, SSD, PCIe NVMe, storage, acceleration, faster resources (server, storage, I/O networks), and protocols	Increased work is done in the same or less time. Lower overhead (CPU, memory, I/O) and rework. Less time spent waiting, more time being productive.	Activity rates (IOPs, TPS, gets, puts, reads, lists, writes, pages per time); data transfer (bandwidth, throughput, MB/KB read/written per time; cache hit rates. Response time latency; queue depths and CPU usage (system, user time). Errors or retransmits.
A	Availability Protection Resiliency Security Consistency	Snapshots, CDP, clones, backup, HA, back-dating replication/mirror RAID/EC/LRC. RAS, BC, BR, DR, and recovery	Information is safe, secure, consistent, and accessible when needed. Protect data infrastructure along with information on various threat risk scenarios or events.	RTO, RPO, and other SLO. AFR, MTBF, some 9's availability, accessibility along with durability. MTTR and other restore/restart times. Uptime planned/unplanned. Errors and timeouts, data loss, damage or corruption, denial of service.
C	Capacity Space Inventory Connections Ports	Tiering, thick or thin provisioning DFR (archive, compress, de-dupe) Space snapshots, reclamation; CPU memory, storage space capacity, I/O bandwidth	Data infrastructure resource inventory is available to support applications and information service delivery. Efficient, productive, and cost-effective resource usage and ROI.	Raw, formatted, allocated, and used, thick, thin, and orphan or ghost space. Reduction ratios, and rates. Usage, free or available resource inventory. Cost per capacity per service level delivered and resources consumed. Hardware, software, services, and licenses usage. How solution can be upgraded.
E	Economic Enablement Effective Energy	One-time and recurring costs of resources such as software, network, or other services; HVAC, facilities, staff, and training	Budget (CapEx, OpEx) impact. One-time or recurring costs. ROI and organizational benefit enablement vs. cost overhead.	TCO, ROI, cost per capacity per given level of service or activity, resources used per level of service per energy used. Various composite (multiple metrics combined) metrics. How solution will be cost-effective in the future for different deployments.

Storage system performance is enabled by the speed of controller nodes, the amount of cache as well as the effectiveness of cache algorithms. Other enablers include the amount of persistent and non-persistent memory for cache, as well as system metadata, along with other optimization algorithms. Number, type, and speed of processor cores, I/O interfaces, protocols as well as software for I/O acceleration, including cache, and micro-tiering are also considerations. Fast software needs fast hardware and vice versa.

Storage system performance and productivity considerations include:

- Effective performance with context of benchmark or indicated metrics.
- Is performance measured or inferred based on the number and type of devices?
- Is indicated performance peak marketing "hero" metric or for normal running mode?
- How is performance impacted with data services enabled under different workloads?

- Can performance claims be easily reproduced with generally available tools?
- Availability of tools for monitoring, reporting, and analysis of performance.
- Storage acceleration (cache) and fast persistent storage for storing data.
- Are additional license fees, software tools, or hardware needed to unlock performance?

> *Tip:* Look beyond the number of processors, ports, amount of DRAM, SCM, NVM, and cache, as well as I/O ports, and devices. Instead, look at what the effective performance of a storage system is that is also relevant to your environment. If your applications are going to be doing large (128 KB to 1 MB or larger) random reads and writes, while small 4 KB sequential reads are impressive as marketing hero numbers, how relevant are they for you?

12.3.1.2. Availability and Accessibility

Availability and accessibility are important attributes of storage systems to protect, preserve, and serve data to data infrastructure along with the applications they support. Accessibility includes being able to access or store data as well as associated metadata. Access includes via logical protocols and associated physical interfaces as well as the storage system personality functionality they support. Access also means that stored data can be accessed (or stored), it is consistent, and it is also safe and secure. Another dimension of availability is durability, or how many copies and versions exist.

Security – Physical access control includes key locks among other measures. Logical security includes encryption of data at rest as well as in-flight (being moved to or from the storage system). Identify access management along with endpoint protection, passwords, access control lists, zoning, and port prohibits, among other items.

Consistency points – Static or dynamic time interval protection includes snapshots, clones, versioning, back-dating, CDP, backups, and checkpoints as well as log switch. The granularity of the consistency point can be application, transactional, system, journals, database, table, file, object, and volume or storage device, among others.

> *Tip:* Some architectures place more emphasis on data and transactional consistency in the storage system, while others rely on upper-level software beyond basic data protection. Additional architectural data consistency features can include data integrity background scans and corrections. These scans can depend on architecture, and implementation can detect as well as correct both data and media errors. Some examples include checking for bit rot, data, and media errors, viruses, and other items proactively.

Access – Keep one or more copies of data (durable) accessible with mirroring and replication along with other RAID options such as parity, Reed-Solomon erasure codes, forward erasure codes and local recovery code, and shingled erasure codes, among others. Other access capabilities include redundant paths, hot-swap controllers, and other components, as well as non-disruptive code updates.

RAS – Reliability, availability, serviceability features include redundant hot-swappable controllers, power supplies, cooling fans, I/O modules (front-end and back-end), drives or devices (HDD and SSDs), mirrored non-persistent DRAM memory cache, persistent cache, and metadata stores. Other RAS features include power-on self-test (POST), redundant east–west (intra-system) communications (fabric or network) for cluster communications, active/active or active/standby failover of nodes, controllers and other devices, logs, and journals.

12.3.1.3. Capacity and Resource Counts

Storage systems and their associated software rely on as well as define, abstract, and manage underlying physical and virtual resources or services. Think of it this way: If there are no available resources, how can you have performance, availability, and capacity to save or access data? Keep in mind that storage systems exist to store, protect, preserve, and serve data along with the programs and settings that transform into information.

Capacity and resources include space capacity including raw, protected, cooked (formatted, file system, or otherwise defined), provisioned and allocated (thick or thin), as well as used (dense or sparse). Other resources include an available processor and volatile memory (DRAM) along with persistent NVM (SSD, SCM) for data cache, storage software, data services, and metadata. In addition to storage space, compute, and memory, other resources include I/O paths such as PCIe bus bandwidth along with physical slots for I/O devices.

I/O devices include front-end and back-end access (i.e., north–south) along with management ports (east–west). While front-end ports of different types, speeds, and protocols support server (or other storage system) access, back-end ports attach to some number of storage devices or storage resource endpoints (other storage systems, cloud, physical or virtual).

> *Tip:* Some vendors will indicate a total amount of cache, for example, 1 TB, for a storage system. Context is important to know if that is 1 TB usable or 1 TB across, say, two nodes or controllers (each with 512 GB). Additional context is needed to understand, of the 512 GB per storage node, how much of that is allocated as a mirror for the peer or partner controller. Another consideration is how much of the cache is set aside for storage system software, features, and data services code, settings, cache or tracking tables, and buffers vs. user data.

Storage system capacity and resource considerations include:

- Number, type (HDD, SSD), speed, capacity, physical form factor, and device interfaces
- Total amount of cache (persistent or non-persistent) and effective usable cache
- Overhead (space capacity and performance) for implementing data protection availability
- DFR capabilities including compress, de-dupe, and thick and thin provisioning
- Number and size of LUN, volumes, virtual disks, sector size that is host accessed
- Number and size of target (back-end) drives and endpoints
- How many and size of file systems, number and size of files or objects
- Expansion capabilities, interoperability, and other hardware and software dependencies

Additional data services features and function capacity considerations include:

- How many total and concurrent snapshots and replications are supported
- Number of depths (children) of snapshots of a parent and how managed
- Optional applications or guest virtual machine and container hosting capabilities
- RAID, mirroring, replication, parity, and erasure code space optimization
- Replication and mirroring modes (synchronous and asynchronous) and topologies
- Number of redundant controllers, I/O modules, power, cooling, and if hot-swappable
- Standards and interfaces, APIs, and management interoperability

12.3.1.4. Economics and Energy

The "E" in PACE is for Economics, which also has a tie to Energy as well as Efficiency (i.e., utilization, such as Capacity) and Effectiveness (i.e., Performance and Productivity). If you are looking for another "E," you can include Enablement via management tools, among others.

As mentioned throughout this book, there is no such thing as a data or information recession due to continued growth, new applications, demand and dependencies on information being kept longer. However, there are economic realities. These realities mean that in some cases budgets are shrinking, resulting in the need to do or enable more with less. In other scenarios, budgets are growing, but not as fast as data and information resource demands are increasing.

The result is that budgets need to be stretched further. Instead of simply looking for ways or areas to cut cost, find and remove complexity or overhead, which in turn will reduce costs. By removing costs without compromising performance, availability, or capacity, the result should be to sustain growth while maintaining or enhancing data infrastructure and applications service levels.

Some economic discussions focus on hardware capital expenditures (CapEx), particularly those focused on moving to cloud-based storage systems and data infrastructures. While CapEx focus is important, so too are recurring operating expenditures (OpEx), not only for hardware but also for software as well as services such as cloud. For example, as part of reducing both CapEx and OpEx, you may look for energy-efficient and effective hardware to lower power and cooling demands, such as using fewer SSDs with better performance vs. many HDDs. Some will focus on energy savings from an environmental or "green" perspective, while others need to support growth with limited available power (the other "green," as in economic as well as effective and productive).

Another example is that you may have a focus on improving hardware utilization to reduce associated CapEx and OpEx costs. However, what are you doing to improve utilization of your software licenses such as databases, hypervisors, operating systems, or applications? Also, what are you doing to improve the effectiveness of your cloud or managed services OpEx costs and recurring fees or subscriptions?

Economics- and energy-related considerations include:

- Metrics such as performance per watt of energy (e.g., IOP/watt, TPS/watt)
- Metrics indicating hardware as well as software usage and return on investment
- Up-front OpEx or CapEx costs and recurring license, maintenance, or subscription fees
- Hidden or not-so-obvious costs and fees for other related services or dependencies

- Costs for peak sustain, normal, and idle or low power and performance modes
- Costs for availability and other overhead tied to a given level of service
- Energy usage and subsequent cooling and HVAC costs
- Exercise caution about comparing storage only on a cost-per-capacity basis

---

Tip: Speaking of environmental concerns, energy efficiency, and effectiveness along with green IT, I happen to know of a good book on the subject, one that I wrote a few years ago called *The Green and Virtual Data Center* (CRC, 2009). Learn more at www.storageio.com/book2.html.

12.3.2. Additional Functional and Physical Attributes

Additional storage attributes include physical as well as logical or functionality.

Management characteristics include:

- Physical connectivity for management and monitoring
- Logical management protocols (IP, HTTP, VNC, SSH, rest)
- Monitoring of temperature, humidity, power consumption

Facilities and environment characteristics include:

- Height, width, depth, and weight of hardware
- Rack, cabinet, or floor space requirements
- Primary, secondary standby, and UPS electrical power
- Electrical power draw for peak start-up, normal, and running
- Power draw in amps, volts, frequency, and number of connections
- Cooling (HVAC) tons to remove BTU heat generated
- Smoke and fire detectors, alarms as well as suppression
- Locking mechanisms for physical security of devices, and access

Physical considerations include what is the physical and logical footprint of various components such as devices, servers, and switches. A normal rack width is 19 in., but some systems and enclosures can be wider, as well as deeper or taller. There are weight and height considerations including for doorways, or for top-loading drive shelves, hot-swapping for replacement, and front- vs. back-of-rack or cabinet access. Also, how will replacements of FRU (field replaceable units), power, cooling, management, I/O interfaces, and drives occur?

12.4. Storage Systems Architectures

There are many different types of environments, applications, servers, networks, and storage devices, and also systems architectures. These various storage architectures align with the different PACE needs of applications, and the data infrastructures that support them.

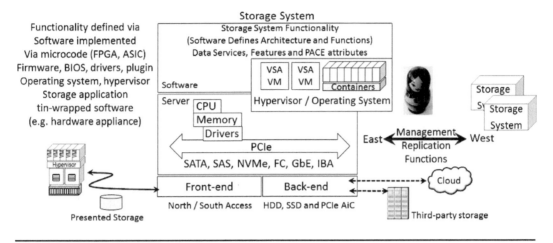

Figure 12.8 General storage system architecture and components.

Different storage system architectures are implemented in various ways, using hardware as well as software with diverse packaging options. Note that all software eventually needs to run on some hardware, and that all hardware needs some software.

Figure 12.8 shows a generic storage system architecture model with physical resources (compute, memory, I/O, and storage devices) along the bottom. Along the top center of Figure 12.8 is the software that defines how the underlying software and hardware function as a storage system. The storage system software can be implemented on a general-purpose server, in a virtual machine or container, as part of a hypervisor, operating system, or a storage application.

Tip: Recall that software can be wrapped inside other layers of software (like the dolls inside of a doll shown in Figure 12.8), as well as simply wrapped in hardware (tin-wrapped) in the form of a storage appliance or system. This means that servers can be defined to be storage systems, appliances, solutions, or services using various storage software. This means there are different views of storage architectures based on size, shape, capacity, performance, functionality, number and type of interfaces, functionality, and packaging, including hardware or software.

Figure 12.9 shows various fundamental storage architectures.

Table 12.3 expands on the fundamental storage architectures shown in Figure 12.9.

Table 12.4 shows various representative examples of architectures in Table 12.3. Keep in mind that this is a simple example; others may be found in the Appendix, and my companion site www.storageio.com/book4.html as well as my www.storageioblog.com.

Note that the different types of storage system architectures shown in Tables 12.3 and 12.4 can be implemented and packaged in various ways. For example, a CI, HCI, CiB solution can be based on type 3 or 4 architectures with aggregated or disaggregated compute, and storage, as well as tightly or loosely coupled software and networking. Similarly, the different architectures shown in Table 12.3 can be used for primary, secondary, tertiary, or protection storage as well as bulk spanning block, file, and object for physical, virtual, and cloud services.

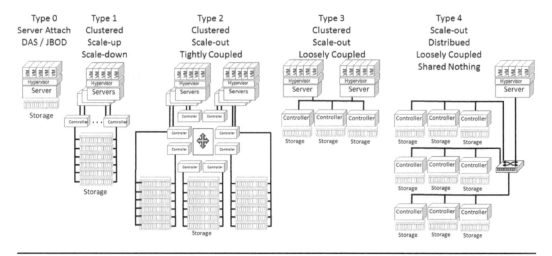

Figure 12.9   Various storage system architectures.

Fundamental storage architecture attributes include:

• Server (internal or external) or storage system (appliance or service) based.
• Various architectures, implementations, and packaging for different needs.
• Architectures get implemented via software and hardware.
• Some architectures are hardware-neutral; others have dependencies.
• Some architectures may be hardware-neutral yet have software dependencies.
• Storage systems ingest (or retrieve) data, process, and store (or access).
• Various data services and functionalities with various types of access (block, file, object).
• Solutions can be propriety, open hardware, and software.

12.4.1. Scaling Up, Down, Out, or Out and Up

Like compute servers and networks, storage systems can scale in different dimensions. This includes scaling of features, the functionality of data services, access, and management. Other scaling attributes include performance, availability, capacity, and economics. The scaling can occur in different ways as shown in Table 12.5 across the block, file, and object storage systems, including software and hardware as well as services including cloud.

For scale-out (horizontal scaling) storage, PACE attributes along with other capabilities are expanded by adding more nodes (compute, storage, management). Performance can vary with scaling, as some solutions have less overhead vs. others. This is also true of space capacity, which can vary by protection overhead. Look for solutions that scale with stability (does not introduce problems with scaling).

12.4.2. Storage Front-End Matters

Storage systems have a "front-end" that provides an endpoint, adapter, connectivity or presentation to the physical, virtual, cloud, or container server or service. The front-end can be

Table 12.3. Fundamental Storage Architectures

Type	Architecture	Characteristics	Pros and Cons	Use For
0	DAS JBOD	Dedicated internal or external with limited or no data services; relies on host server software for functionality. HDD, SSD, and PCIe devices. SAS, SATA, NVMe, PCIe, Thunderbolt attach.	Pros: Dedicated local low-latency storage, potential cost savings. Use software to create storage functionality. Cons: Results in islands of technology, and more management.	Storage close to applications. Low latency with fast SSD. Small-scale storage, or as a node for large scale-out storage
1	Clustered scale-up/scale down	General-purpose focus, multiple protocols (file, block, object). Expansion storage can be added; various data services, shared metadata. Single or dual controllers, strong data consistency. Some systems have good performance, while others focus on lower cost or other features.	Pros: Multiprotocol for dedicated or shared networked access. Adapts to applications. Many vendor options for including hardware or software. Cons: Performance limited by the size of the controller, and software. Scales by adding more systems vs. nodes.	General-purpose as well as primary for many applications, and environments that need flexibility of block, file, or object access. Can be combined with overlay software that virtualizes the hardware for scaling.
2	Clustered scale-out	Multiple controllers with shared metadata, strong transactional consistency, extensive RAS features, functions, and data services, optimized for large mission-critical aggregated applications workloads and service providers.	Pros: Optimized for large-scale PACE, consolidated workloads, large cache for performance. Can support many protocols as well as thousands of HDD or SSD, large cache. Cons: High cost for entry of large system.	Primary storage for mission-critical and other applications. All HDD, AFA, or hybrid. Large-scale consolidation, cloud, virtual, and physical servers, and service providers.
3	Loosely coupled scale-out	Multiple loosely coupled (networked) storage nodes (controllers). Distributed metadata, transactional consistency with various access (block, file, object). Use commodity servers with internal or external storage and software.	Pros: Scale PACE by adding more nodes using cost-effective hardware, and software. Some solutions are CI, HCI, or for bulk storage. Cons: Not all solutions scale equally across PACE, with different overhead.	General-purpose applications, can be primary for some environments. For others, it can be for secondary, and bulk or secondary storage as well as content or video serving.
4	Distributed, nothing shared	Multiple loosely coupled distributed nodes and metadata, nontransactional consistency (handled by higher-level software).	Pros: Suited for bulk lower-cost durable with flexible access. Scale by adding nodes. Cons: Complexity of multiple nodes, and performance.	General-purpose storage for content, and new third-platform big data, bulk, content, object, and data protection.

physical interfaces with different logical protocols, or virtual. These include block, file, object, and API access.

Some storage systems support multiple physical interfaces as well as logical protocols, while others are limited to a single presentation. For example, some solutions support only block iSCSI, while others support NFS, SMB3 direct. Other solutions support a mix of protocols

Table 12.4. Examples of Fundamental Storage Architectures

Type	Architecture	Examples
0	DAS JBOD	Hardware includes various servers with internal or external storage expansion enclosures (SAS, SATA, NVMe, USB, Thunderbolt attached). Internal PCIe cards as well as SSD and HDDs. PCIe external attached using NVMe include Dell EMC DSSD, among others.
1	Clustered Scale-up/scale down	Datacore (software), Data Direct, Dell EMC (VNX, Unity, SC, Datadomain, MD), Fujitsu, HDS, IBM (V7000 & SVC, FlashSystems), HPE (P2000), NetApp (ONTAP & E-Series), NEC, Nexsan, Nimble, Oracle, Quantum, Reduxio, Pure Storage, Seagate/Dothill, and many others.
2	Clustered scale-out	Dell EMC (VMAX, Isilon, XtremIO), Fujitsu, HDS, HPE (3PAR, XP), IBM (DS8000 & XIV), and Oracle, among others.
3	Loosely coupled scale-out	Data Direct, HPE StoreVirtual, HDS HCP, Microsoft SOFS/S2D, NetApp (ONTAP SolidFire, StorageGrid), Nutanix, Seagate (Clusterstor [Lustre]), Simplivity, Tintri, VMware vSAN, and EVO. among many others.
4	Distributed nothing shared	AWS S3, Data Direct, Dell EMC (ECS/Atmos/Centera), Ceph (Fujitsu, Google GCS, Micron, Red Hat, Suse, WD/SANdisk), OpenStack Swift based, NooBaa, OpenIO, Quantum, WD/Amplidata, and many others.

(converged), although they might be all block, file, or object, and some support unified or mixed interfaces and protocols.

Tip: Note that block access options include both open systems as well as IBM Mainframe protocols such as FICON (a Fibre Channel ULP), among others.

Depending on the architecture, packaging, and implementation of the front-end, ports may be on a motherboard, purpose-built front-end adapter card, I/O module, or industry-standard adapter. Some front- ends are hot-swappable as field replaceable units, while others are not. Various host server device drives and multi-pathing software can leverage redundant paths for performance load balancing and resiliency.

Table 12.5. Scaling Storage PACE and Functional Resources

Scaling	Direction	Attribute
Up	Vertical	Bigger, faster PACE resources, and functionality to support larger environments. Larger mid-range, and monolithic enterprise systems.
Down	Vertical	Smaller, perhaps slower, for lower-end cost-sensitive use. Mid-range, entry-level, SMB, and SOHO systems and appliances.
Out	Horizontal	Smaller or similar-size PACE resources to scale beyond limits of a single server, system, or appliance. Scaling out can be done for performance, capacity, availability, or a combination of purposes. Clusters, plexes, grids, rings, stamps, redundant array of inexpensive nodes (RAIN). These can be tightly or loosely coupled with shared or dedicated system-level metadata (state and configuration). Some are shared-everything while others are distributed shared-nothing. Note that not all scale-out boosts performance, capacity, or availability the same way. Some have more overhead vs. others.
Out and up	Horizontal, and vertical	Combines more, and larger, faster servers and storage systems for increased PACE. Best of scale-up and scale-out.

Front-end adapters can support Fibre Channel (SCSI_FCP and or FICON), Ethernet (IP with iSCSI, NBD, NFS, HDFS, CIFS/SMB, or object [S3, swift]), InfiniBand, SAS, Thunderbolt, or PCIe (NVMe), among others. Some systems have adapters that can be defined via software firmware, and device drivers.

This means that some solutions allow adapters to be defined to be a front-end or even back-end to handle different tasks. Some adapters or I/O ports can do both (target and initiator), function as a front-end as well as back-end. For example, some front-ends, in addition to communicating with application host or compute servers, can also be used for east–west traffic to other storage systems or applications. Examples include using a front-end port for communicating with another system for replication, or accessing third-party storage for virtualizing, or accessing a cloud storage service.

Front-end characteristics include:

- Handle north–south storage access from application or compute servers
- Some number and type of physical ports with various speeds, and protocols
- Support for various numbers of buffers, and queues for I/O processing
- Coordination with controller, and storage system software to handle front-end tasks
- Respond to in-band management activity
- Provide optional off-load or preprocessing for storage system
- Front-end tasks can include:
  - In-line compression, de-dupe or other DFR activities
  - Metadata tagging, classification, and management
  - Adding data integrity checks and safeguards
  - Caching to speed up subsequent operations

Depending on storage system capabilities, the front-end can present LUN, volumes, file systems, and objects of various size. Likewise, the front-end presents block storage such as a 512-byte, 4K or larger access. Other functions include support for presentations such as VVOL or other device emulation. Actual features and functions are determined by the storage systems capabilities, discussed later in this chapter.

Tip: Note that metadata (data about data) has different meanings so context is important, particularly with storage systems. Storage systems (hardware, appliances, software-defined, services, or other solutions) store, manage, and protect data including its associated metadata. Storage systems also store system metadata (also known as system configuration) along with telemetry data (event, health, status, and activity information). With storage systems metadata, there are different levels, layers, tiers, or depths of metadata, ranging from general system configuration to volume, file system, object, and other status, and mappings, along with snapshots, replicas, clones, and other data services or functionalities.

12.4.3. Internal Processing and Functionality

At the heart of a storage system architecture is the core functionality to support PACE attributes along with associated data services. Note that some systems rely on external add-on data services implemented in an appliance, gateway, or other location instead of having them integrated as part of the core solution.

Depending on the architecture and implementation, the front-end, back-end, data services and other functionalities and features may be implemented on a single system (node, controller, server) or distributed across two or more for resiliency (and performance). Some of the storage system functionality may be off-loaded to front-end and back-end I/O modules, management to in-band or out-of-band modules, RAID, data footprint reduction (compression, de-dupe) or ASIC, FPGA or special cards.

Core internal processing functionality includes managing the underlying physical and software resources such as cache (persistent and non-persistent). Other processing includes general user data, system as well as metadata and memory management. For many solutions, core processing also includes managing I/O (front and back-end as well as east–west intra-system networks), and storage media (HDD, SSD). In addition to managing the resources, the storage system functionality also includes mapping and abstracting the underlying physical (or cloud, or virtual) resources to the presentation of how applications will access and use it.

For example, mapping objects, APIs, file, and block to underlying storage, end-point, implementing availability, data protection, performance optimization, Flash SSD wear-leveling, RAID, mirroring, replication, and thin provisioning, among other features and functions.

> *Tip:* What will vary by architecture and implementation is where and when the various data services are implemented. For example, some systems implement DFR such as de-dupe and compression post-processing delayed until sometime after data is written. Other systems implement DFR at ingestion, when received. Some systems implement various levels of cache and micro-tiering combined with DFR, logs, and journals for data resiliency, among other features.

Additional functions implemented as part of storage system architecture include maintaining event, activity, and other telemetry data for monitoring, planning, troubleshooting, and diagnostics reporting. Other functions include implementing data services, RAID, erasure codes, snapshots, CDP, replication, and DFR features. Additional functions include performing background data integrity checks and repairs along with snapshot management. The storage systems software can also handle clean-up, performing load balancing for performance and availability, as well as using newly added capacity. Some other functions include shrinking and expanding volumes, and conversion of data as well as volumes. For some storage systems, tasks can also include managing virtual guest machines and containers for storage, data infrastructure, or general applications. Note some storage can also host VMs and containers.

Access and protocol-handling functions include block (SAN, DAS) along with the number and size of LUN, sector size, and buffer queues; file (NAS) number and size of file systems, number, and size of files in a file system; objects, and APIs including the size of objects, any limits on some objects, blobs, buckets, or containers, metadata handling, ingest size, and chunking or sharding.

Other functions include the personality or how the system handles access. Personalities can be optimized for general-purpose, small or large concurrent or parallel access, streaming or random, reads or writes. Additional attributes or capabilities include transactional or data protection, among other characteristics. Some other tasks performed include coordinating attributes for multiprotocol access such as different NAS (NFS and CIFS/SMB) reading and writing common data files. Tasks implemented via software as part of the personality include managing quotas, device or protocol emulation, abstraction, security, and endpoint management. The storage software can also handle back-end device management, including number, type, and size of targets (HDD, SSD, tape) or cloud, and virtual endpoints.

12.4.3.1. Storage System Personalities

Storage systems have personalities defined by and enabled as part of their architectures. Some solutions have personalities for specific roles or what they are optimized to act upon, while others can act or perform multiple roles (application and workload support). The personality is how the storage system (hardware, software, or service) functions and behaves. This includes how the system enables various application PACE attributes for general or specific workloads.

Also, the personality can be general-purpose, optimized for smaller environments, application– or data infrastructure–specific such as a transactional database, file serving, video or content streaming, as well as data protection, among others. Table 12.6 shows various common storage system hardware, software, and services personalities.

> *Tip:* Watch out for confusing a storage system's interface, protocols, and medium (back-end devices) with its personality. For example, a storage system can have 10 GbE physical interfaces supporting NFS or iSCSI with NVMe SSD as back-end storage, yet have a personality optimized for performing backups/restores, replication, archives, or other data protection tasks vs. being a general-purpose file server. Likewise, a file server might be optimized for general access or HPC parallel file streaming.

Data protection or protection storage are used as targets or destinations to store backups, clones, and copies, snapshots, and replicas along with archives. While general-purpose storage can be used for data protection, protection storage has a personality that is optimized for protecting large amounts of data in a compact, low-cost footprint.

Besides supporting different physical interfaces such as FC, FC-NVME, NVMe, GbE, and SAS along with protocols such as iSCSI, NFS, CIFS/SMB, HDFS or object (S3, Swift),

Table 12.6. Common Storage System Personalities

Personality	Characteristic	Use For
Analytics Big Data	Durable, relatively low cost for large capacity, streaming, and metadata	Big data, log, and telemetry file data processing, content processing
Bulk storage	High capacity, relatively low cost, DFR (compress, de-dupe), streaming access	Static and reference data, secondary and content applications, protection
General purpose	Mix of PACE attributes, transactional consistent, various access (block, file)	Block and file, SAN, NAS, file serving, database, VSI, and VDI
Protection storage	High capacity, low cost, streaming, durable, DFR (compress, de-dupe)	Data protection (backup, archive, BC, DR), immutable, compliance
Database Transactional	Transactional, small random I/O with strong transactional consistency	Database and key-value SQL as well as NoSQL systems
VDI, and VSI	Mix of SSD and HDD along with AFA, various access, mixed workloads	General virtualization along with VDI for server consolidation
Video and content	High capacity, streaming access to large files, and objects with metadata	DAM and DEM along with other content, video security surveillance

protection storage solutions have various personalities. These personalities are optimized for storing large amounts of data in a dense, reduced footprint using DFR techniques such as compression and de-dupe among others. In addition to DFR, protection storage can also be optimized for sequential streaming of large amounts of date to support archive, backups, and replication.

12.4.3.2. Aggregated and Disaggregated

Storage systems can be aggregated (tightly or loosely coupled), or disaggregated (separate) from servers (compute functions). In a disaggregated model (on the left in Figure 12.10), compute servers access external storage systems resources via some type of server I/O interface and protocol. A server I/O interface that is external to the server can be dedicated or shared, direct attached or networked, block, file, and object, or API.

Examples of server I/O interfaces and protocols covered in Chapters 4, 5, and 6 include PCIe, NVMe, SAS, SATA, Thunderbolt, Fibre Channel (FC), FC over Ethernet (FCoE), NVMe over Fabric and FC-NVME, Ethernet, iSCSI, InfiniBand, RoCE, NFS, CIFS/SMB, SMB direct, HDFS, and object as well as various APIs, among others.

> *Tip:* With disaggregated solutions, storage and compute PACE resources are scaled independently of each other. What this means is that computing can be scaled independently of storage I/O or space capacity, and vice versa.

> *Tip:* Note that in addition to north–south access, storage solutions such as CI, HCI, and clusters, grids, RAIN, and rings also have some east–west or intra-system networking. The east–west network can be tightly or loosely coupled using PCIe, RapidIO, InfiniBand, or Ethernet, among other technologies.

In addition to aggregated solutions in which compute or server and storage resources are tightly integrated and packaged, and disaggregated solutions, there are also hybrid approaches.

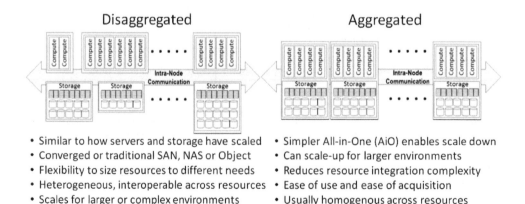

Figure 12.10   Disaggregated and aggregated servers and storage.

Hybrid approaches can have a mix of disaggregated access as well as storage systems (or software) that have disaggregated or aggregated internal architectures. Another hybrid approach is CI, HCI, and CIB that support a mix of aggregated and disaggregated resources.

On the right-hand side of Figure 12.10 is an aggregated approach in which the server and storage PACE resources are scaled in unison with each other. In other words, as more storage is added, compute capabilities also increase, and vice versa. Which type (aggregated or disaggregated) is best depends on your application PACE needs, environment, preferences, what you currently have for hardware and software, as well as future growth plans. You might even discover that a hybrid of aggregated and disaggregated is the right mix for your environment or specific applications.

In addition to servers and storage systems functionality being aggregated or disaggregated, some solutions have distributed architectures. For example, servers can be disaggregated from a storage system such as an SAN, NAS, object, or external solution. The storage system or solution can itself have an aggregated architecture, with one or more controllers or storage processor nodes that have some amount of storage attached. The storage system is disaggregated from the servers, yet internally the storage system is aggregated. Examples of this approach include traditional mid-range and open-systems storage systems, arrays, appliances (physical, and virtual) such as Dell (CS-Compellent, MD-Engenio), Dell EMC (Unity, VNX), Fujitsu Eternus, HDS, HPE, IBM, NetApp (E-Engenio Series, and ONTAP), Nimble, Oracle, and Reduxio, among many others.

Another variation is a storage system that is disaggregated from servers and that also has an internal distributed or disaggregated architecture. Examples include clusters, grid, RAIN, rings, and plexes, among other names.

Storage resources including controllers or storage processors as well as disks can be scaled independently of each other, with caveats. Some solutions are tightly coupled, while others are loosely coupled and can span longer distances—for example, Dell EMC (Isilon, XtremIO), HPE 3PAR, IBM XIV, Oracle, and many others, including ones based on software or CI, HCI. Note that some large enterprise-class systems such as Dell EMC VMAX used to be monolithic frame-based but over time have evolved to a more distributed architecture.

12.4.3.3. Caching, Tiering, and Micro-tiering

Keep in mind that the best I/O is the one that you do not have to do; second best is the one with least overhead and impact on applications. Most applications need to do I/O, so an objective might be to reduce the number of actual I/Os occurring to actual storage systems or devices, as well as their overhead impact. This ties back to data locality of reference, discussed in Chapter 4, where cache and buffers are used to reduce I/O impact. Cache and buffers exist in many different locations from applications to a server, adapters, and controllers cards to network and storage devices.

Cache and tiering are similar yet different, including when, where, and how they are implemented. Both have as objective benefits to improve performance when needed while maximizing space capacity in an economical way. With cache, a copy of data is kept close to where it is being used in a local dedicated or global shared buffer to improve subsequent I/Os while the actual persistent or authoritative copy of data is on some other storage device.

For example, data could be stored on a fast SSD provisioned as persistent storage, or on a slower, high-capacity, lower-cost SSD or HDD. With cache, data is buffered in a persistent (using NVM such as SSD or another form of SSD) or non-persistent (DRAM) media. If an I/O can be resolved from cache (i.e., cache hit), then its impact will be lower, its response faster. Likewise, if data is not in cache (i.e., cache miss), then an actual I/O to the persistent storage occurs.

With tiering, the data gets migrated (promoted, demoted, moved) from one tier to another to speed up performance, or to reduce cost by relocating to slower, lower-cost storage. Different solutions vary in the size or granularity of what is moved, when, and based on different policies. A variation of tiering is micro-tiering, which occurs at a smaller granularity with less overhead vs. traditional tiering. Think of micro-tiering as a hybrid or best of both worlds between cache and tiering. For example, with cache, if you have a 1-TB persistent SSD as the acceleration storage sitting in front of a 100-TB storage tier, your aggregate space capacity will be 100 TB, but with the performance benefit of cache.

Table 12.7. Cache and Tiering Options

	Characteristics	Considerations
Micro tiering	Enables all or part of a volume, file, or object to be promoted (or demoted) from a slow to a fast tier. Unlike cache, which can result in stale reads, micro-tier data is consistent. Micro-tier provides an aggregate capacity of fast and slow tiers.	What is the time delay and system overhead impact when data is promoted to improve performance? QoS options including time- or activity-based, reads, writes, and I/O size. Can files or objects be pinned to cache?
Read cache	As its name implies, data is buffered in a cache to speed up subsequent reads. There can be multiple levels of read cache from applications to the database, file systems, operating systems, and hypervisors, as well as in storage devices.	How will cache be rewarmed (reloaded) after system crash or controlled shutdown? What is the performance impact while the cache is warmed? How are stale reads prevented, for data consistency?
Write-back cache	Data is buffered in persistent or nonpersistent storage memory to speed up write operations and subsequent reads. Cache may be mirrored locally or remotely. Similar to read cache, write-back cache can be implemented in various locations. Various standby power approaches prevent data loss due to power failure.	How is data integrity consistency maintained across multiple storage systems, servers, or nodes for updates to prevent stale cache? Is cache mirrored, and nonvolatile in case of power loss or system crash? Battery backup should be used to prevent data loss should power fail.
Write-through cache	Data is written synchronously in real time for integrity consistency, which could impact write performance. Cache hit improves performance if resolved from the cache.	Performance is not as fast as a write-back cache; however, data consistency is maintained. How will data consistency occur across nodes?
Read-ahead cache	Data is prefetched based on predictive access patterns, then buffered to speed up subsequent requests.	Configuration options for how much data to prefetch, how to eliminate stale reads for data consistency.
Tiering	Data (or a portion of it) is moved from one location (tier, class, or category) of storage to another, automatically, manually, or semi-automatically. Move from expensive fast SSD to lower-cost high-capacity storage or vice versa, or from local to cloud storage for archive or cold data.	Performance impact during the tiering or movement, granularity (LUN, volume, file system, file, or object). Does entire item get moved, or only a portion? What is left behind (i.e., is old data removed)?. System overhead and resources are used for movement.

Likewise, you would have a 100:1 storage space-to-storage cache size ratio. With tiering, you have the same space capacity (e.g., 100 TB), but if the entire volume were moved (promoted) to faster storage, you would need 100 TB of SSD (unless the solution allows smaller granularities). With micro-tiering, you would have an aggregate storage space of 101 TB (i.e., 100 TB + 1 TB), with data promoted and demoted by the tiering engine. Unlike cache, which needs to be rewarmed (i.e., reloaded or refreshed) after a system shutdown/startup, the micro-tier data is persistent, which also eliminates stale reads.

Another aspect of cache is that some systems can do more with less, while others simply use a large amount (they are cache-centric) to support large-scale consolidation or aggregated workloads. Some solutions allow you to pin LUN, volumes, files, or objects to cache so that they are not displaced. Think of pinning as locking the data down into cache for predictable (deterministic) performance.

Cache considerations include:

- High cache utilization simply tells you the cache is being used.
- High hit rate gives you an indicator of cache effectiveness benefit.
- Low hit rate with high utilization may indicate non-cache-friendly data.
- Look at activity (IOPs, TPS) as well as latency to gauge performance benefit.
- Streaming or sequential operations can benefit from read-head cache.
- Look for solutions that prevent stale cache, to avoid cache inconsistency.
- A little bit of cache in the right place with the right algorithms can have big benefits.
- How long does it take for cache to be rewarmed (reloaded) after start-up?
- Does the cache solution require extra hardware or software tools, and drivers?

Table 12.7 shows various cache and tiering options that can be implemented in various locations, from applications, databases, file systems, storage applications, operating systems, hypervisors, drivers, plug-ins, appliances, controllers, adapters, cloud, and storage devices (HDD, SSD). Which cache approach is best depends on your application PACE needs.

Which approach is best? It depends on your application, systems, and other considerations.

12.4.3.4. Servers for and as Storage

In addition to using storage to support server functions and applications, a data infrastructure application can be for a server to be configured as a storage server, storage appliance, storage gateway, storage system, or database appliance, among other roles.

In addition to defining the server with storage software to make it a storage system, various I/O adapters or ports are also included. Other resources include CPU, memory, PCIe bus slots, and storage devices (internal or external).

Tip: Remember that CPU, memory, and I/O resources are needed to implement storage software and associated data services. Many small I/Os (e.g., IOPs or transactions) require more CPU than fewer yet larger I/Os (e.g., bandwidth). Likewise, CPU and memory are needed for performing primary I/O functions as well as background tasks including snapshots, replication, DFR (compress, de-dupe), snapshot management, and data integrity checks, among other tasks.

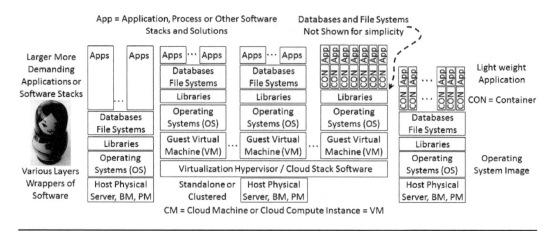

Figure 12.11   Physical, virtual, cloud, and container servers for storage.

Servers can be general-purpose commercial off-the-shelf or engineered and configured for specific uses. For example, a general-purpose motherboard can be engineered into a solution by specifying the type of processor (cores, threads, and features), type and amount of memory, I/O adapters, and ports, among other features. Other enhancements can include custom cards, adapters, ASIC, and FPGA, or even a custom built-to-order controller or server motherboard and chassis.

In addition to physical servers and storage expansion enclosures, storage software and the architecture they implement can be deployed on virtual machines, cloud instances (also virtual machines) as well as via containers, as shown in Figure 12.11.

As shown in Figure 12.11, storage software can be implemented on bare-metal physical servers, via virtual machines, and using containers. The storage software can also be implemented in various layers of the data infrastructure stack to perform various functions. The doll on the left of Figure 12.11 represents the layering, and how storage software and functionality can be wrapped inside other software-defined storage such as virtual storage, VMDK, VVOL, and VHDX.

12.4.4. Storage Back-End Media Matters

Similar to the front-end that handles north–south I/O traffic between application host compute servers while interfacing with controllers running the storage system functionality, back-ends handle the north–south traffic between controllers and actual storage devices. Storage system back-ends are what provide access to actual storage media and devices such as HDD, SSD, or PCIe cards. In addition to component devices located in drive slots or expansion enclosures, back-ends can also be used for accessing other storage systems as well as cloud services, depending on the implementation.

Back-end storage devices can be SAS, SATA, or PCIe NVMe based on various industry standards and custom form-factor packaging. Some systems support attachment of dual-ported HDD and SSD for redundant paths, while others are single attached devices. Note some context here is that a single-ported HDD or SSD can be installed into an enclosure that in turn can be dual-attached to one or more controllers. Most storage back-ends are block (SCSI/SAS, SATA, NVMe)–based, although some can support files or objects.

A variation of objects is Seagate's kinetic key-value drives with an Ethernet interface. Not to be confused with an Ethernet and IP interface for iSCSI, NBD, or NAS, kinetic drives rely on storage server software to implement an API where data is stored using a key-value object implementation. The context here is that "object" does not mean S3 or Swift, among others.

Tip: Some storage software requires that back-end storage devices be in raw or pass-through mode, meaning no abstraction via RAID or other management. For example, VMware vSAN and Microsoft S2D, among others, need back-end storage to be natively accessible as their software handles device mapping and management.

Back-end characteristics include:

- Handle north–south storage access from the controller to storage devices
- Some number of physical ports of given type, speed, and protocol
- Implement protocol handling including buffering
- Support some number of LUN, targets, or volumes of a specific size
- Can implement in conjunction with controller encryption and caching

12.4.5. Common Storage Architecture Considerations

In addition to the various characteristics and attributes already mentioned, the following are some additional common architecture considerations for storage systems. Keep in mind that context is important in that a storage system can be a software-defined cloud, virtual, container, or physical bare-metal (tin-wrapped) implementation. Likewise, storage system architecture functionality implemented in software can be deployed in various ways.

Storage functionality gets implemented via software (algorithms and data structures) packaged in various ways (FPGA, ASIC, firmware, BIOS, device drivers, utilities, hypervisors, operating systems, plug-ins). Software (even cloud, and virtual) needs hardware; hardware needs software. People and software tools define the functionality of hardware, software, and services to be a storage system part of a software-defined data infrastructure.

Legacy storage systems or appliances have relied on a mix of propriety hardware and software, along with open or commodity technologies. For some, a server is used as a storage system defined by software, or perhaps a networking switch defined by software may be new, and revolutionary. The focus of storage systems should not be as much about the architecture, rather, what the solution enables as an outcome to support applications and data infrastructures.

A storage system architecture and its features, data services, and functionalities that define underlying hardware (and software or services) resources can be themselves components of larger architecture or solutions. For example, a storage system appliance or software implementation on a server can be part of a CI, HCI, and CiB solution along with other technologies.

The resulting CI, HCI, or CiB solution can, in turn, be part of a larger footprint, stamp, plex, cluster, or rack-scale solution. For example, multiple servers with internal or external storage defined with storage software to function and have the personality of a storage service can be part of a rack-scale stamp or cluster. The resulting stamp or cluster can then, in turn, be part of a larger data infrastructure deployment making up a site, availability zone, fault domain, or region. More about availability zones, regions, and resiliency later in this chapter.

> *Tip:* When looking at storage system (hardware, software, appliance, service) features, data services, access, PACE attributes, and other functionalities, look beyond checkbox items. Some solutions simply mention speeds, feeds, slots, and watts specifications, while others provide more extensive information. For example, some might say they have GbE ports or IP storage, but not indicate whether those GbE ports are for data access or management. They may not say whether the IP storage is iSCSI, NBD, FCIP, NFS (which version), CIFS/SMB (which version), or object (which object variant).

Another example is that some solutions mention HTTP rest access, but may not be clear whether that is for management or actual data access. Or they may include snapshots and replication without indicating the number or type of snaps, synchronous or asynchronous replication, among other considerations. Concerning PACE characteristics, some vendors specify an IOP or bandwidth number without context, or simply indicate some number of ports at a given speed with SSD, so you assume the solution is fast.

> *Tip:* Some vendors provide additional context, including type and size of benchmarks along with metrics that matter. While it is important to consider speeds and feeds, it is the sum of the parts, how those features work and are leveraged to support your applications and data infrastructure that matters.

Some additional common storage system considerations are dependencies such as the following.

Hardware – Servers with internal or external storage enclosures, storage devices (HDD, SSD, PCIe, or custom components), I/O network, and specialized processing cards (ASIC, FPGA, GPU's custom firmware). CPU, memory, I/O bandwidth, and storage space capacity.

Software – BIOS, firmware, hypervisors, operating systems (or kernel), containers, device drivers, plug-ins, management utilities, and tools.

Network – Hardware, software, and services for bandwidth, and availability.

Cloud or managed services – Combine hardware, software, processes, and policies into a data infrastructure with various applications, services, or functionalities.

Facilities – Physical, cloud, or managed services habitats where data infrastructure technologies live. This includes storage systems, services functionality to protect, preserve, and serve information. It also includes electrical power (primary and standby), HVAC cooling, smoke and fire detection, notification, and suppression, as well as physical security.

In case you were wondering where software-defined storage is or why it is not listed as part of architectures, no worries. We are about to get to that, plus we already covered it. Keep in mind, as has been mentioned throughout this book, that bytes, blocks, and blobs get defined to implement various data structures including to represent software-defined data infrastructure.

Likewise, algorithms are defined to implement the processing, logic, rules, policies, and personalities or protocols associated with the data structures. The result is software-defined storage that gets implemented and packaged in various ways. While there is some architecture that depends on or is designed around some hardware or software dependencies, most are independent of hardware (and software).

Granted, how the architecture gets implemented and deployed can result in technology being perceived as hardware- or software-centric, or even architecturally based. In some cases, marketecture (i.e., marketing architecture) defines how a storage system, solution, or service appears, is packaged, and sold, rented, leased, or subscribed to.

12.5. Storage Solution Packaging Approaches

Storage functionality today is implemented, and exists, is accessed, and used as well as managed via different packaging approaches. To some people, a storage system is a large SAN or NAS sharing a propriety hardware-centric solution (tin-wrapped software). Others will see a storage system as being software in a hypervisor, operating system, third-party software, virtual or physical as well as cloud-based. Still others will see a storage solution as CI or HCI, among other variations.

Figure 12.12 shows how various storage functionalities and data services that make up storage system (solution or service) capabilities defined by software get packaged for deployment. Storage functionality software is often deployed and packaged as "shrink-wrapped" or "downloadable" software to run on a bare-metal physical server. The software can also be prepackaged as a turnkey appliance or storage system solution including a server and associated storage hardware devices (i.e., tin-wrapped). Another variation shown in Figure 12.12 is storage software deployed as a virtual storage appliance running in a virtual machine, container, or cloud instance.

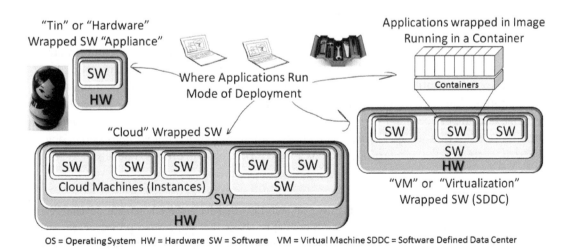

OS = Operating System HW = Hardware SW = Software   VM = Virtual Machine SDDC = Software Defined Data Center

Figure 12.12   How storage functionality defined by software gets wrapped.

Not shown in Figure 12.12 is how storage functionality defined by software, besides being deployed on general-purpose servers, can be packaged as firmware, microcode, or as part of an ASIC or FPGA to handle specific functions (compress, de-dupe, RAID, EC, and LRC, among others).

What is common across all of the above, among others, is some storage capability, functionality, and data services that provide resources to meet data infrastructure and application PACE needs. There are various architectures and personalities or optimizations to support different PACE and application needs. However, the big difference across storage systems, appliances, systems, and services is how they are packaged, deployed, and consumed.

Tip: Storage systems are packaged or wrapped in different ways to meet various needs while protecting, preserving, and serving data. Which is the right or best storage solution or packaging approach? The one that meets or exceeds your application and environment needs.

Additional storage system functionality packaging considerations include:

- Do-it-yourself vs. turnkey solution or appliance
- Open source vs. proprietary software
- Standalone, clustered, scale-out, CI, HCI, CiB
- Open commodity white box vs. engineered or proprietary hardware
- Deploy on bare metal, virtual, container, or cloud instance
- All NVM and SSD, or all HDD, or a hybrid of SSD and HDD
- Additional hardware, software, network, or management tool dependencies
- Aggregated compute and storage or disaggregated solutions
- Freedom and flexibility to use your existing servers and storage

12.5.1. Tin-Wrapped and Hardware Defined

There is a common myth that only newly developed or newly released software implementing storage functionality is software-defined storage. Another myth is that software-defined storage only applies to software that is a virtualization technology, part of an operating system, hypervisor, or third-party tool. Another common myth is that many of today's existing storage systems, and appliances have software tied to specific hardware and are not portable. Then there is the myth that storage software can only be software-defined if it is open-source, object-accessed, scale-out, from a start-up, and has a robust API, among other variations.

Tip: There are still a handful of storage system solutions that have some hardware dependencies such as an ASIC, FPGA, special PCIe card, or adapter. Note that some cloud providers are using ASIC (Google) and FPGA (Microsoft) in their custom-built servers to accelerate and optimize some functions. Some of these custom server designs can be found as part of the open compute project and are also used as platforms for running storage software (e.g., software-defined storage).

While the trend has been for about a decade or more moving away from proprietary main or motherboards as storage controllers, there are still some with custom controller boards. However, most storage systems and appliances are based on standard off-the-shelf server and processor components; granted, some of these are engineered to build to a specific standard, for example, having additional PCIe or other I/O slots or interfaces.

The reality is that with the majority of storage systems in the market today, although they are bundled, packaged, or tin-wrapped with a given vendor's choice of hardware (i.e., an engineered solution), the actual software is portable. Some software is more portable vs. others, for example, able to run on any Intel 64-bit processor with enough memory and I/O devices. Other software may require some number of GbE, PCIe, SAS, NVMe, or other devices as well as CPU processors of at least a specific make, model, speed, and functionality. For example, some VSA will run fine on VMware vSphere ESXi installed on an Intel NUC, while others do a CPU and hardware check for minimum configuration support.

Most vendors have in the past, as well as today and into the future, package or wrap their storage software and functionality in tin or metal as a turnkey hardware appliance for ease of marketing, selling, and deployment. However, many of these same vendors, such as Dell EMC, HPE, NetApp, and many others, also make available the software that was traditionally available only as a hardware appliance. For example, you can buy a turnkey hardware system or appliance from Dell EMC such as VNX or Unity, or you can download the software and run it as a VSA. Another trend is to run the storage software not only on bare metal or a VSA but also in a container or cloud instance.

Likewise, there is new software that implements storage functionality that ends up being tin-wrapped by the customer, service provider, or system integrator for ease of use. The difference between a server running an operating system such as Linux or Windows Server configured with several HDD or SSD, and a storage system or virtualization appliance, can be blurry. The reason is that some storage systems and appliances have a Linux kernel or Windows operating system underneath the specific storage software.

Tin-wrapped simply means storage software that is prepackaged or delivered as an appliance or storage system with hardware included. A vendor may prepackage its storage software on an engineered industry standard server or simply use commodity or "white-box" servers and storage enclosures with their name applied.

Some vendors only offer their software as tin-wrapped, implying that they are hardware-dependent and non-software-defined. Other vendors promote software-defined and hardware-less (no hardware or hardware-independent) yet offer a tin-wrapped appliance for ease of deployment.

Just as there are different types of servers, networks, and storage devices as well as systems, there are also various tin-wrapped (hardware supplied) storage options. Keep in mind that hardware requires software; software needs underlying hardware somewhere in the stack. Also, storage software relies on some server with storage devices attached. With the various types of servers available, there are thus many options for wrapping your software-defined storage with different kinds of hardware.

Figure 12.13 shows how a server with attached storage gets defined with various hardware options and software to provide storage system functionality for various application workloads. Note that some storage system software and functionalities have hardware as well as software or other resource dependencies. Everything is not the same, so there are various options for scaling PACE attributes along with data services enabling various functionalities.

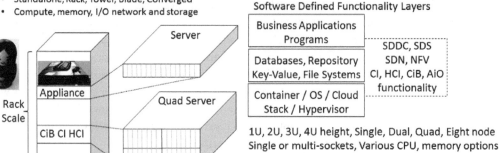

- Supports various application workloads
- Span different types of environments
- Compute node, network or storage platform
- Hardware and software defined functionality
- Commodity white box or custom engineered
- Standalone, Rack, Tower, Blade, Converged
- Compute, memory, I/O network and storage

- Servers also known as appliance, platforms, nodes or systems
- Single or Multiple Servers, Standalone or Clustered or Grid
- Some are disk less others have large number drives (HDD or SSD)
- Various expansion options exist depending on packaging
- Different options for defining how the server platform gets used

Figure 12.13   Server storage hardware defined with software (tin-wrapped).

12.5.2. *Software-Defined Storage*

Software-defined storage (SDS) can mean many different things to different people. The reason is that, as mentioned earlier, all storage systems, appliances, services, and solutions include functionalities implemented in software, which then gets packaged, and delivered in various ways. For some people, SDS means storage functionality that does not exist in a traditional storage system or appliance. Other people may see SDS as being storage functionality that is delivered as software you can run on your hardware, or optionally packaged as a hardware appliance (tin-wrapped). Still others will see SDS as being something new, and not including older storage solutions that were or are in fact hardware-independent yet are packaged with hardware (tin-wrapped) for ease of sale.

> *Tip:* A myth is that some storage systems are software-defined if they are new from a start-up vs. some that are from legacy or other start-ups. Many vendors only offer their storage software tin-wrapped as a system, giving the impression that it is hardware-dependent even if it is not. The reality is that it depends.

At the time of this writing, I can count on one hand the number of storage systems, appliances, and solutions that are truly hardware-dependent. Many vendors who in the past sold their storage software as hardware are now decoupling the software that is portable, and are now making it available via downloads or other means. What this means is that there are many "tin-wrapped" software-defined storage appliances, systems, and solutions in the marketplace.

There are many variations of software-defined storage, including functionality built into hypervisors such as VMware vSAN or operating systems such as Microsoft Storage Spaces Direct (S2D). There are also third-party SDS stacks and solutions that can be installed in or on top of hypervisors as guest VM, as virtual storage appliances virtual storage arrays (VSA), or as part of an operating system, among other approaches including cloud as well as containers.

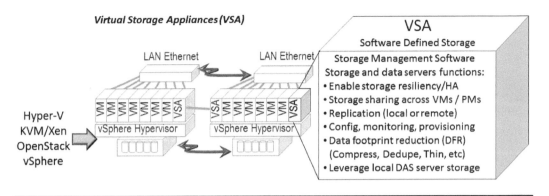

Figure 12.14   Software-defined storage VSA.

Tip: By default container or micro-services have ephemeral non-persistent storage, however they can be configured for persistence. Chapter 14 expands on containers, persistent storage including installing software-defined storage into containers to function as a storage system.

Figure 12.14 shows an example of a VSA. The VSA itself can be many different things, including simply being a virtual machine hosted version of traditional storage software commonly deployed as a hardware appliance or virtual storage solution. Some examples of VSAs include those from Cisco, Datacore, Dell EMC (ECS, ScaleIO, Unity, and others), FalconStor, HDS, HPE, IBM, Microsoft, NetApp, Nexenta, Nutanix, and Starwind, among many others.

Figure 12.15 shows another software-defined storage system implementation with VMware vSAN as an example. The VMware vSAN storage system functionality is implemented within the hypervisor, enabling local direct attached storage on physical servers (VMware hosts) to be protected and shared across vSAN nodes. Various data services functions complementing data protection capabilities that are part of VMware vSphere and related tools. vSAN storage is directly accessible to nodes in a cluster that can scale to 64 nodes (perhaps larger in the future).

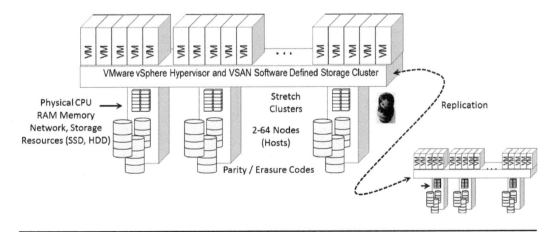

Figure 12.15   Software-defined storage VMware vSAN.

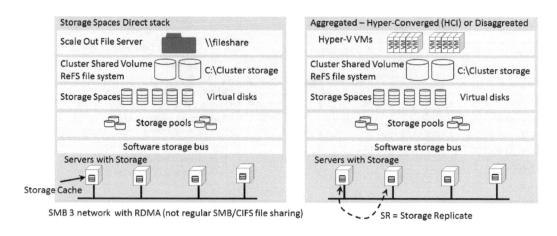

Figure 12.16   Various software-defined storage approaches.

Figure 12.16 shows another software-defined storage system implementation with Microsoft Storage Spaces Direct (S2D) along with Storage Replica and Scale-Out File Server (SOFS) that are part of Windows Server. S2D can be configured for aggregated as well as disaggregated deployment using local direct attached storage. Note that the software storage bus used to connect S2D nodes together uses SMB 3 Direct, which is a low-latency deterministic RDMA and RoCE-based protocol. This is important context: Don't confuse it with traditional SMB IP-based file sharing for the local and wide area.

On the lower left, Figure 12.16 shows storage cache which is implemented in S2D below the file system layer leveraging NVM including SCM. For data consistency, performance data that is written goes into a mirrored NVM journal log that can be used for resolving subsequent read operations. As data cools, becoming less frequently accessed, it is destaged from fast

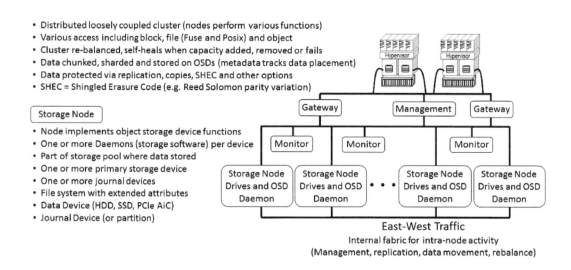

Figure 12.17   Scale-out, bulk, object, loosely coupled software-defined storage.

NVM and SSD tiers to slower SSD and HDD tiers that are protected with parity-based erasure codes. S2D storage can be replicated to another location using synchronous and asynchronous replication, also part of Windows Server 2016. As of Windows Server 2016, ReFS (Reliable File System) is now the default file system instead of NTFS.

Figure 12.17 shows another software-defined storage system, a loosely coupled cluster (grid, RAIN, or ring) for scale-out, bulk, and object storage. The example shown is based on Ceph, but it loosely also applies to many another cloud, object, grid, cluster, RAIN, and scale-out storage providing block, file, object, and API accessible storage. The architecture leverages general-purpose servers with internal or external dedicated storage, and software that defines various node-type functions. Some nodes are defined as management, monitors, and gateways or for actually storing data (i.e., storage).

In Figure 12.17 the storage nodes receive, store, or retrieve data that in turn gets stored on underlying storage devices (HDD, SSD, PCIe AiC, or cloud). The storage software defines the function and characteristics of the storage nodes, as well as the monitoring and management nodes that include determining where to place data, how to protect it, and how to access it.

Figure 12.18 shows another software-defined storage system approach that is used by different solutions from scale-out to CI and HCI (with variations). Storage system controller functionality along with optional data services are implemented (i.e., packaged) as a guest VM (or in parent partition). The storage controller VM communicates and manages underlying physical (virtual or cloud) storage resources, abstracting and presenting storage to other VM.

Depending on the implementation, data and management traffic may flow from VM to the storage controller VM, and then to underlying storage (in-band), or out-of-band. With out-of-band, the storage controller VM gets involved only with management activity, setup, configuration, allocation, failover, or other tasks. The result of out-of-band is management that steps in when needed to help out and then gets out of the way of productive work being done.

Scaling is accomplished by adding additional nodes to support PACE requirements. Storage controller VM or storage node software communicates with other storage nodes via some communication fabric or network. Depending on the implementation, general-use VM (or physical

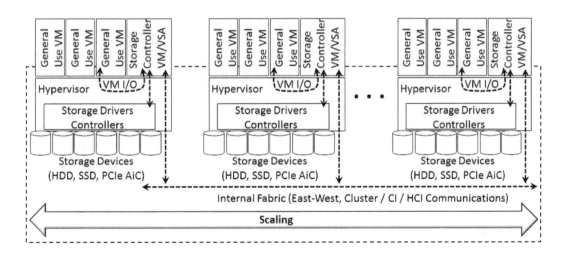

Figure 12.18   Software-defined scale-out, CI, and HCI storage.

bare-metal, cloud, or container) servers can plug and play with the storage controller using a standard block or file access and protocols.

Some implementations require an additional plug-in, agent, driver, or software-defined storage management, and volume managers or file systems to access shared storage. Likewise, some implementations only make the storage accessible within the cluster or given namespace. Meanwhile, others using drivers or plug-ins, can share data outside the cluster, including with non-virtualized servers. Many CI and HCI solutions use a variation of Figure 12.18 as part of their architecture, with implementation packaged in various ways.

12.5.3. Virtual Storage and Storage Virtualization

Storage systems implement storage virtualization enabling virtual storage in many different ways depending on their architecture and packaging. The virtualization functionality varies from simple abstraction to aggregation, pooling and associated endpoint management, and presentation to host servers, as well as more advanced functionalities.

Different implementations provide various features, functionalities, data services, and enablement. Some storage solutions are storage virtualization software deployed as tin-wrapped storage systems or appliances, such as Dell EMC (ViPR), DataCore, IBM SVC, among others that support attachment of third-party storage. As an example, IBM also has the V7000 based on SVC with dedicated storage as a turnkey storage system. IBM also leverages their SVC software running on appliances to add data services functionality to their Flash system SSD or AFA storage systems. Another approach is used by vendors such as Dell EMC (VMAX) as well as HDS VSP G systems and NetApp, among others, whose systems support attachment and abstraction of third-party storage systems.

> *Tip:* In addition to being block, virtualization can also be file or object storage. Different file systems or object and cloud endpoint and namespaces can be abstracted into a common managed namespace.

Figure 12.19 shows on the left a generic example of storage software performing virtualization functions of underlying storage resources. The underlying storage resources can be JBOD, storage systems, or appliances as well as cloud storage, depending on different solutions. The storage software at left center performs management tasks, enabling PACE attributes, data services, and other feature functionalities shown on the right of Figure 12.19.

The top right of Figure 12.19 shows applications and server access (front-end) using block, file, and object as well as API methods of virtual storage, and associated data services. The bottom right of Figure 12.19 shows various underlying back-end storage, including block, file, and object endpoints.

Virtual storage and storage virtualization considerations include:

- Where the functionality is implemented (storage system, appliance, or elsewhere)?
- What virtualization functionalities and data servers are enabled?
- Is there interoperability with third-party storage systems and services?
- Is the virtualization in-band (in the data path) or out-of-band?

- What performance or space overheads are introduced by the solution?
- Are additional host server or client drivers, plug-ins, or software tools needed?
- How is data from external or third-party storage imported into virtual storage?
- Does data have to be unloaded, then re-imported from third-party storage?

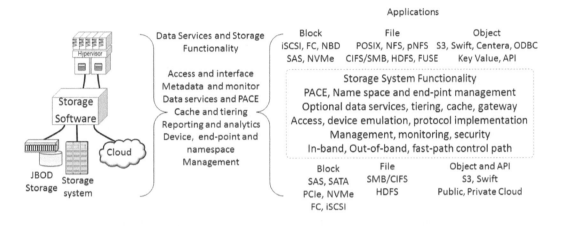

Figure 12.19   Storage virtualization.

12.5.4. Hybrid, All-NVM, All-Flash, and All-SCM Arrays

Different storage systems are defined by their architectures along with personalities to support various data infrastructure and application PACE needs. Some storage systems are focused on general purpose, supporting a mix of HDD and SSD with a balance of PACE attributes (i.e., hybrid), while others are optimized around, for example, all Flash (i.e., all-Flash array or AFA).

Other variations of storage systems are defined by the type of storage media besides all-HDD, SSD, and hybrid, including Flash, NVM, or RAM, among others. Table 12.8 shows various types of common acronyms or categories, and characters of storage arrays, appliances,

Table 12.8. Types of Storage Arrays, Appliances, and Building-Block Components

Name	Characteristics
AFA	All-Flash array, a storage system using all Flash, NVM, or SCM
ANA	All-NVM array, a storage system using all NVM, Flash, or SCM
ASA	All-solid-state array, a storage system using only SSD
Hybrid	A mix of HDD and SSD using Flash, SCM, other NVM combined with storage software running on a server, storage controller, or adapter adding data services to other functionalities to define a storage system, service, appliance, or solution
JBOD	Just a bunch of disks that relies on storage software that is part of a storage system, controller, storage node, adapter, hypervisor, operating system, or layered solution to define the hardware into a storage solution. Note that the disks can be all HDD or SSD or a hybrid mix.
JBOS	Just a bunch of SSD or SCM that functions similar to JBOD except that all storage media are NVM Flash or SCM
VSA	Virtual storage array or appliance that implements storage functionality and data services either in a hypervisor or as a guest virtual machine using underlying physical or virtual storage devices

and building-block components. Which is best? It depends on what your application workload PACE characteristics are along with other requirements of your environment.

Note that JBOS (and JBOD) using only SSD (or SCM), when attached to a server running some storage software, a RAID adapter card, a storage controller, or another solution that implements data services as well as other functionalities becomes an ASA. Likewise, a hybrid array or storage system supports both HDD and SSD, adding data services and other functionalities to those resources.

Also, note that items in Table 12.8 can be combined to create other types of storage solutions. For example, JBOD and JBOS are often combined with software running on a server platform to create a software-defined storage solution that is part of a hypervisor such as VMware vSAN or an operating system based such as Microsoft Windows Storage Spaces Direct (S2D).

Other possibilities include layering storage software on a server with JBOD, JBOS, or other types of storage arrays and attached systems. For example, using software such as Dell EMC ScaleIO, Ceph, DataCore, IBM SVC (available as a tin-wrapped appliance), OpenStack (Swift and Cinder), Swift stack, NooBaa, OpenIO.

12.6. Converged Infrastructure (CI) and Hyper-CI (HCI)

Keep in mind that everything is not the same in IT (or non-IT) data centers as well as organizations with various applications, workloads, size, and complexity, among other diversities. Likewise, not all CI, HCI, CiB, AiO, or software-defined solutions are the same. Some scale down, some scale up, some scale out, some are focused on particular applications, workloads, or market segments. Some solutions are focused on smaller to mid-market (SMB and SME) as well as larger organization workgroup, departmental, and remote office/branch office (ROBO).

HCI does not have to mean hyper-small, hyper-compromised, or for hyper-small environments. Some solutions are architected for smaller environments and specific workloads, while others can scale up, scale out, and scale across different applications workloads without introducing complexity or compromising performance.

> *Tip:* Keep in mind your focus or objective for convergence, for example, converging management functions, data protection, hardware, software, or something else. Are you converging around a particular hypervisor and its associated management software ecosystem such as Citrix/Xen, KVM, Microsoft Hyper-V, OpenStack, or VMware? Other things to keep in mind are how your congruence will scale over time, and whether it will eliminate pools, islands of technologies or point solutions, or create new pockets, pools, and islands of technologies.

Figure 12.20 shows many different aspects of server and storage I/O convergence along with various focus themes as well as dimensions. For example, are you looking for a do-it-yourself solution combining different hardware and software into a solution you define leveraging reference architectures or "cookbooks"? Or are you looking for a turnkey solution, in which everything is integrated—hardware, software, networking, server, storage, management, and applications? Perhaps you need something in between, with some assembly required to customize to fit your environment, and other existing technology as well as management tools.

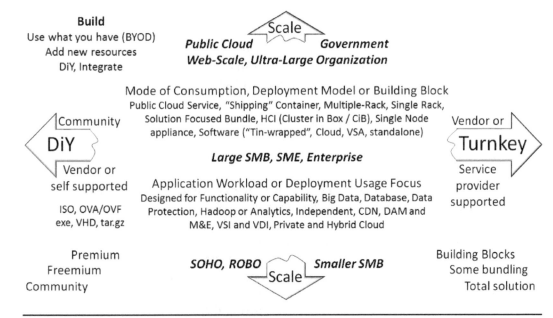

Figure 12.20   Various CI and HCI considerations, and market characteristics.

Also shown in Figure 12.20 are different scales, from small to SMB and ROBO bottom center, to large-scale enterprise, cloud, service provider, or government environments (top center). Other characteristics include, in the center of Figure 12.20, the applications and workloads. For example, is your focus VDI, and general VSI, or database, email messaging, big data, and Hadoop or other forms of analytics, video DAM, and DEM or content servers, among others?

Remember that everything is not the same, from environments to applications and systems. This also applies to CI as well as HCI. Some solutions are focused on larger SME and enterprise, others on larger web-scale, hyper-scale, cloud, managed service provider, or very large enterprise.

Some other differences include turnkey, either application specific or general purpose; others are turnkey with some assembly required, still others are all packaged yet assembly is required, while yet others are reference architectures with cookbooks for doing it yourself.

Likewise, some solutions are heterogeneous regarding what hardware is used as well as allowing bring-your-own-license (BYOL), while others are homogenous but allow new extra or outside hardware even using the same software.

Just as there are different markets, industries, sizes of organizations, application workloads, and solutions, there are also various ways to deploy and use CI, HCI, CiB, and AiO solutions. These include as a consolidation platform, application specific such as VDI or VSI consolidation, and general-purpose workload, to big data and little data database to video, and content systems as well as data protection.

Tip: It is not one size (solution) fits all scenarios when it comes to CI and HCI, although there are some solutions that tend to fit the various markets. Thus, don't limit your view of what CI and HCI are or can be based on what some solutions are limited to; instead, look at the broader market, and how different solutions or approaches fit various needs.

Storage system functionality is implemented either in the hypervisor or as part of the underlying CI, and HCI software functionality. Similar to a traditional storage system, various storage functions and data services are available which vary by vendor implementation. Some solutions today only support a specific hypervisor. Other implementations support different hypervisors on various nodes or in separate clusters, while by the time you read this there may be others that have implemented heterogeneous mixed hypervisor CI and HCI solutions, along with bare-metal operating systems and container support.

CI, HCI, and CiB considerations include:

- What hypervisors and bare-metal operating systems or containers are supported?
- Can you choose your hardware, including external expansion storage?
- Can you use your existing software license as part of a solution?
- What are the granularities of increasing compute, storage, or I/O networking?
- Do you need aggregated, disaggregated, or a hybrid approach for your applications?
- Are specialized nodes such as GPU, compute, or storage-intensive needed?
- What is the east–west intra-system I/O network or fabric?
- Is storage accessible only to virtual machines part of the converged solution?
- What are the data services along with management features of the solution?

12.7. Cloud Instances and Services

There are many different types of cloud storage with various facets, functions, and faces as well as personalities (what they are optimized for). Some clouds are public services; others are private or hybrid. Particular cloud storage may be accessible only from a given cloud service such as AWS, Azure, or Google, among others. What this means is that these cloud storage capabilities, for example, AWS Elastic Block Storage (EBS), can only be native, accessed directly from within AWS. On the other hand, some cloud storage capabilities can be accessed externally, such as AWS Simple Storage Service (S3) objects, Azure Blobs or Files, and Google GCS, among others. Note that these externally accessed storage services and capabilities can also be accessed from within a cloud.

Some cloud storage is accessible as what appear to be storage targets or endpoints such as blocks (iSCSI LUN), files (NFS, SMB/CIFS, and HDFS), or objects (S3, Swift). Other cloud storage appears as an endpoint service, such as SQL/NoSQL, queues, tables, or APIs. Also note that, similar to a physical server with direct attached storage (HDD or SSD), cloud instances or cloud machines which are virtual machines can also have dedicated storage such as SSD.

> *Tip:* Why or how you would use storage that is part of an instance is the same as you would with storage that is direct attached to a physical server. Likewise, cloud instances with dedicated storage can be configured as persistent or ephemeral, leveraging cloud services for protecting storage (backups, snapshots, replicates).

Besides storage endpoints, storage can also be accessed via a service (AaaS/SaaS) such as file or data sharing (e.g., Dropbox or similar from Apple, Google, and many others), backup,

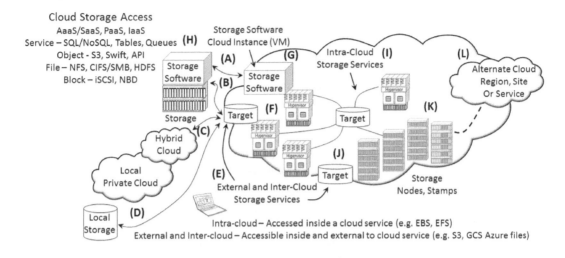

Figure 12.21   The many different facets of cloud storage.

archive or DR, and data protection service, as well as many others. Another variation of cloud storage is a cloud instance where storage software runs on a cloud machine (instance) functioning as a peer node to other like storage systems. Another cloud storage variation is a gateway implemented in a storage system, data infrastructure application such as backup, replication, and archive, or as a stand-alone gateway appliance.

Besides how cloud storage is accessed, its personality or what it is optimized for, and where it is located, there are also different types of media, for example, HDD and SSD, based on various PACE attributes and data services.

Figure 12.21 shows the different faces and facets of cloud storage. Besides being public service, private, or hybrid, cloud storage also has many different facets and personalities. For example, there are storage and associated data services accessible only from within a cloud (intra-cloud), such as AWS EBS or EFS. There are also cloud storage and data services that can be accessed externally (inter-cloud) from a cloud (as well as via internally), such as AWS S3, Google GCS, and Azure Blobs and Files, among others.

Figure 12.21 shows at top left different access methods and personalities of cloud storage. Also shown are cloud services such as for file and data sharing, as well as storage targets. Additional insights into Figure 12.21 and the many different facets of cloud storage include:

A   Local physical or virtual storage system (H) communicates with a peer node (storage server) that is running in a cloud on a cloud instance. For example, storage server node (H) communicates with a cloud-based node (G) and replicates data or performs other functions. The storage nodes function as though they were both local and remote, the difference being that one of the nodes happens to be inside a cloud using cloud resources. The benefit is that the cloud-based storage node (G) functions as a cloud gateway, abstracting cloud resources while appearing as a known peer storage node.

B   Local physical or virtual storage system (H) accesses cloud storage such as AWS S3, Google GCS, Azure, or other services. Cloud storage is treated as a back-end storage media similar to local HDD or SSD. The storage node has a built-in gateway or cloud storage access and management APIs. The local storage node (H) handles endpoint management, identity, and security access control, mapping, and other functions for the cloud storage.

C   Private, hybrid, and virtual private cloud (VPC) implementations access public cloud externally accessible block, file, or object cloud storage. Local private or hybrid cloud storage could be based on OpenStack, Ceph, Microsoft Azure Stack, Springpath, or VMware, among others.

D   Local storage is backed up, protected, archived, copied, synched, or cloned to externally accessible block, file, or object cloud storage.

E   Cloud storage (block, file, object, API) accessible (J) storage and data services (tables, queues) are accessed by mobile devices, servers, and other systems.

F   Compute nodes, also known as cloud instances, access internally accessible (intra-cloud) cloud storage and services, as well as externally accessible storage services.

G   A cloud-based instance is running storage software that is a peer of local storage node (H). Storage for the cloud-based storage node can be a mix of block, file, and object storage among other cloud storage, and data services.

H   Local storage system also accesses cloud storage, and the cloud-based storage node.

I   Intra-cloud storage and services are only accessible from within a given cloud such as AWS, Azure, or Google, among others. For example, AWS EBS storage is accessible to AWS EC2 instances and related services, but is not externally native accessible. External access to EBS requires some software running on an EC2 instance such as a gateway or storage software such as in example G.

J   Storage target, endpoint or service that is accessible from outside a cloud. The storage target can block, such as iSCSI or NBD, file (NFS, SMB/CIFS, or HDFS), object (S3, Swift, Atmos), or API-based.

K   Storage nodes and stamps form a storage system or farm. The storage nodes are made up of storage software on servers with storage attached to those systems.

L   A separate cloud availability zone (AZ), site, data center, region, or a different cloud provider, where data can be protected to, or accessed from, for multi-region data protection.

12.8. Gateways, Appliances, Adapters, and Accessories

Various accessories can enhance or extend the value of storage systems. These include physical and virtual HBAs (Chapters 4, 5, and 6) and associated drivers that can be used in servers to access storage, as well as in storage systems as front-end or back-end to communicate with storage devices. Another type of card is RAID, and cache or I/O acceleration, to off-load servers from performing data protection and performance tasks. There are also acceleration cards for data footprint reduction (DFR) including de-dupe and compression, along with off-loading other data services capabilities.

Besides adapters, physical cables, connectors, and switches, other accessories include gateways, bridges, routers, and distance for handling protocol conversion, or accessing disparate local and remote storage. For example, there are NAS gateways that attach to underlying block (or object) storage and enable them to become accessed file storage. There are also object as well as other types of gateways for accessing local and remote cloud-based storage.

The gateways or their functionalities to implement a protocol or API can be built into the storage system software or may be external physical or virtual appliances. Gateways and access, as well as protocol conversion functions, can also be implemented in other software such as backup, data protection, archiving, and data movement or virtualization technologies.

12.9. Storage Management Software

Context matters with storage management software. Some may consider data protection (archive, backup, BC, BR, and DR) or data and storage migration as storage management. Others may see the software that implements storage system architecture, functionality, data services, and PACE capabilities as management software.

There is software for managing storage systems (services, software, solutions) along with their associated functionalities, as well as tools for enabling and managing data infrastructure storage–related tasks, for example, configuring, monitoring and reporting, event notification, allocation, and other storage-related tasks.

Context and focus, for now, is storage management software to configure, monitor, provision, and allocate storage system resources. Common storage system management tasks include:

- Configuration, maintenance, upgrades, troubleshooting storage
- Provisioning and allocating storage resources and endpoints
- Defining and implementing storage PACE, QoS, and policies
- Data and storage migrations (inter- and intra-system)
- PACE resource usage and event (telemetry data) collection
- Monitoring, reporting, and notification of storage activity
- Performance and capacity planning, analysis, modeling
- Protecting, preserving, securing, and servicing data that is consistent

Storage management considerations include:

- What additional hardware, software, network resources, tools, or licenses are needed
- Plug-ins, drivers, agents, firmware, BIOS, or other needed software
- Supported vendor hardware, and software compatibility list
- Management interfaces including html5 and mobile devices
- Call home, call cloud, NOCs, vNOC, and cloud NOCs
- For fee (one-time or recurring), freemium, or free community software
- Open or proprietary interfaces and tools (GUI, Wizards, CLI, PowerShell, APIs)
- Bundled with hardware, or provided with hardware for turnkey solution
- Federation, interoperable with third-party storage hardware, software, services

12.10. Resiliency Inside and Outside Storage Solutions

Figure 12.22 shows three common storage system deployment approaches balancing performance, availability, capacity, and economics. Which is applicable depends on your environment and needs, including underlying technologies as well as their configuration. On the left in Figure 12.22 are shown multiple servers accessing a shared storage system (a physical array,

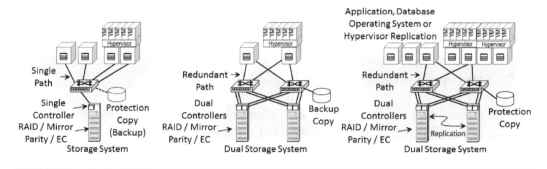

Figure 12.22 Various resiliency options.

appliance, or software-defined server), but with many points of failure. These points of failure range from single I/O network connections from servers to storage, lack of backup, and data protection if it does not exist, and a single storage controller (hardware- or software-based). Other considerations include applicable RAID, mirroring or parity, and erasure code (EC, RS, LRC, SHEC), and snapshots that exist in the same system that is being protected.

The center of Figure 12.22 removes points of failure by introducing servers with redundant paths for load balancing and failover. Also shown are redundant I/O or network paths, and storage systems or appliances with dual controllers, applicable RAID, mirroring, snapshot, and data protection.

On the right, Figure 12.22 shows a more resilient solution that relies on multiple layers of protection. There is application- and database-level replication, and consistency points, redundant hot-swap components and paths, as well as storage systems that replicate to each other or off-site. The off-site copy can be to the same or different storage, or cloud.

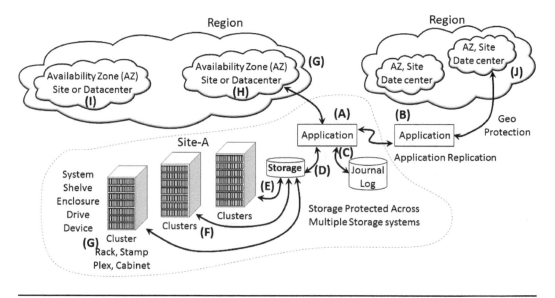

Figure 12.23 Multi-region, multi-site (AZ) storage resiliency and fault domains.

Figure 12.23 shows multiple storage systems as part of a resilient data infrastructure spanning different fault and availability zones, domains, and regions. In this figure, an application (A) is replicated to its peer (B) located in a different availability zone (AZ) in another region (J) for geographical protection. Also shown are the application (A) logging updates to a journal (C) on protected fast storage (SSD), with actual data then being written to primary storage (D). Primary storage (D) has a copy placed in different storage systems, clusters, or stamps (E and F), providing additional durability and resiliency. Resiliency is provided (G) in the case of a storage server software or hardware enclosure, shelf, drive, or cabinet failure. Note that the application (A) is at site-A, which is in fault domain or AZ (H) that is part of region (G), where another AZ or site also exists (I).

General design and configuration considerations for resilient storage systems include:

- Design for system maintenance (planned and unplanned) of hardware and software.
- Plan for failures (hardware, software, or services), and design around them.
- Resiliency should enable applications to continue at run to a given level of service.
- Leverage change control management for hardware and software changes.
- Look for and avoid or design around single points of failure.
- Isolate and contain faults in hardware or software to prevent them from spreading.

12.11. What's in Your Storage Toolbox?

"Back in the day," storage system tools were generally hardware-centric, but today they tend to be more software. However, even with software, a good pair of basic tools such as torx and regular screwdrivers, magnifying glass, flashlight, cables, connectors, and related things come in handy. As to software tools, things I have in my toolbox include SSH and RDP as well as various types of cloud access (e.g., S3 browser, Cloudberry, Cyberduck, curl, s3motion).

Other tools include Clonezilla, gpart, Acronis, diskpart, ping, dd, diskspd, fio, vdbench, iperf, benchmark factory, hammer, sysbench, himon, iometer, Rufus, iotop, esxtop, visualesxtop, and perfmon, among others. Other tools in my toolbox include plenty of USB thumb drives, servers for running different applications, operating systems, and hypervisors.

12.12. Common Questions and Tips

Why not just go all Flash, SCM, NVM, SSD? At some future point in time (perhaps not that far off), the cost of simply deploying all NVM such as Flash, 3D Xpoint, or other types of SSD and SCM will simplify the types of devices for storing data. In the near term, and perhaps for the next decade or so, there will still be a need for HDD in hybrid deployments. Keep in mind that SSD has good performance capability, and improvements have helped it become also cost-effective long-term storage as an alternative to HDD.

What often gets overlooked with NVM such as SSD is that it can be used to buy time. By buying time, I mean time to solve problems that might be associated with poor-running applications or other troubleshooting scenarios. Granted, this is masking the problems vs. finding and fixing them, but if time is of premium value, then using SSD can help buy time. Also,

SSD can help to stretch the life of servers as well as software licenses by getting more productive work out of them.

Is future storage converging around hyper-converged? It depends. For some environments and applications, converged server and storage solution variations are a good fit, while other options are applicable for different environments. Keep in mind that there are different approaches to converged and hyper-converged (i.e., CI and HCI) that address various market, customer environment size, and application PACE needs.

What is the best storage solution? It depends on what your application needs in terms of PACE attributes along with associated data services and management. Where and how do you want your servers and storage to integrate, aggregated or disaggregated or some hybrid? Do you need the storage to be on-site or remote? Are you looking to refresh your servers, or leverage what you already have?

Do clouds need cloud storage? It depends on your definition of a cloud. For example, private clouds can use traditional storage products combined with additional management tools and best practices. Private clouds can also leverage cloud-specific solutions as well as third-party-provided public clouds.

Public clouds often include storage as part of their solution offering or partner with cloud storage services providers. Also, public clouds can access storage located at other cloud service providers as well as on-premise for hybrid scenarios. As an example, I have some data located primarily at one cloud provider but protected to another provider as part of a business continuance and resiliency plan.

Doesn't DAS mean dedicated internal storage? DAS is often mistaken for meaning dedicated internal attached storage in a server. While this is true, DAS can also refer to external shared storage that is directly attached to one or more servers without using switches. For example, some DAS devices and enclosures can be directly accessed over interfaces, including NVMe, NBD, AHCI/SATA, SAS, iSCSI, Thunderbolt, and Fibre Channel, among others. Another name for external shared DAS is point-to-point, as in where iSCSI storage is connected directly to a server via its Ethernet ports, without a switch.

How do you decide which storage system or architecture, and packaging, is best? Step back, and look at your application PACE needs along with other data infrastructure requirements as well as growth plans. What will the storage system need to do: required features, compatibility with existing servers, storage, networks, or management tools? Leverage the various tables, figures, and examples in this as well as other chapters along with the Appendix. You can also find additional tips and recommendations at www.storageioblog.com.

What is a storage system BoM? There are other meanings, but let's assume here that a BoM is a bill of materials, also known as a manifest of components (hardware, software, network, services, tools) that make up a storage system solution or reference architecture. Other components include servers, storage enclosures, controllers, drives or devices, switches, adapters, cabling, routers, and gateways along with associated software, drivers, plug-ins, firmware, or BIOS.

What is a reference architecture? As its name implies, it is an architecture including implementation plans for deploying a data infrastructure resource solution to meet different applications or workload needs. The reference architecture may be generic or include specific BoM components, including how to configure the hardware, software, and network settings for use.

12.13. Learning Experience

Some learning experiences for this chapter are to practice what you learned by visiting a vendor's website (or call them up; I'm sure they would love to talk to you ;). Look at the various types and categories or classes of storage systems, and the solutions they offer. Pay attention to how they group, organize, or categorize their different offerings, particularly those who offer multiple solutions. Also look at their speeds, feeds, and specifications to see how they compare, not just on who has the same features, but also how extensively they discuss them.

Another learning experience, if you have a virtual, physical, or cloud lab, is to download from various vendors' sites their software-defined storage systems to gain some experience. For example, how easy are they to download, install, configure, and provision for use vs. others?

12.14. Chapter Summary

There are many different types of storage systems, ranging from traditional proprietary hardware arrays to software-defined and tin-wrapped open hardware–based appliances, to virtual and cloud services. Storage systems also vary in their architectures, implementation, packaging, and functionalities as well as personality (optimized for different applications or workloads). Also keep in mind when buying (or subscribing to) a storage system solution that how it meets today's needs may not be cost-effective over time as you grow and upgrade. On the other hand, what meets your needs today for one deployment may be applicable for another use in the future if it is cost-effective to do so.

General action items include:

- Hardware needs software, software needs hardware, and other software.
- Everything is not the same with applications or data centers, and also with storage systems.
- The focus of software-defined storage should be the enabling outcome.
- There are many different faces to cloud, virtual, CI, and HCI storage systems.
- What is your focus for CI or HCI, purchasing, hardware, a hypervisor, or something else?
- Keep future upgrades and PACE enhancements in mind for what you acquire today.
- Keep application PACE needs and other data infrastructure requirements in perspective.
- The best storage system is the one that meets or exceeds your requirements.

Bottom line: For now, look beyond the cover or packaging. This includes looking beyond the speeds, feeds, interfaces, protocols, number of HDD or SSD, size of memory and cache, as well as the number of processor sockets (cores and threads). Simply put, how will a storage system, solution, appliance, or service work for you and adapt to your environment as well as applications, instead of you working for it?

Part Four

Putting Software-Defined Data Infrastructures Together

Part Four includes Chapters 13, 14, and 15. This part leverages your tradecraft skills about what you have learned as part of expanding (or refreshing) your software-defined, data infrastructure, server, storage, I/O networking, and related data services. This includes server storage solutions and architectures, management takes, design, optimization, reporting, and deployment options.

Buzzword terms, trends, technologies, and techniques include appliances, applications, benchmark, simulation and testing, block, cloud, containers, converged CI, CIB and HCI, data protection, file, metrics, management monitoring, planning, remediation, reporting, tools, troubleshooting, object, performance, availability, capacity, economics, software-defined, trends, and virtual, among others.

Chapter 13

Data Infrastructure and Software-Defined Management

Having insight and awareness avoids flying blind with data infrastructure management

What You Will Learn in This Chapter

- Knowing what tools to have and how to use them as well as when to use them
- Putting together what we have covered so far
- The importance of metrics and insight to avoid flying blind
- Where to get and how to use various metrics that matter

This chapter converges various themes and topics from the different chapters along with some additional tradecraft, insight, and experiences. The focus is on common data infrastructure (both legacy and software-defined) management topics. Key themes, buzzwords, and trends addressed in this chapter include analytics, metrics, server, storage, I/O network, hardware, software, services, CI, HCI, software-defined, cloud, virtual, container, and various applications, among others topics.

13.1. Getting Started

All applications have some performance, availability, capacity, and economic (PACE) attributes that vary by type of workload, environment, size, scale, and scope, among other considerations. Different environments, applications, subapplications, or application component workloads also vary, having diverse PACE needs along with different service-level objectives (SLOs). Similarly, the applications, data, and metadata along with the configuration of and the data infrastructure resources they depend on also need to be managed.

There are different focuses and domains of interest when it comes to data infrastructure along with related applications and data center facilities infrastructure management. Management spans automation, processes, procedures, policies, insight (awareness, analytics, metrics) and practices, along with people skillsets and experience (i.e., tradecraft). In addition, management includes tools, technologies, as well as techniques across different legacy and software-defined data infrastructure (SDDI) IT focus areas.

Keep in mind that SDDI and software data infrastructures are what are housed in physical (and cloud) data centers and are generically referred to as software-defined data centers (SDDC). Does that mean SDDI = SDDC = SDI? Yes. However, SDDI can also mean leveraging software-defined management applied to cloud, virtual, container, and other environments.

> *Tip:* The context of domain means different things, such as networking and URI, URL domain names for endpoints, or security domains such as with active directory (AD) and domain controllers (DCs). Another use of "domain" refers to areas of focus such as higher-level applications, databases, file systems, repository, servers, storage, or I/O networking, along with facilities infrastructure. Yet another use of "domain" refers to an area of knowledge, expertise, experience, or your tradecraft.

There are domains of focus and tradecraft specialty, such as database administration, storage management, hypervisors, VMware, Microsoft, OpenStack, Docker containers, data protection, security, performance, and capacity planning, among others. For example, "cross-domain management" can mean managing across different network or security domains, but it can also refer to managing across applications, server, storage, and data protection.

As you have read throughout this book, some recurring themes include:

- Everything is not the same in various data centers, data infrastructures, and environments.
- There are different sizes, scopes, and scales, with diverse applications along with workloads.
- While things are different even across industries or applications, there are similarities.
- Management involves different tasks, functions, and domains of focus and tradecraft.
- People define software to manage other software, hardware, and services such as cloud.
- SDDI management spans legacy, virtual, container, and cloud environments.

13.1.1. Management Tradecraft and Tools

In prior chapters we had smaller sections built around toolbox and tradecraft topics; in many ways, this entire chapter is about your toolbox and tradecraft and applying them. Figure 13.1 shows some data infrastructure–related tools spanning hardware, software, and services. However, as we have discussed throughout this book, other tools include your experience, skills, and tradecraft, including what to use when, where, why, and how for different scenarios.

Recall Chapter 1, where I provided a sample of various toolbox items for SDDI and SDDC as well as legacy environments including hardware and software. These items, among others in Figure 13.1, are what define, configure, diagnose, monitor, report, protect, secure, and provision resources, among other data infrastructure functions. Some tools focus on hardware,

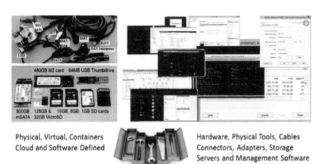

Apt, Yum, Vum, wget, curl, cyberduck, cloudberry, ycsb, rufus, gpart, fdisk, wmic, esxtop, SAP, Hadoop, Oracle, Microsoft, Cassandra, MongoDB, Hbase, vdbench, diskspd, BMF, TokuDB, OpenStack, diskpart, sysbench ,Enmotus-FuzeDrive, Graphviz, caspa, CPU-Z, datadog, DPA, JAM Treesize, blkdid, nttcp, iperf, iotop, htop, lscpu, ntop, netstat, sar, hdparm and ping among others

Data Infrastructure Tool Box

Figure 13.1   Data infrastructure toolbox items (hardware, software, and services).

others on software, services, virtual, container, cloud, file systems, databases, backup and data protection, specific products or vendors, as well as other functionality.

> *Tip:* In addition to object, bucket, and other bulk cloud storage, I have a regular volume (EBS) with an image configured with common tools as well as preferences. When I start an instance, I attach that volume to it. It is like moving an HDD or SSD from one physical server to another for a jump start or other activity. In other words, I have a virtual toolbox with different tools that I keep in the cloud.

In addition to the hardware, software, and services such as cloud found in the data infrastructure toolbox, there are different techniques and approaches for using them in various ways. Companions to the tools and techniques, along with policies and best practices leveraging trade-craft experiences, are people with different roles, focuses, and skills across various domains.

Additional terms and context for this chapter include *landscapes* in the context of different application and workload environments. This includes applications such as SAP, M1, and other ERP, financials, Cachè medical and Dexis Dental healthcare; SAS, Hadoop, and MapReduce-enabled big data analytics. Additional landscapes include Oracle, Microsoft SQL Server, Cassandra, MongoDB, HBase, and TokuDB, among many other databases and repositories. Still another context for landscapes refers to *greenfield* (i.e., starting from scratch, brand-new, nothing exists so start fresh) as well as *brownfield* (i.e., something exists, needs to evolve, be migrated, be converted).

Host is another term that requires some context. Host can refer to a physical server configured with a hypervisor that hosts guest virtual machines; or it can be a nonvirtualized server that hosts different applications, also known as a server. Another variation is that some people use the term for the cloud, the web, or other services hosting, or even as a variation of co-location or where your servers, storage, network, and other data infrastructure resources are housed. Thus context matters for the host.

We have discussed efficiency (e.g., utilization, space optimization, data footprint reduction, and savings) as well as effectiveness (e.g., productivity, performance, doing more work, removing waste) as being different yet complementary. Something similar is the idea of needs and wants, two items that often get interchanged along with requirements.

Tip: Think of needs as what you must have, i.e., mandatory requirements, while wants are what you would like to have, secondary or lower on the requirement list. Also keep in mind what are your needs and must have requirements vs. what are the wants of others (what they want or perhaps need you to have). Where wants and needs come into play is across various data infrastructure management activities including configuration and resource decision making among others.

Another distinction to mention here is vertical and horizontal integration. Vertical integration means that resources, tools, and technologies are aligned, with interoperability up and down the data infrastructure stack, from high to a low level (or altitude) and vice versa. Horizontal integration, on the other hand, as its name implies, spans across different resources tightly or loosely integrated including federated. Context matters in that technology and tools can be vertically (or horizontal) aligned. However, some vendors also have vertical integration. Vertical integration for a vendor or supplier simply means they have streamlined their own resources and supply chain, removing complexity and costs while simplifying things for their customers.

Orchestration and coordination may exist across vertical or horizontal and federated resources and across different application landscapes and environments. Later in the chapter, we will look at some of the sources for data and metrics that matter as well as provide insight awareness. These telemetry data (i.e., metrics) come from various hardware, software, and services with a different focus. Sources of metrics that matter can come from servers, storage, I/O networking, application or data infrastructure, and event logs. Additional insight and metrics sources include shims, sniffers, analyzers, trace, collectors, drivers, hardware, and probes as well as taps. Speaking of telemetry, as a refresher, the term refers to an event, activity, or other data such as logs, performance, availability, access, or security, among others.

Framework, orchestration, dashboards, and key performance indicators (KPI) can align to different resource layers and functionality. In some situations, tools such as Splunk and others are used for enabling big data analytics along with the correlation of telemetry and other data infrastructure as well as broader data center events. Data infrastructure analytics, Artificial Intelligence (AI), deep and machine learning, along with IT service management (ITSM), performance management databases (PMDB), configuration and change management databases (CMDB), along with other data center infrastructure management (DCIM) tools, are essentially the IT systems used by data infrastructures, data centers, and information factories. These various tools and insight metrics track the productivity, resource inventory, costs, quality, on-time, and customer satisfaction of services.

A few more terms and context include *Day one* and *Day two* nomenclature, which refers to the user or customer experience at different points in time. For example, on Day one a technology is delivered, with perhaps initial setup, configuration, and integration along with various integrity and health checks. Day two can refer to what's next: The technology or solution is made ready for use and broader deployment, advanced configuration, provisioning as part of the data infrastructure. Depending on the size, scope, and complexity, some things may be accomplished on Day 1 or Day 2, while others might span additional days, weeks, or months.

Considerations, tips, and recommendations include:

- *Customer* and *user* can mean different things at various levels of the data infrastructure.
- What is below a given level (or altitude) is considered infrastructure to those above it.

- Those above a given altitude can be thought of users or customers of that layer.
- Understand the context of different terms, as everything is not the same.
- Good management and decisions need insight and awareness via metrics.
- The best metrics are those applicable to your needs vs. somebody else's wants.
- Data infrastructures reside in data centers and other habitats for technology facilities.

13.1.2. Data Infrastructure Habitats and Facilities

Data infrastructure is the collection of hardware, software, services, people, processes, practices, and policies that combine to support different application workloads. Data centers, also known as habitats of or for technology, are where data infrastructures reside. If you prefer the stack or layering model, data infrastructures sit below information infrastructures (e.g., business applications), and on top of data center facilities and their associated infrastructures.

Similar to a traditional factory, information factories are the collection of application workloads and their data supported by a data infrastructure that resides in a data center. Often the broad term *data center* is used to refer to data infrastructures, while to others it can be the applications and data infrastructure, yet for others it is a facility physical (or cloud) focus.

> *Tip:* When you hear the term or expression *software-defined data center* (SDDC), its context is the data infrastructure (i.e., SDDI or software-defined data infrastructure). Note that in Figure 13.2, the physical data center (i.e., the habitat for technology) is where data infrastructures are deployed into or hosted from. Collectively, the physical data center, data infrastructure, and applications along with the data they support are *information factories* whose product is information and services.

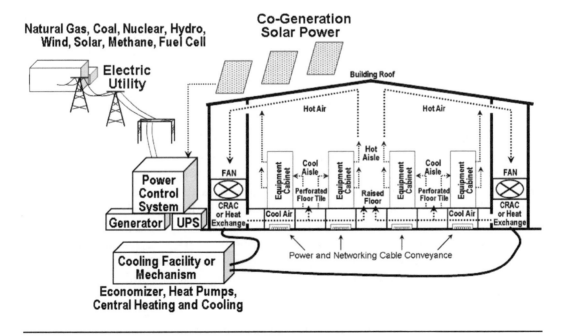

Figure 13.2 Data centers and habitats for technology.

IT, service provider (SP), and cloud data centers are habitats housing data infrastructure (servers, storage, networks, hardware, software, and applications), also known as information factories. These technologies are defined to support various business applications that transform data into information services.

Similar to a traditional factory that transforms material and components using various tools, technologies, and processes housed inside a facility, data centers and their data infrastructure, plus information infrastructures and their applications, enable a similar capability (e.g., function as an information factory to deliver information services). While the goods and services that are delivered differ between traditional and information factories, what is common to either is having insight and awareness. The common insight and awareness is into available resources, productivity (effectiveness), utilization (efficiency), errors and rework, quality, cost, service delivery, and customer satisfaction.

> *Tip:* Factories rely on insight, awareness, and metrics to know whether resources are being used cost-effectively (productive) and efficiently (eliminate waste, remove costs). Having insight also enables knowing how to reconfigure technology, implement new services and processes, as well as boost productivity, deliver services on-time, and return on investment, while improving customer satisfaction.

Habitats for technology include physical, cloud, co-located, shared, your own or others, big or small. Technology topics for data center facilities (i.e., habitats for technology) include primary and secondary electrical power, switching, floor space and floor weight loading, cabinet and rack height, as well as cable (I/O and networking along with power) conveyance. Additional topics include heating, ventilation, and air conditioning (HVAC), smoke and fire detection, notification and suppression, humidity and antistatic resources, and physical security.

Considerations, tips, and recommendations include:

* Understand the relationship between data infrastructure resources and facilities.
* Work with data center and facilities staff to understand each other's needs (and wants).

13.2. Data Infrastructure Management

For some, the focus of management is dashboards, GUI, CLI, SSH, PowerShell, API, and other interfaces for doing and monitoring things. For others, the focus is broader, including what needs to be done, as well as how it is (or can be) done, along with associated tools. Management tasks, tools, techniques, technologies, trends, and topics focus areas include many things that start with "R" and "Re" (among other letters). For example, RAID, RAIN, re-balance, re-build, re-construct, re-cover, re-host, re-inflate, reliance, reload, repair, re-platform, replicate, resiliency, resolve, resources, restart, restore, resume, re-sync, re-transmit, review, revoke, risks, roll-back, roll-forward, and roll-out, among others. Note that data infrastructure management tasks, topics, tools, and techniques are not exclusive to things that start with the letter "R," as there are many other activities and focus areas as well.

Examples of common data infrastructure management activities include:

- Portals, dashboards, service offering menus, and catalogs
- Orchestration and coordination, allocation, provisioning (self and automated)
- Monitor, collect, analysis, insight, report on events, alarms, activity, and usage
- Diagnose, detect, analysis, correlate, isolate, or identify real vs. false problems
- Planned and unplanned maintenance, repair, replace, remediation, and fix
- Configure, setup, modify, change, update, reclaim, and clean up
- Resource, application workload and data assessment, classification, optimize, and tuning
- Upgrades, replacement, expansion, refresh, disposition of technology resources
- Protect, preserve, secure, audit, and implement governance of data and resources
- Maintenance, migration, conversion of systems, data centers, applications, and data

Additional tasks include decision making for the above as well as other areas of focus: strategy and planning, performance and capacity planning, architecture and engineering. Other tasks include establishing best practices, automation and physical policies, templates, workflows, deployment or implementation guidelines, and cookbooks or template recipes.

> *Tip:* Different management tasks have various impacts on the data infrastructure, resources, and applications. Monitoring should not have a negative impact: it should be passive. However enabling additional detailed debug, trace, and troubleshooting telemetry and log data could have an impact on performance while benefiting troubleshooting. Making changes to a component, system, service, or some aspect of the data infrastructure may have adverse effects if it is not done properly. Part of data infrastructure management tasks is understanding the impact of doing something along with what granularity that may provide a benefit versus having an adverse impact (and on what items).

Figure 13.3 shows various focus areas for data infrastructure, both legacy as well as software-defined, physical, virtual, and cloud. Note the different layers from higher-altitude application

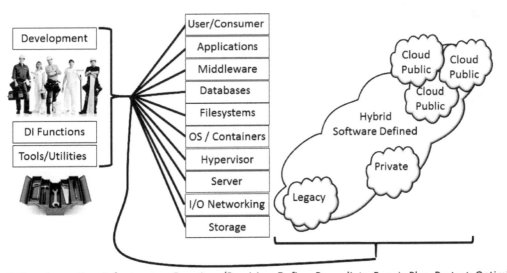

DI Functions = Data Infrastructure Functions (Provision, Define, Remediate, Repair, Plan, Protect, Optimize)

Figure 13.3   Various data infrastructure management focus areas.

Figure 13.4 Dashboard using Datadog software-defined management tools.

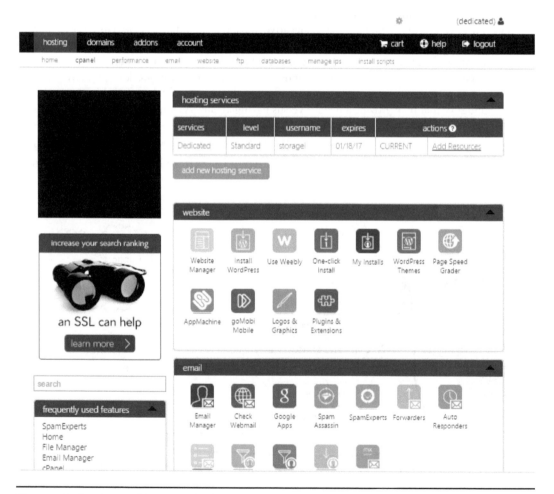

Figure 13.5 Cpanel hosting dashboard.

focus to lower-level data infrastructure resource components spanning different technology domains. Remember that there are different meanings for domains in IT, data centers, and data infrastructures, so context is important.

The next examples move from a higher level to a lower level across technologies.

Figure 13.4 shows a dashboard (Datadog) that spans various layers of data infrastructure resources from lower-level (lower-altitude) virtual machines to higher-altitude (higher-level) applications. With tools such as Datadog, you define the software tool as to what to monitor and collect data about, as well as how and when to display results. In other words, tools like Datadog are software-defined management tools created by people leveraging their tradecraft, skills, and experience to meet the needs of a given software-defined data infrastructure.

Figure 13.5 shows a dashboard (cPanel) for hosting such as websites, blogs, email, and other items. This could be a shared hosting dashboard view, a virtual private server (VPS), or a dedicated private server (DPS). Various management functions from provisioning to configuration, monitoring, reporting, troubleshooting, and diagnostics as new feature enablement can be done from the dashboard.

Figure 13.6 shows another cloud-based dashboard and service catalog, this being Amazon Web Services (AWS). Different regions can be selected, along with account and billing information, health status reporting, as well as resource and service provisioning, configuration, and monitoring—for example, accessing Elastic Cloud Compute (EC2) to set up, start, or access compute instances (virtual machines), or access and use Simple Storage Services (S3) or Route 53 networking, among others.

Other orchestration and management functions that can be done via service dashboards include setting up health and resource checks—for example, setting up a health check to monitor the status of various service endpoints or IP addresses. Should a fault be detected, notification can be sent, and some tools can automatically repair. Other tools can make dynamic changes to DNS as well as work with load balancers among other technologies. Additionally,

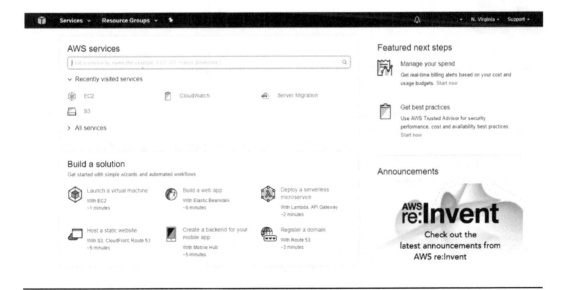

Figure 13.6   AWS service catalog dashboard.

alarms and event notification can be set up for resource usage, activity, performance, availability, capacity consumption, or billing notices. Other tasks include provisioning, remediation, and service management, among others.

> *Tip:* Remember to power down your cloud servers as well as clean up resources when you are done with a project. This can mean taking a snapshot of a server or volume from the primary cloud to object, blob, or other bulk storage. Also remember that if you are done using a database, table, or other resource, clean it up or you might encounter some billing surprises.

Figure 13.7 moves from the cloud to storage, which could be on-site physical hardware-hosted (wrapped) or software-defined. In this example, Dell EMC Unisphere storage management tool accesses and shows the status of a Dell EMC Unity storage array. Note that this particular system is a software-defined virtual storage array (VSA) in which the Unity software is running as a guest virtual machine (VM) on a VMware vSphere host.

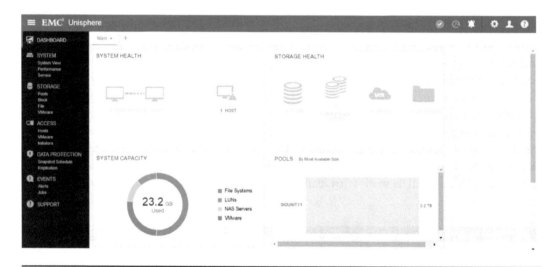

Figure 13.7   Dell EMC Unisphere storage system management.

The Unisphere software management tools enable orchestration of storage system functions, provisioning, allocation and host access, data protection, and monitoring, among other tasks. From an old-school standpoint, Unisphere and tools like it are sometimes called element or device managers. New-school tools such as Unisphere are sometimes called platform-focused orchestration, federated management, or other software-defined management themes.

Management interfaces and focus include:

- Vendor-specific, third-party, and industry standard interfaces
- A single pane of glass (screen) or multiple views or windows
- Web- and GUI-based, including HTML5-enabled and Wizards
- SSH, CLI, shell, PowerShell, and PowerCLI, among other scripts

- HTTP REST and other API-based, SNMP MIBS, and SMIS, agent and agentless
- MQTT and AMQP, among others, for IoT/IoD scenarios
- Large-scale or small-focused, consumer, SMB, SME, enterprise, or other
- On-premise or cloud-based tools (works on premise but from cloud)
- Cross-technology domain and focus (different layers of the data infrastructure stack)

Note that operating systems, hypervisors, file systems, volume managers, databases, software-defined networks, and software-defined storage, among others, are also often considered to be management tools, as well as security, backup, and data protection, among others. Thus context and area of focus for "management software" and tools matter.

Some familiar and perhaps new technologies and tools to add to your toolbox or buzzword vocabulary for various functional areas include Ansible, Apt, AWS cloud watch, Blue Medora, Chef, Curl, Datadog, Docker, dpkg, Komprise, Kubernetes, MaaS, Mesos, Github, Microsoft MAP, OpenStack Horizon and Ceilometer, PowerShell and PowerCLI, Puppet, Saltstack, Splunk, Swarm, System Center and WMI, UCS Director, vRealize, VUM and wget, as well as Yarn and Yum, among others. Besides the tools (programs), there are also patches, updates, installation and other software kits, including .bin, .ova, .ovf, exe, .iso, .img, .vhdx, .pkg, .zip, .vib, and .msi, among many others.

Considerations, tips, and recommendations include:

- Some tools are for fee and require additional dependent software (or hardware).
- Other tools are for fee, but interoperable with no extra costs for plug-ins.
- There are also free and community as well as other open-source software.
- Do plug-ins, drivers, and modules for interoperability or extra functionality exist?
- There are tools for collecting, analysis, correlating, storing, and visualizing metrics.
- Tools can be installed on physical, virtual, container, or cloud resources.
- Tools can be hosted via clouds, as well as used for managing clouds.

13.2.1. Troubleshooting, Problem Solving, Remediation, and Repairs

Another aspect of data infrastructure management includes troubleshooting, problem solving, remediation, and repair. This aspect ranges from simple common problems, with easy fixes for known issues, to more complex and involved issues. Some remediation and repairs are in response to problems, others are proactive as well as to implement changes.

General troubleshooting, problem-solving, and diagnostic questions include: Are other data infrastructure and IT personnel or vendors seeing the same issues? Can the problem be forced to occur, or is it transient and random? Are devices and systems on-line, powered on, or need to be reset? Is the problem isolated and repeatable, occurring at a known time or continuously? Is the issue or problem unique to your environment or widespread and affect others? If the hardware or service is ok, what about various data infrastructure or applications software—Is it on-line, available, or need a restart?

Figure 13.8 shows a simple generic troubleshooting decision-making tree that can applied and extended to meet different environment needs.

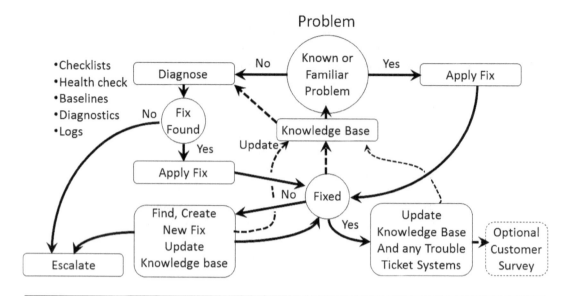

Figure 13.8   Generic simple troubleshooting flow tree.

Additional general troubleshooting and remediation tasks questions include:

- What happens if you force a device rescan or refresh?
- What is working and what is not? Is the problem isolated or widespread?
- Where is the problem? What do your logs and event monitors indicate?
- Who or what is causing the problem? Who is not seeing the problem?
- When did the problem start, or reappear if it has been seen before?
- Are resources up to date with current software, drivers, firmware, and BIOS/UEFI?

Tip: As part of troubleshooting and remediation, as well as being proactive, check with vendors or have them notify you of urgent mandatory updates, as well as recommended upgrades. This should also include resources such as batteries, power supplies, and other not-so-obvious items as well as higher-level software, firmware, and drivers.

General tips and recommendations include: Is this a now familiar scenario, recurring, transient problem, or something new? What are others seeing and experiencing? What is normal vs. abnormal; do you have a baseline to compare with? What changed; when did something break and stop working? Similarly, something may have been working before but not as well as you thought, and something changed to amplify or make the issue known.

What test and diagnostic tools do you have, and how familiar are you (or somebody in your organization) with using them? Are your different systems and components, from servers to storage, networking, and others, set to the proper data and time? Are license keys, security and authentication certificates, and subscriptions valid and up to date or about to expire? Avoid finger-pointing by having good information, insight, and awareness tools to work proactively with others, or use that information to defend your findings.

Additional general tips and recommendations include:

- Keep good notes, cheat sheets, run books, and other documents.
- Leverage bash, shell, or PowerShell, among other scripts, for common tasks.
- Scripts are also a good way to document processes, and to prevent typos.
- Utilize versioning, snapshots, and PIT copies when doing builds and testing.
- When doing installations, pay attention and watch out for typos (yours and others).

Do you have a problem reporting and trouble ticket tracking system? How about checklists and automated scripts to detect or collect information about known or common problems? What do you have for a preventive and planned maintenance schedule for updates (hardware, software, services)? How frequently do you update software and firmware on various devices?

Are your systems kept at current software and firmware release levels? Alternatively, do your systems stay one or more steps behind current release levels? Are hot fixes applied, or do you wait for regularly scheduled updates if possible? How are software and firmware updates tested and verified before deployment? Will a fix or workaround cause problems elsewhere or a problem to return?

Some I/O- and networking-related topics include: Is the problem local to a server, storage, system, service, or network segment? Is it isolated to a site or facility, location, or geography? Is it all network traffic, or specific ports (TCP, UDP), protocols, or types of traffic? Are there bad cables, physical ports, connectors, interposers, transposers, or other devices?

> *Tip:* Set up a centralized event and activity logging repository where logs are collected and protected (backed up) on some specific time interval to enable later restoration. Leverage log insight and other analysis tools including Splunk as well as operating system, hypervisor, and third-party or opensource tools. Create an information base (infobase), knowledge base, or other venue for tracking known issues, symptoms, workarounds, and remediation.

It may seem obvious, but check if affected devices along with their ports (physical and logical), adapters, HBA, NIC, switches, or routers are powered off, or not properly configured. Are there issues with software configurations and definitions, such as missing IP gateways, or DNS or DHCP settings? Are the network and I/O issues consistent or intermittent? Is there a pattern to correlate when errors occur? What are the error logs and other telemetry data indicating? Are MTU and jumbo frames set correctly, as well as firewall and iptables ports properly enabled (or disabled).

> *Tip:* Establish a trouble ticket and tracking system to log, even in a spreadsheet, incidents, what happened, how they were resolved, date and time, along with other information. This log can be used to analyze for trends as well as to plan for future upgrades, remediation, and other changes. Also, establish your health check, status, and diagnostics, including capturing relevant logs, metrics, and configurations to aid in problem resolution. As part of the diagnostics, you can also eliminate what is working and focus on what is not and why the problem is occurring.

Additional I/O and general networking items to look at include:

- What changed, when, and where; what are the event logs indicating?
- What do network traffic displays show or indicate?
- What are your VLAN, VPN, VPC, firewall (or iptables), and load balancer settings?
- For cloud resource access, do you have proper inbound or outbound addresses?
- Are proper security certificates valid and access keys in place?
- Is there a network bandwidth service provider hardware or software issue?

Server-related topics include hardware as well as software configuration, setup and settings, driver or version mismatch, device or driver conflicts, physical and logical connections. Other server considerations include:

- Higher-level software and drivers, including hypervisors and operating systems
- Utilize various hypervisor, operating system, and related tools.
- If a server or VM will not boot, check GPT and MBR, as well as BIOS and UEFI settings.
- Verify software license or subscriptions are valid and have not timed out.

Availability, data protection, security, and backup topics include: Can you restore the contents of a storage device or medium (e.g., SSH, HDD, tape, or optical)? This means can you access the stored data and restore it to some other medium? Are you then able to read and use the restored data, or does it need to be decrypted, security certificates applied? Are permissions, access control, and other attributes preserved and restored to expected settings?

Additional availability and data protection along with security topics include:

- Has a resource to be protected (name, address, endpoint) been moved?
- Is there enough free space on the target device or destination, or is there a quota or limit?
- Are backups or protection copies taking longer than normal due to there being more data?
- Did protection copies run faster than normal even though there was more data?
- Did a backup, snapshot, or other protection task fail due to open files in use?
- Are there network, server, storage, or other issues causing problems?
- What protection and availability tasks worked? Which failed and when?

Can you access files (read and write), local or remote, with firewall enabled or disabled? Can you ping remote devices and endpoints? Can you *nmap–v* a server or storage to see what ports are open or available? What happens if you try to ping the DNS or another known name local or remote? Can you create a local file? If not, can you do more than an *ls* or display of a file or its directory and metadata? What are the file ownership and permissions (i.e., on Linux, do a *ls – al*).

Storage-related problem-solving, troubleshooting, and diagnostics topics span from hardware to software and services, similar to servers and networking. Can you access or see the storage system, device, endpoint, volume, LUN, file system from different systems, or only certain ones? Are you able to log in or access storage systems or software-defined storage software management tools, and what are you able to observe, including any health, status, or error log messages?

Are there any other current issues or problems with networks, servers, or other data infrastructure activities and problems? If the device is not shared, can you reset it, or force a scan from the operating system or hypervisor to see if the device appears? What happens if you reboot the server operating system or hypervisor? Also verify if firmware, BIOS/UEFI, drivers and other software are up to date.

Other considerations include: Is there a device name, address, letter, ID, or mount point conflict? Is there a missing mount point, directory, or folder preventing the device from mounting properly? Is there a problem with incorrect security or permissions? What changed? Did the storage work before? Is the device (hardware or software-defined) stuck in a state that prevents it from being re-initialized, or is software enabled as read-only? (See the Appendix for some possible workarounds.) If an operating system command does not work, what happens if you run in Administrator mode (Windows) or sudo (Linux)? Are permissions and access control as well as ownership configured as expected and needed?

Additional storage-related items include:

- If it is a new device, has the volume or file system been initialized, prepared, exported?
- NAS- and IP-related: Is the DNS working? Can you ping and do other things?
- What is appearing in event and error logs?
- What about access controls and certificates—Are they valid or missing?
- Various tools such as Windows diskpart and Linux fdisk and gparted can be useful.
- How much free space exists on the different storage devices?
- If a Linux command does not work, try running in sudo mode.

Cloud-related items include:

- Do you have a proper endpoint or address for service or resources?
- Do you have the proper access keys and security certificates?
- Are network ports and firewall settings configured for inbound and outbound?
- Are you using HTTP or https as part of your endpoint and applicable ports?

File systems, databases, and other data repository issues include:

- Verify that the database server instance and database are running.
- Check whether network access ports, named pipes, and endpoints are configured.
- Review local and remote access settings, including firewall and user authentication.
- Are the tempdb and other database items full, or on a device that is full?
- Are the main database tables and log files on a slow device or one that is full?
- Does the database server have enough memory allocated to its buffers?
- What are the logs and other telemetry data indicating?
- If you have test SQL queries, what results do they show?

Figure 13.9 shows various data infrastructure and application focus areas as well as layers on the left. On the right of Figure 13.9 is a focus on hypervisors with VMware vSphere using various tools to gain insight into performance, availability, and capacity along with configuration. Different tools provide various levels of detail and insight as well as help correlate what is causing issues, where the problem exists, or perhaps where it does not exist.

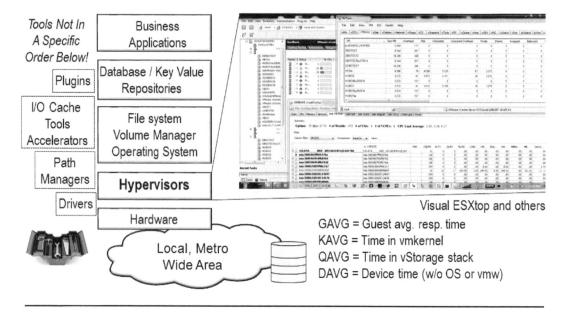

Figure 13.9   Data infrastructure focus layers.

Tip: Sometimes, quickly identifying where the problem does not exist, or where there is no impact, can streamline problem resolution. Eliminating where there is no issue or problem can help bring focus to the area where the problem exists, or what is causing or amplifying it to be an issue. Also pay attention to moving problems—when you fix in one location and something happens elsewhere.

Figure 13.10 shows, on the left, status information for a four-node Ceph software-defined storage cluster using commands such as *ceph health, ceph osd tree,* and *ceph–s.* On the right of Figure 13.10 is a status display from a Microsoft Failover Cluster health and configuration check. The Microsoft Failover Cluster tool checks hardware and software configuration,

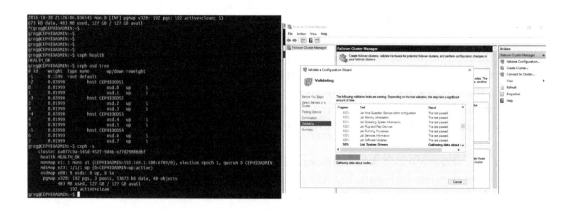

Figure 13.10   Ceph Status (left) and Microsoft Windows Cluster Health check (right).

Figure 13.11   VMware vSAN health and status.

pointing out issues, problems, and configuration items to address before creating a cluster (or that might prevent a cluster from being created).

Figure 13.11 shows a VMware and vSAN health status display. Note that all of the alerts and alarms shown point out abnormal conditions such as when something fails (or introduces a fault for training, or learning). Various alarm conditions via vCenter for a VMware vSAN are shown—for example, where the nodes (hosts) have been disconnected (hence why the errors) due to a network issue.

> *Tip:* Are devices properly plugged into electrical power and I/O network connections? Double-check the obvious to make sure components are properly connected or determine loose and weak connections. Sometimes a device may be plugged into power or a network connection and may need to be simply unplugged for 15–30 seconds or a minute, then reconnected—magically, things may start to work (though not always).

Do you have basic and advanced troubleshooting scripts and diagnostics tools? Tools that can check and verify connectivity to services and resources? Ability to investigate performance responsiveness against known behavior? For example, scripts that can access management ports or services to get health, status, and event activity logs or information to help isolate and determine problems. Another example is whether you can ping a device, system, or service locally as well as remotely to verify network connectivity, DNS and routing, as well as firewalls and iptables configurations.

> *Tip:* Make sure not to unplug the wrong device! In addition to making sure power cords, I/O, and network interfaces are securely connected, are they attached to the proper ports, devices, adapters, or network components? Also remember to label cables and devices so you know what they are for.

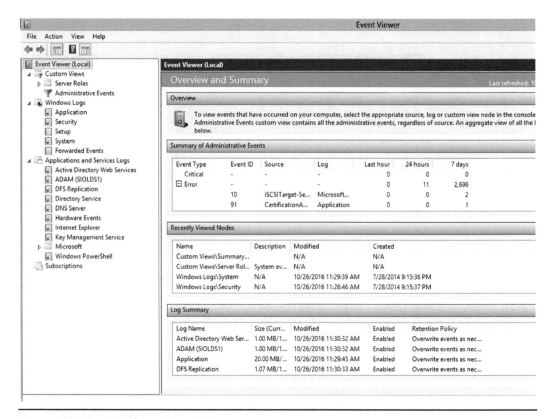

Figure 13.12   Windows Event Viewer.

```
[greg@centio02 ~]$
[greg@centio02 ~]$ ls /var/log
anaconda  cron       lastlog             pcsd       secure             wtmp
audit     cups       libvirt             pluto      speech-dispatcher  Xorg.0.log
boot.log  dmesg      maillog             ppp        spooler            Xorg.0.log.old
btmp      dmesg.old  messages            qemu-ga    tallylog           yum.log
chrony    gdm        openlmi-install.log sa         tuned
cluster   glusterfs  pcp                 samba      wpa_supplicant.log
[greg@centio02 ~]$ sudo tail /var/log/messages
Oct 29 17:13:02 centio02 dbus-daemon: dbus[707]: [system] Activating via systemd: servi
ce name='net.reactivated.Fprint' unit='fprintd.service'
Oct 29 17:13:02 centio02 dbus[707]: [system] Activating via systemd: service name='net.
reactivated.Fprint' unit='fprintd.service'
Oct 29 17:13:02 centio02 systemd: Starting Fingerprint Authentication Daemon...
Oct 29 17:13:02 centio02 dbus[707]: [system] Successfully activated service 'net.reacti
vated.Fprint'
Oct 29 17:13:02 centio02 systemd: Started Fingerprint Authentication Daemon.
Oct 29 17:13:02 centio02 dbus-daemon: dbus[707]: [system] Successfully activated servic
e 'net.reactivated.Fprint'
Oct 29 17:13:02 centio02 fprintd: ** (fprintd:26337): WARNING **: fprint init failed wi
th error -99
Oct 29 17:13:02 centio02 systemd: fprintd.service: main process exited, code=exited, st
atus=157/n/a
Oct 29 17:13:02 centio02 systemd: Unit fprintd.service entered failed state.
Oct 29 17:13:02 centio02 systemd: fprintd.service failed.
[greg@centio02 ~]$
```

Figure 13.13   Centos Linux event logs.

Tip: Error and event logs for Windows systems can be found in the *%SystemRoot%\system32\ winevt\logs* directory folder and accessed via the Event Viewer via Control Panel or the command line using *C:\ eventvwr.* Logs and events for the local system are displayed, as well as access to remote Windows computers activity using the Action Tab from Event Viewer. Note that you may need to do some firewall and other configuration changes to use remote capabilities. Figure 13.12 shows an example of Windows Event Viewer; note the Action Tab at top left, which can be used to access remote computers. Additional troubleshooting activities include:

- Has storage or memory space been exceeded (no free space, or quota limits)?
- Did a resource move to a new location and some task is looking in the old place?
- Are timeouts occurring while a service or resources wait on something?
- What are the symptoms or issues? Is one person or system affected, or everybody?
- What changed, when, where, and by whom?
- What's different now vs. before the problem?
- Is the problem known and isolated, or yet to be determined?
- Is the fault isolated, or is it being triggered by some other event?
- What are your indicators (logs, event notification, error counters) showing?
- Do you have optional debugging and trace modes that generate more telemetry (log and event) data that you can enable for troubleshooting (remember to disable when done)?

Tip: On Linux systems, look in the /var/log directory folder (and subfolders) for various logs. You can use commands such as (among others) *less, grep, more, cat, ls,* and *tail* of a specific log file. For example (unless they have been moved), if you have MySQL installed, look in /var/log/mysql, or for OpenStack, compute /var/log/nova, or /var/log/ceph for Ceph. Figure 13.13 shows a directory listing "*ls /var/log*" of various logs and subfolders, followed by a "*tail /var/log/messages*" to display some log event activity.

Some network tasks for gaining insight, awareness, and testing configurations include:

- What are your event and activity logs indicating?
- Can you ping your known resources by name or IP address?
- Can you ping your DNS servers or another known address such as 8.8.8.8?
- Windows *ipconfig /all* along with *ipconfig /release* as well as *ipconfig /renew*
- On Linux systems, instead of using *ipconfig,* use *ifconfig,* among other tools
- Some additional Linux tools for testing and troubleshooting networks include:
  o Perform an arp scan on a network interface such as eth0: *arp-scan –I eth0 –l*
  o Display network mapping info on a network: *nmap –sP 192.168.0.0/20*
  o *Netstat* along with *lshw –class network* or *arping –I eth0 –c 5 192.168.1.1*
  o *nmap -sS -sU -PN -p 1-500 66.33.99.11* to see open ports on a device

Tip: If you are going to ping a resource, make sure that it allows being pinged, to avoid chasing a false error. In other words, if you ping something and it does not support ping (or ping is disabled), you may in fact not have an error, or the error is occurring elsewhere.

As an example, suppose there is a performance problem with the server and storage system such that the application slows down. Someone might say that he has a problem with his database storage server and that he has Cisco, Dell, HPE, IBM, Oracle, or some other. Is he referring to a server, storage system (or appliance), hardware or software, network, or other data infrastructure resource? Context matters!

Let's say some additional questions and insight determines that the application is a database on Linux, and the storage is dual-pathed. However, the application slows down. Ask some questions, such as what type of storage, SAS, NVMe, iSCSI, or FC. Sometimes the answer might be a specific vendor, make, and model, while at other times the answer might be generic such as a Dell EMC, Fujitsu, HDS, HPE, IBM, NetApp, Nimble, Nutanix, Oracle, Pure, Quantum, Seagate, or product from another vendor. If the vendor has only one product, that narrows things down. On the other hand, if, for example, somebody says it is a Dell EMC, then it could be ECS or Centera, SC (Compellent), Datadomain, PS (Equllogic), Isilon, MD, VMAX, VNX, Unity, or XtremIO, among others. Another example is if somebody said it is a NetApp system, then that could mean FAS/ONTAP, E (Engenio), or StorageGrid, among others. Likewise, somebody might say the storage is fiber attached to a server; instead of assuming it is Fibre Channel, verify whether it is Gb Ethernet (or 10 GbE or faster) supporting iSCSI or NAS, or something else.

When running on one path, things are ok; however, when running on both paths, things are not ok. So, what path driver is being used? What happens if you failover to the other path: Can you run on path A ok, or path B, but not on path A and path B at the same time? What do you see in the system and storage error logs? Are IOs being shipped to a different node or controller for processing? What's normal behavior?

Tools, technology, and techniques include:

- Note that, in addition to automating tasks, scripts also serve to document activity.
- Use tools including hardware and software probes, sniffers, tracers, and analyzers
- Leverage change management to fall back or off from a change.
- Use automated tools and scripts including health checks to verify current status.
- Revert to a previous version, PIT snapshot, copy, or backup:
  - Different granularity from application to system or device
  - Application and database checkpoints and consistency points
  - Snapshots and other PIT copies or recovery-point protection

---

Tip: Creating a baseline of your environment can be as simple as having a report, graphic, image, or other information of what is normal vs. abnormal behavior. This can be from an application and workload perspective, such as how many users are normal, amount of CPU, memory, and storage space consumed, network activity, and I/O performance.

Other baseline indicators include normal number of errors or other log entry activity, response time for different application functions, and configuration. Response time can be measured by applications or data infrastructure tools, or as simply as with a stopwatch (real or virtual). The point is that you can start simple and evolve into a more complex baseline. A more complex baseline would include scripts that collect and compare logs, events, resource usage, and activity. For databases, this can include diagnostic or other transactions that can be run to see if they work, and how long they take.

General considerations include:

- Establish a baseline to compare normal and abnormal behavior of the environment.
- Implement test and diagnostics tools along with scripts for automated fault isolation.
- Leverage planning and analytics tools to test and simulate workload changes.
- Utilize automated tools for proactive and reactive problem isolation.
- Do root-cause analysis: what caused the problem, how to fix it, and how to prevent it from reoccurring in the future.

Considerations, tips, and recommendations include:

- Have a test system and lab environment that can be used in support of problem-solving.
- Start with the simple and basic; leverage automation where possible.
- Compare with known experiences; utilize your knowledge base and tradecraft.
- Conduct a failure or fault postmortem to capture insight to prevent future incidents.
- Create a knowledge base, wiki, or cheat sheet of troubleshooting hints and tips.

13.2.2. Availability, Data Protection, and Security

Threats can be physical or logical, such as a data breach or virus. Different threat risks require multiple rings or layers of defense for various applications, data, and IT resources, including physical security. Data infrastructures rely on both logical and physical security. Logical security includes access controls or user permissions for files, objects, documents, servers, and storage systems, along with authentication, authorization, and encryption of data. Another aspect of security is to "bleach" (wipe) servers and storage with digital secure erase that does deep cleaning. Digital erase should also provide verification along with optional documentation for compliance, like a digital certificate of death for the device and your data.

> *Tip:* Know and understand legal and regulatory requirements for your organization as well as applications and data. This includes what applications and data can be located in different geographies for data privacy, among other regulations. Likewise, are there applicable regulations that require you be able not only to demonstrate having copies of applications and data off-site, but also the ability to demonstrate a restoration in a given amount of time.

Compliance and governance apply to availability, protection, and security to determine who can see or have access to what, when, where, why, and how. This might require solutions where systems administrators who have root access to systems are blocked from access to read or access data or even metadata in secured systems. RTO and RPO can be part of regulations and compliance for applications and their data, as well as retention of data, some number of copies, and where the copies are located.

Additional availability, data protection, and security topics include subordinated and converged management of shared resources including AD. Remember to keep mobile and portable media, tablets, and other devices in mind regarding availability as well as security. How will your solutions work with encryption and data footprint reduction tools such as compression, de-dupe, and eDiscovery.

Also keep in mind: Do you have orphaned (forgotten) storage and data, or zombie VM (VM that are powered up but are no longer in use)? Have you assessed and classified your applications along with their data in terms of data protection, availabity, and security SLO alignment needs? Do you know the number of and diversity of telemetry and log files to monitor as well as analyze, how long will they be retained, and who has access to them?

Considerations, tips, and recommendations include:

- Have international and multilanguage support via tools and personnel.
- Implement proactive intrusion detection software (IDS) and systems.
- Manage vendors and suppliers as well as their access or endpoints.
- Utilize secure erase (digital bleach) solutions to wipe clean devices for disposition.
- Track physical and logical assets with printed barcodes as well as RFID and QR codes.
- Leverage heath checks, automation, proactive data integrity, and virus checks and repairs.
- Implement Data At Rest Encryption (DARE) via servers, storage, devices, and clouds.
- Talk with your compliance and legal experts about what's applicable for your environment.

Additional availability, data protection, and security considerations include:

- Utilize multiple layers of protection, from application to lower-level component.
- Find a balance between higher-level application and lower-level component protection.
- Protect your data protection, security, and availability configurations, settings, and keys.
- What are you looking to protect against and why (refer to Chapters 9 and 10)?
- What are your application PACE needs and RPO, RTO, SLO?
- Keep the 4 3 2 1 rule in mind for your different applications.
- Everything is not the same; do you have the same FTT, RTO, RPO, and other SLO?

13.2.3. Analytics, Insight, and Awareness (Monitoring and Reporting)

Recall from Chapters 1 and 2 that all applications require some amount of performance, availability (durable, secure, protected), capacity (space, bandwidth), and economics (PACE) attributes. Different application PACE requirements depend on corresponding data infrastructure resource (servers, storage, I/O, hardware, software) needs that must be balanced.

Data infrastructure metrics provide insight into application PACE and their corresponding resource usage, service levels, and activity. Some metrics focus on activity usage and utilization, others on resource saturation or starvation, and others on errors and events. Other metrics involve metadata about systems, software, applications, and actual data.

Fast applications need fast data infrastructures, including servers, networks, and storage configured for optimal PACE. To enable optimization, do you have timely, accurate insight and awareness into your data infrastructure, or are you flying blind? Flying blind means that you lack proper instruments, impacting confidence and control to make decisions without situational awareness of your current or soon-to-be status. In other words, you could be headed for unplanned performance degradation problems or worse.

Do you have data infrastructure insight into how and where:

- Your application and resource bottlenecks currently exist, or where they may move to
- How Flash SSD and NVMe technologies can speed up applications
- Availability (or lack of it) impacts application productivity
- Performance degrades during recovery or failure modes and workload impact
- Application changes behave when deployed in production and productivity impact
- Additional resources needed (forecasting) and when

Figure 13.14 is a reminder that all applications have some variation of PACE attributes. Some applications have and need more performance, while others focus on availability, capacity, or low-cost economics, and yet others use various combinations.

| Performance | Availability | Capacity | Economics |

Figure 13.14 All applications have some PACE attributes.

Tip: Informed decision making spans from day-to-day operational and tactical administration to strategic architecture, engineering, planning, forecasting, and procurement, among other tasks. Informed, effective decision making, by humans or machines, requires timely, accurate, and relevant information. The lack of good information can result in garbage in, garbage out decision making. That is, garbage information (or lack of accurate insight) can result in garbage or poor decisions based on incomplete or inaccurate situational awareness.

Figure 13.15 shows different layers of resources and focus areas for data infrastructures on the left, and aggregated views for analysis, AI, deep learning, correlation, reporting, and management on the right. On the right can be a collection of standalone or integrated systems, depending on the size and scope of your organization. For example, smaller organizations might manage their environments with Excel spreadsheets, simple databases, or other solutions. Larger organizations may have IT service management (ITSM) platforms, CMDB, PMDB, or various other DCIM-related systems.

Having timely, accurate, and actionable data infrastructure insight awareness facilitates:

- Avoiding flying blind, either without instruments or using bad data, to prevent bad decisions
- Informed problem resolution, proactive prevention, agile and smart decision making

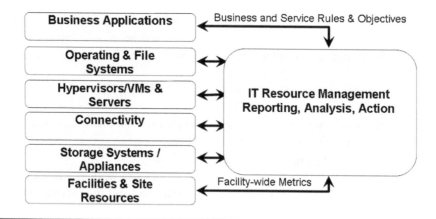

Figure 13.15   Various metrics that align with different data infrastructure layers.

- Establishment of performance QoS baselines to know what is normal and what is abnormal
- Maximized investment in data infrastructure resources
- Application productivity performance and availability
- Putting the customer instead of the vendor in control
- Addressing data infrastructure, application, and service questions

Figure 13.16 shows sources of information, insight, and awareness across data infrastructures layers accessed by and fed into different data infrastructure management tools. Depending on the size of your organization, your approach may be much simpler, or more involved.

Data center infrastructure management (DCIM) and IT service management (ITSM) tools allow monitoring, reporting, tracking, and maintaining an inventory of resources including

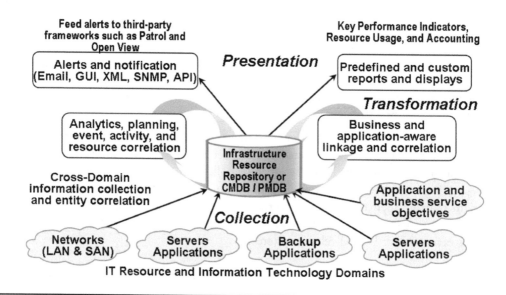

Figure 13.16   Various data infrastructure and IT information systems.

servers, storage, I/O, networking hardware along with software, and services. These assets can be tracked in configuration management databases (CMDB) and their activity monitored via performance management databases (PMDB). Some of the tools focus on lower-level DCIM facilities, while others are more end to end in focus, from higher-level business and information applications to how they use and consume data infrastructure resources and services.

Performance depends on availability; availability relies on performance to support processing, movement, and storing of data along with associated applications. The cost of hardware continues to decrease. However, more of it is needed (servers, storage, networks) to support growth, new applications along with software-defined data infrastructures.

There is a tendency to focus on physical capital costs such as hardware, networking, and facilities. However, there are also opportunities to optimize operational expenses. Besides staffing, maintenance, and services fees, are you getting the maximum value and ROI out of your software? For example, are you getting the highest utilization and effective productivity out of your AWS, Citrix, Cisco, Epic, Meditec, Dell EMC, Desix, EPIC, HP, IBM, M1, Microsoft, OpenStack, Oracle, SAP, SAS, and VMware, among other software investments?

Considerations, tips, and recommendations include:

- Leverage automation tools for the collection, analysis, and reporting of information.
- Organizations rely on applications to keep them informed about how they are running.
- Data infrastructures need applications to keep them informed about how they are running.
- The right metrics are the ones that have meaning in your environment.
- Avoid flying blind and making decisions based on guesses: Be informed.

13.2.3.1. Metrics That Matter (and Where to Get Them)

To address data infrastructure challenges, you need to be able to obtain applicable metrics promptly. This means metrics that are relevant, real-time, and do not require extensive manual effort to extract. There are many different tools and resources for obtaining metrics that matter.

What is your area of focus, applications, database and file systems, management tools, operating systems, drivers, hypervisors, servers, I/O networking, and storage or cloud resources, among others? In addition to focusing on your area or technology domain, are you looking for metrics to troubleshoot and perform diagnostics for problem solving, service management, planning, or design? Are you looking for performance, availability, capacity, security, and cost information?

Figure 13.17 shows various data infrastructure points of interest as well as sources for metrics. These metrics can be used to support various data infrastructure management activities. In Figure 13.17, moving from left to right, there is the user (client, consumer of information services), there are devices including mobile, tablets, laptops, and workstations, access networks sending packets, frames, IOPs, gets, puts, messages, transactions, files, or other items. In the middle of Figure 13.17, there are networking devices including switches, routers, load balancers, and firewalls. On the right side of Figure 13.17 are servers supporting different applications that have various storage devices attached. Metrics can be collected at various points for different time intervals such as per second, and results can be correlated and reported for different purposes.

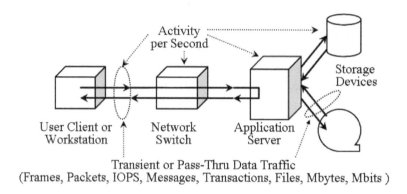

Figure 13.17 Data infrastructure metrics points of interest.

There are many different types of metrics that matter for data infrastructures. Recall that in Chapters 1 and 2 some of these were outlined with reference to application workloads and PACE characteristics (as well as energy). Additional metrics were discussed in Chapter 5, including server I/O activity, among others. In the data services chapters (Chapters 8–11), additional metrics were discussed covering PACE, data footprint reduction, efficiency and effectiveness, SLO, RTO, and RPO, among others.

Recall that metrics can be standalone, such as activity per time interval, e.g., I/O operations per second (IOPS), including block, file, and object gets or puts, transactions per second (TPS) or API calls, among others. There are also compound metrics with multiple items combined to provide a multidimensional (composite) view. For example, IOPS or TPS per watt of energy used, or cost per terabyte, among others.

Also recall that there are averages, peaks, standard deviation, and sustained metrics that are measured as well as derived or calculated. There are rates of activity and data movement along with ratios of change; some are server-based, others arise from the I/O network, storage, cloud, hypervisor, or component. Metrics also include availability, reliability, failures, and retries, as well as bandwidth, throughput, response time, and latency.

Another aspect of metrics is that they can be a higher-level summary or detailed for different time intervals. Figure 13.18 (left) shows an example of workload activity by month

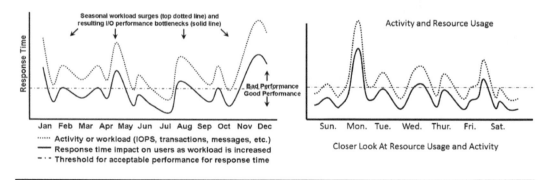

Figure 13.18 Server I/O activity: monthly view (left) and daily view (right).

along with a response time threshold, response time, and activity. In the example, as activity increases, response time also goes up (lower is better), with the threshold indicating an expected or required service-level objective.

Figure 13.18 (right) shows a drill-down or detail, looking closer at a specific week of a given month to see what is occurring on days of the week as well as an indication of the time of day. Further investigation into details of a specific day or hour time range can be seen in Figure 13.19, which shows workload across different physical or virtual servers.

Just as there are many different types of metrics for various purposes representing applications and data infrastructure, along with data center facilities, there are also multiple sources of metrics. These include different tools, some built into operating systems, hypervisors, drivers, file systems, networks, storage, and databases, among other points of interest. Some of the tools are available for a fee and are tied to a particular technology or focus area; others are free or open-source and community editions. Then there are neutral, independent tools and services for fees.

Tip: In this age of digital, IoT, and IoD technology, a convenient, time-tested (pun intended) tool and technique to gauge the performance and response time of something is a stopwatch. This can be an analog or digital watch or wall clock, the clock on your computer display, or simply adding time stamps at the beginning and end of a script or command to see how long it ran.

Common to different sources of metrics is some focus on collecting telemetry and other activity or event data, either in real time or over time. Some specialize in analysis and correlation, while others concentrate on visualization, presentation, or export to other platforms. Then there are tools, solutions, and services that combine all of the above functions with different interoperabilities.

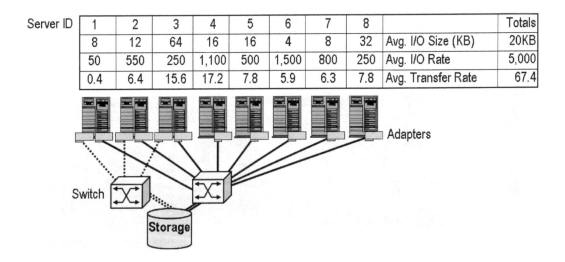

Server ID	1	2	3	4	5	6	7	8		Totals
	8	12	64	16	16	4	8	32	Avg. I/O Size (KB)	20KB
	50	550	250	1,100	500	1,500	800	250	Avg. I/O Rate	5,000
	0.4	6.4	15.6	17.2	7.8	5.9	6.3	7.8	Avg. Transfer Rate	67.4

Figure 13.19   Server and storage I/O metrics.

> *Tip:* A metric-related term is *test harness*, associated with simulation, testing, and benchmark discussions. One context of test harness is a physical cable or connector for collecting electrical power or other telemetry data. Another context is a software wrapper, kit, shell, shim, or script package for a particular tool to make it easier to work with, as well as tailored for a given workload scenario. For example, there are several test harnesses that wrap around Vdbench to perform different tasks. These include the SNIA Emerald green and power tools, VMware HCIBench for testing standalone, CI and HCI including vSAN nodes, among others.

Additional examples of tools, scripts, services, and technologies for gaining metrics include those from or apart from AWS, Apple, Broadcom, Cisco, Dell EMC, Fujitsu, Google, HDS, HPE and Huawei, IBM and Intel, Juniper, Lenovo, Mellanox, Micron and Microsoft, NetApp and Nutanix, Oracle and OpenStack, Promise, Quantum, Rackspace and Redhat. Additional tools are available from Samsung, Seagate, Servicenow and Simplivity, Teamquest, Ubuntu, Veritas, VMware and WD, among others.

Some other examples of insight and reporting awareness tools as a source of metrics include solutions or services from Bocada, Caspa (network metrics), Cacti, Cloudkitty, Collectd, CPUID CPU-Z (server CPU metrics), Dashboard, Datadog (dashboard for various metrics from application to servers and cloud), Datagravity, and DPA (Dell EMC Data Protection Advisor).

Additional sources for metrics and insight include Firescope, Enmotus FuzeDrive (micro-tiering and server storage I/O metrics), ExtraHop, Freeboard, Graphviz (visualization), Hyper I/O (Diskranger and HiMon [windows server metrics]) and JAM Treesize (space usage). Some Linux tools include (blkdid, cat [/proc/cpuinfo], df, dstat, fdisk, iotop, hdparm, htop, lscpu, lspci, lshw. Other tools include Manageiq, Microsoft MAP, Nagios, Ntopng, sar, sysmon, netstat MonitorDB as well as Nttcp (workload generator). Others include Openit, Opvizor, RRDtool, stashboard, Statsd, and Solarwinds, among others.

Figure 13.20 shows using Linux *lspci* command to display PCIe LSI devices (adapters), along with performing a disk read performance test to device /dev/md0 (a software-defined RAID set). Refer to the Appendix for additional examples of workload simulation, benchmark, testing, and metrics.

More awareness tools include Perfmon (various Windows and related metrics), Ping (along with ipconfig, ifconfig, Ibping, Fcping among others), Jperf (visualization), Quest (Foglight,

```
greg@DOCKI005:~$
greg@DOCKI005:~$ sudo lspci | grep LSI
00:10.0 SCSI storage controller: LSI Logic / Symbios Logic 53c1030 PCI-X Fusion-
MPT Dual Ultra320 SCSI (rev 01)
02:01.0 SCSI storage controller: LSI Logic / Symbios Logic 53c1030 PCI-X Fusion-
MPT Dual Ultra320 SCSI (rev 01)
greg@DOCKI005:~$ sudo hdparm -tT --direct /dev/md0

/dev/md0:
 Timing O_DIRECT cached reads:    516 MB in  2.01 seconds = 257.15 MB/sec
 Timing O_DIRECT disk reads: 7512 MB in  3.00 seconds = 2502.85 MB/sec
greg@DOCKI005:~$ []
```

Figure 13.20   Various Linux status commands.

Spotlight along with other SQL and Oracle as well as AD tools), R (visualization), and RVtools (various VMware-related metrics and configuration information).

Add to the previously mentioned tools, SAS (statistical analysis software) and Solarwinds (various metrics) along with Spiceworks as well as Splunk. Additional tools include Windows Task manager, Turbonomic (server, storage, I/O, and configuration metrics). Tools are also part of other operating system, hypervisor, server, storage, network, and cloud solutions. Other tools include Visual Studio (Microsoft) Web Performance and Load Tests, VMware Esxtop and Visualesxtop (VMware metrics), vSAN observer and Virtual Instruments (server, storage, I/O, and configuration metrics), among many others.

Needless to say, there are many additional tools for gaining insight and awareness for different layers or altitudes about data infrastructure. There are also tools for monitoring application workloads, as well as lower-level facilities as well as habitats for the technology. Among these are Atto tools (various workload and reporting tools), Benchmark Factory (database and other workload generator and reporting), Black magic for Apple platforms (storage and I/O workload generator), Cosbench (object storage workload generator), Crystal (various workload generator and reports) and dd (data copy tool that can be used to generate server storage I/O activity. Other workload generators, benchmark, and simulation tools include Dedisbench (storage workload generator with dedupe), DFSio (Hadoop and DFS), Diskspd (Microsoft open-source server and storage I/O workload generator) as well as DVDstore (mixed-application workload generator).

Additional workload generation, simulation, and benchmark tools include ESRP (Microsoft Exchange workload generator), FIO.exe (storage workload generator), HammerDB (various TPCC workload generator), HCIBench (VMware test harness and wrapper for Vdbench), Hdparm (storage workload generator and configuration display), Iometer (storage workload generator), Iorate (storage workload generator), Ior (for HPC and Lustre large scale vs. general-purpose file systems), Iozone (storage workload generator), and Iperf (network workload generator).

Figure 13.21 shows VMware vSAN metrics with a series of thumbnail (small) performance and health indicators that can be clicked on to expand for further detail.

Besides those mentioned above, other workload, simulation, benchmark, and test tools include Jetstress (Exchange Email workload generator), Loadgen, LoginVSI (VDI and workspace workload generator), Mongoose (various object workload generators), PCmark (various server and I/O workloads), and Ptest (processor performance test. Additional tools include

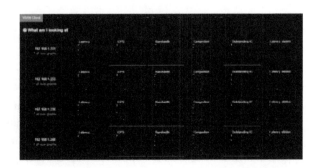

Figure 13.21   VMware vSAN Observer dashboard.

Quest Benchmark Factory (various TPCC and other workload generator, trace, replay and extensive reporting with automation), SharePoint Performance Tests, MOSS and WSS (various workload generator), STAC (workload generator for financial trading), Storage Performance Council (SPC) workload generator, Swingbench (Oracle database workload generator and reporting), and Sysbench (Simulation and workload generator including database).

Then there are Teragen/Terasort (Hadoop and DFS), TPCC (various database transactional, decision support and analytics workload simulation tests), Vdbench (simulation workloads, file and block I/O, various wrappers and test harness including VMware HCIBench, SNIA Emerald among others), Virtual Instruments (Simulation and Analytics), VMmark (Simulation, mixed workload based on DVDstore), Xbench (Apple-focused benchmarks) along with YCSB (Yahoo Cloud Serving Benchmark workload generator), among many others. Some of the tools can be found at download sites such as Git Hub.

Following is an example of a YCSB workload (update heavy, 50% reads, 50% writes) against a local MongoDB database. Other YCSB workloads include B (95% reads, 5% writes), C (100% reads), D (read latest inserts), E (short ranges of queries), and F (read modify writes or updates to existing records).

```
./bin/ycsb run mongodb -s -P workloads/workloada -p recordcount=10000000 -threads 2
-p measurementtype=timeseries -p timeseries.granularity=2000 >> results.txt
```

In addition to the above, there also various vendor solutions that are part of servers, storage, network, and cloud services, including examples in the Appendix.

Considerations, tips, and recommendations include:

- The most important metric and benchmark are the ones that matter for your environment.
- What is your objective for measuring and metrics, along with test and simulation?
- Some tools are for fee; others are free, open-source, or available as community editions.
- Focus and level (or altitude) for monitoring, measuring, test, or simulation
- Application, database, API, object, file system, or block, among other focuses
- Keep context of object in mind, as some tools focus on S3 or Swift objects vs. others.
- Tools such as Curl, S3FS, S3motion, wget, Cloudberry, and others can be used to test or validate access and basic functionality of cloud and object storage resources.

13.3. Data Infrastructure Decision Making

Keep in mind that there is what you need (your must-have requirements and criteria) and then there are what you would like (your wants or nice-to-have items). Also keep in mind that there are what you need and must have, and there are what vendors, solution, service providers, and consultants or advisors might want (or need you to have).

Watch out for the wants and needs of others becoming your wants and needs if they are not applicable. Likewise, look for opportunities to leverage the different things people are suggesting or recommending, as they may become something of value. Table 13.1 shows requests for information, proposals, and tenders. These include bids that form the basis for tenders, which are formal bid-and-proposal processes.

Table 13.1. Request for Information and Proposals

What	Meaning	When and Why
RFI	Request For Information	Early in a decision-making process to assess the market, vendor, and technology offerings that might be applicable now or in the future. This is part of a learning and discovery process. Can be generic or specific to a set of requirements. Use as a means to meet with and learn about vendors, technology, and solution offerings in advance of a formal RFQ or RFP process. Information from RFI can be used to help prepare formal RFQ and RFP. RFI can be more general and may include broad or rough "for planning purposes" pricing costs.
RFP	Request For Proposal	Similar to an RFQ. You may have more general criteria requirements, leaving the actual solution up to a vendor, solution, or service provider to respond how they would address cost and terms. Both RFQ and RFP can be common for government-related projects, but also for commercial and private activities, large and small.
RFQ	Request For Quote	Similar to an RFP. However, you may know what you want in general terms, provide vendors or solution providers with your criteria for them to respond with a solution quote including pricing. Both RFQ and RFP are typically more formal activities with rules, processes, and guidelines along with timing specified on how the winner will be chosen. Likewise, there are usually criteria on how disputes will be addressed.
RFT	Request For Tender	Similar to an RFP or RFQ, vendors are invited or offered to respond to a formal tender (set of requirements and process) as to how they would provide a solution, along with costs.
SOW	Statement of Work	Summary (or expanded) description of what the work is to be done, by whom, when, what the general as well as specific expectations are. The SOW can include project milestones, goals, and objectives.

RFI, RFP, RFQ, RFT, and SOW, among other processes, can vary in detail and scope. Generally, they include a due date for the submission, rules, processes, and procedures, as well as whom to contact, or whom not to contact. Details can include what the requirements are, compatibility, interoperability and other technical as well as business considerations.

Business or commercial considerations include costs and fee schedules for up-front as well as recurring maintenance, licensing, and subscriptions as applicable. Vendor or supplier background information, including technical, product, marketing, sales, service, and support, as well as financials, are other items. There may be questions to be addressed by the respondent, which may be simple yes, no, or more detailed. Responses may be specified to be general or very strict, with non-marketing-type answers.

Questions might also be ranked both regarding their weight of importance as well as encouraging the respondent to put a rank or weight on their solution. There might also be sections for competitive perspective from the respondents. Other items include terms for delivery, installation, and configuration and acceptance criteria. Acceptance criteria might include specific performance results validated with a benchmark or other proof of validation, as well as functionality and interoperability demonstration as proposed. Warranties, service, and support, including who will do what and when, may be included, along with other contractual terms. For example, what are the vendor or service providers remedies to make things right if the solution does meet its contractual and SLA terms?

What are your constraints regarding budget (CapEx and OpEx) now and into the future? What are your power, cooling, primary, and standby electrical power, as well as floor space, floor weight loading, and other data center (habitat for technology) constraints? What are your networking capabilities, local and remote, including for cloud access. Do you have any legal, regulatory, or other requirements to address? What compatibility and interoperability do you need with current hardware, software, networks, and services, as well as tools, staff skills, and procedures? What is your timeline for decision making, installation, acceptance, and deployment?

Understand what you currently have, how it works (or not) to meet your workload, as well as growth expectations. Also, what do you like or dislike about what you have, what do you need and want, as well as current strengths and weakness? What will be the cost to migrate to new technology, including applications along with data movement (and migration). Also, understand if there are any new technologies to learn, new processes as well as procedures to adopt. What will you do with old or existing technology and solutions; is there any residual value in some other role within your data infrastructure, or do the costs favor disposing of the item?

> *Tip:* Look for consistency in what others recommend to you to use. Do they have a good history of being consistent in their advice and recommendations, or do they simply push whatever is new or whomever they are currently working for. Trust, yet verify.

A good salesperson's job is to remove barriers for his or her employer and to look out for the customer's long-term interest so that he or she can come back and sell you something again in the future. On the other hand, some salespeople are just looking for, or being told what to sell, at the current time and moment and may also set barriers for other vendors. Make sure that the salesperson and his or her organization are looking to partner and bring value, so you both win.

> *Tip:* When you ask questions, be specific with context and follow-up detail as needed to avoid simple answers. For example, a question such as "Do you support snapshots?" might be answered with a yes, but does that tell you how extensive the snapshot features are.

If you have a concern or problem with a vendor or their solution, ask them to be candid, even if you have to go under a non-disclosure agreement (NDA) to get the information. It is better to know that something is broken or not working properly or supported so you can work around it than not to know what the issues are. Likewise, if you go under an NDA, honor it.

Watch out for apples-to-oranges (apples-to-pears) comparisons, where different things are compared that are not the same, fair, or relevant. Not everything will be apples to apples or the same. However, there are similarities, and different criteria can be used as needed. Also look for context, as that can help normalize the differences between apples-to-oranges comparisons.

> *Tip:* Balance the pro's and con's, feature benefits and caveats of open vs. closed, vendor propriety or commodity solutions, technology as well as how they work for you, your environment, and achieve what you need to do. Are you making a change or selection because it is what everybody is talking about, or will it bring actual benefit to your environment? Can you run a proof of concept (POC) or extended test to see how the solution works for your needs and requirements?

Some solutions are touted as having no single point of failure, being totally resilient, even with Internet and YouTube videos of people shooting with a gun or blowing something up while it continues to work. What happens if somebody misconfigures, sets the wrong switch setting (button, dial, knob, parameter), or applies software or firmware the wrong way, or simply unplugs the device. If something wrong can happen, it will, and usually it involves human error.

You might want or somebody needs you to spend a large amount of cash on a massive cache (SSD or SCM) for your server, storage, or network. However, upon investigation, you know or determine that yours is a random environment where a lot of cache is not needed, rather, a smaller amount in one or more locations that is also more effective and cash-friendly. Likewise, you might discover that while SSDs and SCMs can boost performance, the real benefit for your environment can be to reduce or eliminate time spent responding to user calls, then searching for and fixing various problems. In other words, SSDs and SCMs can be a form of IT and data infrastructure aspirin or ibuprofen pain relief.

A solution might look expensive up front as a capital expenditure, but it may have low recurring operating or OpEx costs including licenses, renewals, and maintenance. Likewise, something that looks low-cost up front may end up costing more with annual fees as well as usage and activity costs. Do your homework: Look for value near-term as well as longer-term.

> *Tip:* Leverage vendors' hardware (and software) compatibility lists (HCL) to confirm as well as support remediation for devices, device drivers, patches, versions, and other updates. Also look into and leverage vendors' tools for managing updates, including automated, manual, or scheduled downloads and implementation. Vendors also have lab resources along with knowledge base, wiki, forums, and other resources to assistant with testing, troubleshooting, diagnostics, comparison, and remediation activities.

Considerations, tips, and recommendations include:

- Do you need multiple vendors? Are you comfortable with single or sole-sourced?
- Ask what might seem like obvious questions, as sometimes you can get unexpected answers.
- Likewise, ask about assumptions, dependencies, interoperability, and other requirements.
- If something is too good to be true, it might be. Do your homework.
- Find a balance between analysis paralysis (stuck in evaluations) and quick decisions.
- Trust yet verify when it comes to decision making, bids, proposals, and tenders.
- Keep in mind your needs and requirements vs. the wants (or needs) of somebody else.
- Manage your vendors, service, and solution providers while enabling them to do their jobs.

13.3.1. Comparing Data Infrastructure Components and Services

There are many different aspects and considerations when it comes to comparing data infrastructure components for legacy and software-defined environments. This includes looking at, comparing, evaluating, making decisions on what to buy, rent, lease, subscribe to, as well as how to configure for use. The chapters in this book lay out the various characteristics, attributes, and considerations in comparing different hardware, software, services, tools, techniques, and technologies for data infrastructures.

> *Tip:* People often look for something with the lowest cost, and smart vendors or providers know this. Vendors may offer to price "as low as" or "from" or "starting at" along with additional fees for optional services and features. On the other hand, vendors and providers also offer premium services or products at premium fees all-inclusive.

There are also value offerings, which often get confused with cheap or low cost. Rather, for a higher fee, you get more services or capabilities. The value offering might be more expensive than the "starting" price of a basic product or service. However, when you add up the extra cost of the optional items vs. what's included in a value offering, the bundled solution may have a better overall cost. This is where comparing costs on apples-to-oranges basis comes into play.

For example, one server might be priced at $499.99 USD while another is at $2,999.99, causing somebody to ask why spend the extra $2,599.99? If all you need are the capabilities of the lower-cost server, that might be sufficient and the better deal. However, if the more expensive server has more processing performance, storage, memory, network ports, redundant power supplies and cooling, operating systems, and other bundled software that fits your needs vs. that might be the better value vs. a lower-cost model that you have to spend to upgrade.

Keep in mind that economics, while often focused on actual costs, also includes the cost of people's times to do things. While the above example is for servers, there also examples for storage, I/O, and networking, cloud, and data protection, among other things that we look at in each of the respective chapters. Also remember to account for taxes, shipping, installation, insurance, training, and other up-front, as well as recurring expense costs.

For some environments and applications, focus is on performance, for others it is on capacity, for others availability, including data protection, while for others it is all about cost. On the other hand, many environments have different PACE needs even within a given application.

Some use a mixed approach of different dedicated storage solutions or media assigned to an application, task, or role, while others use hybrid technologies. So what is the best type of storage device, component, or media and medium? It depends! What are your application PACE needs and requirements?

Some environments, applications, or sub-application pieces have the need for more performance with less storage capacity, which would probably lean you toward more SSD. However, what if most of those applications requests are reads of data with a high degree of locality, which means adding some local cache? The local cache can be a mix or DRAM for reads as well as SCM or NVM on the server, as well as some shared DRAM and NVM on a storage system, appliance, or converted and virtual device that also could include some traditional spinning-media HDD for hybrid, an all-Flash array (AFA) or all-SSD array (ASA), or an all-NVM array solution. For performance environments, instead of comparing just cost per capacity, which does not reflect the functionality of fast storage, instead compare cost per activity such as cost per IOP, transaction, TPS, file, video, page, or object processed.

Watch out for comparing on a cost per capacity basis vs. cost per performance or productivity basis, or comparing cost vs. availability, accessibility, and durability. In other words, don't just buy capacity at the lowest cost without considering performance, availability, reliability, and cost to replace on a more frequent basis. For example, comparing a 1-TB NVM SSD to 1 TB of low-cost, high-capacity desktop or client-class HDD vs. low-cost, low-performance cloud storage is at best comparing apples to oranges, as they do not reflect the various characteristics and the ability of the different storage to meet different application PACE needs. Remember, everything is not the same, for workloads as well as technologies.

Keep in mind that some technology is older and has known capabilities as well as limits or issues. Likewise, some technology is newer with known features, but issues or limits may not be known yet. How to balance the old vs. the new is also how you can innovate by using old things in new ways, new things in new ways, sometimes even as companions in hybrid ways.

Also, understand your options and alternatives, even if they are not your preference, so that you can articulate why you chose a solution.

Vendor lock-in may not be a bad thing if you get a benefit from it. On the other hand, if you do not get a benefit, then vendor lock-in can be bad. Also keep in mind that the responsibility for managing vendor lock-in is on the customer, although some vendors may help to facilitate the lock-in. Also keep in mind that the golden rule of virtualization and software-defined anything is that whoever controls the management tools and interfaces controls the gold (that is, controls the gold or budget).

Some solutions are low-cost to enter, starting small, but they do not scale without adding complexity and increasing overhead as well as costs. This can apply to smaller servers, storage, network switches, and CI and HCI solutions vs. larger variants that can scale with less cost and complexity. You might pay more up-front for that scaling, but keep in mind what your growth plans are.

> *Tip:* Ask vendors (politely, of course, and with respect) "So What?". So What does something mean, or So What does something do for you or your environment? So What does something work with or require to work. As part of making decisions, do a SWOT (Strength, Weakness, Opportunities, and Threats) analysis of subjective as well as quantitative comparisons.

Beware of focusing on hardware and software solutions that promise near-term cost savings, yet lack the tools to manage and leverage those solutions effectively without added costs. Worse is that the solutions intended to reduce your costs are not used effectively and end up increasing your long-term costs.

If you have a capacity plan and forecast that includes server, storage, I/O and networking, hardware, software, and tools as well as PACE attributes, you can use that to leverage various end-of-the-month, end-of-the-quarter, or year-end deals. Likewise, sometimes vendors have excess product or inventory and offer promotions; if you know about them and can use the resources, you might have a good opportunity.

> *Tip:* Look at how a solution initially deploys and how it can be expanded over time: how disruptive those upgrades and enhancements are. Disruptive means downtime, as well as any other hardware, software, or service and tools dependencies. Disruptive also means to your budget. Are you paying more up-front for lower-cost later expansion, or are you paying less up front and then paying more later.

Watch out for solutions that play to lower cost per capacity based on a given data protection, data footprint reduction (i.e., effective capacity), RAID level including parity and erasure codes to reduce costs. However, that also reduces your performance and productivity for primary workloads. These types of solutions, however, might be a good fit for other scenarios.

> *Tip:* Speaking of cost, while it is important to look at NVM SSD on a cost-per-IOP or productivity and activity basis, which is where they perform well, an opportunity can be missed by focusing just on performance. For example, while an environment or application may not have strong performance needs for SSD, by implementing as much SSD as possible, even at a higher cost per capacity, if those devices magically clear up other application, troubleshooting, and support problems that in turn free up staff to do other things, not to mention making customers happy, then SSD can be a better value even if it is not being measured on a performance basis.

Considerations, tips, and recommendations include:

- Look beyond lowest cost per component: What's the effective enabling benefit value?
- How proven is a technology, solution, service or provider: Who else is using it?
- Understand your total cost of acquisition, install, training, migration, conversion.
- How will technologies work for you today, vs. you having to work for the solution?
- Does the solution require additional hardware, software, services, tools, staff and support?
- Vendor lock-in may not be a bad thing if you get a benefit from it.
- When in doubt, ask others for input or advice (my contact information is at the back of the book); however, do respect other people's time.

13.3.2. Analysis, Benchmark, Comparison, Simulation, and Tests

Analysis, benchmarks, comparisons, tests, and simulations can be used for assessing and validating new or planned technology as part of acquisition decision making. Testing and analysis can be used for problem solving, planning, engineering, validation, and verification of new application or data infrastructure software, hardware, services, and their configuration, as well as whether they function as intended. Other uses of a simulation (aka workload modeling) are to understand how new or changing applications and workloads impact performance as well as availability.

Another use for simulations is to troubleshoot, forecast, and perform what-if analysis to prevent problems. Workload simulation, test, and benchmark can determine what a particular hardware/software combination can do, or can be used to support some number of users, activities, or functions as part of a solution, as either a vendor, solution provider, or customer.

Will you be testing from the perspective of a user of applications (i.e., end to end or E2E), or as a consumer of data infrastructure services or data infrastructure components (server, storage, network, hardware, software or services)? What are the metrics (or insight) you are looking to obtain, what parameters for test configuration (duration, number of workers, workload mix)?

Where and what is your focus and objective?

- Stressing a system, solution, storage, software, or service under test (SUT)
- What will drive, generate, and act as a system test initiator (STI)
- Are you testing to saturation to see when and what breaks (failure scenarios)?
- Is your focus on utilization or at what point response time and usefulness degrade?
- Compare different technologies, tools, and techniques under various conditions.

How will you be testing, with recorded traces replayed against a given data infrastructure layer such as vs. an application, database or file system, operating system, hypervisor, server, storage, network or service? Instead of a trace replay, will you be running a random workload generator or script that runs a known workload that can be repeated with accuracy and consistency?

Will you be using a do-it-yourself (DIY) approach with your own or different scripts and tools, leveraging open-source and freeware, vendor-supplied or services? What workload will you be running—your own, somebody else's, something you found on the web, what a vendor, consultant or solution provider said to use? The biggest issue with benchmarks, tests, and simulations is using the wrong tool for the given scenario, along with the wrong configuration settings. Use the tool and settings that are relevant to your environment and application workloads, unless you can use your actual application. When in doubt, ask somebody for input and advice as well as understand what your application's workloads are.

Tip: Look for data infrastructure performance analytics solutions that not only automate and take care of things for you in an auto-pilot mode but also provide insight and a learning experience. For example, with the continued growth, scale and interdependence, automation helps to manage those complexities.

Does the tool or solution provide the ability to learn, enable analysis, and gain insight into why things are behaving as they are? In other words, can the automated solution help you manage while expanding your tradecraft experience and skill sets, while also applying those to your data infrastructure optimization needs?

Table 13.2 shows a basic five-phase process for conducting a simulation, test, or benchmark study project. With Table 13.2, keep in mind that everything is not the same in the data center, or information factory, as well as across different data infrastructures, and test and simulations will also vary. This means that not only are there different tools and technologies for simulating various workloads; the process steps can vary along with setup and parameter configuration.

You can increase or decrease the number of phases and steps to meet your particular environment needs, scope, and complexity. The basics are the plan, define, set up including practice, execute and analyze results, repeat as needed. This approach can be used for end-to-end exercise of the entire data infrastructure stack along with applications workloads or can be component-focused (i.e., on the server, storage, I/O network, or some other area).

Additional considerations include whether the workload trace, test, simulations (aka so-called benchmarks) accurately represent your specific production environment, or are they relevant to somebody else? Part of developing and expanding your data infrastructure performance, availability, capacity planning, tuning, optimization, and engineering tradecraft skillsets is gaining relevant knowledge to make informed decisions vs. guessing.

Figure 13.22 shows a sample system test initiator (STI) and SUT configuration for testing various storage devices with database and file serving workloads based on an actual simulation. Note that there can be multiple STIs as well as many SUTs as part of a simulation. You can view more detail on simulation configuration and results at www.storageio.com/reports.

In the example in Figure 13.22, the STI performed workload to test the SUT using Transaction Processing Council (TPC) TPC-C (OLTP transactional) workloads. These workloads simulated transactional, content management, metadata, and key-value processing. Microsoft SQL Server 2012 was configured and used with databases (each 470 GB, scale 6000) created and workload generated by virtual users via Quest Benchmark Factory (running on STI Windows 2012 R2).

General workload testing, simulation, and benchmark tips and recommendations include:

- Verify and validate what you are seeing.
- Format, Initialize, thin or thick, eager or full.
- Determine duration and reporting intervals.
- How much ramp-up and run-down time?
- What other workload running at same time?

Table 13.2. General Testing, Benchmark, Simulation Phases and Tasks

Phase	Action	Description, Characteristics, Tips, and Recommendations
1	Plan Learn Pretest Identify Schedule Obtain Tools Skills	The objective of what will be tested, when, how, where, for how long, by whom, using what tools and parameters. Identify candidate workloads that can be run in the given amount of time applicable to your environment or needs. Planning can also involve running some simple tests to learn how or what a tool is capable of to gain insight as well as see types of metrics or reporting available. Part of the planning involves determining an end-to-end or component (layer) focus testing, saturation, or responsiveness and productivity. Determine time schedules and resource availability, and create a project plan. Resources include people along with applicable tradecraft skills (experience) as well as tools and testing. In addition to planning the tests, what workloads, tools, configuration, and settings, also plan for how to analyze, interpret, and use the results.
2	Define Configure Validate Practice Pretest	Leveraging objectives, focus, and requirements for testing along with items from the planning phase, start to refine and define the test. Who is going to do what, when, preparations. This can also involve doing some additional pretest test runs to determine optimal configuration to establish the best foot forward for the actual test. After all, practice makes perfect, plus it is easier and can be more cost-effective to find bugs, issues, or problems with configuration early, before the actual test run.
3	Setup Test Validate Practice Review	Organize your test environment including applicable servers, storage, I/O and networking hardware, software, and tools, along with any cloud or managed services needed. This also means obtaining and configuring simulation, test, trace replay, or other benchmark workload tools or scripts. Initial setup and configuration of hardware and software, installation of additional devices along with software configuration, troubleshooting, and learning as applicable. Conduct any other pretest or "dress rehearsal" for practice runs and fine tuning. This is also a good time to practice using the results and subsequent analysis. Not to mention verify that all of the telemetry data collection, logs, and reports are working. Another consideration is whether you see the same or similar repeatable results with your test runs, or are they varying, and if so, why?
4	Execute Monitor Initial Report Assess	Now is the time to reap the benefits of the work done in steps 1–3. Run or execute the tests per your plan. Verify results to validate they are what you expected, re-run as needed, maintain good notes about configurations, what was done, any other observations on the SUT, as well as on the STI. Perform initial analysis to catch and correct if needed.
5	Analyze Report Disclose	Using the results, review: Are they what was expected? Are there any issues or anomalies (besides not getting what you wanted)? Identify issues, re-run if needed. Analyze and report results, including context about what was done, how done, configuration, and other disclosure information.

Additional tips and recommendations as well as reminders include:

- What other features or data services are or need to be enabled?
- Is data compressed or de-duped? How does that impact testing?
- Number of threads and workers, local or remote (separate from STI and SUT)
- Data size and locality, cache friendly or not
- Reads, writes, random, sequential, large, small
- Long stride, short, bursty, skip sequences

Some context is that a long stride is not to be confused with long or short stroking, discussed in earlier chapters. Instead, a long-stride workload pattern is where many sequential

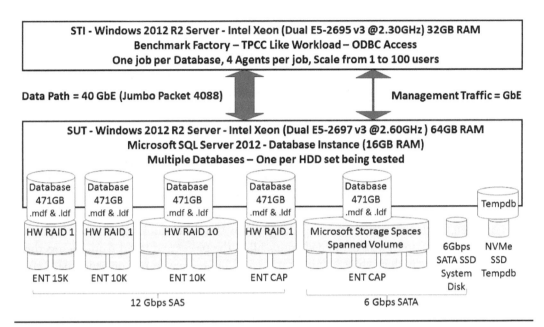

Figure 13.22   STI and SUT configuration examples.

reads (or writes) of various size are done, for example, a partial database table, file system, or another scan. Then a series of short, bursty random reads or writes are done. This could be for scanning an index or metadata to find a match, perform a sort or select, and then jump to another location to retrieve the actual data. Bursty workloads are just that: start, stop, pause, start, stop, and pause with various delays.

> *Tip:* Some workload generation tools such as Quest Benchmark Factory allow you to specify various think and delay parameters as part of configuration to support different test scenarios. For example, to simulate user think time, add longer delays; to simulate a streaming server that sits behind a load balancer with hundreds or thousands of requests (i.e., no think time), the values can be set low or to zero, among other options.

Testing of CI, HCI, CiB, and other converged solutions may at first seem complex and difficult, particularly if you approach from a component or silo focus. However, if you approach from a different focus, that of multiple applications and workloads running on a "collective" or combined, aka converged, system, things might seem easier. For example, if you are used to running iometer, crystal, iperf, or nttcp, among others, you can still run those on different layers.

On the other hand, if you are used to running database simulations with Quest Benchmark Factory (BMF), Hammerdb, Sysbench, YCSB, Jetstress, Sharepoint (WSS, Moss, and others), VMmark or DVDstore, among others, you may be well on your way for running converged workloads.

If all you need to do is put a load on a CI, HCI, CiB, or other converged solution, you can use workload generators such as vdbench powered HCIBench for VMware (both vSAN and traditional vSphere) environments. For VMware environments, there is also VMmark, which is a collection of workloads (DVDstore). There are also tools for testing VDI and workspaces such as those from LoginVSI and Virtual Instruments, among others. For exchange, there is Jetstress (unless you can get your hands on a copy of ESRP). Likewise, there are various scripts for Microsoft Diskspd to exercise Windows and Azure environments among others on the companion site for this book (www.storageio.com/sddi).

You can even cobble together your own workloads using a combination of the above among others and then manually, or, doing some scripting, automate their staging, execution, and results processing. Keep in mind whether you are testing the storage, server, or networking performance, or what the user sees and experiences under a given workload scenario.

Also keep in mind that with converged solutions, the focus for some might be on just the storage, file system, server, hypervisor, or networking. For others the focus might be on the management interfaces, underlying data services functionality, and capabilities, among other points of interest. However, with converged, the focus should be looking at what the entire solution is capable of doing as a combined entity, all of the pieces and layers working together.

What this means is that while it is important and interesting to test or benchmark converged systems storage, or server, or networking capabilities as data infrastructure component layers, also test the entire solution end to end. This means that if your CI, HCI, CiB, or other converged solution is going to be supporting VDI and workspaces, then test and compare with applicable, relevant workloads (besides your own). If supporting SharePoint (Visual Studio test, WSS, and MOSS) or Exchange (Jet), then use other applicable tools.

Tip: If your converged solution is going to be doing transactional database (SQL or NoSQL), then use applicable workloads; if doing data warehouse, analytics, big data, video or other content-focused, again, use applicable workloads. There are different tools for testing general-purpose file systems; then there are others such as IOR for Lustre large-scale file systems. Use the right tool for the given scenario.

General considerations include:

- Keep accurate, good notes documenting initial configuration and parameter settings.
- Perform calibration of any hardware, software, and cloud services, as well as tools.
- Conduct any pretest tuning as part of practice runs before formal tests (best foot forward).
- Validate that test tools, scripts, and traces are working, data is being collected.
- Keep all telemetry (results), event, log, and configuration data along with screen shots.
- Verify that STI or network to SUT are not bottlenecks or barriers.

Considerations, tips, and recommendations include:

- The best benchmark, test, or simulation, is your actual application workload.
- Second best are those that accurately represent your workload.
- Workloads, tests, benchmarks, and simulations should be repeatable and verifiable.
- A successful test, benchmark, or simulation starts with a good plan and objective.
- Know what tools exist, and how to configure and use them for your environment.
- Peaks, averages, standard deviations, sustained and stabilized.

13.4. Data Infrastructure Design Considerations

Data infrastructure design considerations include access, performance, availability, security and data protection, along with capacity optimization, economics, and management. Different environments and even portions of data infrastructures will have various design considerations to meet specific application PACE requirements and SLO. Also keep in mind that there are similarities across environments, so the following are some general considerations.

Your data infrastructure can be software-defined or legacy with software-defined management on-site, off-site, or in the cloud. Likewise, it can be private, public, or hybrid cloud and may leverage virtual, containers, and other technologies. The data infrastructure can exist on a single server, or CI, HCI, CiB, or cluster, as well as on a large scale to meet different needs.

While the technology components are important, keep the application focus and associated data in perspective. This means performance, availability, capacity, and economics (PACE) as well as applicable SLO, along with access and data services that enable some benefit. Also, design your data infrastructure with maintenance, service, support, and management in mind. Having a good inventory of your data infrastructure resources, hardware, software, services, cloud providers, management tools, policies, procedures, and applications can help in making informed decisions.

Common considerations include whether you will need dedicated resources for any application workloads (or data), or whether they can be shared with applicable tenancy, security, and performance. What applies to remote offices/branch offices (ROBO) and work groups might be similar for small–medium businesses (SMB), while larger enterprises may have different needs.

Performance considerations include:

- Application and data locality: Place active data as close to applications as possible.
- Keep storage allocation alignment in perspective for database and virtual environments.
- Fast applications need fast servers, software, and access to data on fast storage.
- Some applications will need support for smaller IOPs or transactions.
- Others will need large sequential streaming bandwidth or low-latency response time.
- Network bandwidth is important, as is latency and reliability.
- A little bit of SSD or cache in the right place can yield a big benefit.

Tip: Keep in mind that performance relies on availability, availability needs performance, and the two are related. Likewise, many configuration decisions have an impact on both performance and availability. For example, faults to tolerate (FTT) and resiliency to provide availability accessibility during failures can vary based on whether performance needs to remain consistent or can be allowed to degrade. Another consideration is the impact on applications during rebuild, repair, or reconstruction of servers, storage, or other data infrastructure resources during a recovery.

Clusters can be used to scale horizontally across multiple servers, storage, network, as well as CI and HCI solutions to increase available resources. Likewise, vertical scaling can be used to scale-up from smaller components, including those that are clustered. In addition to boosting performance and capacity, clustering can also be used to enhance availability with stability.

Availability, RAS, protection, and security considerations include:

- Hot spares, redundant components, failover to standby or active resources
- Fault isolation and containment along with self-healing or repair capabilities
- Leverage dual or multipath for server I/O to storage and general networking.
- Factor in data protection including versions, recovery points, and resiliency.
- Plan for rolling cluster upgrades and ability to have spare servers or storage.
- Will servers, storage, or network devices need to be removed from clusters for upgrades?
- Can you do in-place upgrades, or do clean installs have to be done?

Capacity considerations include:

- How much spare resources for servers, storage, and I/O do you need?
- Spare resources can be used for peak processing, support upgrades, and changes.
- What is your effective network bandwidth vs. advertised line rates?
- Are spare resources on-line, active or standby?

Considerations, tips, and recommendations include:

- Are your software licenses portable; can you move them across different servers?
- Can your software licenses be moved from physical to virtual to cloud, CI, or HCI?
- Leverage diverse network paths that do not have single points of failure.
- Note that you can have different network providers who have common paths.
- You can also have a single provider who has unique, diverse paths.

13.5. Common Questions and Tips

What does scale with stability mean? As you scale up and out to expand resources, this should occur without introducing performance bottlenecks, availability points of failure, or excessive resource overhead and management. In other words, as a result of scaling, your data infrastructure should not become unstable, but rather, more stable across PACE and SLO.

Are chargeback and orchestration required for cloud and software-defined environments? Different types of orchestration tools should be leveraged to help with automation of not only cloud but also other software-defined, virtual, and legacy environments. Depending on your environment needs, a chargeback may be applicable, or a showback as a means of conveying or educating users of those services what the associated costs, as well as benefits, are.

What level of audit trails and logging is needed? Maintain audit trails of who has accessed or made copies of data for event analysis. As important as is what you collect, it is also important how you use and preserve logs for analysis or forensic purposes. Leverage automation tools that can proactively monitor activities and events as well as distinguish normal from abnormal behaviors.

Who should be responsible for security? Some organizations have dedicated security groups that set policies, do some research, and do some forensics while leaving actual work to other groups. Some security organizations are more active, with their own budgets, servers, storage, and software. Security needs to be a part of activity early on in application and architecture decision making as well as across multiple technology domains (servers, storage, networking, hardware, and software) and not just via a policy maker in an ivory tower.

What tools exist that can help collect, assess, and plan for server, storage, and desktop among other migrations to virtual, cloud, or other software-defined data infrastructures? There are many different tools that can collect various data infrastructure and application resourced as well as activity information. Some of these tools focus on the operating system or hypervisor, some focus on enabling migration to different hardware, software, or cloud platforms. Tools can be used for performing inventory, collecting performance baselines, and other activities.

Figure 13.23 shows an example of the Microsoft Assessment Planning (MAP) tool that collects information on various servers, desktops, applications, and resource activity. The collected information can be used for making decisions to move to VDI workspace, virtual servers, or cloud, among others. Others tools are available from AWS, Citrix, Quest, Solarwinds, and VMware, among many others.

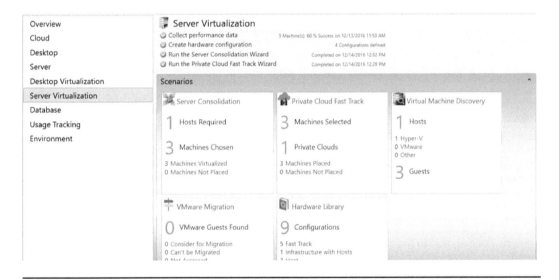

Figure 13.23   Microsoft Assessment Planning tool.

When should legacy vs. CI vs. HCI vs. software-defined cloud or virtual be used? That depends on what you are looking and needing to do. Different applications have various PACE attributes and SLO needs. Hopefully, you can leverage your expanded or refreshed tradecraft experiences from the various chapters in this book to determine what to use when, where, why, and how. Chapter 14 provides some data infrastructure deployment considerations.

13.6. Chapter Summary

Having metrics and insight enables awareness to avoid flying blind with your data infrastructure, including decision making and other management tasks.

General action items include:

- Gain insight and awareness of your application workloads and data infrastructure.
- Leverage new and existing technology in new ways.
- Balance optimization for efficiency utilization vs. productivity effectiveness.
- Know your tools and how as well as when to use them in different ways.
- Decision making applies to acquisition, planning, configuration, and troubleshooting.
- Leverage your expanded or refreshed tradecraft to know what to use when, where, and why.

Bottom line: The best metrics are those that apply to your environment, applications, and data infrastructure to enable informed decision making (and avoid flying blind).

Chapter 14

Data Infrastructure Deployment Considerations

Where do you want to deploy vs. what your applications need.

What You Will Learn in This Chapter

- The roles of different techniques, technologies, and tools
- What to look for and consider in comparing features
- What approach to use for different scenarios
- How to apply your data infrastructure essential tradecraft skills

This chapter converges various themes and topics from the previous chapters combined with some additional tradecraft, insight, and experiences. The focus is on common data infrastructure (both legacy and software-defined) deployment topics. Key themes, buzzwords, and trends addressed in this chapter include server, storage, I/O network, hardware, software, services, converged infrastructure (CI), hyper-converged infrastructure (HCI), software-defined, cloud, virtual, container, database, big data, scale-out, grid, cluster, and various applications, among other related topics.

14.1. Getting Started

In Chapter 13 the focus was on managing data infrastructures including legacy, software-defined data centers (SDDC), software-defined data infrastructure (SDDI), and software-defined infrastructure (SDI). This included various application, technology, and SDDI life-cycle management (LCM) areas of focus. Now let's shift our focus to deployment considerations for various applications, as well as SDDI-enabling technology, tools, and techniques. Keep in mind that all applications have some type of performance, availability, capacity, and economic (PACE) attribute that vary by type of workload, environment, size, scale, and scope, among other considerations. Different environments, applications, or application component workloads also vary, having diverse PACE needs and service-level objectives (SLO).

14.1.1. Deployment Tradecraft and Tools

In prior chapters, we provided smaller sections devoted to toolbox, context, and tradecraft topics. However, as part of nearing the wrap-up of this book, the tradecraft is sprinkled throughout the chapter. In other words, building on the previous chapters, let's leverage what we have covered, filling in some blanks by looking at some common data infrastructure deployment considerations. This will include a mix of legacy, proprietary, open-source, cloud, virtual, containers, and platforms.

14.2. Applications, Tips, and Learning Experiences

Many different application landscapes and workloads have been mentioned throughout this book. A few more buzzwords and buzz terms in addition to those already mentioned, include email; messaging and collaboration using Axigen, Exchange, Exim, Google, O365, SharePoint, and others; big data and analytics, including Apache, Cloudera, Datameer, SAS, DataStax, Elastic search, Exasol, Hadoop and MapReduce, Hortonworks, Kafka, MapR, Pivotal, SAP HANA, Splunk and Tableau, among many others.

Additional applications, tools, and middleware include SITA, CDN, and CMS including Drupal, WordPress along with other Linux, Apache (HTTP) or Nginx, MySQL, PHP (web-focused), EPIC, Meditech, and Desix healthcare. Others middleware includes Tomcat, JBOSS, J2EE, IIS, and SAP, as well as Magneto and redis, among many others tools and applications.

While everything is not the same across environments and even the same applications, given the number of users and corresponding workload, as well as features or functions used, there are similarities. Likewise, all applications have some PACE attribute and SLOs. As has been discussed throughout this book, part of expanding or refreshing your essential data infrastructure tradecraft skills is gaining experience and understanding various applications that are part of the organization's business infrastructure and IT in general.

By having an understanding of the applications and their corresponding PACE attributes, you can avoid flying blind when it comes to making data infrastructures decisions. These decisions span from deciding what technology, architecture, and service trends and techniques to apply when and where to how to use different application workloads or environments.

Where applications can or should be deployed:

- On-site, off-site, or with another cloud service
- As physical (legacy, CI, HCI, CiB), virtual, container, or hosted
- On a single physical server with internal or external dedicated storage
- On multiple clustered physical servers, each with dedicated storage
- On multiple clustered physical servers with shared networked storage

- On aggregated servers, storage, and networks, or on disaggregated resources
- What type of I/O networks (local and remote) as well as storage to use

How applications, settings, metadata, and data are protected:

- Basic RAS and HA, FTT and FTM modes
- Mirroring, replication, and clusters for availability
- Point-in-time (PIT) copies, checkpoints, snapshots
- Logical and physical security, intrusion detection

Considerations, tips, and recommendations include:

- Turnkey service, buy solution, subscription, build DIY, some assembly required
- The application workload and PACE attributes along with SLO
- What are your design premises and priorities?
- Design for resiliency, performance or capacity, and cost savings
- Design for fault isolation, containment, or rapid failure-tolerant modes
- How resource failures are handled by applications including caches

14.2.1. Software-Defined, Virtual, Containers, and Clouds

Applications have different PACE attributes SLOs. Likewise, applications depend on servers (compute), local and remote I/O networking, and storage resources. These resources are managed as well as enabled and defined via software. Software defines hardware, other software, as well as various services to support different needs.

Figure 14.1 shows various software-defined and legacy data infrastructures supporting different application landscapes, services, and modes of consumption or use. These span from

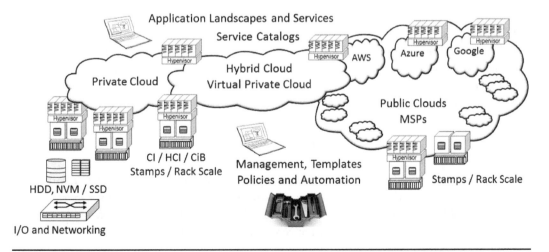

Figure 14.1 Software-defined and legacy data infrastructures.

legacy "brownfield" environments to "Greenfield," newly emerging, on-premise to all in the cloud. In addition to legacy, public and private, also shown are hybrid infrastructures adapting the best of available technologies, techniques, tools, and policies to different needs. Also shown are traditional compute servers, software-defined virtual and containers, CI, HCI, and rack-scale, along with hardware and software storage, networking, and management tools.

Benefits and attributes of SDDC, SDI, and SDDI environments include:

- Elastic to scale up, out, and "burst," or grow rapidly to meet application needs
- Growth includes compute, storage, I/O networking, and PACE-related resources
- Flexible and scalable to meet different and changing needs
- Resilient, available, durable, depending on how configured or used
- Cost-effective, enabling you to pay for what you use, when you use it
- Multi-tenancy for privacy, security, and protection while sharing resources
- Track and provide accounting reporting, showback, chargeback, and insight
- Leverage new technology in new ways at a faster cadence (speed of doing things)

SDDC and SDDI utilize or are implemented using:

- Clouds—public, private, and hybrid
- Cloud stacks such as Microsoft Azure and OpenStack, among others
- Containers. Docker Linux and Windows
- Virtual machines (VM) on legacy as well as CI, HCI, and CiB solutions
- Workspace and virtual desktop infrastructures (VDI)
- Bare metal as a service (MaaS), such as dedicated private servers (DPS)
- Virtual private cloud and hybrid, spanning sites, locations, and services

General considerations include:

- How will you protect, preserve, and secure applications, settings, and data?
- Who are the users? Where do they access different applications?
- What are those applications' information services and data infrastructure dependencies?
- Balance lower costs with increased enablement, resiliency, and flexibility.
- Large scale vs. medium size vs. remote office/branch office (ROBO) or small office needs
- Management tools, interfaces, and interoperability with what you already have

Considerations, tips, and recommendations include:

- "Greenfield" (brand new, from scratch) for your applications or data infrastructure
- "Brownfield" (existing applications or environment) to move, migrate, or re-host
- Who and how will resources and applications be managed as well as supported
- Where and what is your focus for convergence and management
- Control, security, privacy, data protection, and application resiliency
- What conversions of applications, data, virtual machines are needed
- What additional tools, migration, and conversion utilities are needed

14.2.1.1. Clouds: Public, Private, and Hybrid

Clouds have accounts with an owner, optional billing, and related information. Different clouds have various focuses, data services, feature functionalities, and service offerings. Activity and usage reporting can be tied to an account or users, as well as alerts, alarms, and health check status notifications. Billing and spend alerts can be set up to let you know when costs reach or exceed a particular point.

> *Tip:* Pay attention to what region or location your applications and data are placed in, making sure they follow any applicable legal or regulatory privacy and compliance requirements.

Resources are assigned and owned by an account. These resources can be in different regions (geographical locations) with one or more sites. A site consists of one or more data centers that might be dedicated, or co-located in a multi-tenant secure facility. Note that different regions and sites can be used for enabling availability and resiliency.

> *Tip:* Verify that cloud resources are disabled when you are done, or to turn off when you are not using them. With traditional resources, you might forget about them although they continue to be in use, without generating an invoice, chargeback, or showback report. With a public cloud, however, you might be surprised to see how those resources can run up an unexpected bill.

Management tools, capabilities, and interfaces vary, but they usually include, among others, REST/HTTP, HTML5-based GUIs, Wizards, templates and policies, XML and JSON formatted files. Other tools and capabilities include workflow managers, orchestration and service catalogs, dashboards, event, activity, resource, and cost use threshold alerts. Additional tools include import and migration along with conversion for moving applications, systems, and data to or from cloud environments.

Different cloud services have various management tools and interfaces including GUI, Wizards, and CLI, along with API. Also, there are various orchestration and dashboard tools as well as services that can span different cloud services as well as private and hybrid clouds.

Some of these tools also support on-site virtual, containers, and other software-defined data infrastructures along with legacy. Tools vary from simply monitoring, reporting, insight, and awareness to provisioning, workflow, and resource orchestration. Examples of tools include Ansible for orchestration, management, and automation, mesos, swarm, Kubernetes, Tectonic, cloudkitty, collectd, dashbuilder, Nagios, openit, rrdtool, Solarwinds, OpenStack, VMware, and Microsoft, among many others.

In the past, there has been time sharing along with service bureaus for supporting access to different resources and applications. Some of those were (or are) general, while others are application- or industry-specific. Likewise, some of today's clouds are general-purpose, while others are industry-, application-, or function-specific. For example, there are government-focused public clouds such as AWS GovCloud. However, there are also private government clouds (or services) providing services to financial security trading and airline reservations systems, among other businesses, institutions, and agencies.

Another approach for supporting applications and data infrastructures is outsourcing and managed services such as those supported by IBM and ASC, among others. In some cases the entire application along with support (and possibly developers) is transferred to the outsourced or managed service provider. In other words, they become employees of the service provider: The applications and associated resources are owned and operated on behalf of the customer to specific SLO, among other requirements. There are also scenarios where applications and data infrastructure resources remain physically where they are, but the service provider takes over the space and management of the various entities.

Table 14.1 shows various cloud, hosting, and co-location venues and characteristics.

Table 14.1. Cloud, Hosting and Co-location Venue Characteristics and Considerations

Venue	Characteristics	Considerations
Cloud service provider (CSP)	Provides various services and functionalities from application to resources, different granularities of computing, storage, network, development, and other tools. Features vary; some are very large, such as AWS, Azure, GCS, and Softlayer, among others. Some are small and regional. Similar to hosting firms who have expanded into the cloud, virtual private servers (VPS), and dedicated private servers (DPS), some traditional cloud providers have added DPS and hosting capabilities as well as other hybrid enabling capabilities.	Can be used to move your applications, production workloads, development, analytics, testing, or other types of processing for compute, storage, or network functions. Different levels of tenancy, as well as resources and cost along with SLO and SLA. Can also be used for a hybrid to co-exist with your legacy environment.
Hosting	Traditionally the web, email, DNS, domain hosting, WordPress blog, and related hosting services. Some offer cloud services and VPS or VM in addition to DPS, for example, Bluehost, and GoDaddy, among many others.	Can be used to offload specific functions and operations to free up resources for primary or core applications. Different levels of service, support, SLO, SLA, and feature functionality.
Co-location (co-lo)	Some supply servers, storage, and networking hardware or software. Others provide safe, secure, and optimally managed space for your technology. Context is needed for management, as it can mean simple monitoring or more extensive hands-on-support. You have control of your technology, how it is used, configured, managed, and maintained, along with troubleshooting. You are responsible for your technology, hardware, software, and its configuration. Understand the security (physical and logical), primary and standby power, as well as networking capabilities. Examples include Datapipe, Equinix, and Level 3, among others.	Use to supplement your existing data center or as an alternative to your own. Enable setup of new small sites where needed. Who has access to your technology and when? Space can range from a cabinet cage to a portion of a data center. Some have "direct connect" high-speed, low-latency connectivity into AWS or other clouds, bypassing normal front-end Internet-based access. Some are large; some are local.

In addition to public cloud services, hosting, co-lo, and managed service providers, you can also build and deploy your own private cloud. Likewise, you can connect your private cloud or legacy data infrastructure environment to public or other services for a hybrid environment. This also means configuring (defining) hybrid clouds that span different providers such as AWS, Google, and Azure, among others. Cloud stack software includes Cloud Stack, Microsoft Azure, NooBaa, OpenStack, and VMware, among others that you can build yourself (DIY).

Tip: On-board, or cloud instance, DAS, like traditional server DAS, can be useful as cache, scratch, or high-speed local storage, but pay attention to whether it is persistent or ephemeral.

DIY ranges from "some assembly required" to complete build-it-yourself from reference architectures, cookbooks, and other resources. There are tutorials and how-to guides from many different vendors as well as organizations, such as OpenStack, Ubuntu, RedHat, Suse, and others.

Tip: Similar to physical (or virtual) servers with dedicated DAS (HDD or SSD), cloud instances have similar feature in addition to persistent external storage. Pay attention to whether the instance is "on-instance" (i.e., DAS) and if that storage is persistent across reboots as well as system shutdown. Some services support persistent "on-instance" storage across reboots, while others do not. Likewise, some are persistent across reboots but ephemeral (your data gets deleted) across shutdown, stop, and restarts. Leverage some software to protect data across server shutdowns of ephemeral storage, including snapshots, backup/restore, and sync, when you need persistent data on cloud instances.

There are also turnkey solutions where you focus on the configuration, definitions of policies, templates, and other deployment tasks. An example of a prepackaged OpenStack solution software appliance is VMware Integrated OpenStack Appliance (VIOA). Another example is Microsoft Azure stack, and there are others. Which is best for your SDDC, SDI, and SDDI environment depends on what you are trying to accomplish or need to do.

Tip: With some services and stacks you can create an instance with attached volumes, build and define your environment with applicable software, and create configurations just as with traditional server images or builds. Once you have created your system, you can create snapshots for creating new volumes and systems. Likewise, you can also *detach* a volume and *attach* it to another instance, similar to removing a HDD or SSD from one server and moving it to another.

Volumes for cloud compute instances can be expanded in size for Windows and Linux systems similar to with physical and virtual deployments. Check with your particular cloud service or software stack solution provider as to what is supported and the steps involved, then leverage your tradecraft skills (or learn new ones) to do the operating system–specific expand steps.

Tip: Different public cloud services and private cloud software stacks have various features, data services and extent of functionality, as well as cost considerations. Some public and private cloud stacks provide tools, while others have open APIs along with SDKs to assist partners and developers with interfacing to those platforms. For example, leveraging various APIs, tools such as DataDog, RightScale, and many others can collect telemetry data to provide insight into activity, events, usage, and configuration information.

General considerations, tips, and recommendations include:

- There are similarities, including management and resources with different granularities.
- Cloud compute instances are similar to traditional servers (i.e., cores, memory, I/O).
- There are various types of storage for different tasks (i.e., block, file, object, tables, queues).
- Various networking options, from general Internet to high-speed direct connect.
- Leverage various tools for monitoring to gain insight and awareness.
- Monitor resource usage, activity, and events along with applicable costs.

Tip: Some cloud services and stacks allow you to launch an instance using preconfigured images on different types (tiers) of servers, storage, network, and security templates. Likewise, some allow you to set up customized instances including using your software images. In addition to providing instances with pre-licensed images, some allow you to bring your software licenses instead of using the service provider's.

Additional considerations, tips, and recommendations include:

- Pay attention to persistent and ephemeral storage along with instance settings.
- Verify that you have persistent settings for instance stop, shut down, and terminate.
- Leverage your tradecraft experience across physical, virtual, and cloud.
- Don't be afraid of clouds; be prepared and do your homework to learn about them.
- If you can list your cloud concerns, you can then classify, rank, and address them.

14.2.1.2. Public Cloud Services

Use public clouds when you need or want flexibility, agility, ability to pay as you go, and scale, along with leveraging off-site capabilities. Public clouds can also be used to move or relocate your IT and data infrastructure resources along with associated applications to, or, in a hybrid deployment, complement what you are doing on-site. There are many different types of public clouds along with managed service providers that have various areas of focus.

Tip: Public clouds can be used to replace on-premise or complement hybrid on-site data infrastructures. Other uses for public clouds include a place to do development, big-data and other analytics, high-performance computing, content rendering and aggregation, testing, and integration, among others.

Uses of public clouds include as a target or destination for copying, replicating, syncing, backing up, or archiving data in support of data protection activities (BC, DR, and BR). Public clouds can also be used as a place to tier data for applications that reside on-premise, as well as a place from which applications can be hosted.

Tip: Having concerns about clouds and public clouds, in particular, is not a bad thing in my opinion, as long as you can document what they are. If you can identify and list what your cloud concerns are, as well as why, then you can also indicate the severity or priority of those issues. From there you can work with others to identify how to resolve and address the concerns, find workarounds, or, in some cases, verify that cloud is not a good fit for a given scenario. However, if you cannot identify what your cloud concerns are, then how you cannot address them.

Various cloud services have different data services, management, features, functionalities, and personalities. Some clouds have personalities optimized for computing, others for specific applications, still others are general-purpose. Likewise, the extent of feature functionality also varies, so look beyond the basic checkbox of what is supported. Costs also vary, with some offering all-inclusive service, and others offering lower à la carte prices with various fees for different functionalities. Look into the service provider, the offerings, and the terms, including SLO, SLA, and fees as well as quotas and limits.

A sampling of public cloud and managed services providers includes Alibaba, Amazon Web Services (AWS), Apple, AT&T, Bluehost, Box, BT, Comcast, CoreSite, CSC, Datapipe, Dell EMC Virtustream, Dimension Data, Dreamhost, Dropbox, and DynDNS. Others include Equinix, Ericsson, FibreLight, Firehost, GoDaddy, Google Cloud Engine, IBM Softlayer, InterCloud, Joyent, Level3, Microsoft (Azure, Office 365, Xbox), NTT, Oracle, Orange, Peak, Rackspace, Salesforce, SMART, Tata, Telkom, Telstra, Telx, Verizon, Virtustream, Vodafone, VPS, VMware, Zayo, and many others. Table 14.2 shows a summary of some common service offerings from AWS, Azure, and Google.

Amazon Web Services (AWS) includes various applications, management, compute from instances (EC2) to Lambda micro services (containers), lightsail (VPS/DPS), block (EBS), file (EFS), and Simple Storage Service (S3) object. S3 includes policy-based standard, reduced redundancy (RR), infrequent access (IA), and glacier (cold for inactive archives) storage as well as relational database services (RDS) and table services, including RDS, Aurora, DynamoDB among others (SQL and NoSQL). Others include IoT hubs, management, SDK, and related tools. General analytics, Route 53 networking, development tools, and much more. Different levels of services include performance, availability, durability, and capacity to meet various economic and SLO needs.

Table 14.2. Summary of Common AWS, Azure, and Google Features

	AWS	Azure	Google
Resource	aws.amazon.com/products/	azure.microsoft.com	cloud.google.com/products/
Compute	EC2 (instances), Lambda micro services (containers). Various size instances, types of images (Windows, Linux, applications), management options, LightSail private servers. Instance-based including SSD (i.e., DAS as part of the instance).	VM and container services. Various size instances, types of images (Linux, Windows, applications), management options. Instance-based including SSD (i.e., DAS as part of the instance).	Various compute engine granularities, sizes, and types (Windows, Linux) of instances and services. Instance-based including SSD (i.e., DAS as part of the instance).
Database	Various services from tables to SQL and NoSQL compatible. Database instance images also for EC2.	SQL databases, data warehouse, tables, DocumentDB, redistributed database on Azure VM..	Cloud SQL, Cloud Bigtable, NoSQL cloud datastore.
Block storage	EBS HDD and SSD	Disk (HDD and SSD)	Persistent (SSD)
Bulk Big data Object Content Archive Log and Snapshot Storage	S3 standard, RR (reduced redundancy, lower cost), IA (infrequent access, lower cost, slower access). Glacier (cold, very slow access [hours], low cost).	Blobs, various regional and geographical protection schemes. Various big data services, resources, and tools.	Object multiregional; regional, near-line, and cold are off-line. Various big data services, resources, and tools.
File storage	EFS (within AWS)	Azure File within Azure, external Windows SMB 3	
Networking	Route 53 DNS, load balancing, direct connect, VPC (virtual private cloud), Edge locations	Virtual networking, CDN, load balancer, gateways, VPN, DNS, ExpressRoute (direct connect), management	Networking, CDN, load balancing, virtual network, DNS, and interconnect

Figure 14.2   AWS EC2 instance and setting network access.

AWS has a large global presence in regions around the world, with multiple availability zones and data centers in each region. Figure 14.2 shows on the left an AWS EC2 cloud instance (VM) that has been provisioned and is being configured. On the right side of Figure 14.2, network security access for the EC2 instance is being configured for different ports, services, and IP addresses. AWS, like many other providers, has import/export capabilities, including where you can ship storage devices to compensate for slow networks, or where there is too much date to move in a short amount of time even on a fast network.

In addition to basic services, AWS also has services called Snowball (smaller appliance) and Snowmobile (semi-truck sized) that can scale from terabytes to petabytes of data moved via self-contained appliances. With these appliances, you simply plug them into your environment and copy data onto them (with encryption and other security). The appliances are then shipped back to AWS (or another provider of similar services), where the data is ingested into your account.

Figure 14.3 shows an AWS EC2 Windows instance on the left. Note that as part of the instance access, the password can be obtained with the proper security credentials and keys, along with an access script such as for RDP. The Windows Server instance desktop is accessed on the right of Figure 14.3. Note in the top right corner of Figure 14.3 that there are various AWS instance settings, including instance ID, type of processor, and instance, among other information.

Dell EMC Virtustream provides public, private, and hybrid cloud storage solutions. Google Cloud Services has large-scale compute, storage, and other capabilities in addition to their consumer-facing and email solutions, as well as collaboration and workspace applications. IBM Softlayer is another public cloud service with many different geographic locations.

Microsoft Azure is another large public cloud service that also supports interoperability with on-premises Azure Stack (Azure in a box at your site) as well as Windows Server environments. There are many different Azure data centers in various geographical areas, with new ones being added, along with increased functionality for compute, storage, networking, management, development, production, CDN, and analytics among others. Microsoft also has the Xbox as well as O365 clouds, and don't forget about LinkedIn, Skype, and others.

Rackspace is another cloud and managed service providers offering both traditional dedicated and shared hosting, DPS, VPS, and cloud instances. There are also many other small to medium-size local, regional, and international cloud and managed service providers; some are Telco's, some are hosting or co-lo, among others.

Figure 14.3   Connecting to AWS EC2 Windows instance.

VMware is known as a supplier of SDDC and software-defined data infrastructure (SDDI) technology for organizations of various sizes (from small to large). However, it also supplies service providers. While OpenStack and other software are commonly thought of regarding the use of the cloud and managed service providers, some have found VMware (and Microsoft) to be cost-effective.

14.2.1.3. Private and Hybrid Cloud Solutions

Use private clouds when you need or want more control over your environment. These can be deployed on-premise at your location, or in a co-lo, or other managed service provider environment. Your choice of going private cloud may also be to establish a hybrid environment spanning your existing, new and software-defined, as well as public data infrastructure environments. Reasons to have more control include cost, performance, locations, availability, or security along as well as privacy among others.

Private cloud and cloud software includes Cloudian, Cloud Stack, Dell EMC Virtustream, Microsoft Azure Stack, OpenStack, Pivotal, Swiftstack, and VMware, among many others. Figure 14.4 shows an OpenStack dashboard display CPU compute (i.e., hypervisor). The

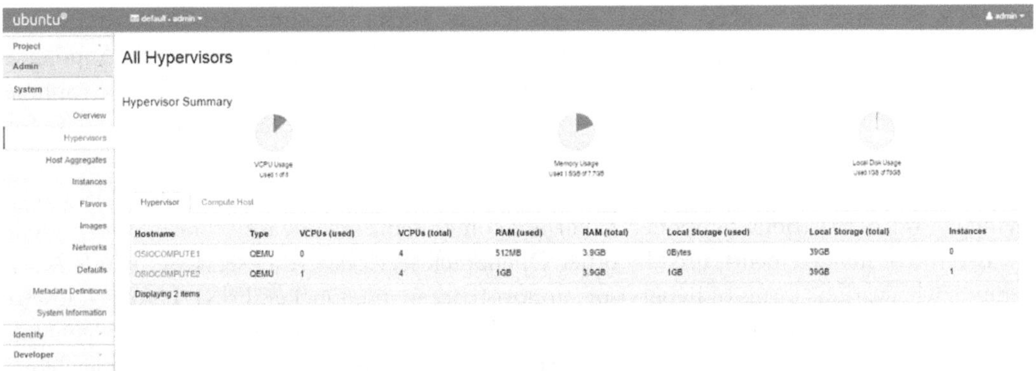

Figure 14.4   OpenStack Horizon dashboard showing compute resources.

image in Figure 14.4 is a multi-node OpenStack deployment I did using Ubuntu and available OpenStack deployment guides.

Tip: As part of protecting an OpenStack environment, make sure to have a backup or other copy of your various databases, as well as snapshots of critical servers or nodes. When something happens, being able to restore all or part of a configuration database with metadata, or configuration files or a point-in-time snapshot can mean the difference between availability and downtime.

OpenStack has emerged as a popular software-defined cloud stack for implementing SDDC, SDI, and SDDI (as well as cloud) environments on both large and small scales. It is used by large service providers who customize it as well as by smaller organizations in different ways. Keep in mind that OpenStack is a collection of data infrastructure components (aka projects) that address different functions and have different functionalities. These projects or components feature well-defined APIs and are highly customizable, from functionality to appearance. Core functionality components include Keystone (security and access), Horizon (management dashboard), Neutron (Networking), Nova (compute), Glance (image management), Cinder (block storage), and Swift (object storage), among many others. Note that you can run compute and dashboard (along with network and security) without Swift object storage, and Swift without compute capabilities.

OpenStack leverages databases such as MySQL or one of its various variant engines as a repository for configuration, security, users, roles, endpoints, access controls, and other metadata. Various configuration files define functionality, as well as support for plug-ins or tools to use, for example Rabbit vs. other messaging software.

Tip: The benefit of doing an install from scratch, such as with OpenStack is to gain insight into what the pieces and parts are, and how they work. Also, it will give you an appreciation for the customization and extensiveness of a solution, along with what will be involved to maintain a production environment vs. using bundled solutions.

Besides OpenStack install from scratch using various cookbooks and tutorials available from OpenStack and other venues, there are also automated build scripts you can find with a quick Google search. Some tutorials and scripts are for a single node, which might be good enough to get some "behind the wheel" or "stick time" with the technology.

There are also some appliances such as the VMware Integrated OpenStack Appliance (VIOA). You can download VIOA from VMware as an OVA and then deploy on a vSphere ESXi system. Likewise, you can download different pieces of OpenStack such as Swift for object storage from OpenStack, as well as others such as Swiftstack.

Figure 14.5 shows a generic OpenStack architecture with (top left) a Horizon dashboard having interfaces to various components. Various components include Keystone (security, identity assessment management), used by other OpenStack services. Other services include Nova compute with various forms of hypervisor support (bare metal, Docker, ESXi, Hyper-V, libvirt [Xen and LXC], Qemu, and XenServer, among others) along with Glance image management.

Note that there is also Magnum (i.e., containers). Software-defined storage includes block provided via Cinder, with files via Manila and objects with Swift or using other plug-ins. Also shown is Neutron (SDN and NFV) networking, facilitating private east–west management or intrasystem communications, as well as intersystem communications with the outside world.

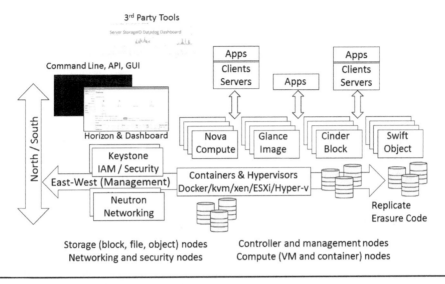

Figure 14.5   OpenStack generic architecture.

There is also north–south (aka northbound, southbound) access of resources shown, as well as other functionalities within networking. Other networking functionalities can include border gateways, DNS, and DHCP, among other network functions virtualized.

Depending on the size and scope of your OpenStack, deployment will involve a different number of nodes (servers) for the various services. For high availability (HA) and resiliency, set up two or more controller nodes, and distribute different functions across nodes depending on workload and performance. Smaller environments can consolidate to fewer nodes, or even containers, using persistent Docker storage (more on this later in the chapter).

Note that OpenStack can overlay VMware vSphere ESXi as well as Microsoft Hyper-V hypervisors (among others), which mean that it can be used as a point of management and orchestration convergence. On the other hand, for simplicity, OpenStack can be deployed with VMware or Microsoft, among others, as a management layer along with other hybrid approaches.

General implementation tips and considerations for private clouds include:

- Implement data protection, including basic as well as HA for resiliency.
- Implement recovery point objective (RPO) protection with snapshots and backup.
- Protect the various databases that contain configuration, authentication, and metadata.
- Deploy redundant nodes and configuration, replication, and other techniques.
- Compute can have local storage similar to a physical server with DAS.

Tip: Besides OpenStack, third-party tools, extensions, and plug-ins can be added to enhance management and orchestration. Some third-party (and open-source) tools for OpenStack include Ansible (policies, templates), Awnix, cacti, Chef, Cloud kitty (billing, invoicing), Collectd, Datadog, Dell EMCcode, Freezer, Kubernetes, Nagios (monitoring), RedHat, RRDtools, Swiftstack (object storage management), Storpool, Suse, Trilio and Ubuntu (metal as a service [MaaS]), and many others. Some other useful tools to have include Cloudberry, curl, Cyberduck, gsutil, S3 Browser, S3FS, S3motion, ssh, wget, and others.

Considerations, tips, and recommendations include:

- Protect your private cloud with resiliency and point-in-time data protection.
- Everything is not the same with cloud, hosting, and managed service providers.
- Some services and features work inside (intra-) a cloud, others are external (inter-).
- Balance the value of your time and budget with DIY vs. kit or turnkey solutions.
- Understand how you will customize and then manage those changes through new versions.
- Do proof of concept (POC), understanding what's involved with DIY for various solutions.
- Leverage POC to determine production considerations including support and maintenance.
- Utilize reporting to monitor resource usage, activity, and costs as well as alerts.

14.2.1.4. Docker, Containers, and Microservices

Compute containers are a form of computing that provides more tenancy (private isolation) for a process, task, or program instance than on a normal-running shared server environment, without the overhead of a virtual machine.

> *Tip:* Some context is needed here to distinguish a standard intermodal shipping container (Figure 14.6), commonly seen on railcars, ships, and trucks, and a Docker (or other) compute server microservice. The reason context is important here is that some large-scale environments leverage variations of standard physical shipping containers to "stuff" with data infrastructure components, including servers, storage, networking, and other items.

Further context is needed in that while data infrastructure components are shipped in physical containers from factory to data centers or another habitat for technology, some are

Figure 14.6   An intermodal shipping container.

also configured as mini-data centers or data center modules. Here, "modules" means that the physical containers are preconfigured with data infrastructure components that "plug in" to data center power, HVAC, and networking, and then are defined with software to perform various functions.

The other context, which is the focus here, is that of a compute service that is a subset of a traditional server (i.e., a microservice). Thus context is important when it comes to containers.

In Chapter 4 we looked at platforms from bare metal (BM) physical machines (PM) to hosts, hypervisors, and containers. Docker is a popular container management platform, along with tools including kubernetes, mesos, and warm, along with "back-end" engines that execute (i.e., host) as microservices running container instances (images).

Tip: There are Linux-based engines as well as Windows. Key context here is that when you hear "container," and in particular "Docker," you need to know whether it is referring to the management or the engine. This context is important because containers built on and for Linux engines can only run on a Linux engine. However, they can be managed from Windows, Linux, or others. Likewise, a Windows-centric container only runs on a Windows-based engine although it can be managed from different platforms. In other words, like programs or applications, "binary" bits or executables built for a specific operating system or hardware architecture, containers have a similar affinity.

Containers are more lightweight, resulting in carrying less overhead baggage than a virtual machine or requiring an entire PM or BM. Containers essentially rely on certain libraries or tools being in place on the host engine, so they do not have to carry that extra baggage.

Think in terms of taking a trip and not having to pack items that you know will be at your destination, such as irons, ironing boards, blankets, pillows, and towels. The result is that you need fewer things in your suitcase, roll-aboard, or backpack, making it lighter, easier to travel with, and will adapt as well as a plug-in to where you are going.

Tip: There is a myth that containers are only ephemeral (e.g., stateless) and cannot be used with or for persistent (e.g., stateful) applications that need to maintain state with persistent storage, or backing store. This is true by default. However, it is also true you can configure containers to use persistent local (not a good idea, as it reduces portability), shared network, or volumes mapped to the networked shared cloud or other storage. Later in the database section of this chapter, there is an example showing MySQL running in a container as well as Dell EMC ECS (Elastic Cloud Storage) images.

Good deployment scenarios for using containers include when you do not need an entire PM or BM, nor the overhead of a VM or full cloud instance. This can be to run in production, development, test, research, operational support, DevOps, among others, or a specific application, program, module, process, stream, or thread.

Another deployment scenario is where you need to stand up or provision very quickly an environment that can be torn down just as quickly. This includes developers, testers, production support, and students, among others. Keep in mind that containers are stateless by default, so for persistent data, they need to be configured accordingly.

Another good use for containers is where you need to launch very quickly images that may have short life-cycle management durations. By being lighter weight, they start up (and shut down) faster, using fewer resources. For example, on public cloud sites, compute instances or VM are usually billed on an hourly basis of granularity. Container services, on the other hand, may be billed on a second, microsecond, or even smaller granularity. What this means is that if you are using only a fraction of a cloud instance, consider deploying into a container or container-based service. Likewise, if you find your monthly compute invoices are getting too large, that might be a sign you need a cloud instance, or a cloud instance configured with multiple containers.

> *Tip:* Various management tools facilitate the creation and container build management, versioning, and other common tasks including scheduling. You can use various tools, settings, and templates to set the amount of CPU, memory, and other resources used, as well as track their consumption as a compute service. Your storage options include ephemeral nonpersistent, persistent local to host, shared network, and portable access using container volume management tools such as Dell EMC RexRay, among others.

Containers by default may be immutable (fixed) with ephemeral (nonpersistent) storage for stateless applications. This is useful for fixed or static applications that perform a given set of tasks, processes, or functions without being stateful. For those types of applications where statefullness is needed, they can connect to a database in another container, VM, cloud instance, or bare metal to store persistent data.

> *Tip:* Display the last 15 Docker log entries for an instances using the `--tail` option in a command such as `Docker logs --tail 20 ecsstandalone`.

Those applications can also access persistent storage via Docker or container volumes, local host, or NAS shared storage. The caveat with using local host storage is that it negates the portability of the container, creating affinity to a given server. Using storage and volume plug-ins, containers can access different shared storage as a persistent backing store while maintaining portability. An example of a plug-in is the libstorage-based open-source RexRay.

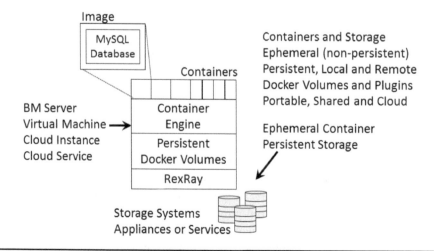

Figure 14.7 Containers using persistent shared storage.

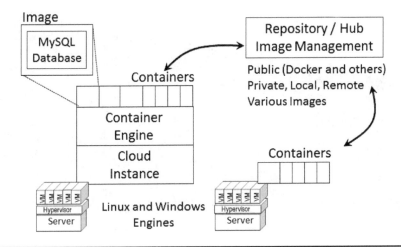

Figure 14.8   Containers live in repositories and run in various places.

Tip: An important value proposition of containers is light weight and portability. Just because you can launch a container to use local storage on a host for persistence does not mean that you should. By doing so, you create an affinity that restricts that container to running on that host. Instead, if you need the flexibility to move containers and run on different hosts, then use shared networked storage (block, file, and object) or a volume plug-in that provides similar functionality.

Figure 14.7 shows a MySQL database container image using persistent storage via a container storage volume plug-in such as RexRay.

Note that Docker supports many different drivers, overlays, and plug-ins that have different characteristics. Which is best for you will depend on your workload PACE requirements and other preferences. Figure 14.8 shows container images running on various platforms including cloud-based services, cloud instances, virtual, and PM or BM. Also shown are repositories or hubs where containers are stored and accessed from, which can be public or private. Public services include Docker among many others.

Figure 14.9 shows a *Docker ps −a* command displaying running Docker container images with information including ID, name, and status, among other items.

Figure 14.10 shows an example of containers using persistent Docker volumes leveraging pluggable volumes with the persistent backing store. In this example, various container images are running on a cloud host (an instance such as AWS EC2 or Azure) that also has libstorage-based RexRay installed as a storage orchestration for volume-handling management. The Docker (another container) engine then uses the persistent storage managed by RexRay for accessing and storing data on shared storage as a backing store. Keep in mind that while many applications or software-defined storage can access object storage natively, or via gateways, some still need file access support. Thus, in this example, access to shared cloud file-based storage is shown.

Figure 14.11 shows, on the left, Docker images and containers, including one for Dell EMC ECS, whose dashboard is shown on the right. The point is to show that containers can be used to perform tasks often relegated to VM or even PM.

```
greg@DOCKIOS5 -$ docker run --name contmysql -e MYSQL_ROOT_PASSWORD=$PSS -v $PWD/local_my.cnf /etc/my.cnf -v /mnt/nd0/docker/contmysql /var/lib/mysql -d mysql/mysql
r
aa5bf6499ee22effa3bd52e350ef27d1d31297a7a96dbfd141140a0c944f56ed
greg@DOCKIOS5 -$ docker ps -a
CONTAINER ID      IMAGE              COMMAND            CREATED         STATUS              PORTS
aa5bf6499ee2      mysql/mysql-server "/entrypoint.sh mysq 3 seconds ago   Up 3 seconds        0.0.0.0 32817->3306/tcp, 0.0.0.0 32816->33060/tcp
ysql
4340f713744a      ubuntu             "/bin/bash"        13 months ago   Exited (0) 13 months ago
r_hodgkin
5c958cbc98a9      ubuntu             "bash"             14 months ago   Exited (0) 14 months ago
d_fermi
ed99a73ae377      hello-world        "/hello"           14 months ago   Exited (0) 14 months ago
ng_bartik
3c4070543f28      hello-world        "/hello"           14 months ago   Exited (0) 14 months ago
r_goodall
3108d66c01cc      hello-world        "/hello"           14 months ago   Exited (0) 14 months ago
_bardeen
2684349ea0cb      enccode/s3motion   "/bin/bash"        21 months ago   Exited (0) 21 months ago
y_stallman
0b97244d7fa4      enccode/s3motion   "/bin/bash"        21 months ago   Exited (0) 21 months ago
rate_mccarthy
cd0d6f605b46      enccode/s3motion   "s3motion --REST"  21 months ago   Exited (143) 21 months ago
y_fermat
a31f729ee5be      enccode/s3motion   "/bin/bash"        21 months ago   Exited (127) 21 months ago
ve_carson
6bd1261fd999      enccode/s3motion   "s3motion --REST"  21 months ago
estorf
a2b0df91f341      enccode/s3motion   "s3motion --REST"  21 months ago   Exited (143) 21 months ago
_bohr
greg@DOCKIOS5 -$
```

Figure 14.9   Displaying Docker container ID and image information.

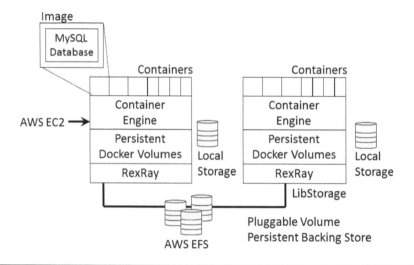

Figure 14.10   Docker containers with persistent storage.

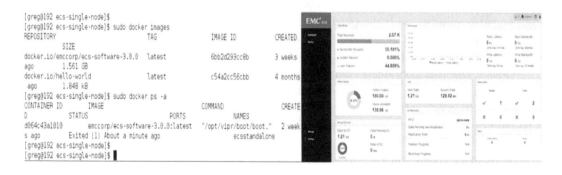

Figure 14.11   Docker images and containers.

> *Tip:* Watch out for container creep, which is similar to VM sprawl; that is, containers, like VM and other technology techniques, are good for some things, but just because you can use them for others, ask yourself if that is the only option, or only what you want to use.

Some vendors, services, solutions, and tools include Apache, Apcera, Atlantis, AWS (EC2, Lambda compute, EFS), Azure File Service, Ceph RBD, Convoy, and CoreOS. Others include Dell EMC (RexRay), Docker, Flocker, Github, Kubernetes, Libstorage, Linux (LXC/LXD), Mesos, Mesosphere, Microsoft Azure and Windows, NetApp, NFS, OpenStack Nova and Magnum, Panzura, Portwk, Storpool, Swarm, and VMware (VIC, Photon), among others.

Considerations, tips, and recommendations include:

- Place logs for active applications, containers, or other busy workloads on fast storage.
- Containers (or their contents) need to be made persistent (stateful) somewhere.
- While containers are immutable and can be ephemeral, protect the hubs they live in.
- Protect the storage or backing store where persistent container data volumes live.
- Just because you can do almost anything with a container, that may not mean you should.
- Leverage containers as another resource allocation tools similar to PM, BM, and VM.

14.2.1.5. Workspace and Virtual Desktop Infrastructure

Virtual desktop infrastructure (VDI), also known as virtual workplaces, complements virtual and physical servers and provides similar value propositions as VM. These value propositions include simplified management and reduced hardware, software, and support services on the desktop or workstation, shifting them to a central or consolidated server.

VDI or workspace benefits include simplified software management (installation, upgrades, repairs), data protection (backup/restore, HA, BC, DR), and security. Another benefit of VDI, similar to VM, is the ability to run various versions of a specific guest operating system at the same time, similar to server virtualization. In addition to different versions of Windows, other guests, such as Linux, can coexist.

Depending on how the VDI or workspace is deployed, for example, in display mode, less I/O traffic will go over the network, with activity other than during workstation boot mainly being display images. On the other hand, if the client is running applications in its local memory and making data requests to a server, then those I/Os will be placed on the network.

Figure 14.12 a workspace or VDI environment being served via (lower left) private on-premises SDDC and SDDI environments. The top right shows workspaces being delivered via cloud services. On-premise options include traditional, CI, and HCI solutions for small and medium-size or ROBO environments. Scale-up and scale-out CI and HCI, as well as rack-scale systems, are options for larger-scale environments.

If many clients boot up at the same time, such as after a power failure, maintenance, or another event, a boot storm could occur and cause server storage I/O and network bottlenecks. For example, if a single client needs 30 IOPS per second either in a normal running state when it is busy or during boot, most servers and networks should support that activity.

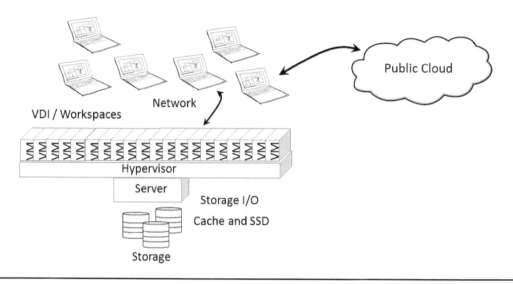

Figure 14.12 VDI and workspace environments.

Tip: Boot storms are a common focus for VDI, which tend to be read-intensive, and good candidates for linked clones as well as SSD-backed read-cache (as well as read-ahead). However, also keep shutdown activity in mind, where many users may need to save their data before going on break or leaving for lunch as well as wrapping up their work day. Shutdown storms result in more data being written, so plan accordingly, with write-back and write-through caches that provide persistence and data integrity as well as performance acceleration.

The above simple example does not factor in the benefit of caching and other optimization techniques. However, it does point to the importance of maintaining performance during abnormal situations (boot and start-up as well as shutdown storms). Note that storms are the flurry of activity resulting from everybody starting up their workstations (lots of reads), or shutting down (lots of writes to save data). This means that for performance-sensitive application workloads, there needs to be consistency within normal running periods and during disruptions or outages (planned and unplanned).

Tip: How many IOPs, gets, puts, reads, writes can an HDD or SSD do? That depends, although there are many old rules of thumbs and dated metrics floating around. Likewise, there are some dated methods that underestimate performance based on revolutions per minute (RPM) of drives that do not take into consideration read as well as write cache. Today's HDD and SSD vary in read and write performance across different types of media and capacities (refer to Chapters 5 and 7). Table 14.3 shows four corners (random, sequential, reads, writes) for 8K across a 5.4K 4TB HDD and an PCIe AiC NVMe SSD using vdbench with 128 threads.

Another consideration for VDI and workspace I/O performance is the size and type of I/O, for example, small 4K or 8K, larger 64K and 128K, or even larger 1MB-sized I/Os, which may be random or sequential reads and writes, cached or noncached. Note that during a typical operating system start-up sequence, there are different types of I/Os, some small random reads, some large sequential, as well as writes to logs to plan for.

Table 14.3. 8K Four Corner (128 threads) IOPs vdbench Result (NVMe and HDD)

	(IOPs) i/o rate	MB/sec 1024**2	bytes i/o	read pct	resp time(ms)	read resp(ms)	write resp(ms)	resp max(ms)	resp stddev	queue depth	cpu% sys+u	cpu% sys
8K 100% Seq. 100% Write												
4TB 5.4K HDD	1570.07	12.27	8192	0.00	80.992	0.000	80.992	305.655	38.656	127.2	5.1	1.7
NVMe PCIe SSD	44069.48	344.29	8192	0.00	2.878	0.000	2.878	1205.039	11.187	126.8	41.8	24.4
8K 100% Seq. 100% Read												
4TB 5.4K HDD	59184.16	462.38	8192	100.00	2.156	2.156	0.000	1624.718	10.984	127.6	32.9	24.1
NVMe PCIe SSD	91562.29	715.33	8192	100.00	1.391	1.391	0.000	1369.334	8.392	127.4	58.8	43.1
8K 100% Random 100% Write												
4TB 5.4K HDD	56.17	0.44	8192	0.00	2240.453	0.000	2240.453	5912.785	1493.667	127.6	1.2	0.3
NVMe PCIe SSD	39462.39	308.30	8192	0.00	3.227	0.000	3.227	1464.869	17.303	127.3	42.3	25.5
8K 100% Random 100% Read												
4TB 5.4K HDD	653.25	5.10	8192	100.00	194.842	194.842	0.000	2016.777	256.913	127.6	1.8	0.6
NVMe PCIe SSD	94770.49	740.39	8192	100.00	1.339	1.339	0.000	1512.827	7.676	126.9	60.5	46.3

> *Tip:* For more server storage I/O performance, refer to www.storageio.com/performance, where there are links to various performance-related tools, results, and related information in addition to those shown elsewhere in this book.

Part of VDI assessment and planning should be to understand the typical storage I/O and networking characteristics for normal, boot, and peak processing periods to size the infrastructure appropriately. Solutions include those from Citrix (XenApp and XenDesktop), Microsoft, VMware Horizon, and View, among others, including those from cloud providers.

Considerations, tips, and recommendations include:

- Understand the different applications and workloads being hosted and your resource needs.
- Use GPUs and increased video RAM (real or virtual) for graphic-intensive applications.
- Leverage linked clones to optimize performance as well as capacity.
- Remember that aggregation or consolidation can cause aggravation (bottlenecks).
- Utilize read/write cache solutions to speed up performance and eliminate bottlenecks.
- Where will the workspace and VDI services be delivered from (on-site, via cloud)?
- What are the applications being used? Do they require extra compute, I/O resources?
- Do the applications require graphics acceleration or other special resources?

14.2.2. Microsoft Azure, Hyper-V, Windows, and Other Tools

Microsoft SDDC, SDI, and SDDI capabilities include traditional workstation, desktop, laptop, tablet, and mobile solutions. In addition to the desktop and workspaces along with various applications such as Exchange, SQL Server, SharePoint, and Visual Studio, and development tools, among others, there is also Windows Server. Windows Server can be deployed with a desktop GUI user interface or in non-GUI mode, on bare-metal physical, virtual, and cloud compute server instances. Note that there are 32- and 64-bit versions of Windows.

Windows Server, along with some newer versions/editions of Windows desktop, supports Hyper-V virtual machines. As a hypervisor, Hyper-V naturally supports Windows-based operating systems, as well as various Linux solutions including Linux Integration Services (LIS). There is also Hyper-V server as well as Storage server, among other offerings.

New is Nano, which is a lightweight version, even lighter and physically smaller than previous non-GUI versions of Windows Server. Nano differs from traditional Windows Server by removing all 32-bit code components in addition to the GUI user interface. The result is a very resource-friendly (less memory, disk space, and CPU needed) Windows server platform for doing more work in a given footprint.

Figure 14.13 shows various Microsoft data infrastructure–related tools and technologies from Azure cloud services on the left, Azure stack (a private on-premise version of Azure you can run on your hardware), Windows Server, Hyper-V, and other tools. Also shown in Figure 14.13 are Microsoft management tools including System Center, Active Directory (AD), Server Manager, Hyper-V Manager, SQL Server, Visual Studio, SharePoint, and Exchange, among others. Additional Azure tools and services include those to support IoT, including hubs, gateways, tables, queues, databases, IAM, and management.

In Figure 14.13, Storage Spaces Direct (S2D) is Microsoft's software-defined storage (SDS) solution, which can be used to create aggregated (HCI) and disaggregated (CI) scale-out clusters. Besides being on Windows Servers, Microsoft has also made SQL Server along with associated management studio and tools available on Linux (and containers). Microsoft Windows (and Azure) supports virtual hard disks (VHD) and extended (VHDX) for VM and physical machines.

Microsoft has also increased its presence in open source on github, as well as being a member of the Open Compute Platform (OCP) group. As part of the OCP, Microsoft has continued their reference design for high-density servers used in their Azure public cloud. Various other vendors are partnering with Microsoft to support Azure Stack in addition to Windows software on their server platforms. Most public clouds, as well as hypervisors, also support various Windows server and desktop editions.

Figure 14.14 shows a Windows folder (left) with various subfolders. Also shown in the center of the left-hand image are three disk or storage-shaped images which are VHDX virtual disks. These are mounted in an existing folder path as opposed to being given drive letters. On

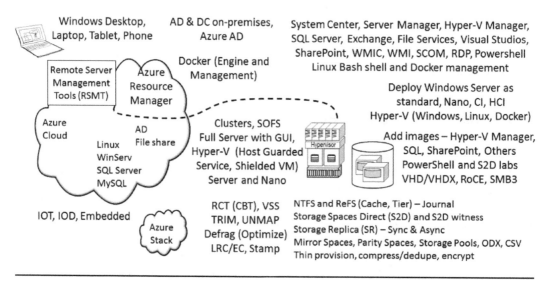

Figure 14.13   Microsoft topics including Azure and Windows.

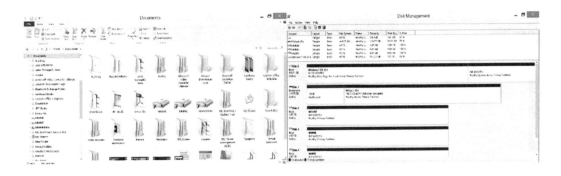

Figure 14.14 Accessing Windows storage without using drive letters.

the right-hand side of Figure 14.14 is the VHDX, shown mounted in the lower portion of the display. Instead of accessing contents of a storage device with a drive letter path, an existing folder path is used. The storage contents simply become part of the folder tree.

The default file system for Windows servers (and workstations) for several years has been NT File System (NTFS), and File Allocation Table (FAT) for portable devices. With Windows Server 2016, the default file system is now Microsoft's Resilient File System (ReFS), which is a journaling solution including integrated data integrity checks. NTFS will continue to be supported for many years, as will FAT for interoperability.

In addition to traditional local storage, Windows Server 2016 builds on Windows Server 2012 Storage Spaces by adding new software-defined storage functionality with Storage Spaces Direct (S2D). Besides local and direct attached storage, Windows servers also support access to storage via block iSCSI and Fibre Channel, among other interfaces and protocols. SMB file sharing along with NFS are also supported among other options, including clustered shared volumes (CSV) and SMB3 direct.

Windows servers can be deployed as standalone or clustered, as disaggregated CI and aggregated HCI using S2D and clustering. Likewise, Windows Server and Workstation support Hyper-V and guest Windows and Linux systems, as well as being a guest itself on VMware and other hypervisors as well as cloud instances.

Other tools include *Disk2vhd* (for migrating from physical disk to VHDX), *Diskmgr*, *Diskpart* (for managing disk partitions and volumes), and *Diskspd* (benchmarking, test, and simulation). Additional tools and utilities include *fsutil* to create large empty files for testing, *Manage-bde* (BitLocker command line), Netsh (for network-setting configuration), and PowerShell for local and remote, SQL Studio, and SQL configuration utility.

Other tools include Task Manager (enables disk status via the command line with *diskperf –y*), Virtual Machine Migrate (for converting from physical to virtual, including VMware to Hyper-V), and WMIC (displays various configuration and device information). See the Appendix for some additional Windows-related tips and tricks.

Figure 14.15 shows, on the left, a Microsoft Azure dashboard provisioning and accessing a cloud instance (virtual machine) instance on the right, which is a Windows Server 2016 instance. Microsoft has been working on enhancing the management and user experience of working with and across traditional Windows platforms as well as Azure. This means that Azure and Windows are compatible and work together, yet can also be used independently.

Figure 14.15   Azure compute access (left) and Windows instance (right).

Microsoft Azure cloud service exists in several different geographical regions, with more being added similar to how AWS is growing. Services which continue to evolve include compute instances such as virtual machines, naturally Windows as well as many different Linux flavors along with containers. Besides compute, Azure also supports development environments and tools, analytics for big data, media, and CDN. Other components include identity and access management (IAM) and security, SQL and NoSQL databases and table services, a data warehouse, queues, and Redis cache and storage. Besides table and database storage (and

Table 14.4. Microsoft Compute Server and Storage Platforms

Resources	Characteristics, Attributes, Tips, and Recommendations
Azure public cloud	Public cloud services including applications, resources, compute instances and containers, analytics, big data, little data, bulk storage and CDN, off-site distributed resources, block, file, and object storage, tables, queues, database, development, and testing. Leverage different geographical locations. Use as a replacement for on-site, on-premise resources or as a complement including for HA, BC, DR, and elasticity.
Docker and Windows container	Docker management of Linux-based containers hosted on Hyper-V as well as Windows-based containers. Use for micro services, threads, and processes that do not need an entire VM to perform specific tasks.
Azure Stack private or hybrid cloud	On-premise version of Azure public service that allows you to select different hardware options for deployment. Leverage as a private cloud to move beyond what you might do with traditional Windows servers, as well as for deploying hybrid clouds. Complement or alternative to OpenStack.
Hyper-V virtualization	For hosting guest virtual machines including various versions of Windows as well as Linux as servers, desktop, workspace, VDI, database, or other applications. Supports consolidation of resources and management or as part of deploying SDDC, SDI, and SDDI environments.
Windows Nano	Streamlined Windows operating system for Hyper-V or bare metal. Workloads where a user interface is not needed; uses less CPU and memory.
Windows Server	For traditional bare metal as well as on virtual servers and cloud instances as the operating system for various applications including Active Directory (AD), Domain Controllers (DC), networking, storage, SQL Server, Exchange, SharePoint, and many others. Nano version for lightweight non-UI–based management, full GUI desktop for user interface–based management, Hyper-V for hosting guest virtual machines.
Workstation desktop	Where a server version is not needed on laptops, physical workstations, tablets, or mobile devices. Note that, depending on version edition, there are common management tools across server and desktop. Some editions also support Hyper-V virtual machines as well as management access.

services), Azure also supports file access including via remote Windows servers, as well as blobs (object storage) and non-blob blocks.

In addition to the Azure public cloud, Azure Stack, Windows Server, and traditional desktop workspace applications including Microsoft Office, Microsoft also has the Office 365 (O365) cloud-based service. This capability allows you to decide on using hosted (cloud) subscription-based desktop productivity tools such as Excell, Word, PowerPoint, and email, among others, as well as traditional on-site hosted (physical or virtual) tools.

Table 14.4 shows suggestions on when and where to use different Microsoft SDDC, SDI, and SDDI solutions to address different needs.

Considerations include:

- Where and how will your IoT devices be managed?
- Will you use the same services for general event management as well as IoT?
- What are the applications? Where are users accessing resources from?
- Do you have an all-Windows, all-Linux, or mixed environment?
- Do you need legacy storage support of Fibre Channel, iSCSI, or NAS using storage spaces?
- Software-defined storage with Storage Spaces Direct (S2D) and replication
- Local and remote networking needs and requirements
- Licensing models including enterprise license agreements (ELA)
- Different environments from SOHO to ROBO to SMB to enterprise
- Consider context of SMB vs. CIFS/SAMBA file sharing
- SMB3 can mean direct connect low latency deterministic, or general file and storage sharing

14.2.3. VMware vSphere, vSAN, NSX, and Cloud Foundation

VMware SDDC, SDI, and SDDI capabilities range from virtual server infrastructure (VSI) hypervisors such as vSphere ESXi to Horizon virtual desktop infrastructure (VDI) and desktop workspace client (View). Other pieces of the VMware software-defined portfolio include virtual storage (vSAN) integrated with vSphere, along with NSX software-defined networking (SDN), many management tools, and solution suites.

In addition to having solutions that can be purchased for a fee along with annual maintenance and licenses, there are also subscription models and enterprise license agreements. VMware also has free software such as basic vSphere ESXi, which provides hypervisor functionality without additional management tools such as vCenter, vRealize, and others.

Figure 14.16 shows a software-defined data infrastructure (SDDI) implementation in the form of VMware Software-Defined Data Center architecture, solutions, and suites. VMware supports virtual desktop infrastructure (VDI) and workspace with Horizon (virtual server) as well as View (clients). Various management tools and suites are available as well as virtualized compute (VM and containers), storage, and networking.

The various VMware components can run on-premise on BM and PM (hosts), nested (on virtualized servers), as well as on some cloud platforms such as AWS. VMware vSphere ESXi hypervisor can also be managed and used by other stacks or frameworks such as OpenStack. Likewise, there is a VMware Integrated OpenStack Appliance that runs on the VMware software-defined data infrastructure.

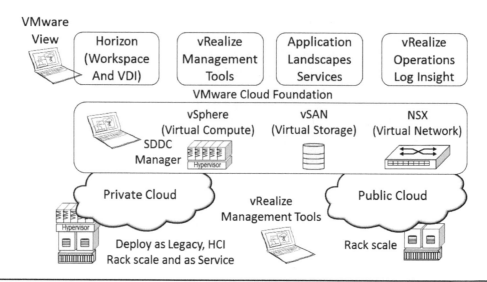

Figure 14.16   VMware Software-Defined Data Center.

VMware management tools such as vCenter are packaged as an appliance, meaning they can run on top of and leverage underlying tools while enabling simplicity of deployment. In addition to vCenter, other VMware tools are also available as preconfigured virtual appliances, as well as close hypervisor kernel integration, for example, with vSAN.

The different VMware technologies can run standalone or in conjunction with other components leveraging different packaging. Different packaging and deployment include legacy server, CI, HCI, and rack scale, as well as the cloud.

Table 14.5 shows the three primary VMware data infrastructure software-defined building blocks converging virtual server compute (vSphere ESXi hypervisor), storage (vSAN), and software-defined networking (NSX). Management is via vRealize suite, among other tools.

Table 14.5. Core VMware Software-Defined Building Blocks

Component	Characteristics, Attributes, Tips, and Recommendations
vSphere ESXI compute	Server virtualization hypervisor that runs on physical servers, and can also be nested installed on top of itself. Supports various physical hardware devices as well as guest operation systems including Windows and Linux. Virtualized CPU, memory, I/O storage, and networking devices along with video among other key resources. VMware File System (VMFS) file system overlays underlying storage resources; VMDK is virtual disks. Also, virtual volumes (VVOL) are supported. Can be deployed standalone without vCenter or other tools as well as clustered. Dynamic Resource Scheduler (DRS) facilitates optimization across clustered hosts, policy and automation along with storage QoS. Data at rest encryption (DARE) support for virtual machines.
vSAN storage	Virtual storage integrated with underlying vSphere hypervisor for HCI deployments scaling from 2 to 64 nodes with various FTM options to meet different FTT needs. Hybrid and all-Flash configurations, support for RAID mirroring, replication, and parity erasure codes along with encryption, compression, de-dupe and thin provisioning. Tools include vSAN Observer and vSAN Console along with vSAN Health Check, among others.
NSX network	Software-defined networking (SDN) including micro-services and segmentation enabling finer granularity as well as enhanced scalability of virtual resources.

APIs include VAAI (VMware API for Array Integration) and VASA (VMware API for Storage Awareness). VASA enables providers (storage systems, appliances, and software) to convey insight up the stack to VMware vSphere, so it has insight into capabilities and configuration of storage resources. Also, VASA combined with VVOL enables vSphere to have finer granular access and insight for optimizing snapshots and other functions.

Another vSphere API is VAAI, offloading some functions from the VM to a compatible storage system that supports specific features. VAAI functionality includes granular SCSI lock management on shared storage volumes or LUN, hardware-assisted VM cloning or copy, and data migration for VM movement—for example, offloading some copy and resource initializes, among other tasks. For storage systems that do not support VAAI, those functionalities (zero copy move, SCSI enhanced locking) are performed by the hypervisor server.

Another vSphere functionality to help address scaling of multiple busy VM on a common PM is Storage I/O Control (SIOC), which enables a proprietary scheme for load balancing. With SIOC, those VM that need I/O performance can be tuned accordingly to prevent bottlenecks from occurring. In addition to VASA and VAAI, other VMware APIs include those for data protection (formerly known as VADP), which facilitates optimization for various data infrastructure tasks. VAIO (vSphere API for IO filtering) supports third-party data services.

Table 14.6 shows various VMware tools and technologies along with deployment as well as consumption models spanning server, storage, networking, software-defined virtual and cloud as well as VDI workspace along with management tools.

Other tools include vSphere web client, VMware converter (supports P2V migration), as well as a workstation and desktop variants of vSphere such as Workstation for Windows, and Fusion for Apple-based systems. Workstation and Fusion enable a vSphere-type environment to be run on a desktop for training, support, development, and testing, among other activities. Workspace ONE is for mobile application management, SSO, and identity access control.

Tip: Leverage 10 GbE (or faster) network interfaces for east–west management traffic including monitoring, cluster heartbeat, logging, server and storage vMotion, as well as vSAN activity.

Proper configuration of virtual video RAM may seem intuitive for graphics and VDI types of application workloads. However, in some scenarios, where the database or other applications appear to be running slow, increasing the video RAM might help if the bottleneck is not the database, but rather the buffering and display of results.

Virtual machine optimization considerations include specific hardware support such as Intel VT assist and 64-bit capabilities. Management tools include metrics and measurements for accounting, tracking, capacity planning, service management, and chargeback. Hypervisors also support APIs and management plug-in tools such as Hyper-V working with VSS writers with application or guest integration. Keep in mind that virtual resources and functionality need memory to hold their software-defined data structures as well as programs and data.

Tip: Standard built-in drivers such as those for networking and storage I/O (HBA), among others, work well out of the box. However, VMware supports optimized "para-virtualized" drivers for networking and storage I/O, among others. These drivers along with VMtools can boost effective performance and in some cases also reduce CPU or other overhead.

Table 14.6. Sampling of VMware Tools and Technologies

Resources	Characteristics, Attributes, Tips, and Recommendations
VMware Cloud Foundation	Unified SDDC platform for hybrid cloud leveraging vSphere, vSAN, NSX, and SDDC manager for automation of data infrastructure operations. Use for turnkey scenarios where faster deployment is needed, leveraging various VMware partner hardware solutions deployed on-site. Integrated with other VMware tools. Replaces EVO as a converged deployment solution.
AWS Management Portal for vCenter	vCenter plug-in that enables migration and conversion of VM from VMware environments to AWS EC2 instances. In addition to migration of VM, the plug-in also provides management integration with EC2 for coordinating AW resources from vCenter. This includes Active Directory (AD) and other identity tools also enabling role-based access controls (RBAC).
VMware Cloud on AWS	VMware tools such as vSphere, vSAN, NSX, and associated management tools running on AWS private servers. Note that these are not normal EC2 instances, rather the equivalent of dedicated private servers. This service enables VMware environments to leverage AWS to seamlessly move or migrate (vMotion) servers, storage, and other resources in a hybrid manner across different software-defined data infrastructures.
vCenter Appliance	vCenter is a central platform, place, and tool for managing, orchestrating, and performing common VMware SDDC and SDDI tasks with various plug-ins. Configure multiple vCenters for scaling performance, availability, or capacity.
Horizon View	Horizon is a set of tools and servers for configuring, orchestrating, composing, and delivery of VDI and workspaces to View clients.
OVA OVF	Open Virtual Appliance (OVA) and DMTF Open Virtualization Format (OVF). These are standards and protocols for packaging for deployment virtual appliances and simplifying virtual application deployment instead of building from scratch using ISO or other image builds.
vMotion	Tools for moving server VM and storage, either inactive or while in use.
vCloud Air	VMware public cloud offering to use as an off-site companion to your private cloud or another VMware deployment, or for compatibility migrate VMware-based software-defined data infrastructure, VM, and applications too.
VIC	VMware Integrated Containers with Photon lightweight Linux kernel. Speeds time of deploy with integrated VMware management for containers hosts.
HCIBench	Test harness tool leveraging vdbench for driving various workloads and reporting metric results on traditional vSphere servers as well as vSAN.
VIOA	VMware Integrated OpenStack Appliance is a self-contained, easy-to-deploy and easy-to-use method of doing a quick proof of concept, support development, and other experiences without having to build from scratch.
vApp	A collection of one or more VM, resources, and policies aligned and grouped for ease of management—for example, start-up/shutdown of all VM together as part of an application system collection of resources.

For compatibility with some older devices, there is the E1000 Ethernet driver, while VMX3NET (or higher) provides support for optimized 10 GbE (and faster) devices. Server storage I/O adapters include standard SAS (LSI Logic SAS) and SCSI (LSI Logic Parallel) along with VMware Paravirtual block I/O driver, among others.

Storage I/O Controls (SIOC) functionality enables performance QoS and other automation including resource allocation with policies. Combined with DRS, policies, and other VMware functionality allocation of data infrastructure resources including server, memory, I/O, storage and networking can be orchestrated and automated in an efficient as well as effective manner.

Tip: In addition to officially released and supported software products, tools, and drivers, VMware, like other vendors, has a collection of tools called "flings." These flings are released via VMware Labs and are a collection of tools and add-on software to enhance, extend, and facilitate management of VMware products and solutions. For example, HCIBench, among others, can be found at labs. vmware.com/flings/hcibench.

Considerations, tips, and recommendations include:

- Align applicable FTM mode and durability (copies) to meet FTT needs.
- Establish performance baselines using HCIBench or other tools for future reference.
- Implement enterprise-class SSDs for write-intensive cache functions and applications.
- Leverage VMware "fling" optional tools to assist with data infrastructure management.
- Refer to the VMware Hardware Compatibility List (HCL) for support configurations.
- Utilize SIOC and another server, storage, and network policy automation.
- Verify that devices and drivers along with other software tool plug-ins are up to date.
- VMware supports 4K page allocations, but make sure applications are page-aligned.

14.2.4. Data Databases: Little Data SQL and NoSQL

Databases and key-value stores or repositories can be used and optimized to handle transactional analytics, also known as data mining. Transactional databases tend to be read-write-update, whereas analytics, references, and data mining, along with decision support, tends to have fewer updates but more reads. Transactional data can be moved into reference and analytics databases or data warehouses at different intervals or frequencies. Databases are also used for supporting websites, blogs, and other applications as well as for handling metadata. Databases can be open-source or proprietary, as well as Structured Query Language (SQL)–based access or NoSQL (non-SQL)–based.

In addition to how databases get used (activity), they also support various types of data, including columnar tabular, text, video, and other entities. Databases and key-value stores can be in-memory, as well as on disk (HDD or SSD) on-premises (PM, VM, containers), or in clouds as a service or instance.

As the name implies, in-memory databases leverage large-memory servers where all or major portions of the database are kept in memory for speed. For availability and durability, some form of high-speed persistent memory or storage is used to load data rapidly, as well as provide consistency of journals and redo logs. The benefit of in-memory is a faster query and processing of certain functions.

Tip: Look into details of your database or key-value repository to see vendor suggestions and best practices for in-memory processing. Some solutions are optimized for queries and others for transactions. Note that some systems may have higher overhead handling transactions even if they are in-memory, due to data consistency, vs. those handling cached queries and reads.

Besides in-memory databases and applications such as MemSQL, Oracle, Redis, SAP, and SQL Server among others, most databases also support various forms of caching with memory, as well as NVM. NVM including storage-class memories (SCM) along with SSD can also be used for journals, transaction and redo logs, checkpoints, metadata and system configuration, system and temporary tables, and scratch space, as well as indices and main data tables.

> *Tip:* Historically, DBAs and others working with databases have been taught the importance of having as many "spindles" or disk drives as possible to spread data out for performance. Using SSD and other forms of SCM and NVM, in some scenarios fewer devices can be used to deliver a higher level of performance than many drives. However, make sure to have enough SSD or SCM to enable availability using a mirror or, when needed, parity-based erasure code RAID. For larger workloads, multiple SSD can be used to further spread out the workload.

Components include a logical database server instance (the database server software), which is installed on a physical, virtual, cloud, or container instance. Note the context that a database server can be the actual server system that runs the database software. However, the database software running on a server can also be called the database server. Keep context in mind. Other components include the server processors (CPU), memory, and I/O network along with storage performance and capacity configured for some level of availability.

Figure 14.17 shows a database server (software) such as Cassandra, IBM, Kudu, MongoDB, MySQL, Oracle, or SQL Server, among many others. Local and remote users, applications, processes, services, and devices access the database, adding, updating, or reading data.

Depending on the particular database solution, there can be one or more instances of the database server, each with one or more databases. For example, on a given PM, BM, VM, container, or cloud instance, there can be a SQL Server database server instance called TESTIO, another called RESULTSIO, and another called STORAGEIO. TESTIO might be used for running various test workloads with different databases. Results such as Benchmark Factory,

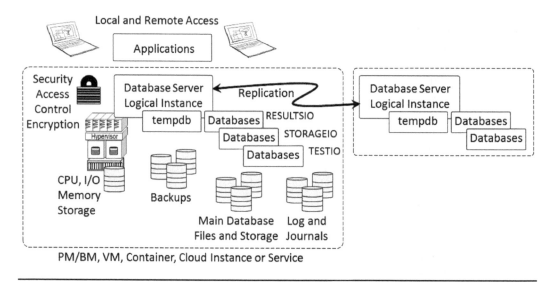

Figure 14.17 Database server instance and database items.

among others, may be stored in databases served by the separate instance RESULTSIO. The general activity might exist in other databases in STORAGEIO.

Also shown in Figure 14.17 are different databases along with a temporary database (tempdb) used for sorting and other temporary functions. The various databases have one or more data repositories (usually a file on or in a file system) that exist on one or more backing stores (storage volumes). Also shown are how databases can be backed up, replicated, as well as unloaded (export or dump) and loaded (import or restore). Resources including CPU, memory, I/O, and storage space capacity. Also shown are security including encryption and access controls to databases, tables, and other resources.

While it is possible to put all databases into a single instance, having different ones enables those instance servers to be tuned and optimized differently. Also, those database instances can be stopped, maintenance performed, and started independently of the others. Work with your DBAs or others responsible for database administration to determine what type of configuration is needed to meet different needs. How and when to protect the databases along with their configuration settings will vary based on PACE and SLO needs.

Additional components include system configuration and metadata for the databases such as a master or system database, and temporary databases for sorting and other functions, also known as tempdb. Then there are specific database instances (databases) that have collections of database entities stored in primary objects along with log and journal files.

Different database systems and their implementations will have various names and locations for these items. Note context here is that a database object is an item or entity and database object storage does not mean S3 or Swift object storage (although for some implementations it might).

Remember that databases store entities and their data structure object items in files as repositories—examples include .ora, .dbf (among others for Oracle), and .mdf for Microsoft SQL Server, among others. For MySQL, MongoDB, and others on Linux systems, you may find the default database files in a location such as /var/lib/MySQL or /var/lib/MongoDB unless they have been moved.

Tip: There may be a temptation to use lower-cost, higher-capacity Flash SSD for databases, which might be ok for mainly read application workloads. However, for update, write-intensive, and cache use, enterprise-class SSD tend to have better wear or durability endurance measured in TB written per day (TBWD), lasting longer and can also be faster on writes. Keep in mind that Flash SSD will wear out over time, so use the applicable type of NVM aligned to your specific workload and PACE needs. Some database and key-value repositories also support Direct Access Extension (DAX) leveraging underlying NVM and SCM.

General management activities include:

- Allocate and provision space, expand space
- Compress, compact, and optimize (reorganize)
- Import, load, and export, unload
- Log and journal file shipping or copies
- Tune resource parameters, buffers, threads, and sessions
- Protection with consistency points and checkpoints

Tip: There are different tools for performing database unload (export) and load (import) in support of data protection, maintenance, and optimization or data migration. Some tools process data in a serial fashion, while others create multiple parallel data streams to move more data in a shorter amount of time. In addition to using tools that support different data import and export methods, also make sure to have enough temporary or scratch space for data being moved.

Additional database and repository management function include:

- Move and migrate, detach, attach
- Performance tuning and troubleshooting
- Relocated backing stores files and storage
- Clean-up, purge, compression, and reorganization
- Protection, backup, clone, replication, checkpoint
- Data load (import) and unload (export)

Tip: Gain insight into data performance activity and resource usage using vendor-supplied as well as third-party tools, for example, Oracle automatic workload repository and performance status, as well as those from IBM, Microsoft, Quest, Solarwinds, and many others. Having insight into resource use, bottlenecks, and constraints can enable smart decision making such as when and where to use SSD along with how much for cache or storing data.

Availability and protection includes:

- Access (local and remote) control and encryption
- Checkpoint and consistency point frequency and retention
- Backup, snapshots, and point-in-time copies
- Replication and mirroring (local and remote)
- Storage placement and configuration
- Redo logs and journals

Tip: Belt-and-suspenders data protection includes using database replication for higher-level strong transactional consistency, along with checkpoints, snapshots, and other point-in-time copies to meet RPO requirements. Lower-level hypervisor, operating system, or storage-based replication can also be used with different frequency and granularity, along with redundant cluster nodes and components with failover capability.

A hybrid approach leverages server or database replication for long-distance synchronous and lower-level storage asynchronous replication. Some local databases can also be replicated not only to other physical or virtual database system servers, but also to some cloud services.

Optimize and allocate resources:

- CPU (cores and threads) along with speed
- Memory (lots and fast) for buffers, cache, applications, and data
- I/O for storage and network access and connectivity

- Storage space (primary, indices, scratch, logs, journals, protection)
- Vary the number of threads and sessions supported per database instance

Tip: To ensure performance and availability, proper database maintenance should be done, including purging, archiving, optimization, or reorganization, even if deployed on SSD. Also make sure to have adequate free space on scratch, work, and primary backing store or storage devices, and that proper security authentication is in place for local and remote access. Also, make sure that your system tempdb or equivalent space used for sorting and other functions has enough space to avoid job termination or other errors.

Troubleshooting tips include verifying that the database physical and logical server instance are on-line and functioning. Verify that the database server has adequate storage space for specific databases and there are main as well as log files. Also check to see whether there is adequate system temporary space for sorting, work, and other tasks and that proper permissions or quotas are in place. Double-check network connectivity, firewalls, and network security as well as whether associated tasks are running. Check your various error logs to see whether resources such as memory, CPU, storage, or threads and sessions have been exhausted or there is some other problem.

Figure 14.18 shows (1) a standalone server with DAS that can be a hybrid mix of HDD and SSD, or all-Flash, or all-NVM. Also shown in (1) are local backups made to disk, tape, or cloud along with checkpoints and other protection techniques. Storage in (1) can be protected for availability using RAID 1 mirroring or other variations including RAID 10 as well as various

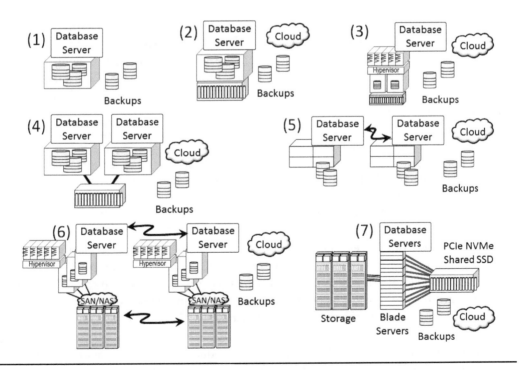

Figure 14.18 Various database deployment scenarios.

erasure and parity code checks. The example shown as (1) is good for smaller environments or those with specialized needs.

At the top center in Figure 14.18, (2) shows a standalone system with additional dedicated or shared DAS. The top right (3) shows a standalone server with the hypervisor and VMs supporting database workloads. This scenario can be used for application and server consolidation. To increase resiliency or scale resources, additional hosts can be added.

> *Tip:* Keep in mind that there is a difference between virtualizing a database system or application and consolidating. Databases can be virtualized, but keep performance in perspective using different QoS and policy parameters vs. simply consolidating as many servers as possible. To meet performance needs, some databases may need their host for most of the time, but you can leverage virtualization for agility, maintenance, availability, load-balancing, and other data infrastructure operations. When the database does not need all the performance, additional VM or containers can be load-balanced on the host.

In Figure 14.18 in the middle on the left (4) are multiple host servers (non-virtualized or virtualized) sharing direct attached storage such as for a simple small availability cluster. In the middle of Figure 14.18 on the right (5) are CI and HCI clusters that are replicating to other CI and HCI systems. Replication can be done via the database, server, or CI and HCI data services. An approach such as shown in (5) is useful for larger SMB or larger ROBO and workgroups that need resiliency across locations.

In Figure 14.18, (6) shows an example of rack-scale and scale-out disaggregated compute servers with SAN, NAS, or other software-defined storage configured for availability and performance for larger-scale or larger-size environments.

The lower right of Figure 14.18 (7) shows a hybrid approach using shared direct attached SSD or SCM attached via low-latency PCIe and NVMe, along with general-purpose storage (HDD or SSD). The shared SSD is used for tempdb, scratch, indices, commonly accessed tables, or other items. Locking and consistency are provided by database software such as Oracle, among others. General-purpose storage is used for less frequently accessed data, checkpoints, import/export, and other scratch space.

> *Tip:* There might be a temptation to use erasure code and some parity-based protection approaches to boost storage space capacity (efficiency), but keep performance in mind. Erasure codes have a higher overhead in terms of CPU usage and longer latency (time required for software to think about how to optimize) to store data and thus should not be used for active-update data. On the other hand, backing stores for backups or archives can be an option for using erasure codes.

Data protection for the scenarios above is a combination of mirror and replication for residency, as well as point-in-time copies for recovery and consistency point protection. Also shown are protecting data locally as well as to clouds, as well as having databases in the cloud. Which is best depends on your environment, applications PACE, and workload, where applications are located (with data close to them), along with size, scale, SLO, and other considerations.

> *Tip:* Configuration files can usually be found in (or in a subdirectory) of */etc* such as */etc/mongod.conf*. These configuration files specify where database files are located and other parameters that can include some sessions, memory for buffers, and other items. Likewise, logs can usually be found in */var/log*.

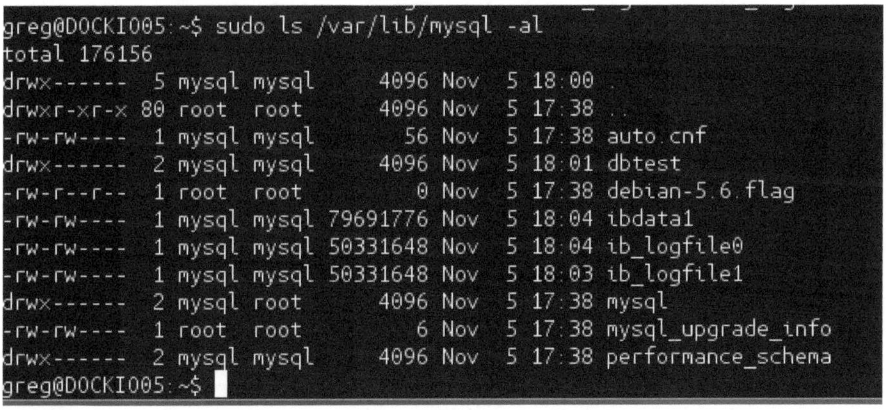

```
greg@DOCKI005:~$ sudo ls /var/lib/mysql -al
total 176156
drwx------  5 mysql mysql     4096 Nov  5 18:00 .
drwxr-xr-x 80 root  root      4096 Nov  5 17:38 ..
-rw-rw----  1 mysql mysql       56 Nov  5 17:38 auto.cnf
drwx------  2 mysql mysql     4096 Nov  5 18:01 dbtest
-rw-r--r--  1 root  root         0 Nov  5 17:38 debian-5.6.flag
-rw-rw----  1 mysql mysql 79691776 Nov  5 18:04 ibdata1
-rw-rw----  1 mysql mysql 50331648 Nov  5 18:04 ib_logfile0
-rw-rw----  1 mysql mysql 50331648 Nov  5 18:03 ib_logfile1
drwx------  2 mysql root      4096 Nov  5 17:38 mysql
-rw-rw----  1 root  root         6 Nov  5 17:38 mysql_upgrade_info
drwx------  2 mysql mysql     4096 Nov  5 17:38 performance_schema
greg@DOCKI005:~$
```

Figure 14.19 Default MySQL data file locations.

Figure 14.19 shows an example of default MySQL data file locations.

Figure 14.20 shows a MySQL database installed on a Docker container on an Ubuntu Linux system using default, ephemeral (nonpersistent) storage. In the example, a container named contmysql is started and a MySQL *show databases* command executed to display databases in the database server instance. Next, a MySQL *create database gregs* command is executed, followed by a *show databases* that displays databases. To accommodate the nonephemeral storage (by default for containers), the container is stopped and removed, then restarted using

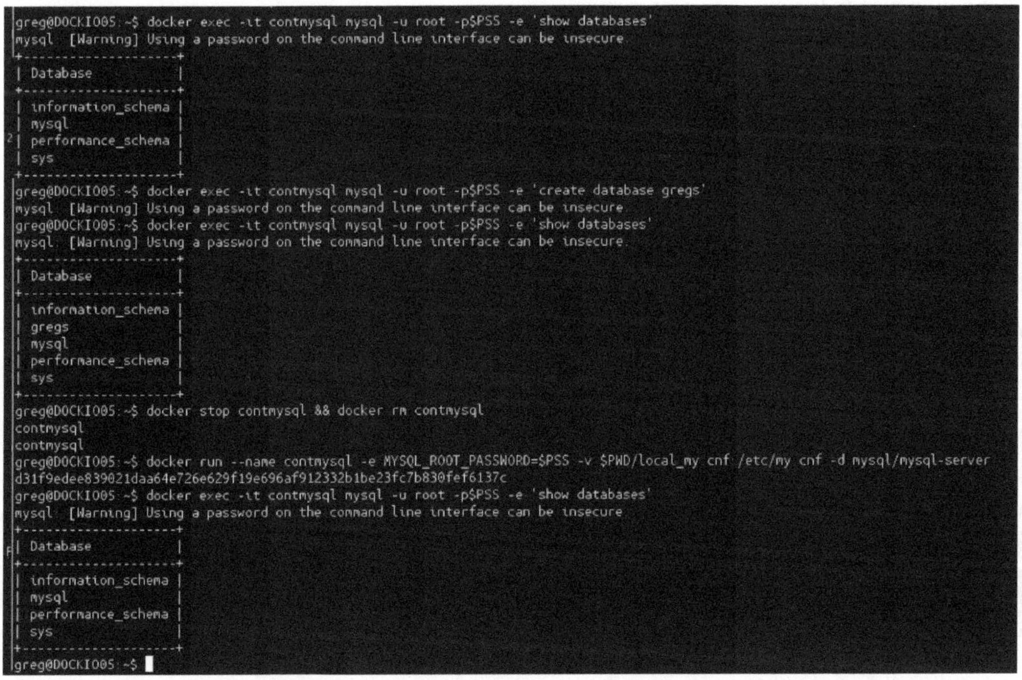

```
greg@DOCKI005:~$ docker exec -it contmysql mysql -u root -p$PSS -e 'show databases'
mysql: [Warning] Using a password on the command line interface can be insecure.
+--------------------+
| Database           |
+--------------------+
| information_schema |
| mysql              |
| performance_schema |
| sys                |
+--------------------+
greg@DOCKI005:~$ docker exec -it contmysql mysql -u root -p$PSS -e 'create database gregs'
mysql: [Warning] Using a password on the command line interface can be insecure.
greg@DOCKI005:~$ docker exec -it contmysql mysql -u root -p$PSS -e 'show databases'
mysql: [Warning] Using a password on the command line interface can be insecure.
+--------------------+
| Database           |
+--------------------+
| information_schema |
| gregs              |
| mysql              |
| performance_schema |
| sys                |
+--------------------+
greg@DOCKI005:~$ docker stop contmysql && docker rm contmysql
contmysql
contmysql
greg@DOCKI005:~$ docker run --name contmysql -e MYSQL_ROOT_PASSWORD=$PSS -v $PWD/local_my.cnf:/etc/my.cnf -d mysql/mysql-server
d31f9edee839021daa64e726e629f19e696af912332b1be23fc7b830fef6137c
greg@DOCKI005:~$ docker exec -it contmysql mysql -u root -p$PSS -e 'show databases'
mysql: [Warning] Using a password on the command line interface can be insecure.
+--------------------+
| Database           |
+--------------------+
| information_schema |
| mysql              |
| performance_schema |
| sys                |
+--------------------+
greg@DOCKI005:~$
```

Figure 14.20 MySQL database nonpersistent storage example.

```
Docker run --name contmysql -e MYSQL_ROOT_PASSWORD=$PSS \
-v /home/user/local_my.cnf:/etc/my.cnf \
-d mysql/mysql-server
```

Next, a MySQL *show databases* command displays known databases, and *gregs* no longer exists.

Note that the MySQL configuration file in the container is overridden by an external (local to the host server) configuration file.

Figure 14.21 shows an example of how persistent storage can be used for containers with applications such as databases that need stateful data. The difference between Figures 14.20 and 14.21 is that the container has persistent storage.

When the container is started, in addition to using an external configuration file, the container default storage path for MySQL *(/var/lib/mysql)* is mapped to an external persistent storage location (*/mnt/md0/Docker/contmysql*). Recall from Chapter 9 that */mnt/md0* was created as an example of a Linux software RAID volume.

```
Docker run --name contmysql \
-e MYSQL_ROOT_PASSWORD=$PSS
-v /home/user/local_my.cnf:/etc/my.cnf \
-v /mnt/md0/Docker/contmysql:/var/lib/mysql \
-d mysql/mysql-server
```

Tip: In Figure 14.21 the persistent storage points to a local storage volume that also ties to or creates affinity with the container to a specific host. Other options include mapping the backing store (the storage location) to a network file share, iSCSI volume, or Docker volume, using plug-ins such as RexRay to map to various persistent storage.

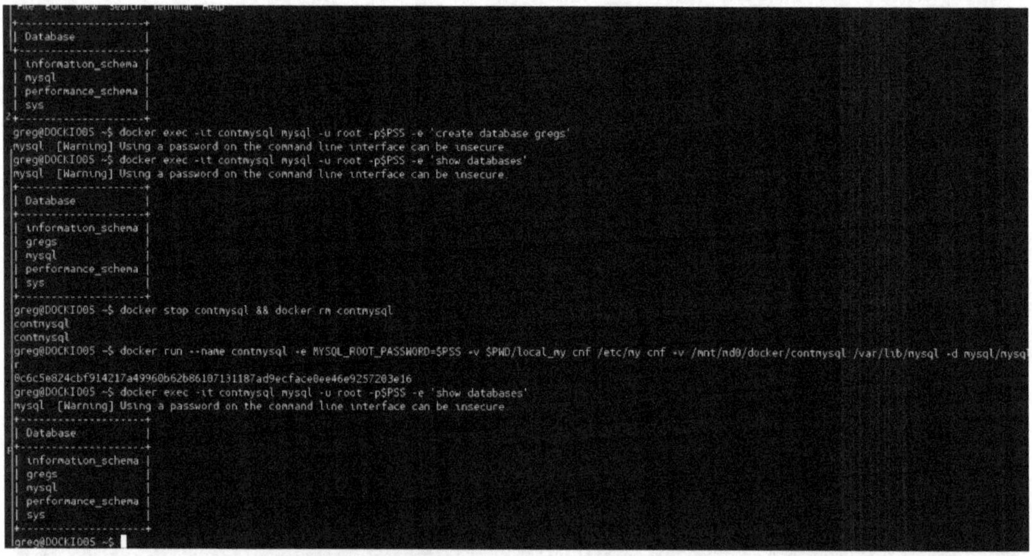

Figure 14.21   MySQL persistent container database storage.

In Figure 14.21, similar to Figure 14.20, the databases are shown, then a database *gregs* is created, the container is shut down and removed, then restored. Unlike in Figure 14.20, however, in Figure 14.21 the container database storage is now persistent, as the database *gregs* exists across reboots.

> *Tip:* If you stop a database server (software) and then copy the associated data files from one location to another (make sure you have a good backup, just in case), make sure you have the security and authentication parameters correct. This includes ownership, access, and other permissions on the files, folders, directories, and other items. For example, if your database does not start after moving it, check your logs to see whether there are permission errors, then check that your destination directory from the root on down to see what is different from the source.

Databases (SQL and NoSQL) and key-value data warehouses and other repositories include Aerospike, AWS (RDS, Aurora, and others), Blue Medora, Basho/Riak, Cachè, Cassandra, ClearDB, Couchbase, Greenplum, and HBase. Ohters include IBM (DB2/UDB and Netezza), Impala, Inodb MySQL engine, Kudo, MariaDB MySQL engine, Memcache, MemSQL, Microsoft (SQL Server [Windows and Linux], various on Azure), and MongoDB. Additional databases and key-value stores include MySQL, OpenStack Trove, Oracle, Parquet, Postgress, Redis, SAP, Spark, Splunk, Sybase, Teradata, TokuDB, and Unisys, among others.

> *Tip:* Use DBA tools such as those that are part of databases and repositories along with third-party and open sources to assess, analyze, report, and gain insight into resource needs, usage, bottlenecks, and optimization opportunities. Tools include those from Microsoft, Oracle such as AWR, MySQL workbench, Quest DBAtools, and many others.

Tools include various studios and workbenches, mysqldump and mysqldumper, among others. Database-related transactions and workload generators such as TPC as well as other workloads include Hammerdb, Quest Benchmark Factory, Loadgen, Sysbench, Virtual Instruments, and YCSB, among others. Analytic, processing, and visualization tools include Hadoop and Kafka, MapReduce, MapR, M1, Pivotal, R, SAP, and SAS, among many others.

> *Tip:* Establish a performance baseline with queries and test data that you can run when problems occur, to compare with known results to verify as well as isolate problems.

Considerations, tips, and recommendations include:

- Databases and repositories are also used in support of big data, including for metadata.
- Everything is not the same, such as applications and their databases or key-value repositories.
- Know your applications, their workload, and PACE requirements, as well as database needs.
- Place data as close to the applications as possible for performance.
- Leverage enterprise-class NVM, Flash, SCM, and SSD for write-intensive applications.
- A little bit of cache in the right place can have a big benefit impact.
- A lot of cache or Flash or NVM SSD and SCM can also have a big benefit.
- Fast applications and fast databases need fast data infrastructure resources.

14.2.5. Big Data, Data Ponds, Pools, and Bulk-Content Data Stores

In addition to those already mentioned in this chapter other applications and workloads in various industries or organizations are also resource intensive. These include those that are CPU compute, memory, I/O, networking, and or storage space capacity intensive. Some of these application and workload characteristics include large amounts of compute using multiple server instances (PM, VM, cloud instance, or container); others are memory focused.

For some of these applications, many lower-performing, lower-cost processors in a grid, cluster, or redundant array of independent nodes (RAIN) scale-out topology are used; others leverage general-purpose multi-socket and multi-core systems. Other applications leverage GPUs for graphics and other compute-intensive activities, including analytics, along with graphics or their rendering. For applications such as genomic sequencing, specialized processors, or cards with ASIC and FPGA such as Dragen, can reduce the time required to process large amounts of data.

Figure 14.22 shows various data repositories including (top left) database, (top middle) data warehouse, (top right) data lake, (bottom center) bulk, and other shared storage or data services. at the top left are SQL and NoSQL databases as well as key-value repositories used for operational, transactional, look-up, metadata, and related functions. Characteristics are highly-structured (software-defined data structures), with applications defined to use a schema of how data is stored. Other attributes include strong consistency, protected, secure, governance, and large and small items accessed by various applications or devices. Examples include Cassandra, MongoDB, MySQL, Oracle, SQL Server, and many others.

At the top middle in Figure 14.22 is data warehouses used for various purposes. Whereas the databases on the left are used for operational or frequent updates or read activity, the data warehouse, as its name implies, is larger, with more structured data for reporting, look-up, and storing other back-end data. Examples include Teradata, SAP, and others.

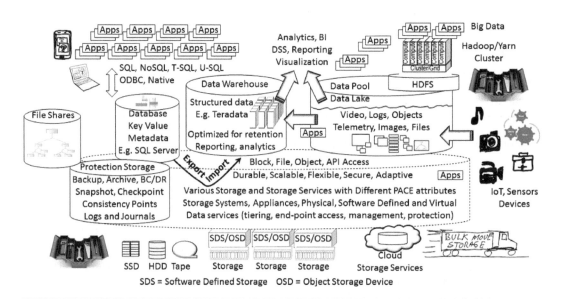

Figure 14.22   Various data infrastructure data repositories.

At the top right in Figure 14.22 is data lakes, with various analytics and applications supporting unstructured, semistructured, structured, and raw data. Different data flows along with data protection and management are shown in Figure 14.22. Keep in mind that everything is not the same, and different environments use databases, data warehouses, and data lakes in various ways. Some environments may not have a data warehouse, instead focusing on data lakes, or vice versa, which in turn are used in different ways to meet their various application and business needs.

At the bottom center in Figure 14.22 is storage that could be one or more systems, solutions, or services including cloud or other repositories that are hardware- and software-defined. Also shown across the bottom of Figure 14.22 are various devices (HDD, SSD, and tape), clusters of converged servers and storage, along with other resources to support the data infrastructure layers.

Other characteristics, as discussed in Chapter 2, include the velocity (how fast—the speed of data arrival) of small or large data items. There are also high volumes of small and large data along with different characteristics, reads, writes, and corresponding metadata. Recall that different applications have diverse PACE needs.

Also recall that there are variations in performance (small or large I/O, IOP or bandwidth, latency, read, write, random, sequential stream, queues) that apply for block, file, object, and APIs. Examples of various applications and workloads that are resource-intensive include big data, big fast data, and bulk data, data warehouse and business intelligence (BI). Additional examples include enterprise data hubs (EDH), click, stream, and other analytics (real-time, postprocessing) along with data transformation and tagging, IoT management and orchestration, log filtering, machine learning, and artificial intelligence.

Tip: Big data means many different things to different people, with diverse application usage scenarios and views. Some focus is for data scientists or other researchers, some for business analysts and business intelligence, some for modern-day executive information systems (EIS) and decision support, an extension of a traditional data warehouse or data marts.

Additional examples include preventive, predictive, and planned maintenance of fleets and other devices, places, and things. Remote monitoring, management of things and places, utilities and other SCADA, active archives, libraries, and reference data repositories, factory and process automation including CNC, robotics, ans sensors, among other devices, stream and telemetry processing, and visualization, among other activities.

Tip: Context with the acronym BD is important, as it can mean big data, backup device, backing device (where data is stored, or stored in), as well as business development aka BizDev (i.e., sales or creating sales). Another context item is EMR, which can mean Elastic Map Reduce, an AWS big data analytics Hadoop cluster service. However, EMR can also refer to electronic medical records.

In addition to the workloads above, applications that are resource-intensive (compute, I/O, and storage) include those for advertising and actuary, energy and mineral exploration, engineering, financial, trading and insurance, as well as forecasting; gaming, government, healthcare, life science, medical, pharmaceutical, and genomics; HPC, IoD, and IoT; logistics, AI, machine learning, manufacturing, media and entertainment (audio, video, photos, and images); seismic, scientific and research, security and surveillance, simulation and modeling, social media, supercomputing, travel and transport, as well as the web, etail, and shopping.

Common to the applications mentioned above are a need for large amounts of:

- Server compute for processing algorithms, programs, threads, tasks, analytics
- Network and storage I/O to move, read, write data, and communicate with devices
- Ephemeral storage and memory during processing of data and data streams
- Bulk persistent storage for storing large amounts of data and metadata
- Scalable, elastic, flexible access, various tenancies, durable, cost-effective resources
- Resiliency leveraging multiple lower-cost components and data copies

The above applications and landscapes workloads and data infrastructure needs have in common large amounts of unstructured and semistructured as well as some structured data needs. Some of this resulting data is large (big actual files, objects, images), others are small in size but with large quantities (many small files or objects). Another variation is telemetry and log files or streams, where the event, activity, status, and other data whose records, fields, lines, or other data structures are relatively small, yet there are very large numbers of them.

> *Tip:* Fast content applications need fast software, multi-core processors (compute), vast memory (DRAM, NAND Flash, SSD, and HDDs) along with fast server storage I/O network connectivity. Content-based applications benefit from having frequently accessed data as close as possible to the application (locality of reference). Also, to prevent bottlenecks, place frequently accessed (read or write) logs and other data onto fast storage repositories, services, and devices.

The resulting log file may be very large and being appended to, or consist of many smaller files or objects. Some applications such as web click analysis or other stream processing can monitor telemetry and other log activity data in real time to spot trends and patterns and enable machine or software-enabled learning. For example, clock analysis running in real time can detect what the user is looking at, and present relevant ads based on predictive analysis and machine learning about the user's browsing behavior. The result is that it might seem like magic when a website starts to display ads or content that might be of interest to you, which in turn causes you to pursue it to learn more.

> *Tip:* Keep locality of reference in mind; that is, place applications and their compute resources near the data they use, or place the data close to where the servers process the date. This can include placing some processing capability at or near where data is ingested for stream and other processing, as well as elsewhere for subsequent postprocessing or analysis, among other tasks.

Besides performance, availability varies from basic RAS to HA, different SLO, and FTT determining FTM configuration or durability, resiliency, RPO and PIT protection, along with logical as well as physical security. The applications among others need large amounts of resource capacity (server, storage, I/O, and network) that is cost-effective. Table 14.7 shows various resources and attributes.

Table 14.7. Data Infrastructure Resources and Considerations

Resource	Attributes and Considerations
Compute memory	Servers physical and software-defined cloud and virtual as well as containers for running code, algorithms, programs, threads, tasks, and analytics. Analytics and other algorithms for processing data can be done during ingest, including tagging, transformation, and filtering, as well as sending copies to different downstream applications and/or data stores and services or tasks. Scale-out compute nodes and instances combined into grids, clusters, RAIN. Specialized cards, ASIC, and FPGA off-load graphics and other task-specific functions. Security using role-based access control (RBAC) defined via different software at various layers.
I/O	Some of the applications and workloads mentioned in this section are sequential streaming (parallel or high concurrency) of data (reads or writes), large or small I/Os. However, there are also random small I/O (IOPS, gets, puts, reads, and writes) of metadata, key values, and related look-up–type activities. Unstructured data including logs and videos, among others, is written sequentially to some backing store, data lake, bulk, object, file system, or another repository. Metadata is added to various SQL, NoSQL, key-value, and other repositories for subsequent processing and use. Applications and services on servers as well as data on storage is accessed via endpoints, protocols, and interfaces.
Network	Various network resources, local and remote, are used for device to servers, servers to servers, servers to storage, storage to storage, and site to site. The networks rely on various underlying mediums, interfaces, and topologies, legacy and software-defined, as well as network function virtualization. Data services include encryption, rights and access control, VPN, firewalls, VPC and VLAN, and micro-segmentation as well as IPv6, among others.
Storage	Bulk (large capacity, low cost, durable), scalable, elastic and flexible, cost-effective, local on-premise, cloud-accessed from on-site, and cloud-accessed from within the cloud. Various access, including block, file, object, API, queues, and tables, among other storage services. Various PACE needs and storage service personalities. Different endpoints, protocols, and interfaces for accessing of storage, storage applications, and data services.

Various solution approaches include:

- Turnkey solutions, services, subscription, DIY, or some assembly required
- On-site, cloud, hybrid cloud accessed from on-site
- Scale-out clusters, grids, RAIN, and software-defined scale-out
- Compute and storage clusters, aggregated or disaggregated
- Various application platforms, middleware, processing software
- Different repositories, databases, SQL (T-SQL, SQL-H, U-SQL), NoSQL, key value

Applicable data infrastructure technologies, tools, and services include databases (SQL and NoSQL) for metadata and other management of data, clusters, grids, rain, tightly and loosely coupled (see Chapter 12); Blue Medora, Cassandra, ClearDB, Couchbase, HBASE, Kafka, Kudu, MemSQL, MongoDB, MySQL, Oracle, Redis, Riak, SAP, and SQL Server (Windows and Linux); AWS RDS, Azure, Google, OpenStack, Oracle, Softlayer, Virtustream, and VMware; Cloudfoundry, Cloudera, Hadoop and YARN, Hive, Hortonworks, Map Reduce, MapR, Morpheus, Pentaho, Pivotal, R, SAS, SAP HANA, Spark, Splunk, and TIBCO, along with various log insight and analysis tools; Docker, Kubernetes, Mesos, and RexRay, among others.

Other technologies, tools, services, providers, and solutions include Blockbridge, Alluxio, Adobe, Apple, Atto, Avere, Avid, Axis, Broadcom, Ceph software-defined storage, Cisco, Cloudian, and Cray; DataStax, DDN, Dell EMC ECS and Isilon, HDS HCP, HPE, IBM

Cleversafe, Informatica, Intel, and Inerplay; LTFS (Linear Tape File System) and LTFS EE+, LSF, and Lustre clustered file system; Maxon, Mellanox, Microsoft, NAS and file access (NFS, pNFS, SMB3, HDFS), Nasuni, NetApp StorageGrid, Nexenta, and NooBaa; OpenStack Swift and SwiftStack, Panzura, Quantum, Supermicro, Stornext, Storpool, Seagate Clusterstor, Sony, Spectralogic, Teradata, Tiger, Vmware, and WD, among others.

> *Tip:* Internet-of-things topics touch many different aspects of data infrastructures. While the IoT devices may be outside the data center and data infrastructures, they have various dependencies and impact. Similar to how users of information services have (and continue) to interact with applications supported by data infrastructures, so to do IoT devices. In some cases, the IoT devices are consuming information or simply being monitored as well as managed via applications supported by data infrastructures.

IoT devices are supported by data infrastructures including on-site, on-premise, as well as cloud and hosted. Another group of IoT devices, in addition to being monitored or managed, also create data along with log, event, and other telemetry data streams. For example, a video camera, in addition to capturing video, also logs time, location, and other metadata. The net result is a need for data infrastructure compute, I/O, and network along with storage resources. IoT data infrastructure resources and components include management service hubs (not to be confused with a physical or virtual network device) and application security authentication and authorization. Additional IoT data infrastructure topics include Protocol gateways, device registry, and rules engines along with end-point management. Other resources and components include developer tools, rules engines, and SDKs. There are various providers that have different IoT-related data infrastructure services, including AWS, Azure, B-SCADA, Google, and Softlayer, among others.

Some IoT protocols, standards, and tradecraft topics include Advanced Message Queueing Protocol (AMQP), Bluetooth and Bluetooth Low-Energy (BLE), Cellular (GSM, 3G, LTE, 4G), Constrained Application Protocol (CoAP), and Device twin or shadow (virtual device with current device state even if turned off). Device to cloud, cloud to device, command, control, communication (C3), LPWAN, NFC, WiFi, WebSocket, PPP, Ethernet, IP, TCP and UDP, JSON, HTTP, and REST. Other IoT topics and terms include MQ Telemetry Transport (MQTT), SCADA and telemetry, along with X.509 public key certificates.

Bulk storage repositories are large, scalable, elastic cost-effective places where structured and semistructured data is stored. This includes a file, object, and other techniques along with various storage mediums (SSD, HDD, tape), locally and in the cloud. Another marketing name for bulk storage is data pools, data ponds that have streams of data flowing into them.

Various applications work with and process data that can then be added to the repository. Metadata about the data can be stored in various SQL and NoSQL databases and key-value stores. Bulk repositories and data lakes can be accessed with flexibility using file, object, APIs, and other means. Security and regulatory data services functionality vary by implementation, solution, or service, including WORM and immutable, litigation hold, multi-tenancy, and subordinated management along with logging, secure digital destruction, and encryption, among others.

Data lakes are storage repositories for storing large amounts of unstructured and semistructured data with durability. A data lake is an implementation of a bulk storage pool or repository and can be a single system, cluster, grid, or federation collection of multiple systems, as well as cloud resources. Some solutions and implementations also provide tiering across different endpoints and namespaces, enabling global namespaces spanning local and cloud, object, and file systems. Different solutions can be optimized for ingest of streaming data, playback of large streams, or processing many small files, among other personality characteristics.

Considerations, tips, and recommendations include:

- Workload simulation tools include Cosbench, vdbench, Mongoose, STAC, and YCSB.
- What availability and data protection, including security, needs to be enabled?
- How will you migrate large amounts of data to clouds (bulk moves)?
- Different bulk and object storage have different performance personalities.
- Do your application workloads need or benefit from GPUs or special devices?
- What data services and functionalities are required for security and regulatory needs?
- Are there geographical location and regulatory concerns for where data is placed?
- Audit, reporting, logging, and access controls

14.3. Legacy vs. Converged vs. Hyper-Converged vs. Cloud and Containers

What is the best platform, deployment architecture, solution, technology, or technique approach for your data infrastructure, or a specific set of applications or functionality? The "converged" folks will generally say that it is CI, HCI fans will say hyper-converged, legacy lovers will say—surprise—legacy, the cloud and container crowds usually say containers and clouds.

Likewise, there are the aggregate vs. disaggregated, open source vs. proprietary, software-defined and less hardware, turnkey, DIY, or some assembly required. In some scenarios, for different applications, workloads, and environments as well as preferences, any of the above can be applicable. However, just as everything is not the same, there are similarities, and while legacy, CI, HCI, cloud, container, and other approaches can be used for many different scenarios, they may not be the right fit for everything.

Of course, if you are a solution provider, or focused on a particular approach, that will often be the focus of what is needed. When I get asked what's the best, my short answer is "It depends." It depends on the application, workload, PACE requirements, SLO, and other needs, environment, constraints, existing technology and tools, as well as preferences, tradecraft skillsets, and experience, among others considerations. Leveraging your tradecraft skillset and experience helps to determine what is applicable, or how to compare/contrast on an informed decision basis, or justify a decision.

Let's start wrapping up with Table 14.8. Table 14.8 provides general characteristics and recommendations of various data infrastructure options. Table 14.8 is not the definitive cheat sheet for decision making; it is a guide to combining with your software-defined data infrastructure essential tradecraft to define your data infrastructure to meet application needs.

Keep application PACE resource needs and SLO requirements in mind. This includes knowing who the users are and where they access the applications, as well as what the resource needs are and the availability (data protection, security, RAS) requirements. Applications should be located on a data infrastructure close to where users access it, or via fast, resilient networks to enable productivity. Data should also be located close to where applications use it (locality of reference) for good productivity.

Table 14.8. General Considerations of What to Use When and Where

Approach	Characteristics, Considerations, When and Where to Use
Standalone BM, PM, host specialized server	Small environment, SOHO, ROBO, workgroup, department, lab, or specialized role or function. Standalone or part of the network, BM host operating system or hypervisor with VM. Many different applications. If BM, consider installing a hypervisor and/or containers to increase flexibility and granularity along with multi-tenancy (security) for applications as the first step to an SDDI, SDDC, and virtual environment. Pay attention to special device needs as well as software licenses as well as data protection.
Networked	Multiple data infrastructure resources (servers, workstations, desktops, routers, storage) networked for communication, application access, storage and data sharing, cloud as well as mobile access. Local, remote, and cloud access. Protect with redundant components, firewalls, access controls.
Cloud	Hybrid. Best of both worlds (public and private), adapt to your needs. Provide elasticity, flexibility, agility, pay for what you use as you consume it, different access modes and granularity of service. Public for pay-as-you-go and off-site, leverage economies of scale and size. Complement your on-site and legacy with hybrid, or go all in, moving everything to the cloud. Private cloud for when more control or large-scale costs preclude using public.
Containers	BM, VM, or cloud hosted provides more isolation of workloads vs. standard non-virtual machine. Less overhead than a virtual machine or of a full cloud instance. Good for small applications, tasks, or components of a larger application that can be compartmentalized. Keep data persistence and ephemeral containers and volumes in mind, as well as portability.
Converged CI, disaggregated,	Scale compute and storage independent converged, hybrid with legacy, SAN, NAS, object, and servers. CI for ease of deployment and management.
Data lake	Unstructured and semistructured bulk data, big data, large-scale file serving
Hyper-converged, HCI aggregated	Small scale, SMB, ROBO, SME, workgroup, a special application for converged management, resources, scaling resources together, ease and speed of deployment. Large enterprise, cloud, and service providers leverage rack-scale CI on a hyper-scale for converging management and deployment. Dedicated to workloads such as VDI/ workspace, general VSI, the web, and general applications including databases, among others.
Cluster, grid (small, large)	Big data, little data, and databases, VM, compute, the web, and others where larger scale-out is needed to meet various application PACE resource needs.
Rack-scale	Legacy, CI, HCI, and software-defined at scale with various software.
Software-defined virtual	Flexibility, extensibility, on-premise or cloud, various applications, different architectures and functionalities for the server, storage, and network. From small to large scale, SOHO and ROBO, to enterprise and cloud scale.
Small cluster *Tip:* Combine with PIT data protection.	Two or more nodes, basic HA removing the single point of failure with one device(s), active/active or active/passive based on applications and data infrastructure software capabilities. Applicable for databases, file and content serving, the web, virtual machines, and many other uses.
Protection	Enable availability, reliability, resiliency, and PIT protection for the above. Leverage various software, services tools, and above approaches.

Data infrastructure management, decision making, and deployment considerations include:

- Abstraction and integration layer software along with middleware
- Aggregate (HCI) and disaggregated (CI or legacy) resources
- Acquisitions and selection of technologies or services
- Are you able to bulk unload (export) and load (import) data, metadata and policies?
- AI, analytics, and automation leveraging policies and templates
- Can you bring your hardware, software, licenses, and policies?
- Data protection, including availabilities, durability, PIT protection, and security

Additional considerations include:

- Health check, troubleshooting, diagnostic, and automated remediation
- Monitoring, notification, reporting, and analytics
- Orchestration of service delivery and resource deployment
- Performance, availability capacity planning, chargeback, show back
- Manual, self- and automated provisioning of resources and services
- Special devices such as GPU, video and other cards, FPGA or ASIC
- System, application, metadata, and data migration or conversation

General data infrastructure considerations include:

- What tools do you have, or will you need to assess and analyze your environment?
- What, if any, geographical regulations apply to where your applications and data can be?
- Will applications be kept available, including for updates, during migrations?
- Does your inventory of resources include applications, databases, middleware, and tools?
- Do you have insight into what device drivers or software update dependencies exist?
- Can you bring your license to cloud or software-defined environments?
- What are your contingency plans for applications, data, data infrastructure, and organization?

There are many different tools that support performance along with capacity and availability planning, optimization, reporting, assessment, inventory, and migration. Some of these tools are focused on particular destinations (cloud, container, virtual server, or desktop workspaces), others are more generic. Tools from various vendors and service providers include AWS, Citrix Project Accelerator, among others, Google, Liquidware Labs VDI Assessment, LoginVSI, Microsoft Assessment Planning (MAP) toolkit along with Azure, among others tools, OpenStack, Quest (various), Solarwinds (various), Turbonomic, Virtual Instruments (insight, assessment, simulation), and VMware (various), among others.

Note that in addition to the above tools and those mentioned elsewhere, many cloud service providers have tools and calculators to estimate your costs for services. However, the key to using those tools, along with those for on-premise, is having good insight and awareness of your environment, applications, and data infrastructure resources, as well as configurations, policies, and your tradecraft experiences.

14.4. Common Questions and Tips

Is a data lake required for big data? Big data does not require a data lake per se, but some repository or place to store large amounts of data is needed. Likewise, data lakes, ponds, pools, and repositories including data warehouses are not exclusive to big data. Of course, that will also vary depending on what your view or belief of what is or is not big data.

Is big data only for a data scientist? There are some big data–centric solutions, approaches, tools, technologies, and techniques that are aligned as well as focused on data scientists. However, big data, in general, is not exclusive to data scientists and is inclusive of everybody from business intelligence (BI) to decision support systems (DSS), AI, business analysis, and many others. Just as there are different types of big data, applications, and tools, so too are there many different types of users and focus areas.

How do you decide between buy, rent, subscribe, or DIY (build) solutions? That comes down to your environment and available resources including staffing, amount of time, and budget. If you have more staff and time than budget, that might push you toward DIY, build, or some assembly required, using open-source or other technologies. On the other hand, if time is money, you might be able to buy time via leveraging technology acquisition either turnkey or with some assembly required. Likewise, for different sizes and types of environments, things will vary.

What are the most important factors when deciding where to site an application, workload, or data (cost, performance, security, scalability, ease of control)? Understanding the application and organizational objectives, requirements, and PACE characteristics all come into play. Also, where are the users of the services of the applications that will be hosted by the data Infrastructure? Do those applications need to be close to the users, and data close to the applications? What applicable threat risks need to be protected against, and what are the cost considerations?

What is the difference between structured, semistructured, unstructured, raw, and data streams? Table 14.9 shows various types of data types.

Can you clarify the difference between database, data warehouse, data lake, data repository, big data, bulk storage, and scale-out file shares and serving? Table 14.10 shows various data and storage stores and repositories. Note that the various items in Table 14.10 can be combined and layered in various ways as part of data infrastructures in support of different applications.

Explain the role and relationships of service catalogs, chargeback, show-back, and how to determine costs of services. Often a picture is worth a hundred or more words, and Figure 14.23 shows various data infrastructure items. Top center is a service catalog of offerings; this can also be accompanied by costs, different types and levels of services, as well as PACE and SLO attributes. Top right is users or consumers of services who subscribe to services, if a public cloud; they get invoices for activity or what is used. For nonpublic clouds, there can be informational or show-back accounting to indicate resource usage and costs. Those costs are tied to the services offerings, left center, which are defined using policies, templates, and bills of materials (what resources are needed) to meet different PACE and SLO.

Table 14.9. Various Types of Data Types

Type	Description and Characteristics
Structured	Data defined into a particular schema such as a database or fixed format. Examples include SQL and NoSQL databases and data warehouse.
Semistructured	Data contains metadata, tags, or other descriptors that help define the data or how to use it and its attributes. Examples include CSV, XML, and JSON.
Unstructured	Undefined data such as files, objects, and other entities stored in file systems or other repositories. Note that data can have a defined data structure, but can be stored in a file or object placed in an unstructured storage or file system. Examples include videos, images, word documents, presentations, PDFs, text files, log files, and telemetry data.
Raw	Data feeds or streams from devices or sensors that eventually ends up in structured, semistructured, or unstructured format.
Data stream	Raw data from IoT and other devices including telemetry.

Table 14.10. Various Data and Storage Repositories

	Description and Characteristics
Commodity storage	General-purpose storage hardware (also known as white box, commercial off the shelf [COTS]), often installed inside or as part of a server that gets defined via software into a storage system, solution, or service.
Big data	Category and class of data along with associated applications stored into data lakes, data repositories, bulk storage, and scale-out file shares or other stores.
Databases, key-value repository (or store)	Focused on handling structured well-defined data by various applications and devices. You can think of unstructured and semistructured data as big data and little data as what goes into databases that are structured. Databases tend to focus on operations, activities, or transactions, whereas data warehouses, as their name implies, house large amounts of reference and other back-end data. Some environments use data warehouses for analytics and reporting as well as a central repository for active and inactive data.
Data lake	Can be a large repository looking and functioning like a very large file share/server (VLFS) for semistructured and unstructured data including big data. Another variation is that data lakes can be focused on a particular application or functionality to support big data, big fast data such as video, stream processing, analytics, or IoT, among other activities.
Data repository	A place where big data, little data, and other is stored in a structured, semistructure, unstructured, or raw format with persistence. Databases, data warehouses, key-value stores, file systems, data lakes, and storage pools are all places where data is stored and thus are data repositories.
File system	Data repository for storing and organizing files with basic metadata attributes including ownership, create, and modify date, time, size, and other information.
Backing store	Persistent place, service, system, a solution where data and an underlying storage resource are combined to protect, preserve, secure, and serve data.
Bulk storage	High-capacity general-purpose storage used to support data repositories, data lakes, file systems, and other places where large amounts of data are stored. Bulk storage includes high-capacity, low-cost, durable SSD, HDD, tape, and optical media and devices, systems, and solutions as well as services. Can be used for general-purpose storage, data protection, big data, and other functions.
Scale-out file shares and serving	A very large file share/server/system, as its name implies, is a place, solution, appliance, or service that supports storing, serving, sharing, and protecting millions, billions, or more files of various size and types of data. Data lakes and other bulk data repositories or forms of data management can be layered on top of scale-out, bulk, or other file servers with common or different namespaces.

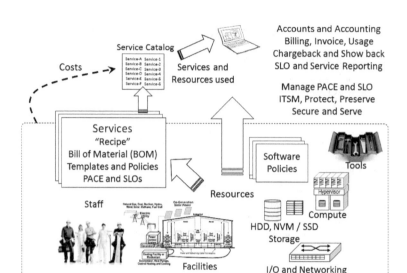

Figure 14.23   Service catalog offerings and costs.

In the center of Figure 14.23 are how resources are packaged and presented as services defined on the left. Bottom center is various resources that also have costs. Those costs include physical (or cloud) facilities, electrical power, cooling, floor space, staffing, insurance, hardware (servers, storage, network, power distribution, desktops), software, and management tools, among others. The collective costs (dotted box around resources) of a given service is the proportion of resources (and their costs) needed to support a given service offering and its subsequent consumption.

Clarify service catalog and orchestration. Like its name implies, a service catalog (top center of Figure 14.23) is a list or menu of service offerings, starting from the lower level, raw or virtual infrastructure as a service (IaaS) such as compute, storage, network, or other resources. Service offerings can be higher level for platform and development such as PaaS, along with even higher-altitude applications along with software as a service (AaaS and SaaS). Those offerings can vary in granularity, scope, and scale as well as cost and SLO. The orchestration is the management tools, processes, policies, templates, and automation that all come together to enable service delivery.

Where do server and applications focus (or domain) start and stop vs. storage, I/O, and networking activities? That depends on your environment, area of focus, and perspective, among other factors. For a heavily siloed or compartmentalized perspective, the lines of demarcation between server, storage, I/O, networking, hardware, software, hypervisors, operating systems, file systems, databases, middleware, applications, and data protection, among others, is well established. However, from other, more converged perspectives, the lines are blurred from technology, task, activity, technology, and management perspectives. If you are a server person, it is time to know more about storage, I/O networking, hardware, software, virtual, cloud, and containers. If you are a storage person, expand or refresh your tradecraft on servers, applications, cloud, virtual, containers, and other related data infrastructure topics. The same applies to networking, applications, database, facilities, and other focus areas.

Keep in mind that servers need storage, storage needs servers, servers and storage need I/O and networking, networking and I/O exist to provide communications, hardware needs software, software needs hardware—they all are used to create and support data infrastructures.

14.5. Chapter Summary

Everything is not the same across different environments, applications, data Infrastructures, and resources. However, there are similarities in how data infrastructure resources can be aligned to meet different application PACE as well as SLO needs. Use new and old tools, techniques, and tips in new as well as hybrid ways to adapt to your needs and environment requirements. Different solutions, services, hardware, software, and tools have various attributes beyond simple cost per resource. Likewise, different applications workloads or services need various approaches to support them: One size does not fit all scenarios.

General action items include:

- Understand your objective and business application focus to determine data infrastructure needs.
- Avoid flying blind by having accurate insight awareness into applications and resources.
- How will you define hardware, software, and services to meet your PACE and SLO needs?
- Balance PACE efficiency (utilization) with effectiveness (productivity).

Bottom line: Evolve and scale today with stability for tomorrow.

Chapter 15

Software-Defined Data Infrastructure Futures, Wrap-up, and Summary

You can't always get what you want, but if you try sometimes well you might find you get what you need.

– The Rolling Stones

What You Will Learn in This Chapter

- Review and recap of what has been discussed in the various chapters
- Emerging techniques, technologies, tools, topics, and trends
- What to do next, where to learn more

This chapter converges various themes and topics from the different chapters along with some additional tradecraft skills, context, toolbox, insight, and experiences. Key themes, buzzwords, and trends addressed in this chapter include data infrastructure, trends, tradecraft, and software-defined, among other topics.

15.1. Getting Started on the Wrap-up

Let's take a quick step back for a moment and look at where we have been to help set the stage for what's next. Figure 15.1 summarizes the organization of this book, which also aligns with various common data infrastructure and software-defined focus areas. This includes different layers and altitudes of technologies, techniques, topics, and tradecraft experience spanning legacy, cloud, virtual, and containers.

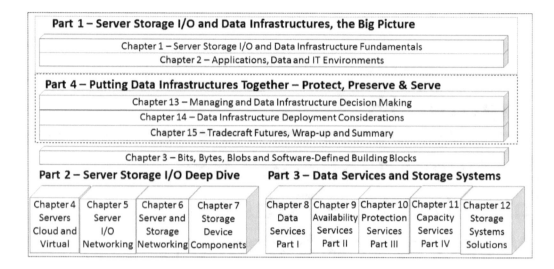

Figure 15.1   Summary of software-defined data infrastructure essentials.

Table 15.1. Chapter Summary and Key Themes

Chapter	Key Themes and Considerations
1	Everything is not the same. However, there are similarities. There are different types and sizes of environments with various applications, trends, and demand drivers.
2	Data infrastructures support, protect, preserve, secure, and serve information. It's about the applications which are programs. Recall: programs = algorithms (rules, policies, code) plus data structures (how organized). Data volume and value.
3	Data structures are defined by software, accessed and used by software, how bits, bytes, blocks, blobs, buckets, and objects are organized and used. Block, file, object and API access, file systems, endpoints to support different application needs.
4	Physical machine (PM) and bare metal (BM) along with virtual machines (VM), containers, converged and cloud machines, memory and I/O including PCIe.
5	Server I/O to devices and storage. The best I/O is the one you do not have to do, second best has least overhead impact. Various protocols and interfaces.
6	Local and remote server I/O to devices, storage, servers, and services, including clouds, spanning different interfaces, protocols, and technologies.
7	NVM including SCM, Flash, and SSD, among others, is in your future. Some storage is for performance; other storage is for capacity or for both. Tiered storage and mediums.
8	Data services. Access, multi-tennancy, performance, and management.
9	Data services. Availability, reliability, and resiliency (cluster, RAID, erasure code).
10	Data services. Availability and recovery points (backup, snapshots, RPOs).
11	Data services. Capacity and data footprint reduction (compress, de-dupe, thin provisioning, tiering) and related topics.
12	Storage software, systems, appliances, solutions, and services.
13	Managing data infrastructure decision making, including gaining insight and awareness, habitats for technology, dashboards, metrics that matter, management tools, common management tasks, remediation and troubleshooting.
14	Data infrastructure deployment considerations across legacy, software-defined virtual, cloud, container, databases, big and little data, OpenStack, VMware, Microsoft, among others.
15	Wrap-up and summary.

Building on Figure 15.1, Table 15.1 summarizes key themes and take-away tips along with considerations covered in the various chapters. In addition to the items in Table 15.1, please refer to the various Appendixes found at the end of this book for additional content, along with companion reference material. Having reviewed where we have been, what's next? Let's continue the wrap-up by filling in a few tradecraft, toolbox, tips, and considerations.

15.1.1. What's in Your Server Storage I/O Toolbox?

Throughout the chapters in this book as well as in the various Appendixes, various tips, tools, examples, and techniques have been mentioned along with terms that have different context and meaning. These are all items that are part of your tradecraft skills, knowledge, and experience. Your data infrastructure tradecraft toolbox should consist of software tools and hardware devices, along with skills and the knowledge to know what to use when, where, why, and how. Another part of your tradecraft toolbox should include resources such as books like this (among others), websites, blogs, videos, podcasts, webinars, conferences, forums, and other sources of information.

Toolbox tips, reminders, and recommendations:

- Create a virtual, software-defined, and physical toolbox.
- Include tip sheets, notes, hints, tricks, and shortcuts.
- Leverage books, blogs, websites, tutorials, and related information.
- Implement a lab or sandbox to try things out (proof of concept [POC]) and gain experience.

15.1.2. Tradecraft and Context Topics

Throughout the book, various tradecraft tips, recommendations, and context matters have been covered. As we continue our wrap-up, here are a few more context matters to keep in mind.

Your data infrastructure tradecraft is (or should be)

- Essential skills and experiences spanning different technologies and focus areas
- In addition to knowing the trends, tools, and technologies, is also aware of techniques
- Knowing various techniques to use new and old things in new as well as hybrid ways
- Expanding awareness into adjacent areas around your current focus or interest areas
- Leveraging comprehension, understanding application of what you know
- Evolving with new knowledge, experiences, and insight about tools and techniques
- Hardware, software, services, processes, practices, and management
- From legacy to software-defined, cloud, virtual, and containers

Some reminders and additional context include the fact that *client* can have different meanings, including software or applications that communicate with a server or service, local or in the cloud. Another variation of *client* is a type of device, such as a tablet, laptop, mobile device or phone, as well as a workstation with varying software for accessing different data infrastructure as well as applications. Another context for *client* is the user, person, or thing such as IoT that accesses and interacts with client software or server and services of application or data resources. Yet another context for *client* is a consumer of lower-level data infrastructure resources or higher-level applications services.

Developers can refer to those who create lower-level (low-altitude) data infrastructure and resource-centric applications, drivers, shims, plugins, utilities, operating systems, hypervisors, FPGA, BIOS/UEFI, and firmware, among other forms of software (and their packaging). At a higher altitude up the data infrastructure and IT stack is developers who create, implement, customize, or support business-focused applications. Those applications can be developed in-house, customized from a third party, or using various tools and platforms including cloud to implement as well as deploy. Context matters as to what type of developers are being referred to.

Another context topic is *DevOps*, which can refer to developers and operations converged, or to developers who are working on software for operations and infrastructures. The latter has ties to what in the past were known as systems programmers (aka sys progs), technical services developers, among others, working on applications that support or enable data infrastructure as well as upper-level applications (and developers).

Another context matter is *software engineer*, which can refer to a system programmer or DevOps specialist or developer, which in turn can mean higher-level applications, or lower data infrastructure, or working for a vendor on one of their solutions. Context is also needed for the term *applications*, including lower-level data infrastructure–related, or higher-level business-focused.

Fabric can also have different context and meaning. It can refer to a network of switches, directors, and routers tying together servers, storage, bridges, gateways, and other devices, but it can also be associated with higher-level application functions, or a cluster of servers or services, as well as data. Keep context in mind regarding *fabric*: whether it is referring to lower-level physical and logical networks, or applications and data, among others.

Roles have different context meaning, including what the role or function a person does with data infrastructure, for example, as an architect, engineering, administrator, operations, or other function. Software applications and utilities also have different *roles* in terms of what they do or pertain to. This includes adding roles or feature functionalities including data services for different hypervisors, operating systems, or data infrastructure software. Context is needed when discussing the roles of people and software, as well as access authentication of people and things to use resources.

A couple more context items are *greenfield*, referring to a new environment, applications, data infrastructure, or facilities, with *brownfield* indicating that something is existing. The platform can refer to a physical or software-defined server, as well as data infrastructure applications such as hypervisors, operating systems, or databases.

Also, *platform* can refer to different development tools and environments, among others. In general, platform 1 refers to legacy mainframe-type processing environments and platform 2 indicates a client server, 2.5 being a hybrid between current client service and next-generation platform 3. Platform 3 refers to the cloud and web-based among other software-defined services, including container tools, technologies, and techniques, while platform 4 is still to be defined (at least as of this writing).

Scale with stability means not introducing complexity or loss of performance and availability. *Scale-up* (vertical) means using larger-capacity, faster resources to scale PACE attributes. *Scale-out* (horizontal) combines multiple resources to support scaling. *Scale-up and -out* uses larger-capacity, faster resources and more of them. *Scale-down* is the inverse of scale-up. *Scale to tomorrow* means future-proofing, enabling your environment to scale with stability for today and tomorrow to meet various needs, without increasing complexity (and costs).

> *Tip:* Keep in mind the software-defined anything fundamental is that applications (lower-level data infrastructure and higher-level business, among others) are software programs. Application software programs can be packaged and delivered in different ways (download, cloud, plugin, tin-wrapped appliance, embedded, ASIC or FPGA, BIOS among others. Also recall from Chapter 2 the software-defined essential that *Programs = algorithms + data structures*. This means that the algorithms are the software that act upon data structures (and data) that define and enable SDDC, SDI, SDDI, SDS, SDN, NFV, IoT, data protection, and other data infrastructure–related management topics.

Remember that *tradecraft* is skills, experiences, tricks, and techniques along with knowing what as well as how to use various related tools as part of what it is that you are doing. At the beginning of this book, I gave the example of an automobile technician who, after years of experience, might become a master mechanic. Another example is engineers and architects (non-IT) who have basic design along with engineering discipline tradecraft, as well as specialties such as mechanical, electrical, HVAC, or environmental, among others. They can leverage their basic tradecraft while extending and enhancing it by gaining insight as well as experience in adajcent areas of focus.

There are many other examples, including salespeople who have the tradecraft of selling, including account as well as relationship building along with the ability to learn new tradecraft related to the trade or items they are or will be selling. And then there are pre-sales and systems engineers, technical marketing, product and program management, test and development, R&D engineering, IT and technology architects, among many others.

Throughout this book, there are examples to expand (or refresh) your tradecraft skills, experiences, and insight. For some of you, this might be the start of a journey for which you decide to take the next step and go deeper into the cloud, containers, virtual, or other software-defined server, storage, I/O networks, applications, and development, among other topics. That might mean going deeper into a particular cloud, hypervisor, container, server, storage, network, vendor, or stack. On the other hand, it might mean focusing on data protection, performance, availability, capacity planning, or another related topic.

15.2. People, Process, and Best Practices

Become familiar with the business or what your organization does, and the information that it depends on via applications along with the underlying data infrastructures. This means understanding what the organization does, how it works, why it uses various applications, and data infrastructure needs.

When I worked at a railroad that was easy, as I liked trains and related topics, so learning more about the business was fun; plus, it helped to be able to speak the language of the business, such as waybills, lifts, moves, and other functions. That, in turn, enabled translating those business functions into data infrastructure resources needs and vice versa. While working at an electrical power generation and transmission company, going out into the power plant helped me to understand what needed to be done when writing preventative maintenance systems, among other business applications.

> *Tip:* If people, processes, politics, or organizational practices are barriers to convergence and productivity, then address those instead of using new technology in old ways. Also keep in mind that while software defines other software, hardware, and services, people define software via programming, configuration, policies, and related activities.

Some general tips, considerations, and recommendations include:

- Ask for and provide context to help clarify what is meant during discussions.
- Pay and play it forward; help others learn; be a student and teacher.
- Give credit where credit is due, including attribution of sites and sources.
- Be respectful of others' time when you ask them for help or how to do something.
- Leverage reference architectures, templates, designs, and tutorials.
- Learn the tricks of the trade that may not be spelled out in tutorials or other materials.

> *Tip:* Get some hands-on, behind-the-wheel time with various technologies to gain insight, perspective, and appreciation of what others are doing, as well as what is needed to make informed decisions about other areas. This also means learning from looking at demos, trying out software, tools, services, or using other means to understand the solution.

Additional tips and considerations include:

- Expand your social and technical network into adjacent areas.
- Get involved in user groups, forums, and other venues to learn and give back.
- Listen, learn, and comprehend vs. simply memorizing to pass a test.
- Find a mentor to help guide you, and become a mentor to help others.
- Collaborate, share, respect and be respected; the accolades will follow.
- Evolve from focus on certificates or credentials to expansion of experiences.
- Connect with me and others via LinkedIn, Twitter, Google+, and other venues.

15.2.1. Skills Development

Learn some programming, including languages and tools as well as how to use them. Once you learn the fundamentals of programming, requirements, analysis, design, and test and troubleshooting, you should be able to transfer those tradecraft skills to using different languages and tools on various platforms.

If you focus on just the tool or technology, that could become your point of lock in or single tradecraft domain experience. While becoming a master, expert, ninja, or unicorn in a particular tool or technology area opens opportunity doors for that domain focus, it can also become an inhibitor. Instead, look to expand your tradecraft skill into other related and adjacent tools, technologies, and domain focus areas. For example, if all you know is .net, Java, C+, Cobol, and Python, then learn another programming (software defining) language. However, also move beyond the language to expand on what can be done with those tools, such as the language or programming environments (e.g., the tools) and expand what you can do with them in new ways (e.g., increase your tradecraft) and what you might become tied to. If your focus is servers, then expand your tradecraft learning about storage, I/O network, hardware, software, and related topics. If your focus is on business applications, learn about the business as well as the underlying technology. And vice versa, if your tradecraft domain focus is lower-level data infrastructure, such as hypervisors, devices, and cloud, among others, learn about the business side of things.

> *Tip:* Some tools and technologies have long-lasting lifecycles, particularly when used in new and different ways. Likewise, fundamental tradecraft skills, techniques, tricks, and experiences can span different generations of technologies and environments. This means knowing how to learn new technologies and apply your tradecraft to use them effectively, without making old mistakes.

Project planning, along with project as well as program management, is another skillset to add to your tradecraft. This can be as simple as understanding how to put together a simple project plan on the back of a napkin together with a statement of work (SOW) involved. Take a class or tutorial, watch a video, read a book, or talk with somebody who has project and program management experience to learn more.

Assessments are done in various layers (altitudes) across data infrastructures, from high-level to low-level. The assessments can be done to discover what applications and associated resources are needed, or data protection including backup coverage, among others. Examples of assessments include gaining requirements and fact finding for design or configuration, along with decision making of new technology, or changes; and health and status checks of applications, services, and other resources during normal operations, as well as part of troubleshooting. Other activities involving assessments and gaining insight about data infrastructures include forecast, planning, design (or decision) validation, tendering (RFP, RFI, RFQ), and assets inventory, among others.

Tasks related to development, design, analysis, planning, and project and program management, among others, are job task analysis (JTA). A JTA is an assessment of what jobs, tasks, steps, or procedures are done, or need to be done, to support something, or complete a given piece of work in support of something. In the case of data infrastructures, JTAs can involve understanding what is being done, as well as what needs to be done to improve service, remove complexity, or reduce costs, among other initiatives.

Keep in mind:

- Your tradecraft is evolving, expanding, and perhaps even refreshing over time
- How to apply experiences, reasoning, and comprehend different scenarios
- Moving beyond the tools and technologies to techniques for various situations
- Track and manage your tradecraft skills with a résumé as well as online venues
- What you know and don't know, and learn how to learn

15.3. Emerging Topics, Trends, and Predictions

Before we wrap up, let's take a few moments to look at some emerging topics, trends, perspectives, and predictions about software-defined data infrastructures and related technologies, techniques, and topics.

Some environments will evolve quickly to all-software-defined cloud, converged, container, or virtual. Other environments will evolve more slowly in hybrid ways to support past, present, and future applications workloads, software, tools, and other technologies. Likewise, some environments will move quickly to all-Flash or other form of NVM technology, while most will continue to increase the use of SSD in various ways. Note that there are different categories (tiers) of NVM and SSDs, similar to HDDs.

Servers, storage, and I/O networking devices will continue to do more work in a smaller physical footprint; there is also more work to be done, data to move and store, applications to run needing additional resources. Some servers will become physically smaller. Likewise, higher-density modular and fat or thick blades continue to be popular, as well as fat twins and quads.

Safe (perennial favorite) predictions and perspectives include:

- Data infrastructures continue to evolve in hybrid ways to support different environments.
- Data's value will change over time; some will find new value, some will lose value.
- Hardware will need software, software will need hardware, both need to be defined.
- More data will find its way into unknown value for longer periods of time.
- Applications and data infrastructures will enable data with unknown value to have value.
- There is no such thing as a data recession, as more information is created and used.
- Business, IT, application, and data infrastructure transformation will continue.

There is more going on in and around data infrastructures today from a software-defined along with legacy topic, technology, trends, and news focus than in the past. Flash, SSD, and other forms of NVM memory including SCM are in your future in some shape, form of packaging, location, quantity, and use. There will be some new additions to the zombie list (i.e., declared dead, yet still alive, in use, and being enhanced) of technologies, including some fairly recent items to make way for other new things. Keep in mind, however, that most techniques and trends are on the zombie list, primarily from a marketing or media buzz perspective. Perennial zombie list favorites include HDD, tape, IBM Z mainframes, PCs, and Microsoft Windows, among others. Even SSD, cloud, and OpenStack, among others, have been declared dead to make room for new things to talk about. Granted, there are technologies that do end up eventually being phased out over time.

NVM and persistent memory, including NAND Flash among other SSD and SCM, have become popular, affordable, and cost-effective storage for use internal as well as external to servers. With the continued decrease in cost and increase in capacity and durability of NVM, SCM, and SSD, HDD also continue to follow similar improvement curves. The amount of data being generated and stored for longer periods of time, along with protection copies, will increase demand for cost-effective storage.

This is where, at least until late in this decade (2019) or early in the next (2020), HDD will continue to be a viable and cost-effective option for storing infrequently accessed, bulk, cold, and secondary or tertiary protection copies. Granted, SSD will continue to increase in deployment and handle more workloads, with faster adoption in some environments. For other environments, budget and capacity constraints, including for cloud service providers, will lead to a hybrid mix of resources. Hybrid resources include a mix of HDD and SSD (tiered storage). Other tiered resources include different types of servers allocated as containers, VM, and cloud instances, as well as Metal as a Service (MaaS), PM, BM, and DPS. Also, to compute servers and storage, tiered networking resources that are also hardware, as well as software-defined, will be leveraged. Another industry trend is that data infrastructure resources will continue to be enhanced with more capability into physically smaller footprints.

Some other emerging enhancements include new I/O and memory access as new applications are developed. Meanwhile, existing memory and I/O models continue to adopt object while also using file, block, and other API-based storage services. Part of I/O and memory enhancements includes leveraging storage-class memories (SCM) for persistence along with new application libraries. These libraries, such as DAX among others, allow programs to leverage SCM and other persistent memories without having to use traditional object, file, or block I/O protocols. Part of enabling SCM and other persistent memories is the media themselves, as well as processors, chipsets, and motherboards along with associated software to leverage them.

In addition to server, storage, and I/O enhancements, network function virtualization (NFV) along with software-defined networks (SDN) will continue to evolve. Similar to how server micro-services such as Docker (mesos or kubernetes) and other containers enable finer grain to compute resource allocation, micro-segmentation for networks will also continue to gain awareness, adoption, and deployment. An example of micro-segmentation–enabled solutions includes VMware NSX among others.

There is also renewed re-awareness of ASIC and FPGA along with GPUs and other specialized processors that are software-defined to offload general-purpose CPUs. Keep in mind that FPGA are Field Programmable Gate Arrays, meaning that they can be redefined with new software, similar to BIOS, firmware, or other packaging and implementations.

Even though the industry trend remains toward using general-purpose, off-the-shelf commercially available servers (both x86/x64/IA, ARM, and RISC among others), solutions from small scale to web scale are leveraging ASIC and FPGA to wrap software that define those devices' functionality.

For example, Microsoft leverages servers with custom FPGA in their Azure and other clouds that enable them to implement new algorithms via a software update vs. physically replacing servers. The FPGA enable Microsoft to offload certain functions including networking from the general processor, which is then able to handle other work. Google is using custom ASIC in its servers which enable offloads as well as some performance advantage, but without the flexibility of re-programming. Which is better, ASIC, FPGA, general processor? That depends on your needs and PACE among other criteria, since everything is not the same.

Keep in mind that there are different timelines for technology that vary by domain focus area. Researchers are on a different timeline than IT organizations that eventually deploy new technologies. For example, you might hear of a breakthrough announcement of new technology and technique in the research and development labs. However, it might take months, years, or more before it can work its way through production labs to commercially viable to mass production in a cost-effective way. From there the technology moves into customer labs for initial testing.

The context here could be a customer in the form of an OEM vendor who gets a new hardware component or software stack to test and integrate into its solutions. From there, the resulting solution or system moves into IT or cloud organizations labs for initial testing as well as integration (or debugging). Eventually, the solution is integrated into a data infrastructure as part of an IT stack so that users of applications can consume information services.

Figure 15.2 shows on the horizontal axis various technology-relevant adoption and deployment timelines. On the vertical axis are different focus areas of adoption and deployment, starting on the bottom (low-level) with research and development, moving up to higher levels. The point of Figure 15.2 is to show different timelines for awareness, adoption, and deployment across various audiences.

Figure 15.2 is an example where the time from a researcher announcing a new breakthrough technology until mass production can be years. By that time, those at the early side of the timeline may see what they worked on in the past as dead or old and time to move on to something new. Meanwhile, others who are getting exposed to the new technology may see it as just that, something new, while others further downstream (or higher up in altitude) see as still being future technology. The industry press, media, analyst, bloggers, and other pundits tend to like to talk about new things, so keep timelines in context as well as what temporal phase of a technology, trend, or tool is being discussed.

> *Tip:* Context is important to understand when something new is announced, particularly what level or audience it is ready as well as being relevant for. Just because something is announced at a lower level does not mean that it is ready for deployment at a higher level. Likewise, something being used, updated, enhanced, and still in use may be declared dead by those who have a focus on something new.

Another context meaning is around *customer*. For example, when a vendor of a component such as a server processor, network or I/O chip, or other device says it is shipping to customers, what is the context of customers? The customers might be original equipment manufacturers (OEM) and system builders to test in their labs, or prepare beta systems for their customers to

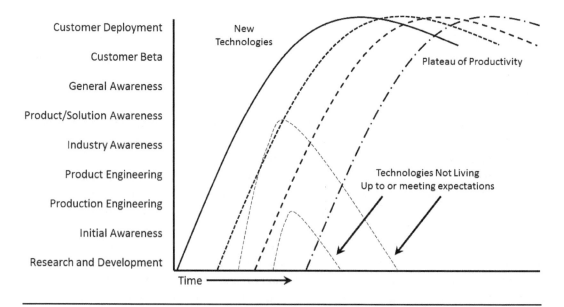

Figure 15.2   Different adoption and deployment timelines.

in turn test and trial in their labs. To clarify, an OEM uses components from other manufacturers such as a processor, I/O, networking, or other components, including memory, drives as well as software and combines them into a solution.

Containers (Docker, Kubernetes, Mesos and other tools, technologies) will continue to gain popularity as data infrastructure host platforms. This includes for some environments ephemeral and persistent storage-based containers to meet different application needs. For some environments there may be a push to put everything into containers, others will align containers to the task or application needs, similar to when and where to use VM, cloud instances, PM and DPS at a cloud or managed service provider.

Automation and orchestration across different technology and data infrastructure layers will also rely on policies, workflows, and processes to be defined by templates, plays, playbooks, or other rules for implementation via automation tools. In addition, automation and orchestration across cloud, software-defined, and hybrid environments will include convergence at various levels. Convergence includes hyper-converged (HCI) and hyper-scale-out for aggregated as well as disaggregated solutions.

Tip: Start your automation focus on common recurring tasks, events, and activities that have well-defined behaviors and solutions, or processes for resolution. This may not be as exciting as taking on something more newsworthy, with big budget impact, and sophisticated. However, whenever you can find and remove complexity, even if small, on a large scale it adds up quickly, perhaps even into a large impact. Also, by automating common tasks, responses, and remediation, you can build successful tradecraft experiences to leverage for the next round of automation. This theme also applies to artificial intelligence (AI), analytics, and related activity.

As a general trend to support large-scale applications and amounts (volume) of data of various size (little data and big data), new techniques are needed for using new and old tools as well as technologies. Instead of simply adding faster servers, NVM storage such as SCM and SSD, along with 40/50/100/400 GbE networks to support growth, new techniques are needed. For example, instead of spending the time to scan files, objects, volumes, databases, and other repositories for data protection, checking for virus or malware, data classification, data footprint reduction, data integrity and consistency checks along with proactive repair, more of those functions need to be done at ingest. When data is ingested, it is tagged (i.e., metadata information added or stored), protection copies made (backup, replica, snapshot, archives), security and access control applied, data footprint reduction (DFR) applied (compress, de-dupe, SIS, thin provisioning), data integrity and checksum enabled. The result is that less time is spent later handling the processes mentioned above, enabling faster protection and management with less overhead.

15.4. Chapter and Book Summary

Key themes, buzzwords, and trends addressed have included server storage I/O tradecraft, data demand drivers, fundamental needs and uses of data infrastructures, and associated technologies.

Our journey and conversation have spanned architecture, hardware, software, services, solutions, tools, techniques, and industry trends along with associated applications, management, and deployment across different environments. This has included IoT, big data and little data, data protection, performance, software-defined, PM, VM, CI, HCI, CiB, cluster, cloud, containers, SSD, SCM, RoCE, and much more.

Depending on when you read this, some of the things discussed will be more mature and mainstream. Meanwhile, other topics, technologies, and trends may be nearing or have reached the end of their usefulness. Some new things will be emerging, while others will have joined the "Where are they now?" list, some of which may currently be being touted as the next "big" thing.

General tips, considerations, and recommendations include:

- Avoid flying blind, leverage insight, become aware to make informed decisions.
- Clouds, containers, CI, HCI, SSD, SDS, SDN, SDDC, and SDDI are in your future.
- Leverage automation with good insight, and manage the automation policies.
- Software-defined does not eliminate hardware; it shifts the location and usage.
- The best server, storage, and I/O technology depends on what your needs are.

Key themes to remember include:

- Context matters regarding data infrastructure and software-defined topics.
- Data infrastructures include servers, storage, network hardware, software, and services.
- Data infrastructures protect, preserve, secure, and serve applications and data.
- Hybrid is the home run for legacy and brownfield data centers and data infrastructures.
- NVM including SCM, Flash, SSD, and others are in your future.
- The best server I/O is the one that you do not have to do (locality of reference).
- Things below your area of focus are considered infrastructure.

We have come to that point in any journey where it is time to wrap up this particular discussion. For those of you who travel via different mediums (plane, train, boat, bus, automobile), it is time to wrap up and stow your things in relative safety as we begin preparations for arrival. Where we are about to arrive may be your final destination or endpoint for this particular journey, or perhaps simply somewhere you are making a connection or layover as part of a longer journey.

There is no such thing as a data or information recession. Society is addicted to information and its availability and reliability, people and data are living longer, and getting larger, which means increased needs as well as demands placed on data infrastructures.

A convenient way of thinking about balancing needs and wants is "You can't always get what you want, but if you try, sometimes you might find you get what you need." What this means is that you should know and understand your wants, needs, and options to address them.

Keep in mind that the fundamental role of data infrastructures is to protect, preserve, secure, serve, and enable applications programs (algorithms), data structures, and data to be transformed into and used as information. This means to design, configure, and deploy flexible, scalable, resilient software-defined (and legacy) data infrastructures that are also efficient and cost-effective. It also means enabling data infrastructures that are easy to manage, able to expand with elasticity to support growth as well as being managed and maintained with minimal disruption.

There is an old saying along the lines of ". . . there are the known and unknown, in between there are the Doors . . ." (the legendary rock band). What I offer is an updated quote applied to the server, storage, I/O networking, hardware, software, cloud, container, CI, HCI, SDDC, SDDI, SDS, SDN, virtual, and software-defined data infrastructure: of ". . . there are the known and unknown, in between there is your tradecraft skills and experience. . . ." How's that for using something old and new in a new way?

Bottom line: This is not the end, it is part of a continuing journey that might be beginning, transitioning, refreshing, or evolving. However, this particular leg, trip, or segment as part of an evolving journey about IT, data center, and data infrastructure has now landed.

Ok, 'nuff said, for now. . . .

Cheers,
GS

Greg Schulz. Server StorageIO® www.storageio.com and www.storageioblog.com
Student of IT, the Software Defined and Data Infrastructure game
Microsoft MVP (Cloud and Data Center) and VMware vSAN vExpert
Twitter @StorageIO and LinkedIn https://linkedin.com/in/schulzgreg

Appendix A

Learning Experiences

The following is a summary of learning experiences from the chapters in this book. Refer to the specific chapter to learn the answers to these and other related items. Review Appendix B for additional learning experiences and tips; Appendix G provides a list of frequently asked questions along with chapters where answers may be found. Keep in mind that a common theme in the learning experiences, along with tips and frequently asked questions, is to help you to understand the content and context, rather than simply memorizing answers.

Chapter 1

1. Can you clarify storage characteristics besides cost per capacity?
2. What are some things that are different in most IT environments?
3. Where does a server store things, including an operating system, applications, and data?
4. What are some common similarities between different applications?
5. What is server storage and I/O tradecraft? How do you acquire and maintain it?
6. Why is context important for server storage I/O topics, trends, terms, and technologies?
7. What is creating the need for more servers, storage, and I/O resources?
8. Articulate the role and importance of server storage I/O hardware and software.

Chapter 2

1. Explain the impact of increasing data volume.
2. What is the difference between data volume and velocity?
3. How do you determine the value of data?
4. What do programs and applications have in common?
5. Who is Niklaus Wirth, and why is he relevant to this chapter?
6. What is the relationship between applications and data?

7. Why is understanding applications and data important?
8. What's the impact on servers, storage, and I/O of latency and bandwidth?
9. What is the importance of sparse vs. dense, random vs. sequential data?
10. How do you decide what data infrastructure resources to use when and where?

Chapter 3

1. Look at your computer and determine how much total storage capacity is reported vs. free space (or in use) and how that compares to what you think you have or bought.
2. Describe the difference between a partition, a LUN, and a volume vs. a file system.
3. Explain why a 64-bit processor running a 32-bit operating system is not able to access or use as much as memory as a 64-bit operating system.
4. What is more important, having performance, availability, capacity, or low cost for the server, storage, and I/O networking?
5. Is S3 a product, service, protocol, software tool, or standard?
6. Is tiering just for capacity, or can it be used for performance?
7. How much raw usable capacity in gigabytes would an SSD advertised as 500 GB have?

Chapter 4

1. On a Linux system, use the commands *lshw –c cpu or lscpu or cat /proc/cpuinfo* or, to determine whether processor virtualization technology exists, *egrep '(vmx|svm)' /proc/ cpuinfo*. To display memory information, use *lshw –c memory*, and for PCIe devices, *lshw –c pci* or lspci combined with your favorite grep or other script output filtering techniques.
2. For Windows, from a command line, you can use the wmic command that displays information ranging from the motherboard and BIOS to network, CPU, disk, memory, and more—for example, to find CPU, *wmic CPU list brief*, and memory, wmic memory chip list brief or wmic memphysical list brief and wmic disk drive list brief.
3. You can also download the Microsoft devcon tool, which is a command-line alternative to the GUI device manager. Also, Windows defines environment variables that you can access via echo, for example,

```
C:\>echo %PROCESSOR_ARCHITECTURE%%PROCESSOR_IDENTIFIER%
x86 x86 Family 6 Model 37 Stepping 2, GenuineIntel
```

Chapter 5

1. Use a tool such as *ipconfig* (Windows) or *ifconfig* (Linux) to display networking information including IP addresses that can then be pinged. Visit www.storageio.com/ performance, where there is an evolving list of tools, technologies, techniques, and tips about server storage I/O networking performance, benchmarking, and related topics.

Chapter 6

1. Try pinging a web or blog site using site tools such as http://tools.pingdom.com/fpt. Use the ping command on your server or system to test access and latency with another device. Ping some device IP address such as 8.8.8.8. Use iperf to test network performance between different servers. For example, set up one system to function as an iperf server responding to other systems acting as iperf clients. Multiple iperf clients can access the iperf server.

2. The following is an example of an iperf client command to perform I/O, specifying TCP window and buffer sizes with results shown in gigabytes for 300 iterations and displaying results every 2 sec. Download iperf from www.iperf.fr.

```
iperf -c ip.of.server -m -fG -d -t 300 -i 2 -w 64k -l 128k
```

3. An alternative to iperf is the Microsoft NTttcp testing tool. Some other commands include ethtool, *iplink, lspci | grep –i eth, netstat –I,* or *ifconfig –a,* among others.

Chapter 7

1. Some useful Linux commands to display disk (HDD or SSD) information include hdparm to display information, *dd* (read or write to a device), *sudo lshw –C disk* (displays disk information), gparted and fdisk (device partitioning), *mkfs* (create a file system), and gnome-disks (device configuration, SMART metrics, and basic benchmark test tools). Windows commands include wmic, diskspd, diskpart, and perfmon, among others.

2. If you need to enable disk performance counters on Windows Task Manger, simply close Task Manager, then, from a command prompt as Administrator, enter *diskperf –Y,* then reopen Task Manager; you should see disk performance stats. Figure 7.11 shows examples of two different tools to monitor HDD and SSD performance.

3. The Windows fio.exe command below (as a single line) performs 70% reads and 30% writes, random with a 4K I/O size with 4K block alignment to a test file called TESTfile. tmp on device X: with a size of 100,000 MB. Some of the other parameters include 32 jobs with an I/O depth of 16, ramp-up time of 3600 sec, a run time of 14,400 sec, and results written to a file called FIO4K70RWRAN.txt in the local directory.

```
fio --filename="X\:\TESTfile.tmp" --filesize=100000M --direct=1
--rw=randrw --refill_buffers --norandommap --randrepeat=0
--ioengine=windowsaio --ba=4k --rwmixread=70 --iodepth=16
--numjobs=32 --exitall --time_based --ramp_time=3600
--runtime=14400 --group_reporting --name=FIO4KRW --output
FIO4K70RWRAN.txt
```

4. Here is an example (in a single line) of using Microsoft *Diskspd* to test device file X:TESTfile.tmp for 14,400 sec, display latency, and disable hardware as well as software cache, forcing CPU affinity. The test does 8 KB random 100% reads with I/Os to the test file sized 100 GB with 16 threads and 32 overlapped I/Os.

```
diskspd -c100g -b8K -t16 -o32 -r -d14400 -h -w0 -L -a0,1,2,3
X:\TESTfile.tmp
```

5. The following is a bit more involved than the previous example and shows how to run a synthetic workload or benchmark. This uses vdbench (available from the Oracle website) as the workload generator and a script shown below. You can learn more about vdbench as well as other server, storage I/O performance, workload generators, and monitoring tools at www.storageio.com/performance.

6. With vdbench downloaded and installed, the following command (as a single line) calls a vdbench file called seqrxx.txt (shown below) with these parameters. Note that this workload assumes a file exists that is larger than 200G (i.e., the number of spaces specified) on the volume being tested. Also, note that this assumes a Windows device; for Linux or Unix, other parameters would be used (see Vdbench documentation). The results are logged to a folder called NVMe_Results (or whatever you call it) at intervals of 30 sec, with each iteration or step running for 30 min. The steps are for 8K, 128K, and 1M using 0, 50, and 100% random of 0, 50, and 100% reads.

```
vdbench -f seqrxx.txt dsize=200G tthreads=256 jvmn=64
worktbd=8k,128k,1m workseek=0,50,100 workread=0,50,100
jobname=NNVME etime=30m itime=30 drivename=\\.\N:\TEMPIO.tmp -o
NVMe_Results
```

where seqrxx.txt is a file containing:

```
#
hd=localhost,jvms=!jvmn
#
sd=sd1,lun=!drivename,openflags=directio,size=!dsize
#
wd=mix,sd=sd1
#r

d=!jobname,wd=mix,elapsed=!etime,interval=!itime,iorate=max,fo
rthreads=(!tthreads),forxfersize=(!worktbd),forseekpct=(!worksee
k),forrdpct=(!workread),openflags=direction
```

Chapter 8

1. Identify data services for Access, Performance, Availability, Capacity, and Economics.
2. Where are data services implemented, and what are the best locations for different scenarios?
3. Describe when is it best to do P2V, V2V, P2C, C2C, and V2P migrations?
4. Discuss where and how to use tools such as those from Acronis, AWS, Google, Microsoft, Quest, Starwind, and VMware, among others.
5. Explain different metrics for Performance Availability Capacity Economics (PACE), including compound metrics for different server, storage, I/O, and network scenarios.

6. The following is an example of Windows PowerShell commands displaying a VHD attached to a Hyper-V virtual machine HVIOWS2012R246 and then increasing the size of one of the VHDs to 250 GB. The first command displays the VHD, its controller type, number, and location, which are used in the second command to resize the volume.

```
Get-VM "HVIOWS2012R246" | Get-VMHardDiskDrive

Get-VM "HVIOWS2012R246" | Get-VMHardDiskDrive -ControllerType
SCSI -ControllerNumber 0 -ControllerLocation 2 | Resize-VHD
-SizeBytes 250GB
```

7. The following is a variation of the above to get and display a specific VM VHD access path.

```
$x = Get-VM "HVIOWS2012R246" | Get-VMHardDiskDrive
-ControllerType SCSI -ControllerNumber 0 -ControllerLocation 1

write-host $x.Path
```

Chapter 9

1. Do you know what legal requirements determine where data can exist?
2. What is copy data management (CDM)?
3. What does the number of 9's indicate, and how would you use it?
4. Why do you want to implement availability and protection at different layers?
5. Do you know how much data as well as storage you have?
6. How many versions as well as copies do you have in different locations (4 3 2 1 rule)?
7. Can you identify when your data was last updated or modified as well as the last read?
8. Explain the role and where to use RTO and RPO as well as FTT and FTM.
9. When would you use mirror and replication vs. parity and erasure code?
10. What's the difference between hardware- and software-based RAID?
11. What's the difference between a gap in coverage and an air gap?
12. Why is RAID (mirror, replication, parity, erasure code) by itself not backup?
13. How or what would you do to leverage some form of RAID as part of resiliency, data protection, and an approach to address backup needs?

Chapter 10

1. What are your organization's terms of service for accessing resources?
2. Have you recently read the terms of service for various web services and resources?
3. Are your protection copies, associated applications, settings, and metadata, protected?
4. Does your organization meet or exceed the 4 3 2 1 backup and data protection rule?
5. How can you improve your data protection and security strategy and implementation?
6. Does your organization treat all applications and all data the same, with common SLO?

7. Does your RTO consider components or focus on end-to-end and the entire stack?
8. How many versions and copies of data exist in different locations, both on- and off-site?
9. Can you determine what the cost of protecting an application, VM, database, or file is?
10. How long will it take you to restore a file, folder, database, mailbox, VM, or system?
11. How long will it take to restore locally, using off-site media or via a cloud provider?
12. Can you restore your data and applications locally or to a cloud service?
13. Using Tables 10.3 and 10.4 as a guide, update them with what's applicable from a regulatory, legal, or organization policy viewpoint for your environment.
14. What regulatory, industry, or organizational regulation and compliance requirements does your organization fall under?
15. For your organization, are there any geographical considerations for where your data can be placed (or cannot be placed or moved to)?
16. When was the last time a backup or protection copy was restored to see if it is good and consistent, as well as that applicable security permissions and rights are restored?

Chapter 11

1. Do you know what data infrastructure capacity resources you have, as well as how they are being used (or allocated and not used)?
2. Do you know what your data footprint impact is, why it is what it currently is, as well as opportunities to improve it?
3. What are your capacity forecast plans for the server (CPU, memory, I/O, software licenses), I/O networking, and storage (capacity and performance), along with associated data protection?
4. Using the tools, technologies, tips, and techniques along with tradecraft from this as well as Chapters 8 and 9, can you address the above, among other questions?

Chapter 12

Some learning experiences for this chapter are to practice what you learned by visiting a vendor's website (or call them up; I'm sure they would love to talk to you). Look at the various types, and categories or classes of storage systems, and solutions they have. Pay attention to how they group, organize, or categorize their different offerings, particularly those that offer multiple solutions. Also look at their speeds, feeds, and specifications to see how they compare, not just on who has the same features, but also how extensively they discuss them.

Another learning experience, if you have a virtual, physical, or cloud lab, is to download from various vendors' sites their software-defined storage systems to gain some experience.

Chapters 13, 14, and 15

These chapters in their entirety are learning experiences, so there are no individual exercises. Refer to the common questions and tips as well as the text in Chapters 13 through 15 for additional learning experiences to expand your tradecraft.

Appendix B

Additional Learning, Tools, and Tradecraft Tricks

The following are a collection of tool and tradecraft tricks for some of the items covered in various chapters. Some of these commands, tools, and tricks are dangerous in that they can cause loss or damage to data; you are responsible for your use. Server StorageIO and UnlimitedIO LLC as well as publisher CRC/Taylor & Francis assume no responsibility for your use or misuse of any items shown here; they are provided as learning to use at your own risk.

Configuration file locations can vary with different *nix implementations, for example, some will have network device settings in /etc/network/interfaces, while others will have those items in a directory such as /etc/sysconfig/network-scripts/ifcfg-xxxx, where xxxx is a device. Also, or in place of DNS, server names and IP addresses are usually found in /etc/hosts, with the name of the local server, device, or system in /etc/hostname.

For system start-up and configuration options on some systems, /etc/rc.local is a place to look, and for storage mounting, /etc/fstab with logs in /var/logs, among other locations. In addition to having different locations for configuration and resource files, folders, and objects, device names can also vary with different *nix implementations.

Other useful commands include lshw, dd, dstat, ntop, iotop, htop, sar, yum, wget, curl, ssh, ping, ifcfg, iperf, df, scp and ssh-keygen, gparted, mkfs, and fdisk, among others.

Do you need to resize an Ubuntu Linux system device? First, it should go without saying, but make sure that you have a good backup that has been tested before doing the following, just to be safe. Using Table B.1, with your system on a new or large disk that has plenty of free space (e.g., you did a dd or other copy from old to a new device, or it is on a virtual machine), the table shows the steps. This assumes the system disk device is /dev/sdx1; verify what your actual device is.

For Windows environments, suppose you need to find a physical name of a mounted devices for configuration? For example, if you have a tool that wants the physical name for your C: drive, that might be \\.\PhysicalDrive0. No worries, use the command *WMIC DISKDRIVE LIST BRIEF*.

Table B.1. Resizing a Linux System Volume

Ubuntu Command	Description and Comments
fdisk /dev/sdx1	Use what your applicable device is.
p	Note the starting block number.
d	Removes partition (breathe deep).
n	Create a new first partition (breathe deep). Make sure partition #1 starts with the same block/cylinder number as noted above.
p	Validate what you have done; quit if you are not 100% sure.
w	If you are 100% sure, 100% correct, breathe deep, and enter w to write the partition. If not 100%, abort and try again.
reboot	Do a df and note size as well as free space.
resize2fs /dev/sdx1	Resize the volume, then do a df again and note change.

Need more detail about the devices beyond what is shown above? Then use *WMIC DISKDRIVE LIST* or, as in the above example, direct the output to a file with the results shown below (scroll to the left or right to see more detailed information). WMIC also provides server, storage, network, and other info.

Is SATA SSD TRIM enabled? If you have a SATA SSD, the TRIM command is a form of garbage collection that is supported with Windows 7 (SAS drives use the SCSI UNMAP). Not sure if your system has TRIM enabled? Try the following (Figure B.1) command as administrator. If you see a result of "0," then TRIM is enabled, while a value of "1" means that it is disabled for your system.

Figure B.1   Verify Windows SSD TRIM enabled.

Have an issue or problem collecting your system statistics, or, when running a benchmark workload generation tool such as vdbench, getting "Unable to obtain CPU statistics"? Try the *Lodctr /R* command (as administrator) shown in Figure B.2.

As its name implies, if you do not have this tool, you can download it from Microsoft not only to delete files and folders, but also to write "0" patterns across a disk to secure-erase it. You can specify the number of times you want to run the write "0" patterns across a disk to meet your erasure requirements.

There is also another use for *Sdelete*: If you need or want to precondition an SSD or another device such as for testing, you can run a preconditioning pass using *Sdelete*. Some command options include -p #n, where "n" is the number of times to run; -s, recursive to process subdirectories; -z, to write "0" or zero out the space on the device; -c, for clean; -a, to process read-only attributes. Note that other tools such as those from Blancco, among others, can erase at a lower technology layer (e.g., go deeper) to digital clean (wipe or digital bleach). For example,

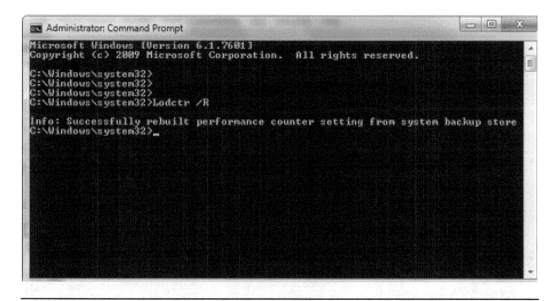

Figure B.2   Windows rebuild CPU statistics.

some technologies only erase or digital wipe, clean, or bleach at a higher-level file system layer. Other technologies can go down deeper into hard to reach persistent memory cell locations in SCM and NVM, such as NAND flash.

Ever have a Windows storage device or system that failed to boot, or a problem with a partition, volume, or another issue? Diskpart can be used to help repair storage devices that have boot problems, or for clearing read-only attributes among other tasks. Don't be scared to use Diskpart, but do be careful. If you prefer GUI interfaces, many of the Diskpart functions can also be done via the Disk Management interface (i.e., Control Panel → All Control Panel Items → Administrative Tools → Computer Management → Storage → Disk Management). Note that Diskpart (Figure B.3) will need to be run as administrator.

In the Diskpart example in Figure B.3, the LIST DISK command shows what disks are present (on-line or off-line), which means that you may see devices here that do not show up elsewhere. Also shown is selecting a disk and then listing partitions, selecting a partition and showing attributes. The Attribute command can be used for clearing read-only modes should a partition become write-protected.

> *Tip:* Ever have a device that had VMware (other volume format) installed on it, you move it to Windows and try to reformat it for use and get a read-only error? If so, look at Diskpart and the Attribute commands. BE CAREFUL and pay attention to which disk, partition, and volumes you are working with, as you can easily cause a problem that will result in finding out how good your backups are.

How about running into a situation where you are not able to format a device that you know and can confirm is ok to erase, yet you get a message that the volume is write-protected or read-only?

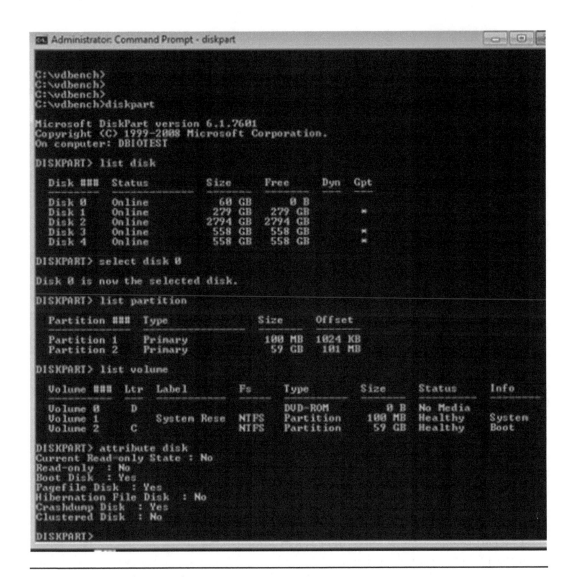

Figure B.3   Windows Diskpart.

Diskpart is a handy, powerful, but potentially dangerous tool if you are not careful, as you could mistakenly drop a good volume or partition (again, the importance of having good backups!).

While the above applies to Windows, here's an approach for Linux. Note that the following will cause damage and loss of data, as well as require sudo or *su* mode. In the example, /of=xxx is a fictional device that could be replaced with an actual device name. Depending on the size of the device, the 102400 value can be increased to write across more of the device.

```
dd if=/dev/zero /of=xxx bs=512 count=102400 notrunc
```

Figure B.4 shows an example of running iperf server on one system and clients on others, generating network activity as well as measuring performance.

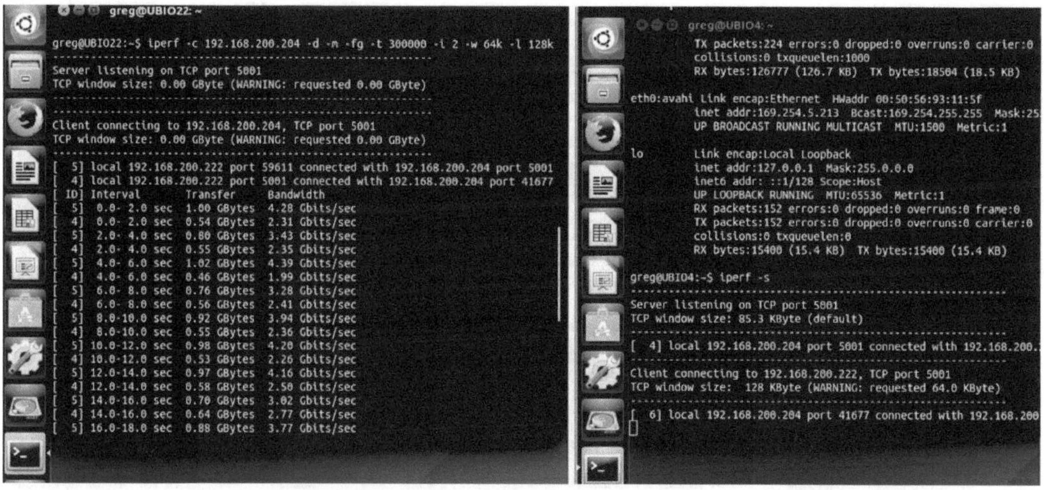

Figure B.4   An iperf client (left) and iperf server (right) for network testing.

The following is a quick example of using sysbench to test a Linux MySQL database. Using MySQL, create a test database:

```
create database dbtest;
```

Tip: Check MySQL config parameters in /etc/mysql/my.cnf, which might point to a default storage location, which could be /var/lib/mysql unless it is mapped to some other location.

Prepare and populate the database:

```
sysbench --test=oltp --oltp-table-size=1000 --mysql-db=dbtest
--mysql-user=user --mysql-password=password prepare
```

Run the workload:

```
sysbench --test=oltp --oltp-table-size=1000 --oltp-test-
mode=complex --oltp-read-only=off --num-threads=6 --max-time=60
--max-requests=0 --mysql-db=dbtest --mysql-user=user --mysql-
password=password run
```

While the workload is running, Figure B.5 shows left *iotop* and right *htop* activity display.

For Oracle environments in which the emphasis is not on exercising the entire data infrastructure stack from application through database and down into lower-level resources, there is Silly Little Oracle Benchmark (SLOB). Tools such as Benchmark Factory, hammer, sqlio, swingbench, sysbench, and ycsb exercise the entire data infrastructure stack. Meanwhile, tools such as diskspd, crystal, fio, iometer, ior, iorate, iozone, SPC, and vdbench, among others, focus below database or upper-level applications. SLOB is able to focus on driving Oracle and underlying data infrastructure harder per CPU cycles consumed. The result is a more

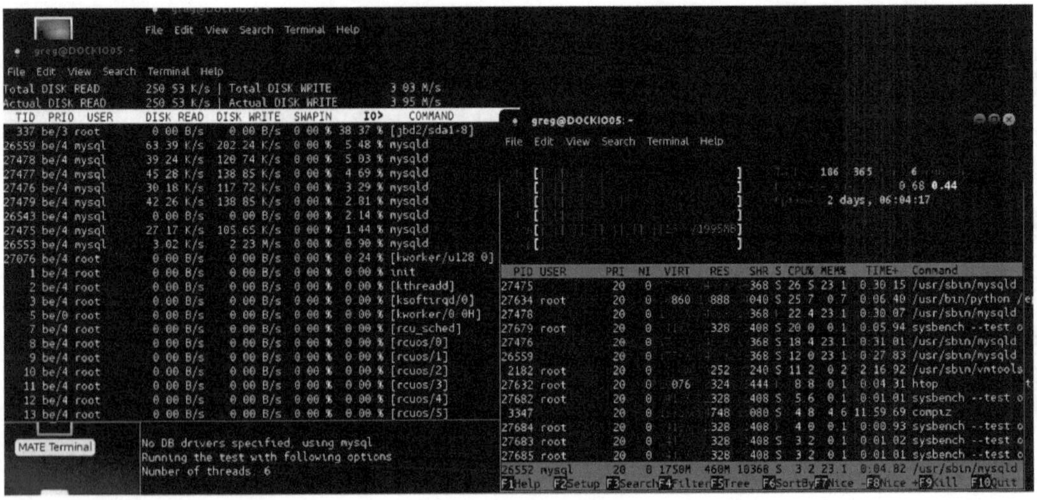

Figure B.5   iotop (left) and htop with CPU (right).

realistic Oracle workload on a data infrastructure for testing. Learn more about SLOB at www.kevinclosson.net/slob and at www.storageio.com/performance.

Need to display lines of error log for a Docker instance? Use the Docker logs command and specify a value as part of the tail option for a specific instance such as shown in Figure B.6.

Figure B.7 shows results from a sysbench command.

```
[root@centio01 greg]#
[root@centio01 greg]# docker images
REPOSITORY                              TAG          IMAGE ID       CREATED
               SIZE
docker.io/emccorp/ecs-software-3.0.0    latest       774f0a57dcf4   5 weeks
ago            1.575 GB
[root@centio01 greg]# docker ps -a
CONTAINER ID       IMAGE                            COMMAND            CREATE
)             STATUS          PORTS          NAMES
3629a1214199        emccorp/ecs-software-3.0.0:latest   "/opt/vipr/boot/boot."   20 hou
rs ago        Up 3 hours                     ecsstandalone
[root@centio01 greg]# docker logs --tail 10 ecsstandalone
12/26/16 17:33:33: Initializing configuration properties
12/26/16 17:33:33: Generating configuration files
12/26/16 17:33:38: Generation failed
/etc/init.d/storageos-dataservice: line 100: /proc/sys/kernel/core_pattern: Read-only f
ile system
Warning: Found stale pidfile(s) (unclean shutdown?)
Setting up SSL certificates  ...done
Starting ViPR services..done
service: no such service cron
Starting nginx service
..done
[root@centio01 greg]#
```

Figure B.6   Docker images, ps, and logs.

```
Running the test with following options:
Number of threads: 6

Doing OLTP test.
Running mixed OLTP test
Using Special distribution (12 iterations,  1 pct of values are returned in 75 pct cases)
Using "BEGIN" for starting transactions
Using auto_inc on the id column
Threads started!
Time limit exceeded, exiting.
(last message repeated 5 times)
Done.

OLTP test statistics:
    queries performed:
        read:                           646296
        write:                          230806
        other:                          92323
        total:                          969425
    transactions:                       46159   (769.27 per sec.)
    deadlocks:                          5       (0.08 per sec.)
    read/write requests:                877102  (14617.45 per sec.)
    other operations:                   92323   (1538.62 per sec.)

Test execution summary:
    total time:                         60.0038s
    total number of events:             46159
    total time taken by event execution: 359.7427
    per-request statistics:
        min:                                2.17ms
        avg:                                7.79ms
        max:                              594.59ms
        approx.  95 percentile:            16.29ms

Threads fairness:
    events (avg/stddev):            7693.1667/144.02
    execution time (avg/stddev):    59.9571/0.00

areg@DOCKTOO5:~$
```

Figure B.7   sysbench OLTP complex test results.

After running the sysbench test, perform clean-up:

```
sysbench --test=oltp --mysql-db=dbtest --mysql-user=user --mysql-
password=password cleanup
```

Access MySQL and remove the test database, for example:

```
drop database dbtest;
```

Figure B.8 shows an example MongoDB with YCSB.

Note that the following are not intended to enable you to be a database administrator, but to develop, expand, or refresh your data infrastructure tradecraft, as well as gain insight and perhaps enable converged conversations with DBAs or others by understanding what they do, as well as how things work.

Figure B.9 shows, top left, various databases for a particular database server instance; top right, properties for the database server instance, specifically memory allocation; top center,

```
> show dbs
dbioit   0.000GB
local    0.000GB
ycsb     11.643GB
> use ycsb
switched to db ycsb
> db.stats()
{
        "db" : "ycsb",
        "collections" : 1,
        "objects" : 10000000,
        "avgObjSize" : 1167.8795472,
        "dataSize" : 11678795472,
        "storageSize" : 11967320064,
        "numExtents" : 0,
        "indexes" : 1,
        "indexSize" : 534544384,
        "ok" : 1
}
> exit
bye
greg@DBIO10:~/YCSB$ df
Filesystem          1K-blocks       Used Available Use% Mounted on
udev                  2009804         12   2009792   1% /dev
tmpfs                  404144       1448    402696   1% /run
/dev/sda1            30831612   20035176   9322636  69% /
none                        4          0         4   0% /sys/fs/cgroup
none                     5120          0      5120   0% /run/lock
none                  2020712        152   2020560   1% /run/shm
none                   102400         36    102364   1% /run/user
/dev/md0              8119736      18420   7665812   1% /mnt/md0
/dev/sdh             41153856   14207932  24832388  37% /mnt/db1
//SIONAS2/SIOXFR 2822269248 2444269244 378000004  87% /media/SIOXFR
```

Figure B.8   Mongo show database and YCSB benchmark.

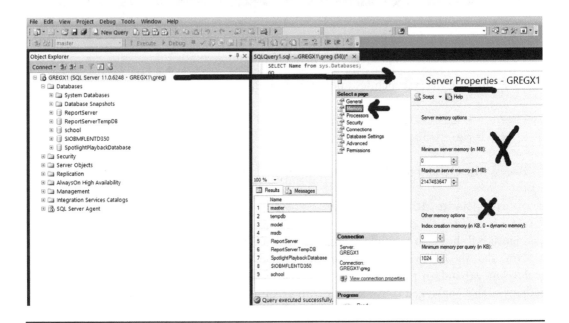

Figure B.9   SQL Server Management Studio.

a SQL command to list all databases shown bottom center. The SQL Server command to list the databases is

```
SELECT Name from sys.Databases;
GO
```

Note that by decreasing the amount of memory (something you normally would not want to do) for testing purposes, the database server can be forced (or encouraged) to do less buffering. This might be useful for putting more workload on an underlying file system, operating system, hypervisor, storage solution, or I/O network. Normally you would want to look at increasing the values for memory to improve buffering and reduce the amount of extra disk I/O.

The following is an example of SQL Server restoring a database (tpcc) from a backup device (W:), placing the MDF main database file on volume S: and its LDF log file on volume I:.

```
RESTORE DATABASE tpcc FROM
DISK = 'W:\SQLS_tpc_Big393_89G\tpcc300Gbackup020915.bak' WITH
    MOVE 'tpcc' TO 'S:\tpcc.mdf',
MOVE 'tpcc_log' TO 'I:\Ttpcc_log.ldf',
        REPLACE, STATS = 10;
```

The following moves a database server instance tempdb (i.e., where sort and other temporary items are stored or processed) from one location to the E: device.

```
SELECT name, physical_name AS Location FROM sys.master_files
WHERE database_id = DB_ID(N'tempdb');
GO
USE master;
GO
ALTER DATABASE tempdb MODIFY FILE
(NAME = tempdev, FILENAME = 'C:\Program Files\Microsoft SQL
Server\MSSQL11.DBIO\MSSQL\DATA\tempdb.mdf');
GO
ALTER DATABASE tempdb MODIFY FILE
(NAME = templog, FILENAME = 'E:\SQLSERVER_TMP_DB\Tempdb.mdf');
GO
```

The following is an example of calling a script C:\SIO_Items\SIO_BMFDB_Backups.sql from the command line, passing various parameters to back up a SQL Server database with benchmark results.

```
sqlcmd -v varDBname="BMFRES" varBUname="E:\BMFRES_BCK" -S 02 -i
C:\SIO_Items\SIO_BMFDB_Backups.sql
```

The following is the script file called from the above:

```
BACKUP DATABASE [$(varDBname)] TO  DISK = N'$(varBUname)'
WITH NOFORMAT, NOINIT,  NAME = N'$(varBUname)', SKIP, NOREWIND,
NOUNLOAD,  STATS = 10
GO
Exit
```

The following might look familiar if you have done an OpenStack install, as it creates a database called glance, granting privileges and access to local and remote to various users. Note that these settings were in a test environment and may not be consistent with your environment's best practices. Consult with your DBA and security, among other folks, for your systems.

```
CREATE DATABASE glance;

GRANT ALL PRIVILEGES ON glance.* TO 'glance'@'localhost' \
  IDENTIFIED BY '~@#$!+_GDBPAS';
GRANT ALL PRIVILEGES ON glance.* TO 'glance'@'%' \
  IDENTIFIED BY '~@#$!+_GDBPAS';
```

The following MySQL displays various parameters.

```
show variables like "innodb_file_per_table";
```

The following Mongo commands show databases for the particular database server instances, selecting a database called testdb and deleting it.

```
show dbs
use testdb
db.dropDatabase()
```

This Mongo command selects a database called testdb and compresses and compacts it to reduce the amount of internal (in the database) and external (storage and file system) space.

```
use testdb
db.repairDatabase()
```

Figure B.10 shows a SQL Server database instance (GREGX1) along with its various databases including *gregs* via SQL Server Management Studio. Also shown are attributes of database *gregs* including tables. Tables include *dbo.Course*, *dbo.CounselInstructor*, and *dbo.Department*, among others. Also shown are the columns for table *dbo.Course*, including *CounselID*, *Title*, *Credits, and DepartmentID*. Also shown at top right is a query, *use gregs select * from dbo.Course*, that selects all data from database *gregs* table *dbo.Course* and displays the values at the lower right in Figure B.10.

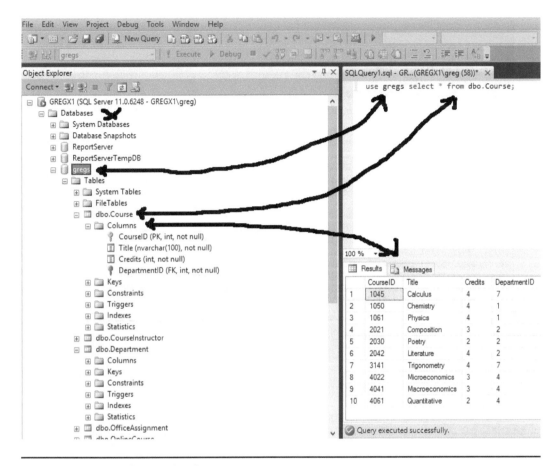

Figure B.10 Displaying database contents.

Hadoop and HDFS include tools for validating or testing installations. These include Teragen, Terasort, and Teravalidate as well as DFSIO, among others. You can use these and other tools to verify that your Hadoop and HDFS environment is performing as expected, as well as to create a baseline for future reference. Then, when there is a problem or change, run the workload and compare to the known baseline. Figure B.11 shows an example of a DFSIO test processing 1000 500MB-sized files for DFSIO.

Not sure whether you may have a slow network, router, browser, application, server, storage, or cloud provider? Besides doing a basic *ping*, *traceroute*, or *nslookup*, among other commands, there are tools such as PingPlotter Pro to have in your toolbox for troubleshooting. Figure B.12 shows an example pinging 8.8.8.8 with PingPlotter running to diagnose where issues may or may not exist.

Review other Appendix sections for additional companion learning material. See examples and related material at www.storageio.com as well as at my blog, www.storageioblog.com.

```
7/29/16 no cache
Logfile:  072916_500GB_100GB_NOCached.log
DFS read loops:  4
DFS file size (MB):  500
DFS number of files:  1000
Using Terasort/validate loops:  4
Using Tera records:  1000000000
DFSIO will use  1000  x  500 MBsized files to create  500000 MB sized DFS space...

dfsio exec 19988.89
16/07/29 22:53:06 INFO fs.TestDFSIO: Total MBytes processed: 500000
16/07/29 22:53:06 INFO fs.TestDFSIO:       Throughput mb/sec: 3.261253350929665
16/07/29 22:53:06 INFO fs.TestDFSIO: Average IO rate mb/sec: 3.404493570327759
16/07/29 22:53:06 INFO fs.TestDFSIO:  IO rate std deviation: 0.7313342751549275
16/07/29 22:53:06 INFO fs.TestDFSIO:      Test exec time sec: 19988.89

16/07/29 23:55:51 INFO fs.TestDFSIO: ----- TestDFSIO ----- : read
16/07/29 23:55:51 INFO fs.TestDFSIO:           Date & time: Jul 29 23:55:51 CDT 2016
16/07/29 23:55:51 INFO fs.TestDFSIO:        Number of files: 1000
16/07/29 23:55:51 INFO fs.TestDFSIO: Total MBytes processed: 498506
16/07/29 23:55:51 INFO fs.TestDFSIO:      Throughput mb/sec: 18.28280363131648
16/07/29 23:55:51 INFO fs.TestDFSIO: Average IO rate mb/sec: 19.959604263305664
16/07/29 23:55:51 INFO fs.TestDFSIO:  IO rate std deviation: 5.840153442531154
16/07/29 23:55:51 INFO fs.TestDFSIO:     Test exec time sec: 3754.115
```

Figure B.11   Example of Hadoop DFSIO output.

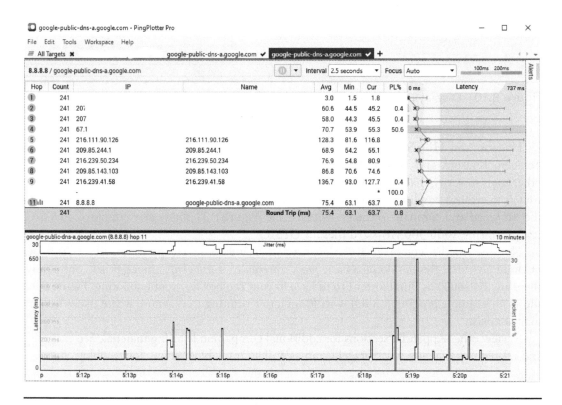

Figure B.12   PingPlotter example of checking for network issues.

Appendix C

Frequently Asked Questions

The following are frequently asked questions along with the chapter in which answers along with related tradecraft learning and refresh material can be found.

Frequently Asked Question	Chapter
Does IOPS apply only to block storage?	3
Do virtual servers need virtual storage?	3
Does a data pond, pool, or lake require a block, file, or object storage?	4
Does having 64 vs. 32 bits mean faster data movement and performance?	4
Does software-defined virtual and cloud eliminate hardware?	4
Does the presence of fiber optics guarantee performance?	6
Doesn't DAS mean dedicated internal storage?	12
Doesn't the 4 3 2 1 data protection rule cause an increase in your data footprint impact along with extra costs?	11
Explain the role and relationships of service catalogs, chargeback, showback, and how to determine costs of services.	14
On a server in an environment where there are a good response time and high I/O rate, a large or high queue depth and a relatively low amount of server CPU time being spent on system time or overhead would indicate what?	2
How can an SSD wear out if there are no moving parts?	7
How can you measure or determine IOP or bandwidth if you know one but not the other?	5
How come you did not say more about backup/restore, archiving, security, using snapshots, and replication along with other data protection topics?	8
How do you decide between buy, rent, subscribe, or DIY (build) solutions?	14
How do you decide which storage system or architecture, and packaging, is best?	12
How do you know what type of data protection technique and technology to use and when?	10
How do you read and compare vendors' data sheets?	8
How important is low cloud cost vs. higher all-inclusive cost, or cost to restore?	11
How is a flat network different from how Fibre Channel SAN are designed?	6
How long does data (primary and protection copies) need to be retained?	10
How many failures do different types of protection and resiliency provide?	10
How many times can a tape, HDD, or SSD be used before it wears out?	7
How to develop tradecraft?	1
I have a heavily utilized server and application that is using Flash SSD already. Could NVMe help reduce the CPU bottleneck?	7
I heard that RAID is dead and bad?	9
If containers are the future, why learn about hypervisors, operating systems, and servers?	4
If the cloud is the future, why spend time learning about servers?	4
Is a data lake required for big data?	14
Is all software plug-and-play today across different servers?	4
Is backup dead, as some people say?	9
Is big data only for a data scientist?	14
Is block storage dead?	3
Is FCoE a temporal or temporary technology?	6
Is future storage converging around hyper-converged?	12
Is latency only a distance issue?	6

Frequently Asked Question	Chapter
Is there a difference between SATA HDD and other drives?	7
Isn't the objective to keep less data to cut costs?	2
Should we use two separate service providers for a redundant remote network connection?	6
Storage is getting cheaper; why not buy more?	11
What are marketing or hero numbers?	8
What are some examples of architectures, solutions, systems, and services that implement various data services?	8
What are some legal considerations regarding data protection, resiliency, and availability?	9
What are the legal requirements for where data can be protected to?	10
What are the most important factors to ask when deciding where to site an application, workload, or data (i.e., cost, performance, security, scalability, ease of control)?	14
What can you do to reduce or manage Flash memory wearing out?	7
What comes next after Ethernet?	6
What does scale with stability mean?	13
What is a de-dupe ratio vs. rate, and how does it compare to compression?	11
What is a reference architecture?	12
What is a server or storage system BoM?	12
What is a tin-wrapped data service?	8
What is a white-box server?	4
What is an engineered server or system?	4
What is effective vs. actual capacity?	11
What is oversubscription?	6
What is the best storage medium to avoid UBER and bit rot?	7
What is the best storage solution?	12
What is the best storage?	1
What is the best technology and location to do encryption?	10
What is the difference between NVDIMM and NVM SSD?	7
What is the difference between efficiency and effectiveness?	11
What is the difference between FIPS 140-2 compliant and certified?	10
What is the difference between structured, semistructured, unstructured, and raw data?	14
What is tin-wrapped vs. software vs. shrink-wrapped software?	4
What level of audit trails and logging is needed?	13
What tools exists that can help collect, assess, and plan for server, storage, and desktop among other migrations to virtual, cloud, or other software-defined data infrastructures?	13
What's the best server I/O protocol storage?	5
What's the best server I/O protocol and network for server and storage?	6
What's the best storage medium?	7
What's the connection between DFR and data protection?	11
What's the difference between regular memory and storage-class memory?	7

Frequently Asked Question	Chapter
What's wrong with using RPM to gauge storage performance?	7
When should I use block, file, or object?	3
When should legacy vs. CI vs. HCI vs. software-defined cloud or virtual be used?	13
When should redundant paths be used in a network or fabric?	5
Where do server and applications focus start and stop vs. storage-, I/O-, and networking-centric activities?	14
Where does hardware stop and software start?	3
Where is the best place to locate or have data services provided from?	8
Who should be responsible for security?	13
Why can't everything simply go to SSD?	7
Why can't we just have one type of data storage such as NAND Flash SSD, or put everything in the cloud?	1
Why cover HDD and tape when the future is all about SSD?	7
Why do storage devices slow down as they fill up with data?	3
Why is data footprint reduction important for clouds or virtual environments?	11
Why not add more memory?	4
Why not just go all Flash, NVM, SSD?	12
Why not just run everything over TCP/IP?	5
Why not keep all data?	2
Why not move everything to the cloud?	1
Why not use de-dupe everywhere?	11
Why not use erasure codes everywhere, particularly with SSD?	9
Why not use wireless everywhere?	6
Why use drive form factors vs. PCIe AiC?	7
Why use SAS or SATA when NVMe is now available?	7
Why would you want to do replication in a storage system vs. in an application, hypervisor, or software-defined storage?	9
Will SCM eliminate the need for DRAM and storage?	4
With a dependence on networks (local and remote) for cloud access, what happens if someone decides to turn off the network?	6
With virtualization, can I get down to having only a single server?	4

Appendix D

Book Shelf and Recommended Reading

Table D.1 is a sampling of various recommended books (besides my own) about different data infrastructure and software-defined focus areas that are on my bookshelf.

Table D.1. Some Suggested Books for a Server Storage I/O Bookshelf

Book Title	Author(s)	ISBN No.	Publisher
Algorithms + Data Structures = Programs	Niklaus Wirth	978-0130224187	Prentice-Hall
Cisco IOS Cookbook	Dooley & Brown	978-0596527228	O'Reilly Media
Docker: Up & Running: Shipping Reliable Containers in Production	Karl Matthias & Sean P. Kane	978-1491917572	O'Reilly Media
Essential Virtual SAN (VSAN): Administrator's Guide to VMware Virtual SAN, 2nd ed.	Cormac Hogan & Duncan Epping	978-0134511665	VMware Press Technology
Hadoop: The Definitive Guide: Storage and Analysis at Internet Scale	Tom White	978-1491901632	O'Reilly Media
Implementing Cloud Storage with OpenStack Swift	Kapadia, Varma, Rajana	978-1782168058	Packt Publishing
Linux Administration: A Beginners Guide	Wale Soyinka	978-0071767583	McGraw-Hill
Seven Databases in Seven Weeks: A Guide to Modern Databases and the NoSQL Movement	Eric Redmond & Jim Wilson	978-1934356920	Pragmatic Bookshelf
Systems Performance: Enterprise and Cloud	Brendan Gregg	978-0133390094	Prentice Hall
TCP/IP Network Administration	Craig Hunt	978-0596002978	O'Reilly Media
The Human Face of Big Data	Rick Smolan & Jennifer Erwitt	978-1454908272	Against Odds Productions
VMware vSphere 5.1 Clustering Deepdive (Vol. 1) Note: Older title, still good fundamentals in it.	Duncan Epping & Frank Denneman	978-1478183419	Independent Publishing Platform

Refer to www.storageio.com/bookshelf for links to the titles mentioned in Table D.1 along with other resources. In addition to the titles mentioned above and those on the Server StorageIO bookshelf, check out the Intel Recommended Reading site. Also check out the various vendor press sites (e.g., Cisco, Microsoft, and VMware Press, among others).

Useful information can be found at BizTechMagazine, BrightTalk, ChannelProNetwork, ChannelproSMB, ComputerWeekly, Computerworld, CRN, CruxialCIO, Data Center Journal

Table D.2. Some Useful Data Infrastructure Sites and Venues

Venue	Description
apache.org	Various open-source software
blog.scottlowe.org	Scott Lowe VMware- and network-related topics
blogs.msdn.microsoft.com/virtual_pc_guy	Ben Armstrong Hyper-V blog
brendangregg.com	Linux performance-related topics
cablemap.info/	Global network maps
CMG.org	Computer Measurement Group (CMG)
communities.vmware.com/	VMware technical community and resources
comptia.org	Various IT, cloud, and data infrastructure certifications
cormachogan.com	Cormac Hogan VMware- and vSAN-related topics
csrc.nist.gov	U.S. government cloud specifications
ethernetalliance.org	Ethernet industry trade group
fibrechannel.org/	Fibre Channel trade group
github.com	Various open-source solutions and projects
http://tinyurl.com/nqf2bcp	Intel-recommended reading list for developers
ieee.org	Institute of Electrical and Electronics Engineers
ietf.org	Internet Engineering Task Force
iso.org	International Standards Organizations
it.toolbox.com/	Various IT and data infrastructure topics forums
Kevinclosson.net	Various databases, including Oracle and SLOB resources
labs.vmware.com/flings/	VMware Fling additional tools and software
nvmexpress.org	NVM Express (NVMe) industry trade group
objectstoragecenter.com	Various object and cloud storage items
opencompute.org	Open Compute Project (OCP) servers and related topics
openfabrics.org	Open-fabric software industry group
opennetworking.org	Open Networking Foundation (ONF)
openstack.org	OpenStack resources
pcisig.com	Peripheral Component Interconnect (PCI) trade group
scsita.org	SCSI trade association (SAS and others)
SNIA.org	Storage Network Industry Association (SNIA)
Speakingintech.com	Popular industry– and data infrastructure–related podcast
technet.microsoft.com	Microsoft TechNet data infrastructure–related topics
vmug.org	VMware User Groups (VMUG)
yellow-bricks.com/	Duncan Epping VMware and related topics

(DCJ), Datacenterknowledge, IDG, and DZone. Other good sourses include Edtechmagazine, Enterprise Storage Forum, EnterpriseTech, Eweek.com, FedTech, Google+, HPCwire, InfoStor, ITKE, LinkedIn, NAB, Network Computing, Networkworld, and nextplatform. Others include Reddit, Redmond Magazine and Webinars, Spiceworks Forums, StateTech, techcrunch. com, TechPageOne, TechTarget Venues (various Search sites, e.g., SearchStorage, SearchSSD, SearchAWS, and others), theregister.co.uk, TheVarGuy, Tom's Hardware, and zdnet.com, among many others. Besides those already mentioned, Table D.2 lists some additional related sites and venues.

Find additional companion material for this book and related topics, tools, techniques, tradecraft, and technologies at www.storageio.com/book4 and www.storageioblog.com.

Appendix E

Tools and Technologies Used in Support of This Book

Moving beyond what's in your toolbox as discussed in the various chapters of this book along with tradecraft, context matters, considerations, recommendations, and tips, here are some tools, technologies, and techniques used as part of this book project. While creating this book (and related projects), I used a hybrid approach for the various examples, figures, images, and other hands-on activities. The approach included on-premise (i.e., at Server StorageIOLAB), off-site, and remote, along with on-line and cloud resources. Resources used included: Acronis (Trueimage), Apache (various software), APC (battery backup UPS), AWS (EC2, EBS, S3 [Standard, IA and Glacier], Route 53 [DNS and IP health alerts] among other services across different regions); Bluehost (DPS) and Broadcom/Avago/LSI (RAID and adapter cards); Centos, Ceph, Cisco (Switches and Routers), Clonezilla, Cloudberry (various cloud and object storage access tools); Datacore (storage software), DataDynamic (StorageX), Datadog (dashboard), Dell EMC (Servers, software including ECS, DPA, VNX, and Unity, among others), and Docker.

Other resources included: Enmotus (FuzeDrive micro-tiering software), Equass (Servers and storage), Google (GCS and other services), HPE (adapters, servers, and software), Intel (NUC and larger servers, NVMe SSD [AiC and U.2]) and Iomega (NAS); Jungledisk (cloud backup), JAM treesize (storage management software), and Lenovo (Servers, laptops, and tablets); Mellanox (HCA and switches), Micron/Crucial (DRAM and SSD), Microsoft (Azure, Windows Server, workstation, desktops, laptops, and tablets, Hyper-V and other software tools), and MSI (Server); NetApp (Software [edge, StorageGrid, and others]), Nexenta (storage software), OpenStack software, Oracle (MySQL, Oracle Database, various management tools, vdbench), and PuTTY (ssh).

Also: Quest Software [Benchmark Factory, Spotlight on Windows/Unix and SQL, Foglight]), Rackspace (cloud), Redhat, Retrospect (backup and clone), and Retrospect (backup and clone); Samsung (various SSD), Sandisk (SSD), Seagate (HDD and SSD), Startech (connectors, enclosures, KVM switch), Supermicro (enclosures), and SUSE; Turbonomic, UBNT (access points, directional antenna), Ubuntu, UpDraft Pro (WordPress backup), Virtustream (cache software), VMware (vCloud Air, VSAN, vSphere, and other software), and WD (HDD), among many others (see various chapter toolbox items).

Appendix F

How to Use This Book
for Various Audiences

The following are some suggested alternative reading flows based on different audiences or needs. Just like IT environments, data centers, data infrastructure, and the applications they support, everything is not the same for different audiences or readers. Figure F.1 provides a visual reference of the various chapters to help you navigate to meet your particular needs.

Some readers will start at the beginning with the Preface and other front matter, then progress sequentially section by section, chapter by chapter, through the Appendix and then the back matter including the Glossary. For others, the flow may be to read parts of the Preface,

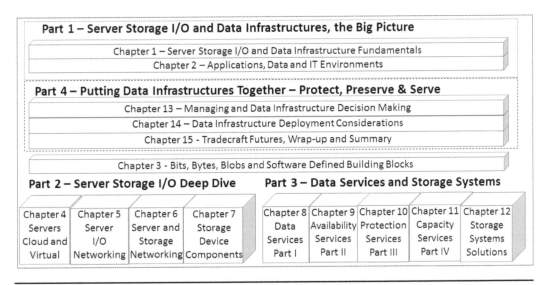

Figure F.1 Book organizations.

take a quick look at how the book is organized, check out Part One (Chapters 1 and 2), and then jump to Chapter 15. Others might jump around the different parts, chapters, or even sections within chapters as needed. The book is laid out from higher level working down into the lower-altitude details of data infrastructures. However, it can also be used by moving around at your pace or to meet your own needs. Table F.1 shows suggested reading sequences to use this book for different purposes. Everyone should read Chapters 1, 2, and 15, plus the following suggested chapters, depending on your focus.

Table F.1. Suggested Reading for Various Focus Areas

Area of Focus or Occupation	Suggested Reading Sequence
Analyst (financial or industry), press, media	All chapters, in order
Application developer, systems analyst	Chapters 4, 8, 13, 14
Break fix, service, repair, and install	All chapters
Call center and help desk	All chapters
Cloud and managed services provider	Chapters 3, 4, 8, 13, 14
College, university, or trade school student	All chapters
Data protection (backup, BC, DR)	All chapters
Facilities and physical infrastructure	Chapters 4, 11, 12, 13, 14
Finance or purchasing	Chapters 4, 6, 7, 12, 13, 14, plus others for future purchases
Instructor, teacher, or professor	All chapters
IT or technical services manager	All chapters
IT, data center, or technical services administrator	All chapters
IT, data center, technical services architect	All chapters
IT, data center, technical services engineer	All Chapters
Networking, data/telecom focus	Chapters 3, 4, 5, 6, 13, 14
Performance, capacity planning, security, and ITSM; sales and marketing professional	All chapters
Server or compute (hardware or software) focus	Chapters 3, 4, 5, 6, 7, 13 and 14
Storage-centric (hardware or software) focus	Chapters 3, 4, 6, 7, 12, 13, 14
Virtualization or hypervisor focus	Chapters 3, 4, 5, 6, 7, 8, 9, 10, 11, 12, 13, 14

Appendix G

Companion Website and Where to Learn More

Table G.1 shows various companion material sites and related content for this book.

Table G.1. Companion Material for This Book

storageio.com/sddi	Companion page for this book with links to related items
storageio.com/converged	Converged topics, technologies, trends, and tools
storageio.com/dfr	Data footprint–related topics (compression and de-dupe)
storageio.com/events	Various industry events, webinars, and activities
storageio.com/lab	Server StorageIO Lab reports and updates
storageio.com/links	Various industry and data infrastructure link resources
storageio.com/news	Various industry news, comments, and trends perspectives
storageio.com/newsletter	Server StorageIO update newsletter
storageio.com/nvme	Various NVMe-related tools, trends, topics, and technologies
storageio.com/performance	Server, storage, I/O performance tools, and topics
storageio.com/management	Server, storage, I/O management tools and related topics
storageio.com/network	Server, storage, I/O networking topics
storageio.com/server	Server-related topics
storageio.com/protect	Data protection, including backup, HA, BC, DR, security
storageio.com/reports	Server StorageIO thought leadership and trends reports
storageio.com/softwaredefined	Software-defined topics, technologies, trends, and tools
storageio.com/store	Various items used and mentioned via Amazon.com
storageio.com/ssd	SSD-related topics, technologies, trends, and tools
storageio.com/tips	Tips, articles, and other resources
storageio.tv	Videos and podcasts, including some for fun

In addition to the items in Table G.1 and elsewhere in the appendix as well as chapters, Table G.2 lists my other solo books that have common themes concerning data infrastructures. Each of the books in Table G.2 has a different color for the cover that also ties in with related themes. Red, green, and blue are primary colors, also known as lambda, that are used in optical networking for local and long-distance communications (I/O networking). Some modern video equipment also introduces a fourth color, yellow, as part of an enhanced visual display. Learn more at www.storageio.com/books.

Table G.2. Books by Server StorageIO Founder Greg Schulz

Book	Title	Subtitle	ISBN 13	Published
Red	Resilient Storage Networks	Designing Flexible Scalable Data Infrastructures	978-1555583118	2004
Green	The Green and Virtual Data Center	Efficient and Effective IT Data Centers (i.e., software-defined data centers before that was a buzz term)	978-1420086669	2009
Yellow	Cloud and Virtual Data Storage Networking	Enabling Your Journey to Efficient and Effective Computing Including Hybrid	978-1439851739	2011
Blue	Software-Defined Data Infrastructure Essentials	Cloud, Converged, and Virtual Fundamental Server Storage I/O Tradecraft	978-1498738156	2017

Glossary

*NIX	Linux, and UNIX
10 GbE	10 Gigabit Ethernet
24×7	24 hours a day, seven days a week
4 3 2 1	Data protection guide rule
4K	4KB sector size, 4K video resolution
512E, 512N, 4K (AF)	512-byte emulated, native, 4KB sector size
9's availability	How much availability in a given period
AaaS	Application as a service, archive as a service
Access method	How applications and servers access resources
ACL	Access control list
Active data	Data being read or written; not idle
AD	Active directory, archive device, active data
AES	Advanced Encryption Standard
AF	Advance Format
AFA	All-Flash array or all-Flash storage system
AFR	Annual failure rate
Agent	Software for performing some tasks or function
Aggregate	Disk or RAID group, virtual or physical device
Aggregated	Consolidated resources; combined server storage
Aggregates	Collection of different aggregates or disk groups
Aggregation	Consolidating resources together
AI	Artificial intelligence
AiC	PCIe add-in card
Air gap	Physical distance between resources for protection
A la carte	Order or select items individually
Altitude	How high or low in a stack of software
ALUA	Asymmetric logical unit access
AMD	Manufacturer of processing chips
AMI	AWS EC2 machine instance
AMQP	Advanced Message Queueing Protocol
Analytics	Applications and algorithms for data processing

ANSI	American National Standards Institute
ANSI T11	Standards group for Fibre Channel protocol
Apache	Open-source data infrastructure software
API	Application Program Interface
Appliance	Hardware or software that performs some function
Application blades	Server, storage, and I/O networking blades
Application	Programs or software that performs some function
Archive	Preserve inactive data to alternative media
ARM	Advanced RISC Machine; Azure Resource Manager
ARP	Address Resolution Protocol
Array	Data structure, virtual, physical storage, enclosure
ASIC	Application-Specific Integrated Circuit
ASM	Automatic storage management
ASP	Application service provider
Asynchronous	Time-delayed data transmission
ATM	Asynchronous Transfer Mode networking
Authoritative copy	Master, definitive version or copy of data
AWS	Amazon Web Services
AZ	Availability zone, site, data center
Azure (Microsoft)	Microsoft cloud and web services
BaaS	Backup as a service
Backing store	Where data gets written to
Ballooning	Dynamic expanding with elasticity
Bandwidth	Throughput, amount of data moved
Base 10	Decimal numbering system
Base 16	Hexadecimal numbering system
Base 2	Binary numbering system
Base 8	Octal numbering system
BBU	Battery backup unit
BC	Business continuance
BE	Back end, where storage attaches to
Benchmark	Workload, real, replay, synthetic for comparison
BER	Bit or block error rate for storage devices
BI	Business intelligence and analytics
Big data	Large volume or amount of unstructured data
Big iron	Large-form-factor, high-capacity hardware device
Binary	1 (on) or 0 (off) numbering scheme, architecture
Bind	Physical or logical connect, concatenate a resource
BIOS	Basic input/output system
Bitnami	Collection of applications and data infrastructure
Blade center	Packaging with blade servers, I/O, and networking
BLE	Bluetooth Low-Energy
Bleach	Digital secure shred (wipe clean) data so unrecoverable
Blob	Binary large object, collection binary data, object
Block	Allocation of memory or storage, sector, or page

Blockage	Congestion, or something causing a bottleneck
Block-based	Raw, cooked, SCSI, NVMe LUN volume access
Block chain	Distributed data for growing list of data blocks
BM	Bare metal or physical machine
BOM	Bill of material
Boot	Starting up and loading software into a server
Bound	Resources combined; bound disk, not a stripe
BR	Business resiliency; business recovery
Bridge	Physical, software to connect protocols, resources
Brownfield	Existing applications and environments
BTU	British thermal unit; heat from energy used
Bucket	Organization of resources, data, or data structures
Bulk	Large quantity use, or amount of resources
Bundled	All-inclusive resources or services, vs. a la carte
Burst	Surge, increase in activity or usage
Buzzword Bingo	Game played with industry buzzwords
BYOD	Bring your own device
C3	Command, control, communication
Cache hit	Data accessed, I/O resolved from cache
Cache warming	Reloading or populating cache after start-up
Capacity	Space or amount of resources, available inventory
CapEx	Capital expense
CAS	Content addressable storage
Cascade	Chain or series of events, flat linear topology
Cassandra	NoSQL database repository
Cat 5/6	Physical cable commonly used with Ethernet
CBT	Change block tracking, computer-based training
CD	Compact disc
CDM	Copy data management
CDMI	Cloud data management interface
CDN	Content distribution network
CDP	Continuous data protection
Centos	Linux operating system
Ceph	Open-source software-defined storage
CERT	Community Emergency Response Team (security)
Certificate	Security access credentials, accomplishment
CFR	U.S. Code of Federal Regulations
CHAP	Challenge Handshake Authentication Protocol
Chargeback	Billing and invoicing for information services
Checkpoint	Consistency point, snapshot, recovery point
Chunk	Allocation size used by RAID, data protection
CI	Converged infrastructure with disaggregated resources
CiB	Cluster-in-box or cloud-in-box
CIDR	Classless interdomain routing
CIFS	Common Internet File System data sharing

CIM	Common information model
CIO	Chief information officer; concurrent I/O
Cipher	Security, crypto, encryption, or locking code
CKD	Count Key Data mainframe data access protocol
CLI	Command Line Interface
Client	Physical, software, or person using resources
Cloud	Public, private, hybrid data infrastructure and service
Cloud access	Hardware, software, protocol for accessing cloud
Cloud computing	Internet or web remote-based application
Cloud instance	Instantiation of an image on a cloud resource
Cloud storage	Storage provided via a cloud service or product
Cloud-in-box	Self-contained private or hybrid cloud
Cluster	Group of resources working collectively
Clustered file system	Distributed file system across multiple servers
CM	Cloud machine, cloud virtual machine, cloud instance
CMDB	Configuration management database
CMG	Computer Measurement Group
CMS	Content management system
CNA	Converged Network Architecture
CNAME	Canonical Name record
CoAP	Constrained Application Protocol
Cold Aisles	Cool-air aisles between equipment cabinets
Cold data	Inactive data, dormant, not being used, archives
Colo	Co-location facility
Compliance	Complying with and meeting requirements
Component	Smaller piece of a larger system or application
Compression	Compact, reduce size of something
CompTIA	Testing and certification organization
Congestion	Busy with activity, possible bottlenecks
Connectivity	Hardware or software connecting resources
Consistency	Data that is as accurate as it is expected to be
Consistency point	Point in time (RPO) where data has consistency
Consumer	User of information services or data infrastructure
Container	Application packaging, repository, micro-service
Contiguous	Data or resources grouped close together
Controller	Hardware, software, algorithm that controls some function
Cookbook	Guide, reference on how to make something
Cooked	Formatted, organized, software-defined; not raw
Cooling ton	12,000 BTU, to cool 1 ton of air
CoreOS	Lightweight Linux derivative operating system
COTS	Commercial off-the-shelf, white box, commodity
Coverage	What is included in scope of being covered
Cpanel	Web and hosting management dashboard interface
CPU	Central processing unit
Crash consistency	Consistency protection in case of fault or crash

CRC	Cyclic redundancy check; name of a publisher
CRM	Customer relationship management
Cross technology	Solutions or tools that address multiple disciplines
CRUD	Create/Read/Update/Delete
CSP	Cloud service provider
CSV	Comma-separated variables for data interchange, Microsoft Cluster Shared Volume
Curl	Web HTTP resource access tool
Customer	User of applications or data infrastructure services
Customer adoption	What customers are doing and deploying
Customer deployment	What customers are installing and using
D2C	Disk (HDD or SSD)-to-cloud copy
D2D	Disk-to-disk snapshot, backup, copy, or replication
D2D2C	Disk-to-disk-to-cloud copy
D2D2D	Disk-to-disk-to-disk snapshot, backup, copy
D2D2T	Disk-to-disk-to-tape snapshot, backup, copy
D2T	Disk-to-tape backup, copy, archive
DAM	Digital asset manager
DAR	Durable availability reduced; lower-cost service
DARE	Data at rest encryption
DAS	Direct attached storage, internal or external
Dashboard	Management tool for monitoring, insight, awareness
Data centers	Habitat for technology, physical facility
Data-in-flight	Being moved between locations or systems
Data infrastructure	Server, storage, network resources and software
Data lake	Large repository, big pool or pond of data
Data pond	Small lake or repository of unstructured data
Data pool	Repository of structured, unstructured data
Data protection	Availability, security, BC, DR, BR, HA, resiliency
Data scientist	Somebody who studies and analyzes data
Data services	Feature functions for data access and management
Data structure	Bits and bytes organized into some defined entity
Data value	Known, unknown, no value, possible future value
Data variety	Different types of data
Data velocity	Speed at which data moves and flows
Data volume	Amount of data; a storage volume for data
Data warehouse	Large-scale repository or place for data analytics
Database	Structured means for organizing and storing data
DAX	Direct-access persistent NVM, NVDIMM, SCM
Day 1, day 2	First and second days of new install, environment
DB loss	Degradation of signal over optics or electric cable
DBA	Database administrator
DBaaS	Database as a service
DC	Direct-current electricity
DCB	Data center bridging

DCE	Data center Ethernet for converged networking
DCIE	Data center infrastructure efficiency
DCIM	Data center infrastructure management
DCPE	Data center performance efficiency
DDR/RAM	Double data rate random access memory
DDR4	Next generation of DRAM memory
De-dupe rate	How fast data is reduced (effectiveness)
De-dupe ratio	How much data is reduced (efficiency, space saving)
De-duplication	Elimination of duplicate data
De facto standard	Widely used convention, not set by a standards body
Delta	Change between points in time, such as for PIT snapshots
DEM	Digital evidence management (e.g., police body cameras)
Demand drivers	Need for application and resource consumption
De-normalize	Optimize, balance for performance and capacity
Dense	Large number of data or resources close together
Depreciate	No longer being developed, planned end of life
Desktop	Workstation or laptop computer, aka PC
Deterministic	Predictable, steady, known behavior
Developer	Creates applications for business, data infrastructures
DevOps	Development operations
DFR	Data footprint reduction (compress, de-dupe)
DHCP	Dynamic Host Configuration Protocol
Dictionary	Reference resource of known items
DIF	Data integrity features
Differential	Difference between time and state of something
DIMM	Dual Inline Memory Module
DIO	Direct I/O operations
Director	I/O and networking large-scale, resilient switch
Disaggregated	Not aggregated, server and storage separated
Disk partition	How storage is organized and allocated for use
Dispersal	Data spread across resources, site, or sites
Distance gap	Space between copies of data (for protection)
DIY	Do-it-yourself
DL	Disk library, storing backup and other data
DLA	Data loss access
DLE	Data loss event
DLM	Data life-cycle management
DLP	Data loss prevention, data leak prevention
DMA	Direct memory access
DMTF	Distributed Management Task Force
DNS	Domain Name System for internet domain names
Docker	Container micro-services and management tools
DoS	Denial-of-service attack
Downtime	Planned, scheduled, or unscheduled nonavailability
DPDK	Data Plane Deveopment Kit

DPM	Data protection management; Microsoft protection software
DPS	Dedicated private server
DR	Disaster recovery
DRaaS	Disaster recovery as a service
DRAM	Dynamic random-access (RAM) memory
Drive	A physical, logical storage device (SSD or HDD)
DRO	Data replication optimization
Dropbox	Cloud data storage sharing, consumer and EFSS
DRP	Disaster recovery planning process
Drupal	Web and content management application
DSL	Digital Subscriber Line
DTDS	Disaster-tolerant disk system
Dual-path	Two or more physical, logical paths to resource
Dual-port	Two or more ports, active or passive, to resource
Dual-socket	Two sockets for CPU processors on motherboard
Duplex	How communication exchange is handled
Durable	Copies exist to protect against data loss, damage
DWDM	Dense wave division multiplexing
DWPD	Disk writes per day (Flash endurance metric)
E2E	End-to-end
East/west	Management, system to system traffic, replication
EBCDIC	Character encoding scheme for IBM mainframe
EBS	AWS elastic block storage
EC2	AWS elastic cloud compute
ECC	Error correction code
ECKD	Extended Count Key Data
EDI	Electronic data interchange
eDiscovery	Electronic search and data discovery
EFS	AWS elastic file service
EFSS	Enterprise File Sync Share
EH&S	Environmental health and safety
EMR	Electronic medical record
Encoding	8b/10b, 64b/66b, 128b/130b
Encryption	Encode data so it is private, secure
Endpoint	Access address of a resource or service
Energy Star	U.S. Environmental Protection Agency Energy Star program
Energy-effective	Increased productivity per energy used
Energy-efficient	Reduced energy usage
EPA	U.S. Environmental Protection Agency
Ephemeral	Temporary, nonpersistent, volatile, short-lived
EPT	Extended Page Translation memory
Erasure code	Reed-Solomon (RS) parity derivative; see also LRC
ERP	Enterprise resource planning
ESCON	Access protocol for legacy IBM mainframe
ESRP	Microsoft Exchange Solution Reviewed Program

Ethernet	Network interface
Eventual consistency	Consistency occurs sometime in the future, async
Export	Volumes, file shares, data exported or unloaded
Extents	Fragments, resource allocations, extended file
Fabric	Network, collection of resources or applications
Fan-in, fan-out	Many into one or few, few expand out to many
Fault domain	Domain, scope of what can fail or cause faults
Fault region	Large-scope, geographically dispersed fault area
FBA	Fixed Block Architecture on HDD and SSD
FC	Fibre Channel
FCIA	Fibre Channel Industry Association
FCIP	Fibre Channel on IP for long-distance movement
FCoE	Fibre Channel over Ethernet
FCP	Fibre Channel SCSI_Protocol
FEC	Forward error correction
Federated	Loose collection of affiliated items, management
FedRAMP	Federal Risk Authorization Management Program
FICON	Mainframe storage access protocol
FIFO	First in, first out
FIO	Benchmark tool for creating test workloads
FIPS	Federal Information Processing Standard
Firewall	Network security device or software
Firmware	Low-level software that gets loaded into hardware
FISMA	Federal Information Security Management Act
Flash	Adobe Flash software, NAND NVM SSD storage
Fog compute	Distributed resources between cloud and data source
Footprint	Physical or logical space, resource used, overhead
Formatted	Organized, software-defined; not raw or uncooked
FPGA	Field Programmable Gate Array chip
Fragmentation	Noncontiguous (scattered) data, nonoptimized
Frame	Physical enclosure, logical data transmission unit
Frequency	How often something occurs
Front end	Where and how resources are accessed by others
FRU	Field replacement unit
FTL	Flash translation layer
FTM	Fault tolerance mode to enable resiliency
FTP	File Transfer Protocol
FTT	Failures or faults to tolerate
FUD	Fear, uncertainty, and doubt
Full-stripe write	All data in stripe written together to reduce overhead
FUSE	File system in userspace
Gap in coverage	Items not covered, skipped, missing, forgotten
Gateway	Hardware, software access to resource or services
Gbe	Gigabit Ethernet
Ghosted	Native file or object not in SharePoint, content database

GHz	Gigahertz frequency, a measure of speed
Github	Place with lots of open-source software and tools
Global namespace	Directory address namespace access resource
Gold copy	Master backup or data protection copy
Google	Search engine; cloud data infrastructure resources
Govcloud	Public cloud optimized for government use
Gpart	Storage partitioning tool
GPS	Global Positioning System; author of this book
GPT	GUID partition table
GPU	Graphics processing unit
Granularity	Coarse, broad focus, or fine-grained detail-centric focus
Green gap	Disconnect between messaging and IT challenges
Green IT	Efficient and effective information services
Greenfield	New applications and environment, clean sheet
Grid	Local or wide-area cluster of resources
Grooming	Logical data clean-up, reclamation, optimization
GT	Giga transfers
Guest	Guest operating system in a VM or LPAR
GUI	Graphical User Interface
GUID	Globally unique identifier
HA	High availability
Habitat	Place where technology lives, such as a data center
Hadoop	Software tool for performing big data analysis
HAMR	Heat-assisted magnetic recording
Hardened	Hardware, software, or cloud resilient to faults
Hash	Computed sum or key for lookup comparison
HBA	Host bus adapter for attaching peripherals
HCA	Host channel adapter for InfiniBand
HCI	Hyper-converged infrastructure
HCIBench	VMware test harness wrapper benchmark testing
HCL	Hardware compatibility list
HD	High-definition broadcast or video
HDD	Hard disk drive such as SAS, SATA, or USB
HDFS	Hadoop Distributed File System
HDMI	High-definition media interface
HDTV	High-definition TV
Heat map	Display of hardware heat; view of active data
Hero number	Benchmark number to show in best way
Heterogeneous	Mixed types of technologies and configurations
High-latency	Slow, long-response wait time for work to finish
Hitech	Health Information Technology Act and Standards
HOL	Head-of-line blocking
Homogenous	Same type server, storage, hypervisor, operating system
Hosting	Facility or service provider that hosts IT services
Hot aisle	Aisle between cabinets for warm air exhaust

Hot data	Data that is actively being read and or written
HPC	High-performance computing
HSM	Hierarchical storage management
HTML5	New version of HTML
HTTP	HyperText Transfer Protocol
Hub	Physical or logical place where things are shared
HVAC	Heating, ventilation, and air conditioning
HVM	Hardware virtual machine
Hybrid	Mixed mode using different technologies and techniques
Hyper-V	Microsoft virtualization infrastructure software
Hypervisor	Virtualization framework
I/O	Input–output operation, read or write
I/O rate	Number of I/O operations (IOPS) in a given time
I/O size	How big the I/O operations are
I/O type	Reads, gets, writes, puts, random or sequential
IaaS	Infrastructure as a service
IAM	Identity access and authentication management
IBA	InfiniBand Architecture
IDS	Intrusion detection software or system
IEEE	Institute of Electrical and Electronics Engineers
IETF	Internet Engineering Task Force
ILM	Information life-cycle management
Image	Photo; operating system, virtual machine, or container copy
Immutable	Persistent, not changeable
In-band	Management done in-line or in data path
Industry adoption	What industry and vendors talk about that is new
InfiniBand	High-speed, low-latency network
Information factory	Clouds, data infrastructure, and data centers that produce information
Ingest	Data being read and processed by de-dupe engine
Initiator	Something that starts or initiates I/O operation
Instance	Occurrence of an item, cloud machine, container
Intel	Large processor and chip manufacturer
Inter-	Across domains, internet, intersystem
Interface	Hardware or software for accessing a resource
Intra-	In domain, intranet, intrasystem, intracabinet
IoD	Internet of Devices
Iometer	Load generation and simulation tool
IOPS	I/O operations per second
Iostat	I/O monitoring tool
IoT	Internet of Things
IOV	I/O virtualization
IP	Internet Protocol part of TCP/IP; intellectual property
IPL	Initial program load (i.e., boot)
IPMI	Intelligent Platform Management Interface

IPsec	IP-based security and encryption
IPv4, IPv6	Internet Protocol versions 4 and 6
ISCSI	SCSI command set mapped to IP
ISL	Interswitch link
ISO	International Standards Organization; data format
ISR	Intelligence surveillance and reconnaissance
ISV	Independent software vendor
IT	Information technology
ITaaS	IT as a service
ITSM	IT service management
JBOD	Just-a-bunch-of-disks, with no RAID
JDEC	Joint Electronic Devices Engineering Council
Journal	Logical log for data consistency, a repository
JRE	Java Runtime Environment
JSON	JAVA Script Notation format
JTA	Job task analysis
Jumbo frames	Large I/O network packet, maximum transmission units (MTU)
JVM	Java Virtual Machine
Kerberos	Network authentication protocol
Key management	Managing encryption keys
Key value	Key that identifies a stored value in a repository
Knowledge base	Where de-dupe metadata, index, and pointers are kept
KPI	Key performance indicator
Kubernetes	Container management tools and framework
KVM	Keyboard–video monitor, kernel-based virtual machine hypervisor
Lambda	A light wave; AWS cloud service offering
LAMP	Linux, Apache (web handler), MySQL, and PHP
LAN	Local-area network
Landscape	Collection of software for a given application
Lane	Physical, logical, virtual path for moving data
Laptop	Portable computer
Latency	Response time or elapsed time
Lazy write	Deferred, delayed, asynchronous, or eventual update
LBA	Logical block address
LBN	Logical block number
LCM	Life-cycle management
LDAP	Lightweight Directory Access Protocol
LDM	Logical disk manager
Legacy	Existing, not new, mature
Library	Collection of software, where things are stored
LIFO	Last in, first out
Lightsail	AWS Virtual Private Server (VPS)
Link	Physical, logical path for data access or address
Linux	Open-source operating system
LIS	Linux Integration Service

Little data	Not big data or database, small files
Load balance	Spread access, optimize across resources
Locality of reference	How far away data or resource is for access
Logical size	Logically represented without data reduction
Loosely coupled	Clustered, aggregated, not tightly coupled
Lossless	No data lost during compression or reduction
Lossy	Some data loss during compression or reduction
LPAR	Logical partition, or virtual machine
LRC	Local reconstruction code, variation of erasure code
LRU	Least recently used
LTFS	Linear tape file system
LTO	Linear tape open
LUN	Logical unit number addressing for storage
Lustre	High-performance, large-scale, big-data file system
LVM	Logical volume manager
LXC	Linux container service
LZ	Lempel Ziv compression
M.2 NFF	Next-generation form factor for NVMe
M2M	Machine-to-machine
MaaS	Metal as a service
MAC	Media Access Control, layer for networking
Magnetic tape	Low-cost, energy-efficient, removable media
Mainframe	IBM legacy large server; large frame-based server
MAN	Metropolitan area network
Management tools	Software to support various management needs
Mapping	Assigning access, making resources available
Marketecture	Marketing architecture
Masking	Hiding or concealing a resource from being seen
MBR	Master boot record
MD5	Hash algorithm
Mdadm	Linux software RAID and storage tools
Media	Medium where and how data stored
Medium	Magnetic, electronic media for memory or storage
Mesos	Container management tools and framework
Metadata	Data describing other data
Metrics	Measurements and results
Mezzanine	Upper level, place, or card that sits above others
MHz	Megahertz frequency, an indicator of speed
MIB	Management information block for SNMP
Micro-services	Container services
Microcode	Low-level software for programming hardware
Micro-tiering	Fine-grained tiering
Migration	Physical, logical move, convert to somewhere else
Mitigation	Prevent or reduce impact of something
MLC	Multi-level cell NAND Flash memory

MMU	Memory management unit
Motherboard	Mainboard of server, storage, network device
Mount	Physical or logical attach and make ready for use
MPIO	Multi-path I/O
MPLS	Multiprotocol labeling WAN network protocol
MQTT	MQ Telemetry Transport
MSP	Managed service provider
MTBF	Mean time between failures
MTTL	Mean time to loss
MTTR	Mean time to repair or replace or rebuild
MTU	Maximum transmission units, jumbo frame
Multi-path	More than one path or access to resource
Multi-protocol	More than one protocol supported
Multi-tenancy	Shared occupancy of a resource
MySQL	Open-source database
Name brand	Apple, Cisco, Dell, HPE, Lenovo, Samsung, or other brands
Namespace	Collection of address endpoints
NAND	Non-volatile computer memory such as Flash
Nano	Small, lightweight Windows operating system
NAS	Network attached storage NFS, SMB data sharing
NAT	Network address translation
NBD	Network block device
NDA	Non-disclosure agreement
NDCA	Non-disruptive code activation, no restart
NDCL	Non-disruptive code load, restart needed
NDMP	Network Data Management Protocol
Near-line	Non-active data that does not need fast access
Nesting	Encapsulating, wrapping item inside of something
Networked storage	Storage accessed via storage or general network
NFS	Network File System (NAS) file and data sharing
NFV	Network function virtualization
NGFF (M.2)	Next-generation form factor for NVMe and SATA
NIC	Network interface card or chip
NIST	National Institute of Standards and Technology
NLM	Network lock manager
NOCC	Network operations control center
Node	Member of a cluster or resource pool, controller
Normalize	Optimize, reduce redundant data resources
North/south	Application or server to resource access
NoSQL	Non-SQL access, database, key-value repository
NPIV	N_Port ID Virtualization for Fibre Channel
NSM	Network status monitor
NUC	Next Unit computing (e.g., Intel NUC)
NUMA	Non-uniform memory access
NVDIMM	Non-volatile DIMM, DIMM with NVM

NVM	Non-volatile memory
NVMe	NVM Express protocol
NVRAM	Non-volatile RAM
OAuth	Open authorization access and security protocol
Object access	How objects are accessed, HTTP REST, S3, Swift
Object storage	Architecture, how data are organized and managed
Object	A place, thing or entity, many different contexts
OC	Optical carrier network
OCP	Open Compute Project
ODBC	Open database connectivity
ODCA	Open Data Center Alliance
OEM	Original equipment manufacturer
OFA	Open Fabric Alliance
Off-line	Data or IT resources not on-line and ready for use
OLTP	On-line transaction processing
On-line	Data and IT resources on-line, active, ready for use
Opacity	No knowledge of data structure in a file or object
OpEx	Operational expense
Optical	Network or storage based on optical mediums
Orchestration	Coordination of resource access, management
Orphaned storage	Lost, misplaced, forgotten storage space or virtual machine
OS	Operating system, VM guest, instance, image
OSD	Object storage device
OSI	Open system interconnection
OU	Organizational unit
Outage	Resources are not available for use or to do work
Out-of-band	Management done outside of data path
OVA	Open Virtualization Alliance
Oversubscription	Too many users have access at the same time
OVF	Open Virtualization Format
P/E	Program/erase, how NAND Flash is written
P2V	Physical-to-virtual migration or conversion
PaaS	Platform as a service
PACE	Performance, availability, capacity, economics
Packaging	Physical, logical how item is wrapped or enclosed
Para-virtualization	Optimized virtualization with custom software
Parity	Extra memory or storage for data protection
Partial file read/write	Portion of a file or object accessed instead of all
Partial stripe write	Some data written, resulting in extra overhead
Partitions	Resource organization and allocation or isolation
Pass-through	Transparent access, raw, no management
PBBA	Purpose-built backup appliance
PC	Personal computer, program counter, public cloud
PCI	Payment card industry; Peripheral Computer Interconnect
PCIe	Peripheral Computer Interconnect Express

PCM	Phase-change memory
PDU	Power distribution unit
Perfmon	Windows performance monitoring tool
Persistent	Data is stateful, non-volatile
Personality	How resources abstracted, act, and function
Phys	Physical connector
Physical volume	A single or group of storage devices
PIT	Point in time
PKZIP	Popular data compression tool
Pl	Probability of loss
Platform 1, 2, 2.5, 3	Platform generations: 1, central mainframe; 2, client server, the web; 2.5, hybrid between 2 and 3; 3, open software-defined and cloud, containers, converged along with IoT/IoD
Plug-in	Software module for additional functionality
PM	Physical machine, physical server or computer
PMD	Poll Mode Driver
PMDB	Performance management database
pNFS	Parallel NFS (NAS) high-performance file access
POC	Proof of concept, prototype, trial test, learning
Pool	Collection of resources or repositories
Port	Physical, logical point of interface attachment
POSIX	Portable Operating System Interface Standard
POST	Power-on self-test
PowerShell	A command-line interface
Protocol	Defined process for how something is done
Protocol droop	Performance degradation due to protocol impact
Provisioning	Allocating, assigning resources or service for use
Proxy	Function representing someone or something else
PST	Microsoft Exchange email personal storage file
PUE	Power usage effectiveness measurement
Puppet	Data infrastructure resource configuration tool
PV	Para-virtual
PXE	Preboot execution environment for network boot
QA	Quality assurance
Qcow	Virtual disk format used by QEMU hypervisors
QoS	Quality of service
QPI	Quick Path Interface used by Intel CPUs
QSFP	Quad small form-factor plug
Qtree	Similar to a subfolder, with additional functionality
Quad Core	Processor chip with four core CPUs
Queue depth	Items, things, I/O, work waiting in queue or line
Rack scale	Scaling by rack or cabinet basis vs. single device
RADIUS	Remote Authentication Dial-In User Service
RAID	Redundant array of independent disks
RAIN	Redundant array of inexpensive nodes

RAM	Random-access memory
RAS	Reliability, availability, serviceability
Raw storage	Storage not formatted with a file system or RAID
RBER	Recoverable bit error rate
RDM	Raw device mapped storage
RDMA	Remote direct memory access
Reduced redundancy	Lower level and cost of durability, AWS S3 RR
Reduction ratio	Ratio of how much data is reduced by
Reference architecture	Example of how to use a technology
Refs	Microsoft Resilient File System
Region	Geographical location, resource location
Re-inflate	Expand, un-compress, restore data to original size; re-expand de-duped data during restore
Reliability	How available a resource or service is
Remediation	Fix, update, repair, bring into compliance
Remote mirroring	Replicating data to a remote location
Reparse point	Object, item in file system to extend its capabilities
Replication	Mirror, make a copy of data elsewhere
Repository	Place for storing information and data
Response time	How fast work gets done, latency
REST	HTTP-based access
Retention	How long data is retained or saved before discard
RF	Radio frequency
RFID	Radio-frequency ID tag and reader
RFP	Request for proposal
RHDD	Removal HDD
RISC	Reduced instruction set computing
RJ45	Connector on Cat 5 and Cat 6 Ethernet cables
RMAN	Oracle database backup and data protection tool
ROBO	Remote office/branch office
RoCE	RDMA over Converged Ethernet
ROHS, RoHS	Restriction of hazardous substances
ROI	Return on investment
Role	What someone, or software, does or is assigned
Root	Base, foundation software, resource organization
Rotation	Migrating usage across different resources
Router	Networking or storage device for routing
RPC	Remote Procedure Call
RPM	Revolutions per minute; RPM Package Manager
RPO	Recovery-point objective
RR	Reduced redundancy, lower-cost storage service
RSA	Crypto, encryption keys utilities; also a company
Rsync	Tool for synchronizing data across resources
RTO	Recovery-time objective
Rufus	Tool for making bootable ISO images

RUT	Rule-of-thumb
S2D	Microsoft Windows Server Storage Spaces Direct
S3	AWS Simple Storage Service; object protocol
S3fs	Plug-in to access AWS S3 buckets from Linux
SaaS	Software as a service; storage as a service
SAMBA	Data access protocol, SMB, Microsoft file share
SAN	Storage area network block storage access
SAP HANA	In-memory database and big data analytics
SAR	System analysis and reporting tool
SAS	Serial attached SCSI; statistical analysis software; shared access signature; server attached storage
SATA	Serial ATA I/O interface and type of disk drive
SCADA	Supervisory Control and Data Acquisition
Scale-out	Expand horizontally across multiple resources
Scheduled downtime	Planned downtime for maintenance
SCM	Storage class memory
Scope	Limits, breadth, boundaries, what is covered or included
Scratch, workspace	Temporary work area for various functions
SCSI	Small Computer Storage Interconnect
SCSI_FCP	SCSI command set for Fibre Channel, aka FCP
SD card	Small Flash SSD for laptops, tablets, cameras
SDDC	Software-defined data center
SDDI	Software-defined data infrastructure
SDK	Software development kit
SDM	Software-defined marketing; software-defined management
SDN	Software-defined network
SDS	Software-defined storage
Sector	Page, block, allocation unit for storing data
SED	Self-encrypting device, such as a SSD, HDD
Semistructured	More organization than unstructured
Serdes	Serializers and de-serializers
Server	Physical, virtual, logical, cloud resource
Server SAN	Shared storage management
Service catalog	List of service offerings
Service port	Physical or logical network management port
SFF	Small-form-factor disk drive, server
SFP	Small-form-factor optical transceiver
SHA	Secure Hash Algorithm
Shard	Chunk or pieces of some data, file, object
SharePoint	Microsoft software for managing documents
Shares	Portion or allocation of resources, data, file shares
SHEC	Shingled erasure codes
Shelfware	Software that is bought but not used
Shell	Physical enclosure or case; software interface
Show-back	Report of resource usage and consumption

Shredding	Physically breaking up a resource, secure digital erase
Shrink-wrapped	Software you buy in a box with shrink wrapping
SIOC	Storage I/O control
SIS	Single-instance storage
Size on disk	Actual space on disk vs. logical size
SLA	Service-level agreement
SLC	Single-level cell NAND Flash
SLO	Service-level objective
SLOB	Silly Little Oracle Benchmark
SMART	Self-monitoring analysis reporting technology
SMB	Small/medium Business; server message block
SMB3	Microsoft data access protocol
SME	Small/medium enterprise; subject-matter expert
SMR	Singled magnetic recording
Snapshot	A picture or image of the data as of a point in time
SNIA	Storage Networking Industry Association
SNMP	Simple Network Management Protocol
SNT	Shiny new thing
SOA	Service-oriented architecture
SOHO	Small office/home office
SONET/SDH	Synchronous optical networking/synchronous
SP	Service provider
Sparse	Blanks, white spaces, allocated but not used
SPC	Storage Performance Council benchmark
SPEC	Performance benchmark
Specmanship	Promoting, competing on speeds and features
SPOF	Single point of failure
SQL	Structured Query Language
SRA	System or storage resource analysis
SRM	Storage/system/server resource management
SRM	Site recovery manager
SRP	SCSI RDMA protocol
SSD	Solid-state device or solid-state drive
SSE	Server-side encryption
SSH	Protocol and tools to access resources CLI
SSHD	Solid-state hybrid device
SSL	Secure Socket Layer
SSO	Single sign-on
Stack	Layers of software and hardware resources
Stale cache	Data in cache that is no longer current and valid
Stamps	Collection of server, storage, network resources
Standalone	Single system, server, storage system, appliance
Stand-up	Setup, configure, install, make available for use
Stateful	Consistent and persistent
Stateless	Non-consistent, non-persistent

Static data	Data that is not being changed, but is read
STI	System test initiator; self-timed interface
Storage system	Hardware and software that provide storage resources
STP	Spanning Tree Protocol
Stripe	Spread data evenly across storage resources
Stripe size	Width of devices across which data is spread
Strong consistency	Synchronous, guaranteed data integrity of updates
Structured data	Data stored in databases, well-defined repositories
Sunset	Retire, phase out, depreciate a technology
SUT	System/solution under test or being tested
SUV	System/solution under validation or verification
Swift	OpenStack storage; object access protocol
Synchronous	Real-time data movement, strong consistency
Tape	Magnetic tape, off-line bulk storage medium
Tar	*nix data protection and backup utility
Target	Destination or location where stored
TBW	TB written per day, metric for Flash durability
TCP/IP	Transmission Control Protocol/Internet Protocol
TDM	Time division multiplex for communications
Telemetry	Log and event activity data from various sources
Template	A pattern, format to define resource into item
Temporal	Temporary, non-persistent, volatile, short-lived
Tenancy	Isolation, separation, shared resources
Test harness	Connector cable, software wrapper for testing
Thick-provision	Space fully allocated, not thin provisioned
Thin provisioning	Virtually allocates, overbooks physical resources
Threat risk	Things that can happen or cause damage
Throughput	Bandwidth; data moved, or data flow rate
Tiering	Moving between tiers
Tiers	Different types and categories of resources
Time gap	Delta or difference in time to facilitate recovery
Time to first byte	Delay from start of I/O until first byte is available
Tin-wrapped	Software packaged with hardware as an appliance
TLC	Triple-level cell
Toolbox	Physical or logical place for data infrastructure tools
Toolchain	Collection of different development tools
TOR	Top of rack
TPM	Trusted platform module for securing resources
TPS	Transactions per second
Tradecraft	Experience, skills of a craft, trade, or profession
Transaction integrity	Ensure write order consistency of transactions, events
Transceiver	Electrical or optical connector for I/O cables
Trim	SATA command for NAND Flash garbage clean-up
Trunking	Aggregating network activity on shared resource
Trust relationship	Secure access, where known trusted partners exist

Turnkey   No assembly required, install, start using
U or RU   Rack unit, 1 U = 1 RU = 1.75 in.
U.2   NVMe SFF 8639 interface used on some 2.5-in. SSD
UBER   Unrecoverable bit error rate
Ubuntu   Linux operating systems support
UDP   User Datagram Protocol, aka TCP/UDP
UEFI   Unified Extensible Firmware Interface
UI   User interface, such as GUI, command line, shell
ULP   Upper-level protocol that runs on top of a network
UNC   Universal Name Convention
Unghosted   File or object abstracted, in a content repository
UNMAP   SCSI command similar to Trim for NAND Flash
Unscheduled   Unplanned downtime for emergency work
Unstructured   Data such as files stored outside of databases
UPS   Uninterrupted power system
URE   Unrecoverable read error
URI   Uniform resource identifier
URL   Uniform resource locator
USB   Universal Serial Bus
V2P   Virtual-to-physical migration or conversion
V2V   Virtual-to-virtual migration or conversion
VAIO   VMware API for IO filtering
VAR   Value-added reseller
Vault   Physical or logical place to store things
Vdbench   Benchmarking and workload generation tool
VDC   Virtual data center
VDI   Virtual desktop infrastructure
Vendor adoption   New things to talk about to get customers to buy
VHDX   Virtual hard disk used by Microsoft platforms
Virtual disk   Software-defined storage located in various places
Virtualization   Tools to abstract emulate and aggregate resources
VLAN   Virtual local-area network
VM   Virtual memory; virtual machine; video monitor; volume manager
VMDK   VMware virtual disk
VMFS   VMware file system
Volume   Physical, logical, virtual, or cloud storage device
Volume manager   Software that aggregates and abstracts storage
VPC   Virtual private cloud
VPN   Virtual private network
VPS   Virtual private server
VSA   Virtual storage array, virtual storage appliance
VSAN   Virtual SAN
VSI   Virtual server infrastructure
VSphere/ESXi   VMware hypervisor
VT   Virtual technology; video terminal; virtual tape

VTL	Virtual tape library
VTS	Virtual tape system
VVOL	VMware virtual volume
VXLAN	Virtual extensible LAN
WAN	Wide-area network
Warm data	Data that has some read or write activity
WDM	Wave division multiplex
White box	Commodity, commercial off-the-shelf, no-name brand
Wi-Fi	Wireless networking for relatively short distances
Windows	Microsoft operating system
Wirth, Niklaus	Coiner of phrase "Programs = algorithms + data structures"
WordPress	Popular blog and content management software
Workflow	Process, flow steps to accomplish task or work
Workload	Activity from applications using resources
Workstation	Desktop PC or laptop computer
WORM	Write once/read many
Write amplification	Extra write overhead resulting from single write
Write grouping	Optimization, group I/Os, reduce amplification
Write-back	Writes are buffered and eventually written
WTC	Write through cache (writes are cached for reads)
WWN,	World Wide Name
WWNN	World Wide Node Name
WWPN	Worldwide port name for addressing
X86	Popular hardware instruction set architecture
Xen	Open-source–based virtualization infrastructure
Xfs, zfs	*NIX filesystems
XML	Extensible Markup Language
YCSB	Yahoo Cloud Serving Benchmark
YUM	RPM package manager installer for some Linux systems

Index